Glencoe
Mathematics
Applications and Connections

Course 1

Glencoe McGraw-Hill

New York, New York   Columbus, Ohio   Woodland Hills, California   Peoria, Illinois

Visit the Glencoe Mathematics Internet Site for
Mathematics: Applications and Connections at

www.glencoe.com/sec/math/mac/mathnet

You'll find:

Chapter Review

Test Practice

Data Collection

Games

 links to websites relevant to
Chapter Projects, Interdisciplinary Investigations, exercises

and much more!

About the Roller Coaster Hologram

The optimum viewing angle for the roller coaster hologram on the cover of this textbook is a 45° angle. For best results, view the hologram at this angle under a direct light source, such as sunlight or incandescent lighting.

Glencoe/McGraw-Hill

*A Division of The **McGraw·Hill** Companies*

Send all inquiries to:
Glencoe/McGraw-Hill
8787 Orion Place
Columbus, OH 43240-4027

ISBN: 0-07-822866-2

6 7 8 9 10 027/043 08 07 06 05 04 03 02

Dear Students, Teachers, and Parents,

Mathematics students are very special to us! That's why we wrote **Mathematics: Applications and Connections,** a math program designed specifically for you. The exciting, relevant content and up-to-date design will hold your interest and answer the question "When am I ever going to use this?"

As you page through your text, you'll notice the variety of ways math is presented for you. You'll see real-world applications as well as connections to other subjects like science, history, language arts, and music. You'll have opportunities to use technology tools such as the Internet, CD-ROM, graphing calculators, and computer applications like spreadsheets.

You'll appreciate the easy-to-follow lesson format. Each new concept is introduced with an interesting application or connection followed by clear explanations and examples. As you complete the exercises and solve interesting problems, you'll learn a great deal of useful math. You'll also have the opportunity to complete relevant Chapter Projects, Hands-On Labs, and Interdisciplinary Investigations. Test Practice, Test-Taking Tips, and Reading Math Study Hints will help you improve your test-taking skills.

Each day, as you use **Mathematics: Applications and Connections,** you'll see the practical value of math. You'll quickly grow to appreciate how often math is used in ways that relate directly to your life. If you don't already realize the importance of math in your life, you soon will!

Sincerely, The Authors

Kay McClain

Patricia S. Wilson

Patricia Frey

Linda Dritsas

Barbara Smith

Jack M. Ott

Ron Pelfrey

Beatrice Moore-Harris

David Molina

Authors

William Collins
Director of The Sisyphus
Mathematics Learning Center
W.C. Overfelt High School
San Jose, CA

Linda Dritsas
District Coordinator
Fresno Unified School District
Fresno, CA

Patricia Frey
Mathematics Department
   Chairperson
Buffalo Academy for Visual
   And Performing Arts
Buffalo, NY

"***Mathematics: Applications and Connections*** *helps students make the connection between mathematics and the real world. Applications lead students through the classroom door into the world of art, geography, science, and beyond.*" —**Linda Dritsas**

Arthur C. Howard
Program Director for Secondary
   Mathematics
Aldine Independent School District
Houston, TX

Kay McClain
Lecturer
George Peabody College
Vanderbilt University
Nashville, TN

David Molina
Adjunct Professor of Mathematics
   Education
The University of Texas at Austin
Austin, TX

Beatrice Moore-Harris
Staff Development Specialist
Bureau of Education and
    Research
Houston, TX

Jack M. Ott
Distinguished Professor of
    Mathematics Education
University of South Carolina
Columbia, SC

Ronald Pelfrey
Mathematics Consultant
Lexington, KY

*"**Mathematics: Applications and Connections** is designed to help middle school students develop mathematical power—the ability to use what they know—and to give them a good start into higher level mathematics. This text also helps students learn reasoning skills, make connections to the real world, become expert problem solvers, and explain their work to others."*—**Jack Price**

*"The strongest focus of any middle school mathematics program should be on problem solving. **Mathematics: Applications and Connections** not only provides such a focus, but it makes the problem solving alive for students through its applications and connections."*—**Ronald Pelfrey**

Jack Price
Professor, Mathematics
    Education
California State Polytechnic
    University
Pomona, CA

Barbara Smith
Mathematics Supervisor
Unionville-Chadds Ford
    School District
Kennett Square, PA

Patricia S. Wilson
Associate Professor of
    Mathematics Education
University of Georgia
Athens, GA

Academic Consultants and Teacher Reviewers

Each of the Academic Consultants read all 39 chapters in Courses 1, 2, and 3, while each Teacher Reviewer read two chapters. The Consultants and Reviewers gave suggestions for improving the Student Editions and the Teacher's Wraparound Editions.

ACADEMIC CONSULTANTS

Richie Berman, Ph.D.
Mathematics Lecturer and Supervisor
University of California, Santa Barbara
Santa Barbara, California

Robbie Bonneville
Mathematics Coordinator
La Joya Unified School District
Alamo, Texas

Cindy J. Boyd
Mathematics Teacher
Abilene High School
Abilene, Texas

Gail Burrill
Mathematics Teacher
Whitnall High School
Hales Corners, Wisconsin

Georgia Cobbs
Assistant Professor
The University of Montana
Missoula, Montana

Gilbert Cuevas
Professor of Mathematics Education
University of Miami
Coral Gables, Florida

David Foster
Mathematics Director
Robert Noyce Foundation
Palo Alto, California

Eva Gates
Independent Mathematics
    Consultant
Pearland, Texas

Berchie Gordon-Holliday
Mathematics/Science Coordinator
Northwest Local School District
Cincinnati, Ohio

Deborah Grabosky
Mathematics Teacher
Hillview Middle School
Whittier, California

Deborah Ann Haver
Principal
Great Bridge Middle School
Virginia Beach, Virginia

Carol E. Malloy
Assistant Professor, Math Education
The University of North Carolina,
    Chapel Hill
Chapel Hill, North Carolina

Daniel Marks, Ed.D.
Associate Professor of Mathematics
Auburn University at Montgomery
Montgomery, Alabama

Melissa McClure
Mathematics Consultant
Teaching for Tomorrow
Fort Worth, Texas

TEACHER REVIEWERS

Course 1

Carleen Alford
Math Department Head
Onslow W. Minnis, Sr. Middle School
Richmond, Virginia

Margaret L. Bangerter
Mathematics Coordinator K-6
St. Joseph School District
St. Joseph, Missouri

Diana F. Brock
Sixth and Seventh Grade Math Teacher
Memorial Parkway Junior High
Katy, Texas

Mary Burkholder
Mathematics Department Chair
Chambersburg Area Senior High
Chambersburg, Pennsylvania

Eileen M. Egan
Sixth Grade Teacher
Howard M. Phifer Middle School
Pennsauken, New Jersey

Melisa R. Grove
Sixth Grade Math Teacher
King Philip Middle School
West Hartford, Connecticut

David J. Hall
Teacher
Ben Franklin Middle School
Baltimore, Maryland

Ms. Karen T. Jamieson, B.A., M.Ed.
Teacher
Thurman White Middle School
Henderson, Nevada

David Lancaster
Teacher/Mathematics Coordinator
North Cumberland Middle School
Cumberland, Rhode Island

Jane A. Mahan
Sixth Grade Math Teacher
Helfrich Park Middle School
Evansville, Indiana

Margaret E. Martin
Mathematics Teacher
Powell Middle School
Powell, Tennessee

Diane Duggento Sawyer
Mathematics Department Chair
Exeter Area Junior High
Exeter, New Hampshire

Susan Uhrig
Teacher
Monroe Middle School
Columbus, Ohio

Cindy Webb
Title 1 Math Demonstration Teacher
Federal Programs LISD
Lubbock, Texas

Katherine A. Yule
Teacher
Los Alisos Intermediate School
Mission Viejo, California

Course 2

Sybil Y. Brown
Math Teacher Support Team-USI
Columbus Public Schools
Columbus, Ohio

Ruth Ann Bruny
Mathematics Teacher
Preston Junior High School
Fort Collins, Colorado

BonnieLee Gin
Junior High Teacher
St. Mary of the Woods
Chicago, Illinois

Larry J. Gonzales
Math Department Chair
Desert Ridge Middle School
Albuquerque, New Mexico

Susan Hertz
Mathematics Teacher
Revere Middle School
Houston, Texas

Rosalin McMullan
Mathematics Teacher
Honea Path Middle School
Honea Path, South Carolina

Mrs. Susan W. Palmer
Teacher
Fort Mill Middle School
Fort Mill, South Carolina

Donna J. Parish
Teacher
Zia Middle School
Las Cruces, New Mexico

Ronald J. Pischke
Mathematics Coordinator
St. Mary of the Woods
Chicago, Illinois

Sister Edward William Quinn I.H.M.
Chairperson Elementary Mathematics
    Curriculum
Archdiocese of Philadelphia
Philadelphia, Pennsylvania

Marlyn G. Slater
Title I Math Specialist
Paradise Valley USD
Paradise Valley, Arizona

Sister Margaret Smith O.S.F.
Seventh and Eighth Grade Math Teacher
St. Mary's Elementary School
Lancaster, New York

Pamela Ann Summers
Coordinator, Secondary Math/Science
Lubbock ISD
Lubbock, Texas

Dora Swart
Teacher/Math Department Chair
W. F. West High School
Chehalis, Washington

Rosemary O'Brien Wisniewski
Middle School Math Chairperson
Arthur Slade Regional School
Glen Burnie, Maryland

Laura J. Young, Ed. D
Eighth Grade Mathematics Teacher
Edwards Middle School
Conyers, Georgia

Susan Luckie Youngblood
Teacher/Math Department Chair
Weaver Middle School
Macon, Georgia

Course 3

Beth Murphy Anderson
Mathematics Department Chair
Brownell Talbot School
Omaha, Nebraska

David S. Bradley
Mathematics Teacher
Thomas Jefferson Junior High School
Salt Lake City, Utah

Sandy Brownell
Math Teacher/Team Leader
Los Alamos Middle School
Los Alamos, New Mexico

Eduardo Cancino
Mathematics Specialist
Education Service Center, Region One
Edinburg, Texas

Sharon Cichocki
Secondary Math Coordinator
Hamburg High School
Hamburg, New York

Nancy W. Crowther
Teacher, retired
Sandy Springs Middle School
Atlanta, Georgia

Charlene Mitchell DeRidder, Ph.D.
Mathematics Supervisor K-12
Knox County Schools
Knoxville, Tennessee

Ruth S. Garrard
Mathematics Teacher
Davidson Fine Arts School
Augusta, Georgia

Lolita Gerardo
Secondary Math Teacher
Pharr San Juan Alamo High School
San Juan, Texas

Donna Jorgensen
Teacher of Mathematics/Science
Toms River Intermediate East
Toms River, New Jersey

Statha Kline-Cherry, Ed.D.
Director of Elementary Education
University of Houston – Downtown
Houston, Texas

Charlotte Laverne Sykes Marvel
Mathematics Instructor
Bryant Junior High School
Bryant, Arkansas

Albert H. Mauthe, Ed.D.
Supervisor of Mathematics
Norristown Area School District
Norristown, Pennsylvania

Barbara Gluskin McCune
Teacher
East Middle School
Farmington, Michigan

Laurie D. Newton
Teacher
Crossler Middle School
Salem, Oregon

Indercio Abel Reyes
Mathematics Teacher
PSJA Memorial High School
Alamo, Texas

Fernando Rosa
Mathematics Department Chair
Edinburg High School
Edinburg, Texas

Mary Ambriz Soto
Mathematics Coordinator
PSJA I.S.D.
Pharr, Texas

Judy L. Thompson
Eighth Grade Mathematics Teacher
Adams Middle School
North Platte, Nebraska

Karen A. Universal
Eighth Grade Mathematics Teacher
Cassadaga Valley Central School
Sinclairville, New York

Tommie L. Walsh
Teacher
S. Wilson Junior High School
Lubbock, Texas

Marcia K. Ziegler
Mathematics Teacher
Pharr-San Juan-Alamo North High School
Pharr, Texas

Student Advisory Board

The Student Advisory Board gave the editorial staff and design team feedback on the design, content, and covers of the Student Editions. We thank these students from Crestview Middle School in Columbus, Ohio, and McCord Middle School in Worthington, Ohio, for their hard work and creative suggestions in making *Mathematics: Applications and Connections* more student friendly.

Mara Flood

Anna Hoza

Neil Gardner

Lareva Summerall

Mike Leonelli

Chris Franklin

Kelly McLoughlin

Jennifer Shoemaker

Maghee Disch

Problem Solving, Numbers, and Algebra

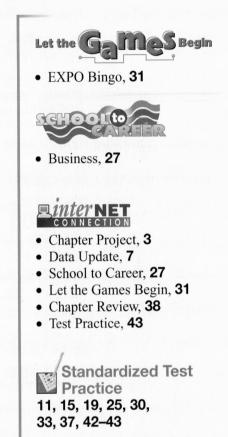

Let the Games Begin

- EXPO Bingo, **31**

SCHOOL to CAREER

- Business, **27**

interNET CONNECTION

Standardized Test Practice

Applications, Connections, and
Integration Index, pages xxii–1.

CHAPTER 2

Statistics: Graphing Data

Let the **Games** Begin

- What's the Average?, **75**

inter NET CONNECTION

Standardized Test Practice

Adding and Subtracting Decimals

Applications, Connections, and
Integration Index, pages xxii–1.

"O" is for Olympics, **128**

• The 1-Meter Dash, **121**

• Aviation, **99**

Standardized Test Practice

MATH IN THE MEDIA

CHAPTER 4
Multiplying and Dividing Decimals

CHAPTER 5
Using Number Patterns, Fractions, and Ratios

Applications, Connections, and
Integration Index, pages xxii–1.

 **Let the Games Begin**

- LCM Spin-Off, **209**

 **School to Career**

- Graphic Design, **185**

 **MATH IN THE MEDIA**

- Peanuts, **205**

interNET CONNECTION

- Chapter Project, **177**
- School to Career, **185**
- Data Update, **196**
- Let the Games Begin, **209**
- Chapter Review, **220**
- Test Practice, **225**

Standardized Test Practice
180, 184, 187, 190, 196,
201, 205, 209, 213, 216,
219, 224–225

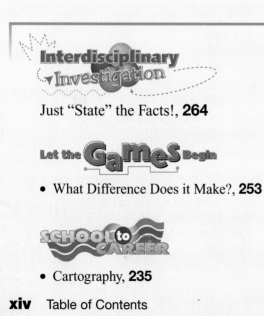

Multiplying and Dividing Fractions

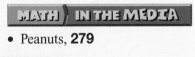
Applications, Connections, and
Integration Index, pages xxii–1.

CHAPTER 8

Exploring Ratio, Proportion, and Percent

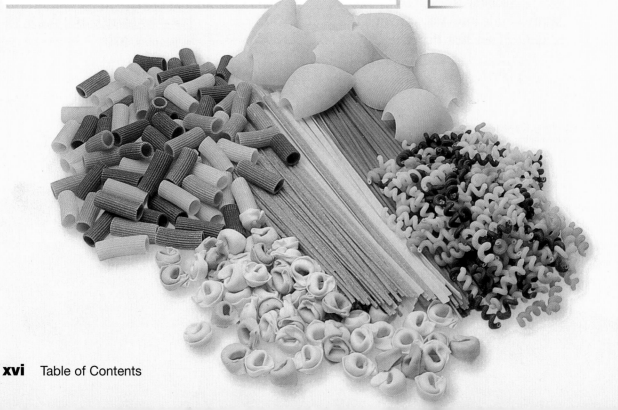

Geometry: Investigating Patterns

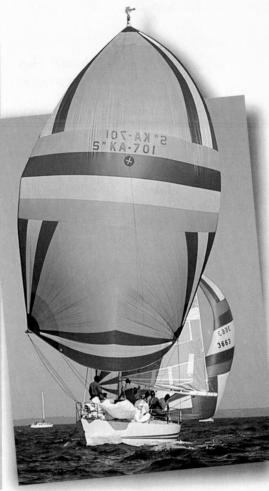

Applications, Connections, and
Integration Index, pages xxii–1.

Interdisciplinary Investigation

That Is One Humongous Pie!, **392**

Let the Games Begin

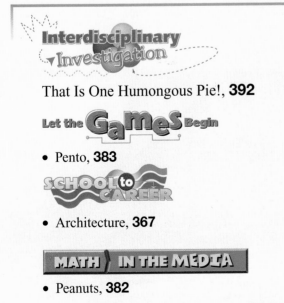

• Pento, **383**

School to Career

• Architecture, **367**

MATH IN THE MEDIA

• Peanuts, **382**

interNET CONNECTION

• Chapter Project, **351**
• School to Career, **367**
• Let the Games Begin, **383**
• Interdisciplinary Investigation, **393**
• Chapter Review, **386**
• Test Practice, **391**

Standardized Test Practice
**355, 357, 361, 366, 373,
378, 382, 390–391**

CHAPTER 10

Geometry: Understanding Area and Volume

Let the Games Begin

- Tiddlywink Target, **409**

interNET CONNECTION

Standardized Test Practice

Algebra: Investigating Integers

Applications, Connections, and
Integration Index, pages xxii–1.

Algebra: Exploring Equations

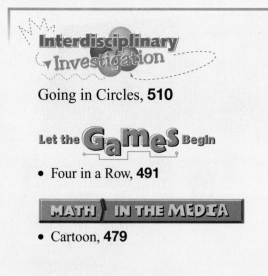

Interdisciplinary Investigation

Going in Circles, **510**

Let the Games Begin

• Four in a Row, **491**

MATH IN THE MEDIA

• Cartoon, **479**

interNET CONNECTION

Standardized Test Practice

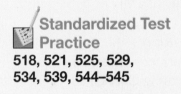

Standardized Test Practice

Applications, Connections, and
Integration Index, pages xxii–1.

Applications, Connections, and Integration Index

APPLICATIONS

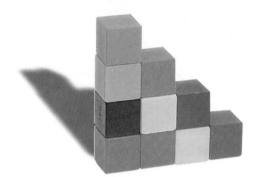

Algebra

Data Analysis, Statistics, and Probability

48%
34%
15%
10%

Popular Pastas

Spaghetti   **Macaroni and cheese**   **Noodles**   **Lasagna**

Source: National Pasta Association

Geometry and Spatial Sense, 366, 389, 452

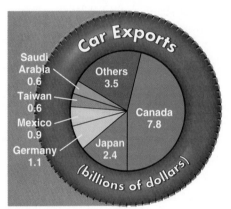

Source: U.S. Census Bureau

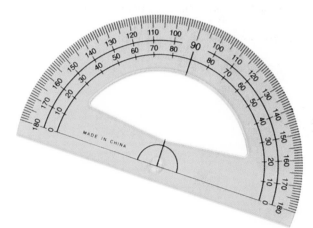

Problem Solving, Numbers, and Algebra

What you'll learn in Chapter 1

- to solve real-world problems using the four-step plan,
- to estimate sums, differences, products, and quotients using rounding,
- to evaluate numerical and algebraic expressions, and
- to solve equations using mental math and the guess-and-check strategy.

CHAPTER Project

CATALOG SHOPPING

In this project, you will use problem-solving strategies, estimation, and algebraic expressions to go on a shopping spree. You can use any catalog you wish for your shopping spree such as a sporting goods catalog or a novelty gift catalog.

Getting Started

- Look at the two shipping tables. Describe the method used by each catalog to determine the shipping charge.
- For each table, find the shipping charge for an item that costs $20 and weighs less than 5 pounds.

Table 1: Sporting Goods Catalog

Shipping Charges	
For the first item	$5.99
Each additional item	$1.99
Rush Orders	
Next day air	$21.99*
Express delivery (2–3 days)	$9.99*
*Plus standard shipping charges	

Table 2: Novelty Gift Catalog

Merchandise Total	Shipping Charges
up to $20.00	$4.95
$20.01 to $30.00	$5.95
$30.01 to $40.00	$6.95
$40.01 to $50.00	$7.95
$50.01 to $75.00	$8.95
$75.01 to $100.00	$9.95
$100.01 to $150.00	$12.95
$150.01 to $200.00	$15.95
1-Day Delivery	$9.99*
3-Day Delivery	$6.99*
*Plus standard shipping charges	

Technology Tips

- Use a **spreadsheet** to calculate the cost of the items and their shipping charges.
- Surf the **Internet** to see if any catalogs have websites.

 **inter**NET CONNECTION Data Update **For up-to-date information on catalogs, visit:**

www.glencoe.com/sec/math/mac/mathnet

Working on the Project

You can use what you'll learn in Chapter 1 to help you go on a shopping spree.

Page	Exercise
15	45
19	23
41	Alternative Assessment

What you'll learn

You'll learn to solve problems using the four-step plan.

When am I ever going to use this?

The four-step plan gives you an organized method for solving problems.

The graph shows the number of years it took from the time a new technology was invented for it to be in one-half of U.S. homes. In what year was the telephone in one-half of U.S. homes? *This problem will be solved in Example 1.*

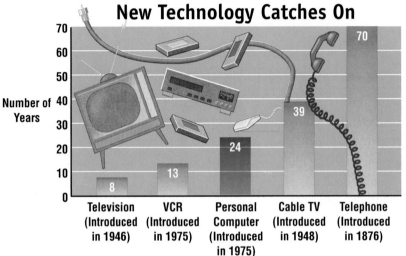

New Technology Catches On

Number of Years

Technology	Years
Television (Introduced in 1946)	8
VCR (Introduced in 1975)	13
Personal Computer (Introduced in 1975)	24
Cable TV (Introduced in 1948)	39
Telephone (Introduced in 1876)	70

You can use a four-step plan to help solve this problem and other problems.

1. *Explore*
- Read the problem carefully.
- Ask yourself questions like, "What facts do I know?" and "What do I need to find out?"

2. *Plan*
- See how the facts relate to each other.
- Make a plan for solving the problem.
- Estimate the answer.

3. *Solve*
- Use your plan to solve the problem.
- If your plan does not work, revise it or make a new plan.

4. *Examine*
- Reread the problem.
- Ask, "Is my answer close to my estimate?"
- Ask, "Does my answer make sense for the problem?"
- If not, solve the problem another way.

1 **Inventions** Refer to the beginning of the lesson. In what year was the telephone in one-half of U.S. homes?

Explore The graph states that the telephone was invented in 1876 and that it took 70 years until one-half of U.S. homes had a telephone. You need to find the year when one-half of U.S. homes had a telephone.

Plan To solve the problem, add the number of years it took until one-half of U.S. homes had a telephone to the year the telephone was invented.

Solve

$$\underbrace{\text{year telephone}}_{1876} \quad \underbrace{\text{plus}}_{+} \quad \underbrace{\begin{array}{c}\text{number of years until}\\ \text{one-half of U.S. homes}\\ \text{had a telephone}\end{array}}_{70}$$

$$1876 + 70 = 1946$$

By 1946, one-half of U.S. homes had a telephone.

Examine $1946 - 70 = 1876$. So the answer is correct.

2 **Work** Tyler delivers newspapers each weekday. If Tyler works every weekday for one year, how many days does he work?

Explore You know that Tyler delivers newspapers each weekday. You need to find out how many days Tyler works in one year.

Plan Multiply the number of weekdays in each week, 5, by the number of weeks in one year, 52, to find the number of days Tyler works in one year.

Solve

$$\underbrace{\begin{array}{c}\text{number of weekdays}\\ \text{in each week}\end{array}}_{5} \quad \underbrace{\text{times}}_{\times} \quad \underbrace{\begin{array}{c}\text{number of weeks}\\ \text{in one year}\end{array}}_{52}$$

In one year, Tyler works 5×52 or 260 days.

(continued on the next page)

Examine There are 2 weekend days each week, Saturday and Sunday. So, in one year, there are 2 × 52 or 104 weekend days. Subtract 104 from 365 to find the number of weekdays. The result is 261. The answer is reasonable. *Why don't the number of weekend days and weekdays add up to 365?*

CHECK FOR UNDERSTANDING

Communicating Mathematics

Read and study the lesson to answer each question.

1. ***Tell*** how each step of the four-step plan for problem solving helps you solve a problem.

2. ***Explain*** why you should check to see whether your answer makes sense for the problem.

3. ***Write a Problem*** with an answer of $500 that uses addition. Ask a classmate to solve your problem using the four-step plan.

Guided Practice

Use the four-step plan to solve each problem.

4. ***School*** Mika studied 224 vocabulary words in two weeks for his French class. If Mika studied the same number of words each day, how many vocabulary words did he study per day?

5. ***Computers*** In 1993, about 5 million computers had CD-ROM drives. In 1997, there were three times as many computers with CD-ROM drives than in 1993. How many computers had CD-ROM drives in 1997?

EXERCISES

Applications and Problem Solving

Use the four-step plan to solve each problem.

6. ***Computers*** A computer disk holds 720K of memory. If three programs are on the disk and they use 27K, 34K, and 52K of memory, how much memory is left on the disk?

7. ***Geography*** On a map of Tennessee, each inch represents approximately 18 miles. Travis is planning to travel from Nashville to Knoxville. If the distance on the map from Nashville to Knoxville is about 10 inches, approximately how far will he travel?

8. ***Earth Science*** On the first day of summer, Barrow, Alaska, has 24 hours of daylight, and Honolulu, Hawaii, has 13 hours and 26 minutes of daylight. How much more daylight does Barrow have than Honolulu?

9. ***Money Matters*** The Anstines want to buy a 27-inch television that costs $390. They plan to make a down payment of $150 and pay the rest in six equal payments. What will be the amount of each payment?

10. *Postal Service* Refer to
    the graph.
    a. How many more cards are
       sent for Valentine's Day than
       Father's Day?
    b. How many more cards are
       sent for Valentine's Day than
       Mother's Day?

11. *Life Science* To estimate
    the temperature in degrees
    Fahrenheit, you can count the
    number of times a cricket chirps in 15 seconds and then add that number to
    40. What is the temperature if a cricket chirps 25 times in 15 seconds?

Holiday Mail

Average number of
cards and letters mailed:

Valentine's Day
925 million

Mother's Day
155 million

Father's Day
101 million

Source: U.S Postal Service, Greeting Card Association

12. *Money Matters* The student council at West Boulevard Middle School is
    planning their annual winter dance. Tickets to the dance cost $3 each.
    There are 345 students in the school. If each student buys one ticket, how
    much money would the student council collect?

13. *Measurement* Mrs. Hall challenged her class to determine the number
    of pennies it takes to reach one mile. If there are 16 pennies in one foot,
    how many pennies are in one mile? (*Hint*: 1 mile = 5,280 feet)

14. *Life Science* An adult male walrus weighs about 2,670 pounds. An adult
    female walrus weighs about 1,835 pounds. How many more pounds does
    an adult male walrus weigh than an adult female walrus?

15. *Critical Thinking* Refer to
    the table. To score in either of
    the last two ways, a team must
    have scored a touchdown on
    the preceding play. Can a
    team score 23 points? Explain
    your answer.

For **Extra Practice,**
see page 557.

Different Ways a Team Can Score in Football	
Touchdown	6 points
Field Goal	3 points
Safety	2 points
2-Point Conversion	2 points
1-Point Conversion	1 point

1-2 Using Patterns

What you'll learn

You'll learn to solve problems using patterns.

When am I ever going to use this?

You can use patterns to make formations in marching band.

Patterns are all around us, even in a bee's honeycomb. Think about your day. What events follow a pattern? Do you always have the same class after lunch? Does your school day always end at the same time?

In this lesson, you will use patterns to solve problems.

Examples

1 Draw the next three figures in the pattern.

In the first figure, the top left corner is shaded. In the figures that follow, the shaded corner moves counterclockwise. The next three figures are drawn below.

2 Draw the next three figures in the pattern.

In the figures above, the triangle moves clockwise. The next three figures are drawn below.

In Examples 1 and 2, geometric shapes were arranged in a pattern. Numbers can also be arranged in a pattern.

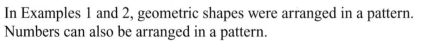

3 Find the next three numbers in the pattern
6, 11, 16, 21, _?_, _?_, _?_.

Study the pattern. How do you get each succeeding number?

$$6, \underbrace{11,}_{+5} \underbrace{16,}_{+5} \underbrace{21,}_{+5} \underline{?}, \underline{?}, \underline{?}$$

Each number is 5 more than the number before it.

21 + 5 = 26   26 + 5 = 31   31 + 5 = 36

The next three numbers are 26, 31, and 36.

4 Find the next three numbers in the pattern 2, 4, 8, 16, _?_, _?_, _?_.

Study the pattern. How do you get each succeeding number?

$$2, \underbrace{4,}_{\times 2} \underbrace{8,}_{\times 2} \underbrace{16,}_{\times 2} \underline{?}, \underline{?}, \underline{?}$$

Each number is twice the number before it.

16 × 2 = 32   32 × 2 = 64   64 × 2 = 128

The next three numbers are 32, 64, and 128.

When using patterns, making a table helps organize the information.

HANDS-ON

MINI-LAB

Work with a partner.   30 centimeter cubes

A "staircase" that is 4 cubes high is shown at the right.
Notice that 10 cubes are needed to build the staircase.

Try This

1. Copy the chart below. Then use centimeter cubes to find the number of cubes needed to build each staircase.

Height of Staircase	1	2	3	4	5	6	7
Number of Cubes Needed	1	?	?	10	?	?	?

Talk About It

2. Describe the pattern you see in the number of cubes needed to build each staircase.

3. Use the pattern to find the number of cubes needed to build staircases that are 9 cubes and 10 cubes high without using centimeter cubes.

Example **5**
INTEGRATION

Geometry The marching band wants to make a triangular formation on the football field. Triangular formations with 2, 3, 4, and 5 students at the base look as follows:

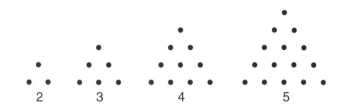

2   3   4   5

How many students are needed to make a triangular formation with 8 students at the base?

Explore   You know the number of students needed to make a formation with 2, 3, 4, and 5 students at the base. You need to find the number of students needed to make a formation with 8 students at the base.

Plan   Make a table to organize the information. Then, study the table to find a pattern.

Solve

Number of Students at Base	2	3	4	5	6	7	8
Total Number of Students Needed	3	6	10	15	?	?	?

+3 +4 +5

The extra students needed increases by 1 each time. Use the pattern to complete the chart.

$15 + 6 = 21$ $21 + 7 = 28$ $28 + 8 = 36$

36 students will be needed to make a triangular formation with 8 students at the base.

Examine   Draw a triangular formation with 8 dots at the base. Check to see if there are 36 dots in the formation.

CHECK FOR UNDERSTANDING

Communicating Mathematics

Read and study the lesson to answer each question.

1. ***Draw*** the first five phases of a geometric pattern. Then write a rule for finding the next shape in the pattern.

2. ***Write*** a number pattern with six numbers. Then write a rule for finding the next number in the pattern.

HANDS-ON
MATH

3. Fold a piece of paper 1 time. Count the number of thicknesses. How many thicknesses are there when the paper is folded 2 times? 3 times? 4 times? Describe the pattern in the number of thicknesses.

Find the next three numbers in each pattern.

4. 0, 8, 16, 24, _?_, _?_, _?_

5. 128, 64, 32, 16, _?_, _?_, _?_

6. *Sports* Jamil and Mark are competing against each other in a tennis match. Their scores are shown below.

	Set 1	Set 2	Set 3	Set 4	Set 5
Jamil	6	2	6	4	6
Mark	3	6	3	6	?

If the pattern continues, what will Mark's score be in set 5?

EXERCISES

Practice

Find the next three numbers in each pattern.

7. 3, 4, 3, 4, _?_, _?_, _?_

8. 7, 10, 13, 16, _?_, _?_, _?_

9. 2, 8, 32, 128, _?_, _?_, _?_

10. 36, 30, 24, 18, _?_, _?_, _?_

11. 7, 8, 10, 13, _?_, _?_, _?_

12. 3, 9, 27, 81, _?_, _?_, _?_

13. Find the next three numbers in the pattern 1, 3, 7, 13, ...

Applications and Problem Solving

14. *Fitness* Marcie is training for the swim team. On the first day, she swims 5 laps. The second day she swims 6 laps. The third day she swims 8 laps, and on the fourth day she swims 11 laps.

   a. If this pattern continues, how many laps will she swim on the fifth day?

   b. On which day will she reach her goal of 33 laps?

15. *Transportation* Isabel needs to take the bus home from the mall. She forgot her bus schedule but she remembers that the bus comes every 20 minutes. She also remembers that there is a bus at 12:30 P.M. What time does the first bus after 5:00 P.M. arrive?

16. *Geometry* Find the next two shapes in the pattern.

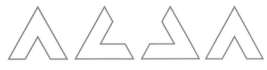

For **Extra Practice**, see page 557.

17. *Critical Thinking* What is the next number in the pattern 1, 2, 2, 4, 8, 32, ... ?

Mixed Review

18. *History* In July 1776, General George Washington had nearly 30,000 troops guarding New York City. After retreating through New Jersey to the Delaware River, only about 3,000 troops remained with him. How many troops did he lose? *(Lesson 1-1)*

19. **Standardized Test Practice** In its first year, Lucas Software had a profit of $2,763,000. The second year, their profit increased by $184,000. How much profit did Lucas Software have in the second year? *(Lesson 1-1)*

   A $2,579,000   **B** $2,763,000   **C** $2,947,000   **D** $3,248,000

1-3 Estimation by Rounding

What you'll learn

You'll learn to estimate sums, differences, products, and quotients using rounding.

When am I ever going to use this?

Rounding can be used to estimate the total cost of items at a department store.

Wild horses can be found throughout the United States. The table shows the number of wild horses living on public lands in five western states. About how many wild horses are living on public lands in these five states?

State	Number of Wild Horses
Nevada	30,798
Wyoming	4,115
Oregon	1,891
Utah	1,884
California	1,745

To estimate the number of wild horses, you can use rounding. Round each number to the nearest thousand. Then add.

$$
\begin{array}{rcl}
30,798 & \rightarrow & 31,000 \\
4,115 & \rightarrow & 4,000 \\
1,891 & \rightarrow & 2,000 \\
1,884 & \rightarrow & 2,000 \\
+\ 1,745 & \rightarrow & +\ 2,000 \\
\hline
& & 41,000
\end{array}
$$

About 41,000 horses are living on public lands in the five states.

Let's review the rules for rounding.

Rounding Whole Numbers	Look at the digit to the right of the place being rounded. • The digit remains the same if the digit to the right is 0, 1, 2, 3, or 4. • Round up if the digit to the right is 5, 6, 7, 8, or 9.

Example

1 **Round 43 to the nearest ten.**

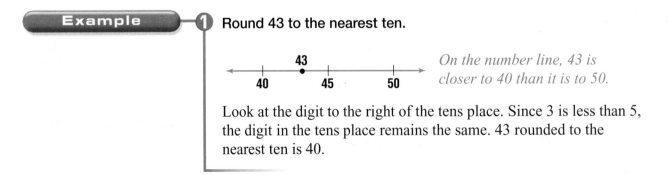

On the number line, 43 is closer to 40 than it is to 50.

Look at the digit to the right of the tens place. Since 3 is less than 5, the digit in the tens place remains the same. 43 rounded to the nearest ten is 40.

2 Round 252 to the nearest hundred.

252
200   250   300

On the number line, 252 is closer to 300 than it is to 200.

Look at the digit to the right of the hundreds place. Since the digit is 5, round up. That is, increase the digit in the hundreds place by one. The number 252 rounded to the nearest hundred is 300.

3 Round 9,973 to the nearest thousand.

9,973
9,000   9,500   10,000

On the number line, 9,973 is closer to 10,000 than it is to 9,000.

Look at the digit to the right of the thousands place. Round up since 9 is greater than 5. Thus, 9,973 rounded to the nearest thousand is 10,000.

Estimation is a useful skill that provides a quick answer when an exact answer is not needed.

Examples

4 Estimate the product of 7 and 64.

$$
\begin{array}{rcr}
64 & \rightarrow & 60 \\
\times\,7 & \rightarrow & \times\,7 \\
\hline
& & 420
\end{array}
$$

5 Estimate the quotient of 112 and 23.

$$
112 \div 23 \;\rightarrow\; 20\overline{)100}^{\,5}
$$

Using estimation, you can also check to see whether an answer is reasonable.

Example

Real World APPLICATION

6 **Money Matters** Eduardo is shopping at Calyan's department store. He purchases a bag of candy, a mechanical pencil, and a package of baseball cards. The total bill is $9.12. He gives the cashier a $20 bill and receives $9.46 in change. Is this amount reasonable?

Round to the nearest dollar amount. Then subtract mentally.

$$
\begin{array}{rcr}
\$20.00 & \rightarrow & \$20 \\
-\;9.12 & \rightarrow & -\;9 \\
\hline
& & \$11
\end{array}
$$

Eduardo should have received about $11. So, the amount of $9.46 is *not* reasonable.

Communicating Mathematics

Read and study the lesson to answer each question.

1. *Tell* how you would round 957 to the nearest hundred by using the number line below.

$$957$$

900   950   1,000

2. *Explain* how you would use rounding to estimate $517 - 192$.

3. *Write* about a situation when you would use estimation.

Guided Practice

Round each number to the underlined place-value position.

4. 3̲5
5. 4̲24
6. 5̲4̲2
7. 2̲96
8. 2̲,088
9. 9̲,922

Estimate. State whether the answer shown is reasonable.

10. $\$6.67 + \$8.99 = \$12.66$
11. $898 - 413 = 485$
12. $215 \times 6 = 1,290$
13. $468 \div 52 = 14$

14. *Food* The average U.S. citizen eats about 47 pints of ice cream per year. The average Italian citizen eats about 22 pints per year. About how many more pints of ice cream does the average U.S. citizen eat than the average Italian citizen?

EXERCISES

Practice

Round each number to the underlined place-value position.

15. 9̲7
16. 5̲2̲0
17. 5̲1
18. 1̲8
19. 4̲4̲4
20. 8̲,1̲96
21. 5̲,5̲00
22. 9̲2̲7
23. 3̲,7̲04
24. 3̲94
25. 5̲,6̲23
26. 3̲,6̲56
27. 9̲,716
28. 4̲,448
29. 1̲,888
30. 7̲,050

31. Round 9,438 to the nearest thousand.

32. What is 6,612 rounded to the nearest hundred?

Estimate. State whether the answer shown is reasonable.

33. $82 \times 3 = 246$
34. $\$3.32 + \$3.34 + \$3.21 = \$9.87$
35. $96 - 18 = 65$
36. $121 \div 2 = 42$
37. $4,901 + 5,002 = 9,903$
38. $392 \times 4 = 12,368$
39. $765 - 234 = 531$
40. $391 \div 23 = 17$
41. $\$5.25 + \$35.27 = \$46.12$
42. $2,914 \times 3 = 8,742$

Applications and Problem Solving

43. *Recycling* The average U.S. citizen uses about 190 pounds of plastic, 55 pounds of aluminum cans, 586 pounds of paper, and 325 pounds of glass per year. If these materials are recyclable, about how many pounds of these materials could be recycled by the average citizen in one year?

44. *Travel* The chart shows the approximate distance, in miles, between some U.S. cities. Suppose you travel from Raleigh, North Carolina, to Nashville, Tennessee. If you average 52 miles per hour, about how many hours will it take you to get there?

Mileage Chart

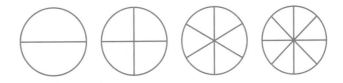

Approximate Mileages	Raleigh, NC	Albuquerque, NM	Nashville, TN	Minneapolis, MN	Portland, OR	Charleston, WV	Dallas, TX
Raleigh, NC		1,760	538	1,215	2,965	299	1,200
Albuquerque, NM	1,760		1,228	1,227	1,377	1,527	654
Nashville, TN	538	1,228		812	2,424	395	668
Minneapolis, MN	1,215	1,227	812		1,748	937	963
Portland, OR	2,965	1,377	2,424	1,748		2,572	2,041
Charleston, WV	299	1,527	395	937	2,572		1,141
Dallas, TX	1,200	654	668	963	2,041	1,141	

45. *Working on the* CHAPTER Project Refer to the tables on page 3.

a. Suppose you place an order from the Sporting Goods Catalog. You order a pair of running shoes for $74.99, a hooded sweatshirt for $35.99, and a pair of shorts for $20.99. Estimate the total cost of the order including express delivery charges.

b. Choose a catalog from which you would like to order. Imagine you have $200 to spend. List the items you would like to order and their prices. Estimate the total cost of the order including shipping charges.

46. *Critical Thinking* You want to purchase three items at a store for $4.43, $5.42, and $6.45. You have $15. To see if you have enough money, you estimate the total cost by rounding the cost of each item to the nearest dollar and adding. Will you have a problem making this purchase? Explain.

Mixed Review

47. *Patterns* Find the next three numbers in the pattern 1, 2, 4, 7, ... *(Lesson 1-2)*

48. *Geometry* Find the next figure in the pattern. *(Lesson 1-2)*

49. *School* At Watson Middle School, the bells ring at 8:50 A.M., 8:54 A.M., 9:34 A.M., 9:38 A.M., and 10:18 A.M. each day. When do the next three bells ring? *(Lesson 1-2)*

50. *Population* In 1998, Denver had a population of 499,000 people, and Atlanta had a population of 404,000 people. How many more people were living in Denver than Atlanta? *(Lesson 1-1)*

51. Standardized Test Practice The Changs' dot matrix printer prints four pages each minute. Hiro Chang has a history report that is 12 pages long. How long will it take the printer to print the report? *(Lesson 1-1)*

A 6 min **B** 4 min **C** 3 min **D** 2 min

52. *School* Jill needs to hand out an equal number of counters to 23 students. If there are 522 counters, how many should she give each student? *(Lesson 1-1)*

For **Extra Practice**, see page 557.

Order of Operations

The San Diego Zoo has more than 3,200 animals. Each year, the animal food bill is more than $685,000. Many people help reduce this cost by adopting an animal. The table shows the costs for adopting various zoo animals. How much would it cost to adopt 1 pygmy hippopotamus, 4 fishing cats, and 3 tapirs? *This problem will be solved in Example 4.*

Animal	Cost ($)
Anteater	500
Spider Monkey	50
Pygmy Hippopotamus	100
Spectacled Owl	50
Tapir	100
Fishing Cat	50

When more than one operation is used, we need to know which one to perform first so that everyone gets the same result. Mathematicians have come up with rules called the **order of operations**.

Order of Operations	1. Multiply and divide in order from left to right. 2. Add and subtract in order from left to right.

Examples

Find the value of each expression.

1 $15 + 7 - 3$

$15 + 7 - 3 = 22 - 3$ *Add 7 to 15.*

$= 19$ *Subtract 3 from 22.*

2 $3 + 6 \times 4 - 2$

$3 + 6 \times 4 - 2 = 3 + 24 - 2$ *Multiply 6 and 4.*

$= 27 - 2$ *Add 3 and 24.*

$= 25$ *Subtract 2 from 27.*

3 $14 \div 7 + 12 \times 3 - 9$

$14 \div 7 + 12 \times 3 - 9 = 2 + 12 \times 3 - 9$ *Divide 14 by 7.*

$= 2 + 36 - 9$ *Multiply 12 and 3.*

$= 38 - 9$ *Add 2 and 36.*

$= 29$ *Subtract 9 from 38.*

Example

CONNECTION

4 **Life Science** Refer to the beginning of the lesson.

a. Write an expression for the cost of adopting 1 pygmy hippopotamus, 4 fishing cats, and 3 tapirs.

b. Find the cost of adopting the animals.

a. *cost for*
    1 hippo   *plus*   *cost for*
                    4 cats   *plus*   *3 tapirs*

$1 \times \$100 \quad + \quad 4 \times \$50 \quad + \quad 3 \times \$100$

b. Find the value of the expression.

$$1 \times \$100 + 4 \times \$50 + 3 \times \$100 = \$100 + 4 \times \$50 + 3 \times \$100$$
$$= \$100 + \$200 + 3 \times \$100$$
$$= \$100 + \$200 + \$300$$
$$= \$600$$

The cost of adopting the animals is $600.

Does your calculator follow the order of operations?

MINI-LAB

Work with a partner. 🖩 calculator

Find the value of $5 + 8 \div 4 \times 3$.

Method 1
Use a calculator.

$5 \boxed{+} 8 \boxed{\div} 4 \boxed{\times} 3 \boxed{=} \ 11$

Method 2
Use paper and pencil.

Follow the order of operations.

$$5 + 8 \div 4 \times 3 = 5 + 2 \times 3$$
$$= 5 + 6$$
$$= 11$$

This calculator follows the order of operations.

Try This

1. Using your calculator, evaluate $3 + 2 \times 5 - 7$.

2. Evaluate this expression using paper and pencil and the order of operations.

Talk About It

3. Compare your answers. Are they the same? If so, why are they the same? If not, why are they different?

Communicating Mathematics

Read and study the lesson to answer each question.

1. *List* each step of the order of operations.

2. *Identify* the first step when evaluating the expression $25 + 2 - 6 \div 3 \times 2$.

Guided Practice

Name the operation that should be done first. Then find the value of the expression.

3. $22 + 5 - 13$

4. $13 - 4 + 7 \times 6$

5. $9 \times 15 \div 5 + 6$

6. $27 - 8 \div 4 \times 3$

7. *Money Matters* Mrs. Jamison's class is planning a trip to the zoo. Tickets cost $6 for children and $15 for adults.

  a. Write an expression for the total cost of 12 adult tickets and 28 children's tickets.

  b. Find the total cost of the tickets.

EXERCISES

Practice

Find the value of each expression.

8. $16 + 5 - 7$

9. $15 \times 8 - 3$

10. $85 \div 17 \times 14$

11. $29 + 56 \div 4$

12. $17 - 3 \times 4 + 6$

13. $21 + 18 \div 6 - 9$

14. $68 + 19 - 7 \times 5$

15. $19 + 45 \div 3 - 11$

16. $26 \div 13 + 9 \times 6 - 21$

17. $9 + 12 - 6 \times 4 \div 2$

18. Evaluate the expression $67 - 21 \div 7 \times 8 + 10$.

19. Find the value of $21 \div 3 + 10 \times 9 - 42$.

20. What is the value of 144 divided by 12 plus 3 times 7 minus 6?

Family Activity

Ask a family member to help you make a list of the steps involved in a project around your home, such as painting a room. Discuss how the order of steps affects the result.

Applications and Problem Solving

21. *Nutrition* The table shows the number of fat grams in one scoop of various ice cream flavors.

  a. Yoko ordered 2 scoops of chocolate chip ice cream and 1 scoop of pralines 'n cream. Robert ordered 3 scoops of chocolate almond. Write an expression for the total number of fat grams in the two orders.

Flavor	Fat (g)
Chocolate Almond	18
Chocolate Chip	15
Pralines 'n Cream	14
Rainbow Sherbet	2

  b. Find the total number of fat grams in the two orders.

22. *Entertainment* The Music Club is planning a production of a play. In order to determine the number of tickets they can sell, they need to know how many seats are in the auditorium. The auditorium has 3 sections, and each section has 24 rows. Sections A and C have 15 seats in each row and section B has 20 seats in each row.

   a. Write an expression for the total number of seats in the auditorium.

   b. How many seats are in the auditorium?

23. *Working on the* **CHAPTER Project** Refer to the tables on page 3.

   a. You decide to order 3 life-size cut-outs for $29.95 each from the Novelty Gift catalog. You also want 2 T-shirts that cost $16.95 each. Write an expression for the total cost of the order including shipping. Find the total cost of the order.

   b. Find a catalog from which you would like to order. Select 3 of one item and 4 of a different item. Write an expression for the total cost of your order including shipping. Find the total cost of the order.

24. *Critical Thinking* Write an expression using four numbers and at least two of the operations +, −, ×, or ÷ in which the value of the expression is 11.

Mixed Review

25. Round 9,046 to the nearest thousand. *(Lesson 1-3)*

26. *Patterns* Find the next three numbers in the pattern 5, 10, 15, 20, … . *(Lesson 1-2)*

27. *Standardized Test Practice* Brandi can run a mile in 7 minutes. At this rate, how long will it take her to run 8 miles? *(Lesson 1-1)*

   A 58 min   **B** 56 min   **C** 54 min   **D** 15 min

For **Extra Practice,**
see page 558.

CHAPTER 1

Mid-Chapter Self Test

Use the four-step plan to solve each problem. *(Lesson 1-1)*

1. Thomas purchased a CD game. The total amount, including tax, was $43. How much change did he receive from $50?

2. On average, an elephant's heart beats 30 times per minute. At this rate, how many times would an elephant's heart beat in 1 hour?

Find the next three numbers in each pattern. *(Lesson 1-2)*

3. 5, 13, 21, 29, _?_, _?_, _?_   **4.** 3, 12, 48, 192, _?_, _?_, _?_

Round each number to the underlined place-value position. *(Lesson 1-3)*

5. 1$\underline{5}$4   **6.** $\underline{9}$,511

7. Estimate 18,392 divided by 317.

Find the value of each expression. *(Lesson 1-4)*

8. $19 - 6 + 24$ **9.** $75 \div 3 - 8 \times 2$ **10.** $132 + 4 \times 9 \div 6 - 53$

COOPERATIVE LEARNING

paper bags

popcorn

1-5A Variables and Expressions

A Preview of Lesson 1-5

In algebra, we often use unknown values. Consider the expression *the sum of three and some number*. The "some number" is an unknown value. When you assign a value to "some number," then you can find the value of the expression.

TRY THIS

Work with a partner.

1 To model the expression *the sum of three and some number*, place three pieces of popcorn next to an empty paper bag. The three pieces of popcorn represent the known value, 3. The bag represents the unknown value. It can contain any number of popcorn pieces.

the sum of three and some number

2 To evaluate the expression *the sum of three and some number*, follow these steps.

- Have one partner assign a value to "some number" by placing any number of the remaining pieces of popcorn in the bag.

- Empty the bag and count the pieces of popcorn that are in the bag. Add this to the other 3 pieces of popcorn. This total number is the value of the expression.

ON YOUR OWN

1. What is the "some number?"

2. In the expression *the sum of three and some number*, replace "some number" with its known value. What is the value of the expression?

3. Find the value of *some number plus eight* if "some number" is 5.

4. *Look Ahead* Find the value of $m + 7$ if $m = 12$.

20 Chapter 1 Problem Solving, Numbers, and Algebra

Work with a partner.

3 To model the expression *two times some number,* place two bags in front of you. Each bag represents the same unknown value.

two times some number

4 To evaluate the expression *two times some number,* follow these steps.

- Have one partner assign a value to "some number" by placing the same number of pieces of popcorn in each bag.

- Empty the bags and count all of the pieces of popcorn. The number of pieces of popcorn is the value of the expression.

ON YOUR OWN

5. What is the "some number?"

6. In the expression *two times some number,* replace "some number" with its known value. What is the value of the expression?

7. How many bags would you need to model the expression *five times a number*?

8. *Look Ahead* Find the value of $4 \times h$ if $h = 6$.

Integration: Algebra
Variables and Expressions

What you'll learn

You'll learn to evaluate numerical and simple algebraic expressions.

When am I ever going to use this?

Knowing how to evaluate algebraic expressions can help you find the miles per gallon achieved by a car.

Word Wise

algebra
variable
algebraic expression
evaluate

Western Park Mall offers a gift wrapping service. The cost for gift wrapping is found by adding the length and the width of the box and multiplying by 5 cents. The table below shows how to find the cost of wrapping the most common size boxes. Study the pattern.

Length	Width	Length + Width	Cost (cents)
10	4	10 + 4	$(10 + 4) \times 5$
12	6	12 + 6	$(12 + 6) \times 5$
16	8	16 + 8	$(16 + 8) \times 5$
20	10	20 + 10	$(20 + 10) \times 5$

In mathematics, **algebra** is a language of symbols. By extending the table, using the letters L and W, the following table results.

Length	Width	Length + Width	Cost (cents)
L	W	$L + W$	$(L + W) \times 5$

The letters L and W are called **variables**. A variable is a symbol, usually a letter, used to represent a number. Any letter may be used as a variable. In this problem, the letter L represents length and W represents width.

The expression $(L + W) \times 5$ is called an **algebraic expression**. Algebraic expressions are combinations of variables, numbers, and at least one operation. $(L + W) \times 5$ is an algebraic expression for the cost of gift wrapping

The variables can be replaced with any number. Once the variables have been replaced, you can **evaluate**, or find the value of, the algebraic expression.

Example

1. Evaluate $14 + c$ if $c = 32$.

$$14 + c = 14 + 32 \quad \textit{Replace c with 32.}$$
$$= 46 \quad \textit{Add 14 and 32.}$$

2 Evaluate $x - y$ if $x = 47$ and $y = 13$.

$x - y = 47 - 13$ *Replace x with 47 and y with 13.*

$ = 34$ *Subtract 13 from 47.*

In algebra, there are several ways to show multiplication.

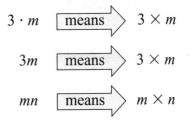

$3 \cdot m$ [means] $3 \times m$

$3m$ [means] $3 \times m$

mn [means] $m \times n$

Be sure to use the order of operations when you evaluate algebraic expressions.

3 Evaluate $3m + 2 \cdot 4$ if $m = 12$.

$3m + 2 \cdot 4 = 3 \times 12 + 2 \cdot 4$ *Replace m with 12.*

$ = 36 + 2 \cdot 4$ *Multiply 3 and 12.*

$ = 36 + 8$ *Multiply 2 and 4.*

$ = 44$ *Add 36 and 8.*

Study Hint

Reading Math The expression $3m + 2 \cdot 4$ is read as *three times m plus two times four.*

4 Evaluate $4 + mn$ if $m = 6$ and $n = 2$.

$4 + mn = 4 + 6 \times 2$ *Replace m with 6 and n with 2.*

$ = 4 + 12$ *Multiply 6 and 2.*

$ = 16$ *Add 4 and 12.*

APPLICATION

Real World

5 **Hobbies** To find distance traveled, you can use the expression $r \times t$, where r represents rate and t represents time. Find the distance a hot air balloon traveled if its rate was 13 miles per hour and it traveled for 4 hours.

Evaluate $r \times t$ if $r = 13$ and $t = 4$.

$r \times t = 13 \times 4$ *Replace r with 13 and t with 4.*

$ = 52$ *Multiply 13 and 4.*

The hot air balloon traveled 52 miles.

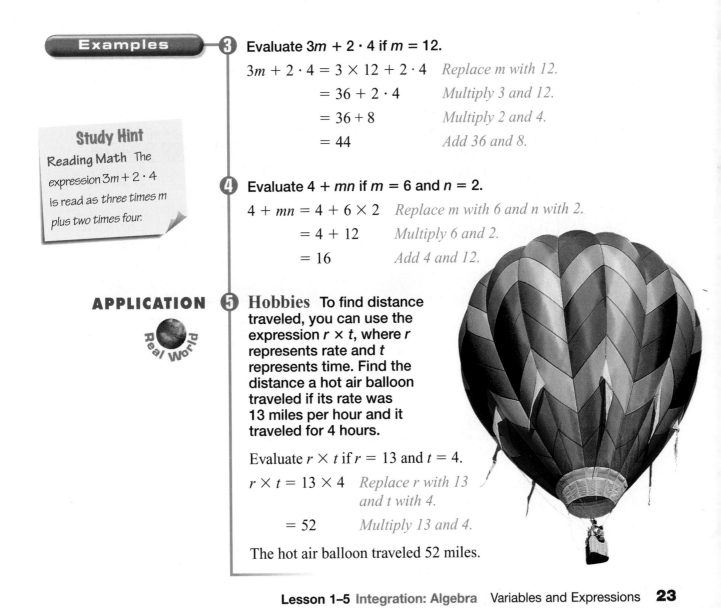

CHECK FOR UNDERSTANDING

Communicating Mathematics

Read and study the lesson to answer each question.

1. *Write* two algebraic expressions that mean the same as $5r$.

2. *Copy and complete* the chart.

Algebraic Expressions	Variables	Numbers	Operations
$4x - 2y$			
$3r - 2s + 5t$			

3. *You Decide* Laura says that the value of $g + h \div 2$ when $g = 2$ and $h = 14$ is 8. Nicholas says the answer is 9. Who is correct? Explain your reasoning.

Guided Practice

Evaluate each expression if $a = 3$ and $b = 12$.

4. $a + 5$ 5. $48 \div b$ 6. $7ab$ 7. $2a + b$

Evaluate each expression if $c = 5$, $d = 2$, and $f = 8$.

8. $f - d$ 9. $32 \div d$ 10. $6d + c$ 11. $c + d + f$

12. *Geometry* The expression $\ell \times w$ gives the area of a rectangle. The letter ℓ represents the length of the rectangle, and w represents the width. Find the area of each rectangle.

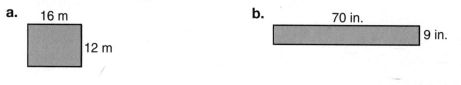

a. 16 m 12 m

b. 70 in. 9 in.

EXERCISES

Practice

Evaluate each expression if $n = 5$ and $p = 7$.

13. $n + 5$ 14. $n \times 0$ 15. $4p - 6$ 16. $12 - n$

17. $3np$ 18. $8 + p$ 19. $2n \times 6$ 20. $n \div 1$

21. $p \times n$ 22. $2p - p$ 23. $3n + p$ 24. $21 \div 3 \times p$

Evaluate each expression if $r = 6$, $s = 12$, and $t = 3$.

25. $s - t$ 26. $s \div 2$ 27. rt 28. $4 + 2r$

29. $t \times 18$ 30. $48 \div s$ 31. $s - r + t$ 32. $rt - s$

33. $3s - 4t$ 34. $4 + 2rt$ 35. $3rt - 2s$ 36. $s - 2r \div t$

37. *Algebra* Find the value of $3z \div y$ if $y = 3$ and $z = 9$.

38. *Algebra* What is the value of $2xyz - 26$ if $x = 3$, $y = 6$, and $z = 10$?

39. *Geometry* To find the diameter of a circle, you can use the expression $2r$, where r is the length of the radius. Find the diameter of a circle whose radius is 6 inches.

6 in.

40. *Travel* To find the miles per gallon achieved by a car, you can use the expression $m = d \div g$ where m represents miles per gallon, d represents the distance traveled, and g represents the number of gallons of gasoline used.

a. Find the miles per gallon for a car that travels 180 miles on 5 gallons of gasoline.

b. Find the miles per gallon for a car that travels 256 miles on 8 gallons of gasoline.

41. *Critical Thinking* Christina and Shawon each have a calculator. Shawon starts at zero and adds three each time. Christina starts at 100 and subtracts seven each time. If they press their keys at the same time, will their displays ever show the same number at the same time? If so, what is the number?

Mixed Review

42. Find the value of $36 \div 9 + 7 \times 2$. *(Lesson 1-4)*

43. Evaluate $2 \times 18 \div 3 + 9$. *(Lesson 1-4)*

44. *Shopping* The graph shows how much consumers spent shopping on-line recently. *(Lesson 1-3)*

a. Estimate the total amount spent shopping on-line.

b. About how much more was spent on travel than on apparel?

Shopping On-Line

	(millions)
$140	Computer products
$126	Travel
$85	Entertainment
$46	Apparel
$45	Gifts/flowers
$39	Food/drink

45. **Standardized Test Practice** What number is missing from this pattern?

$\ldots, \square, 35, 41, 47, 53, \ldots$ *(Lesson 1-2)*

A 29

B 31

C 26

D 24

46. *Business* In one month, Pam collected $51 from the customers on her paper route. If she owed the newspaper company $32, how much was her profit? *(Lesson 1-1)*

For **Extra Practice**, see page 558.

GRAPHING CALCULATORS

1-5B Evaluating Expressions

A Follow-Up of Lesson 1-5

graphing calculator

Graphing calculators observe the order of operations. So, you can enter an expression just as it is written to evaluate it. You can also use parentheses keys on a graphing calculator to indicate multiplication.

The expression appears on the screen as you enter it. On many calculators, the multiplication and division signs appear on the screen as * and /.

TRY THIS

Work with a partner. ✎

Evaluate $3(x - 6) + 1$ if $x = 8$.

3 ⎣(8 ⎣– 6 ⎣) ⎣+ 1 [ENTER]   *You can use arrow keys to go back and correct any error you make by typing over it, or by using the [INS] (insert) key, or by using the* [DEL] *(delete) key.*

If $x = 8$, the value of $3(x - 6) + 1$ is 7.

If you discover that you entered the expression incorrectly, you can use the REPLAY feature to correct your error and reevaluate without reentering your expression. Press [2nd] [ENTRY]. Then use the arrow keys to move to the location of the correction. After you make the correction, press [ENTER] to evaluate. You don't have to move the cursor to the end of the line.

ON YOUR OWN

Use a graphing calculator to evaluate each expression if $x = 4$, $y = 7$, and $z = 9$.

1. $12 - z$ **2.** $x + 9$ **3.** xy **4.** $2(18 - z)$

5. $x \div 2 + 12$ **6.** $\frac{2(z-x)}{(y-2)}$ **7.** $14 + \frac{3x}{2}$ **8.** $x(y + z) - x$

9. Evaluate the expression $5(3 + 8)$ with a scientific calculator and with a graphing calculator. Can you enter the expression the same way on both calculators? If not, explain the difference.

BUSINESS

Gary C. Comer
MAIL-ORDER BUSINESS OWNER

Gary C. Comer is the founder and Chairman of the Board of a major mail-order catalog. When he first founded the company in 1963, there was one catalog in which sailboat hardware and equipment was sold. Today, there are eight different catalogs. Consumers can purchase unique clothing and accessories, and products for the home.

If you are interested in owning your own mail-order business, you should consider taking courses in business management, accounting, and mathematics. A bachelor's degree in marketing and advertising, communications, business management, or accounting is also helpful.

For more information:
The Direct Marketing Association
1120 Avenue of the Americas
New York, NY 10036-6700

interNET
CONNECTION
www.glencoe.com/sec/
math/mac/mathnet

Someday, I'd like to start a mail-order business. I would travel around the world to find unusual products to sell in my catalog.

Your Turn
Design a catalog for a mail-order business you would like to start. Select a name for your company and describe and illustrate the products you want to sell.

Integration: Algebra
Powers and Exponents

What you'll learn

You'll learn to use powers and exponents in expressions.

When am I ever going to use this?

Knowing how to use powers and exponents can help you determine the amount of water needed to fill an aquarium.

Word Wise

factor
exponent
base
power
squared
cubed

The state of Hawaii has an area of more than 10,000 square miles. You can express 10,000 as the product $10 \cdot 10 \cdot 10 \cdot 10$.

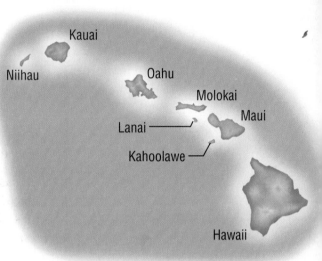

When two or more numbers are multiplied, each number is called a **factor** of the product. For $3 \times 12 = 36$, 3 and 12 are factors of 36.

When the same factor is repeated, you can use an **exponent** to simplify the notation. An exponent tells how many times a number, called the **base**, is used as a factor.

$$10{,}000 = \underbrace{10 \cdot 10 \cdot 10 \cdot 10}_{four\ factors} = 10^4 \leftarrow exponent$$

$base$ 10^4 *is a power of 10.*

A **power** is a number that is expressed using exponents. The powers 12^2, 5^3, and 7^4 are read as follows.

Symbols	Words
12^2	12 to the second power or 12 **squared.**
5^3	5 to the third power or 5 **cubed.**
7^4	7 to the fourth power.

Examples

① Write $8 \cdot 8 \cdot 8 \cdot 8$ using exponents.

The base is 8.
Since 8 is a factor four times, the exponent is 4.
$8 \cdot 8 \cdot 8 \cdot 8 = 8^4$

② Write 4^3 as a product. Then evaluate.

The base is 4.
The exponent 3 means that 4 is a factor three times.
$$4^3 = 4 \cdot 4 \cdot 4$$
$$= 64$$

Since powers are a form of multiplication, the rules for the order of operations need to be expanded to include them.

Order of Operations	1. Do all powers before other operations.
	2. Multiply and divide in order from left to right.
	3. Add and subtract in order from left to right.

Example **3** Evaluate $5 \cdot 3^2 - 8$.

$$5 \cdot 3^2 - 8 = 5 \cdot 9 - 8 \quad \textit{Evaluate } 3^2 \textit{ first. } 3^2 = 3 \cdot 3 \textit{ or } 9.$$
$$= 45 - 8 \quad \textit{Multiply 5 and 9.}$$
$$= 37 \quad \textit{Subtract 8 from 45.}$$

Powers and exponents are often used in algebra.

Examples
INTEGRATION

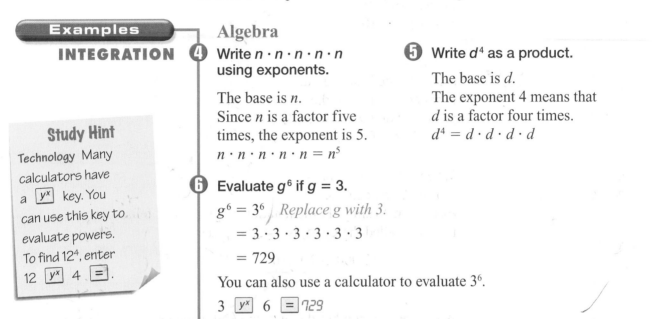

Algebra

4 Write $n \cdot n \cdot n \cdot n \cdot n$ using exponents.

The base is n.
Since n is a factor five times, the exponent is 5.
$$n \cdot n \cdot n \cdot n \cdot n = n^5$$

5 Write d^4 as a product.

The base is d.
The exponent 4 means that d is a factor four times.
$$d^4 = d \cdot d \cdot d \cdot d$$

6 Evaluate g^6 if $g = 3$.

$$g^6 = 3^6 \quad \textit{Replace g with 3.}$$
$$= 3 \cdot 3 \cdot 3 \cdot 3 \cdot 3 \cdot 3$$
$$= 729$$

You can also use a calculator to evaluate 3^6.

3 [y^x] 6 [=] 729

Study Hint

Technology Many calculators have a [y^x] key. You can use this key to evaluate powers. To find 12^4, enter 12 [y^x] 4 [=].

CHECK FOR UNDERSTANDING

Communicating Mathematics

Read and study the lesson to answer each question.

1. *Write* a paragraph explaining the terms *exponent, base,* and *power.*

2. *Demonstrate* how to evaluate 5^8 using a calculator.

3. *You Decide* Esteban says that the value of $3^2 + 2^4 \cdot 6$ is 150. Tamika says the value is 105. Who is correct? Explain your reasoning.

Guided Practice

Write each product using exponents.

4. $2 \cdot 2 \cdot 2 \cdot 2 \cdot 2 \cdot 2$ 5. $m \cdot m \cdot m \cdot m \cdot m$ 6. $3 \cdot 3 \cdot 5 \cdot 5 \cdot 5$

Write each power as a product.

7. c^3 8. 4^7 9. h^6

Evaluate each expression.

10. 2^3 11. 10^5 12. $5^2 + 4 \cdot 2$

13. *Foreign Languages* Mandarin Chinese is spoken by more people in the world than any other language. An estimated 10^9 people speak this language. About how many people in the world speak Mandarin Chinese?

Practice

Write each product using exponents.

14. $11 \cdot 11 \cdot 11$ **15.** $3 \cdot 3 \cdot 3 \cdot 3 \cdot 3 \cdot 3$ **16.** $w \cdot w \cdot w$

17. $3 \cdot 3 \cdot 4 \cdot 4 \cdot 4 \cdot 4$ **18.** $15 \cdot 15 \cdot 8 \cdot 8$ **19.** $r \cdot r \cdot r \cdot r$

20. $y \cdot y \cdot y \cdot z \cdot z$ **21.** $6 \cdot 6 \cdot 6 \cdot 1 \cdot 1 \cdot 1$ **22.** $a \cdot a \cdot b \cdot b$

Write each power as a product.

23. 14^2 **24.** 21^4 **25.** 7^5 **26.** $9^3 \cdot 2^4$

27. 16^6 **28.** a^7 **29.** $x^3 \cdot y^4$ **30.** $b^2 \cdot c^4 \cdot d^3$

Evaluate each expression.

31. 10^3 **32.** 6^3 **33.** 2^5

34. 16^2 **35.** 3 squared   **36.** 4 cubed

37. $4^5 + 3$ **38.** $4^2 \cdot 3^3 \cdot 6$ **39.** $4 \cdot 5^2 + 3$

40. *Algebra* What is the value of $x^3 + y^2$ if $x = 3$ and $y = 6$?

41. Complete the table below.

$3^4 = 81$	$5^4 = 625$	$10^4 = 10,000$
$3^3 = 27$	$5^3 = 125$	$10^3 = 1,000$
$3^2 = 9$	$5^2 = 25$	$10^2 = 100$
$3^1 = 3$	$5^1 = 5$	$10^1 = \underline{?}$
$3^0 = \underline{?}$	$5^0 = \underline{?}$	$10^0 = \underline{?}$

Applications and Problem Solving

42. *Earth Science* A light year is the distance light travels in one year. The Milky Way galaxy is about 100,000 light years wide. Write 100,000 as a power with 10 as the base.

43. *Geometry* The volume of a cube can be found by using the expression $V = s^3$, where s is the length of a side. The aquarium shown at the right is in the shape of a cube. What is its volume? *The volume will be expressed in cubic units.*

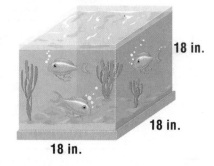

18 in.

18 in.

18 in.

44. *Critical Thinking* Evaluate 10^3, 10^4, 10^5, and 10^6. Explain how you can evaluate 10^{20} without using a calculator.

Mixed Review

45. *Algebra* Evaluate the expression $ab - a$ if $a = 7$ and $b = 3$. *(Lesson 1-5)*

46. **Standardized Test Practice** The Marcos family is going to a football game. Tickets cost $5 for adults and $2 for students. Mrs. Marcos will need 2 adult tickets and 4 student's tickets. Which expression could be used to find the cost in dollars of the tickets? *(Lesson 1-4)*

     A $2 + \$5 \times 4 + \$2$ **B** $\$5 + \$2 + 6$

     C $\$2 + 4 \times \$5 + 2$ **D** $6 \times \$5 + \$2$

     E $2 \times \$5 + 4 \times \$2$

47. Use rounding to estimate $398 + 688 + 241$. Explain whether 1,337 is a reasonable answer. *(Lesson 1-3)*

48. *Patterns* Find the next three numbers in the pattern 7, 12, 17, 22, ... *(Lesson 1-2)*

For **Extra Practice**, see page 558.

49. *School* The number of students at Aurora Junior High is half the number it was twenty years ago. If there were 758 students twenty years ago, how many students are there now? *(Lesson 1-1)*

Let the Games Begin

EXPO Bingo

Math Skill
Powers and Exponents

Get Ready This game is for two to four players.

🖩 calculator

Get Set Each player copies the EXPO bingo playing card onto a sheet of paper. Then each player selects 16 different powers and writes them in any of the upper right hand corner boxes. Choose a person to be the bingo caller.

EXPO BINGO CARD

				5^2	12^2	3^3	2^4
7^2	2^3	8^2	3^2
1^3	4^3	9^2	11^2
6^2	14^2	5^3	15^2
10^3	3^4	13^2	2^2

Go ● The caller reads one power at a time. Each player marks the space by writing the equivalent number in the larger box.

● The first player with four powers listed in any row, column, or diagonal wins.

🖥 *inter***NET** CONNECTION Visit www.glencoe.com/sec/math/mac/mathnet for more games.

PROBLEM SOLVING

1-7A Guess and Check

A Preview of Lesson 1-7

Yvonne and her friend Catalina are putting together a computer generated jigsaw puzzle. They are trying to decide the best way to start the puzzle. Let's listen in!

Yvonne

Do you know what the picture is supposed to look like?

I can't tell. Maybe we should start by finding the piece that will fit in the lower right-hand corner.

All right. Three of the four corner pieces are already in place. Let's look for the fourth corner piece.

It is the one that has a straight edge on the right side and on the bottom.

Yes! That piece looks like it will fit.

Catalina

THINK ABOUT IT

Work with a partner.

1. **Analyze** Catalina's and Yvonne's thinking. Do you agree with their thinking? Why or why not?

2. **Think** of another way for Catalina and Yvonne to start the puzzle.

3. **Choose** the puzzle piece that will fit on the right side of the top left-hand puzzle piece. Explain your reasoning.

4. **Apply** what you have learned from Catalina's and Yvonne's situation to solve the following problem.

   Find an even number between 50 and 60 that is divisible by 2 and 3.

For **Extra Practice**, see page 559.

ON YOUR OWN

5. The second step of the four-step plan for problem solving asks you to make a *plan* for solving the problem. **Tell** how you can use the **guess-and-check** strategy to help you make a plan for solving a problem.

6. *Write a Problem* that can be solved using the guess-and-check strategy. Then ask a classmate to solve the problem.

7. *Explain* how you could use the guess-and-check strategy to answer Exercise 35 on page 36.

MIXED PROBLEM SOLVING

STRATEGIES

Look for a pattern.
Solve a simpler problem.
Act it out.
Guess and check.
Draw a diagram.
Make a chart.
Work backward.

Solve. Use any strategy.

8. *Money Matters* Kamaria earns $5 per hour baby-sitting. Last week, she worked 16 hours. How much money did Kamaria make?

9. *Life Science* Last week, the Mahoning Animal Shelter sent a total of 24 dogs and cats to new homes. There were 6 more dogs than cats. How many of each were adopted?

10. *Entertainment* The chart below gives admission costs to Mount Carmel's health fair. Twelve people paid a total of $50 for admission. If 8 children attended the health fair, how many adults and senior citizens attended?

Admission Costs	
Adults	$6
Children	$4
Senior Citizens	$3

11. *Life Science* Each hand in the human body has 27 bones. There are 6 more bones in your fingers than in your wrist. There are 3 fewer bones in your palm than in your wrist. How many bones are in each part of your hand?

12. *Money Matters* Refer to the chart below. Olivia and Imena each got a perm, a manicure, and a bottle of shampoo. About how much money did they spend?

Alberto's Salon	
Cut/Style	$12
Perm	$56
Manicure	$17
Shampoo	$ 8

13. *Physical Science* Sound travels through air at a speed of 1,129 feet per second. At this rate, how far will sound travel in 1 minute?

14. *Sports* Jamal's basketball team won twice as many games as they lost. They lost 8 games. How many games did they win?

15. *Standardized Test Practice* Jorge purchased a new car. His loan, including interest, is $10,740. How much are his monthly payments if he has 60 payments to make?

A $195

B $179

C $135

D $159

E Not Here

Lesson 1-7A THINKING LAB   **33**

Integration: Algebra
Solving Equations

What you'll learn

You'll learn to solve equations by using mental math and guess and check.

When am I ever going to use this?

Knowing how to solve equations can help you determine the number of new students in your school.

Word Wise

equation
solve
solution

On March 23, 1775, ___?___ said, "*I know not what course others may take; but as for me, give me liberty, or give me death!*"

This sentence is neither true nor false until you fill in the blank. If you answer Patrick Henry, the sentence is true. If you answer George Washington, Harriet Tubman, or any other historical figure, the sentence is false.

In mathematics, an **equation** is a sentence that contains an equal sign, $=$. Equations can be either true or false.

$$34 - 12 = 22 \quad \text{\textit{This equation is true.}}$$
$$45 \times 2 = 100 \quad \text{\textit{This equation is false.}}$$

An equation like $x - 4 = 5$ is neither true nor false until x is replaced with a number.

Study Hint

Reading Math The equation $x - 4 = 5$ is read as *x minus four is equal to five.*

Replace x with 10.

$$x - 4 = 5$$
$$10 - 4 \stackrel{?}{=} 5$$
$$6 = 5$$

This sentence is false.

Replace x with 9.

$$x - 4 = 5$$
$$9 - 4 \stackrel{?}{=} 5$$
$$5 = 5$$

This sentence is true.

When you replace a variable with a value that makes the equation true, you **solve** the equation. The replacement that makes the equation true is called a **solution** of the equation. The solution of $x - 4 = 5$ is 9.

Example

1 Which of the numbers 3, 4, or 5 is the solution of $y + 8 = 12$?

Replace y with 3.

$$y + 8 = 12$$
$$3 + 8 \stackrel{?}{=} 12$$
$$11 = 12$$

This sentence is false.

Replace y with 4.

$$y + 8 = 12$$
$$4 + 8 \stackrel{?}{=} 12$$
$$12 = 12 \checkmark$$

This sentence is true.

Replace y with 5.

$$y + 8 = 12$$
$$5 + 8 \stackrel{?}{=} 12$$
$$13 = 12$$

This sentence is false.

The solution is 4 because replacing y with 4 results in a true sentence.

In Example 1, $y + 8 = 12$ was solved by replacing y with a number until a true sentence resulted.

Some equations can be solved mentally by using basic facts or arithmetic skills you already know well.

Examples

Solve each equation using mental math.

2 $2m = 10$

$2 \times \mathbf{5} = 10$ *You know that $2 \times 5 = 10$.*

$ 10 = 10$

The solution is 5.

3 $7 = 5 + d$

$7 = 5 + \mathbf{2}$ *You know that $5 + 2 = 7$.*

$7 = 7$

The solution is 2.

You can also solve equations by using guess and check.

Example

 **APPLICATION**

4 **School** Last year, the enrollment in Pointview Middle School was 695 students. This year there are 748 students. If n represents the number of new students, the equation $695 + n = 748$ results. Find the number of new students.

Use guess and check.

Try 50.	*Try 52.*	*Try 53.*
$695 + n = 748$	$695 + n = 748$	$695 + n = 748$
$695 + 50 \overset{?}{=} 748$	$695 + 52 \overset{?}{=} 748$	$695 + 53 \overset{?}{=} 748$
$745 = 748$	$747 = 748$	$748 = 748$ ✓
This sentence is false.	This sentence is false.	This sentence is true.

The solution is 53. So, there are 53 new students.

Study Hint

Estimation You know that 695 is about 700 and 748 is about 750. Think: $700 + 50 = 750$. So the answer should be close to 50.

CHECK FOR UNDERSTANDING

Communicating Mathematics

Read and study the lesson to answer each question.

1. **Write** an example of an algebraic expression. Write an example of an equation. How are they alike? How are they different?

2. **State** an equation with a solution of 7.

Tell whether the equation is *true* or *false* by replacing the variable with the given value.

3. $18 + j = 45; j = 29$

4. $y \div 7 = 35; y = 245$

Identify the solution to each equation from the list given.

5. $b + 5 = 12; 5, 6, 7$

6. $36 - h = 15; 21, 22, 23$

7. $72 \div d = 18; 3, 4, 5$

8. $16r = 96; 4, 5, 6$

Solve each equation mentally.

9. $4 + x = 13$

10. $5w = 20$

11. $z - 8 = 2$

12. *Shopping* Ana purchased a helmet and some reflectors for her bike. The cost of the helmet was \$25. She spent a total of \$32.

   a. Let c represent the cost of the reflectors. Write an equation that represents the total cost of the items.

   b. Solve your equation to find the cost of the reflectors.

EXERCISES

Practice

Tell whether the equation is *true* or *false* by replacing the variable with the given value.

13. $p + 11 = 27; p = 16$

14. $4w = 24; w = 6$

15. $t - 15 = 32; t = 47$

16. $57 \div d = 21; d = 3$

17. $93 = 3v; v = 32$

18. $y \div 39 = 6; y = 234$

Identify the solution to each equation from the list given.

19. $8 + h = 23; 14, 15, 16$

20. $t - 6 = 3; 5, 7, 9$

21. $42 = 6a; 7, 8, 9$

22. $23 + n = 56; 31, 32, 33$

23. $b = 65 \div 5; 11, 13, 15$

24. $31 - m = 19; 10, 11, 12$

25. $k + 38 = 45; 7, 9, 11$

26. $76 \div q = 19; 3, 4, 5$

27. $11 = d \div 4; 41, 43, 44$

28. $7 \div g = 1; 6, 7, 8$

Solve each equation mentally.

29. $m - 9 = 17$

30. $3 + j = 18$

31. $12 = d + 5$

32. $22 = 2y$

33. $24 \div w = 8$

34. $4n = 32$

35. $p \div 5 = 45$

36. $a = 4 + 13$

37. $18 \div s = 9$

38. *Algebra* Which of the numbers 3, 6, or 9 is the solution of $7h = 63$?

39. *Algebra* Find the solution of $176 = 11s$.

40. *Banking* Currency rates are calculated daily. If 140*d* represents the number of Spanish pesetas for each United States dollar, represented by *d*, how many pesetas will be received in exchange for $50?

41. *Food* Celina and Marcus went to a fast-food restaurant for lunch. Celina ordered a hamburger with cheese, one order of onion rings, and water. There are 19 grams of fat in one order of onion rings. There is a total of 65 grams of fat in the entire meal.

  a. Let *g* represent the number of grams of fat in a hamburger with cheese. Write an equation to find how many grams of fat are in a hamburger with cheese.

  b. Solve your equation to find the number of grams of fat in a hamburger with cheese.

42. *Critical Thinking* Translate each sentence into an equation. Then solve.

  a. *Eight less than m is equal to 5.*

  b. *The sum of h and 7 is equal to 18.*

  c. *The product of 4 and 16 is equal to n.*

Mixed Review

43. **Standardized Test Practice** How is $2 \times 3 \times 3 \times 11$ expressed in exponential notation? *(Lesson 1-6)*

  A $2 \times 2 \times 3 \times 11$

  B $2^3 \times 11$

  C $2 \times 3^2 \times 11$

  D $2 \times 2^3 \times 11$

44. *Algebra* Evaluate $3x - 2 \cdot 4 + y$ if $x = 7$ and $y = 1$. *(Lesson 1-5)*

45. Find the value of $11 - 3 \div 3 + 2$. *(Lesson 1-4)*

46. *Transportation* The steamer *Alaska* traveled 182 miles in 5 days. Use rounding to estimate how many miles it averaged each day. *(Lesson 1-3)*

47. *Sports* The graph shows the average cost of professional hockey, football, basketball, and baseball tickets. Camilia is planning to attend a hockey and a football game. About how much will she spend on the two tickets? *(Lesson 1-3)*

For **Extra Practice,**
see page 559.

Sports Ticket Costs

NFL $35.74

NHL $34.79

NBA $31.80

MLB $11.13

Source: Team Marketing Report

Study Guide and Assessment

interNET Chapter Review **For additional lesson-by-lesson review, visit:**
CONNECTION **www.glencoe.com/sec/math/mac/mathnet**

Vocabulary

After completing this chapter, you should be able to define each term, concept, or phrase and give an example or two of each.

Algebra
algebra (p. 22)
algebraic expression (p. 22)
equation (p. 34)
evaluate (p. 22)
solution (p. 34)
solve (p. 34)
variable (p. 22)

Number and Operations
base (p. 28)
cubed (p. 28)
exponent (p. 28)
factor (p. 28)
order of operations (p. 16)
power (p. 28)
squared (p. 28)

Problem Solving
guess and check (p. 32)

Understanding and Using the Vocabulary

Choose the correct term or number to complete each sentence.

1. When finding the value of an expression, first (multiply and divide, add and subtract) in order from left to right.

2. $(7, h + 2)$ is an algebraic expression.

3. The base of 4^5 is (4, 5).

4. When the same factor is repeated, a(n) (variable, exponent) can simplify the notation.

5. The factors of 12 are (3 and 4, 6 and 6).

6. Four cubed is written as $(4^2, 4^3)$.

7. $2 \cdot 2 \cdot 2 \cdot 2 \cdot 2$ can be written as $(2^5, 5^2)$.

8. A symbol, usually a letter, used to represent a number is called a (variable, power).

9. If $d = 4$, then the value of the expression $14 - d$ is (10, 18).

10. $(3x = 6, 2x + y)$ is an equation.

11. The solution of $16 - g = 5$ is (21, 11).

In Your Own Words

12. *Summarize* the rules for the order of operations.

Objectives & Examples

Upon completing the chapter, you should be able to:

● solve problems using the four-step plan *(Lesson 1-1)*

Use the four-step plan to solve.
Migina studied 2 hours each day for 10 days. How many hours has she studied?

Explore	Migina studied 2 hours each day for 10 days. Find the total number of hours studied.
Plan	Multiply 2 by 10.
Solve	$2 \times 10 = 20$
	Migina studied 20 hours.
Examine	The answer makes sense and is reasonable.

● solve problems using patterns *(Lesson 1-2)*

Find the next three numbers for the pattern 1, 4, 7, 10, ?, ?, ?.

Each number is 3 more than the number before it.

$10 + 3 = 13$ $13 + 3 = 16$ $16 + 3 = 19$

The next three numbers are 13, 16, and 19.

● estimate sums, differences, products, and quotients using rounding *(Lesson 1-3)*

Estimate the product of 8 and 57.

$$
\begin{array}{ccc}
57 & \rightarrow & 60 \\
\times\,8 & \rightarrow & \times\,8 \\
\hline
& & 480
\end{array}
$$

Review Exercises

Use these exercises to review and prepare for the chapter test.

Use the four-step plan to solve each problem.

13. *Money Matters* The Music Store sells compact discs for $9.98 each. If Lilla buys 4 CDs, how much will he spend?

14. *School* Tickets to the school dance cost $4 each. The dance committee collected a total of $352. How many students bought tickets?

15. *History* In the 1932 presidential election, Franklin Roosevelt won with 472 electoral votes while Herbert Hoover had 59. How many electoral votes were cast?

Find the next three numbers or shapes in each pattern.

16. 3, 6, 12, 24, ?, ?, ?

17.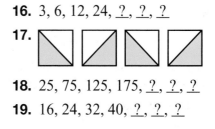

18. 25, 75, 125, 175, ?, ?, ?

19. 16, 24, 32, 40, ?, ?, ?

Round each number to the underlined place-value position.

20. <u>4</u>,650   21. <u>2</u>7

22. 5,3<u>2</u>1   23. 6<u>5</u>4

24. *Money Matters* Chapa purchases a coat for $91, a shirt for $17, and jeans for $24. The cashier asks for $163. Is this amount reasonable?

Objectives & Examples

Review Exercises

● evaluate expressions using the order of operations *(Lesson 1-4)*

Find the value of $6 + 12 \div 3$.

$$6 + 12 \div 3 = 6 + 4$$
$$= 10$$

Find the value of each expression.

25. $9 - 3 \times 2 - 1$

26. $3 \times 7 + 5 \times 9 \div 3$

27. $1 + 8 \div 4 - 1$

28. $4 + 8 - 6 + 10$

29. $2 \times 8 \div 4 + 3$

30. $30 \div 3 + 16 \div 4$

● evaluate numerical and simple algebraic expressions *(Lesson 1-5)*

Evaluate $2a + b$ if $a = 7$ and $b = 6$.

$$2a + b = 2 \times 7 + 6$$
$$= 14 + 6$$
$$= 20 \bullet$$

Evaluate each expression if $a = 4$, $b = 3$, and $c = 12$.

31. $c + a$ **32.** $8 - b$

33. $2a - 1$ **34.** $c - b - a$

35. $c + 3b$ **36.** $3a - 4b$

37. $5a + 2c$ **38.** $2ab - c$

● use powers and exponents in expressions *(Lesson 1-6)*

Evaluate each expression.

$$4^3 = 4 \cdot 4 \cdot 4$$
$$= 64$$

$$6 \cdot 3^2 = 6 \cdot 3 \cdot 3$$
$$= 6 \cdot 9$$
$$= 54$$

Evaluate each expression.

39. 4^6

40. $5^2 \cdot 3^2 \cdot 4^2$

41. x^3 if $x = 5$

42. ten squared

43. Write $t \cdot t \cdot t \cdot t \cdot t$ using exponents.

44. Write 3^6 as a product.

● solve equations by using mental math and guess and check *(Lesson 1-7)*

Solve $m - 2 = 7$ mentally.

$9 - 2 = 7$ *You know that*
$7 = 7$ *$9 - 2 = 7$.*

The solution is 9.

Identify the solution to each equation from the list given.

45. $14a = 42$; 3, 4, 5

46. $b + 23 = 31$; 7, 8, 9

47. $9 = g \div 7$; 61, 62, 63

48. $y - 7 = 12$; 18, 19, 20

Solve each equation mentally.

49. $72 \div x = 8$

50. $m - 13 = 11$

51. $18 = 3w$

52. $12 + j = 21$

Applications & Problem Solving

53. *History* John F. Kennedy began his first term in the House of Representatives in January 1947. In November 1960, he was elected president of the United States. How long had he been in the legislature when he was elected president? *(Lesson 1-1)*

54. *Fitness* Ricardo made a bar graph of the number of push-ups he did in 4 days. If the pattern continues, how many push-ups will he do on the seventh day? *(Lesson 1-2)*

Ricardo's Push-ups

Day	Number
1	8
2	12
3	16
4	20

55. *Travel* Kanya and her family are planning to go to Hawaii with a tour group. The cost of the trip is $964 per person. There are 63 people in the tour group. About how much money is needed for the tour group to go to Hawaii? *(Lesson 1-3)*

56. *Guess and Check* The combined weight of two brothers is 300 pounds. One brother weighs 50 pounds more than the other. How much does each brother weigh? *(Lesson 1-7A)*

Alternative Assessment

Open Ended

Suppose you are planning to buy a television that costs $410. The store has two payment plans. Plan 1 requires you to pay half of the purchase price up front and the remaining cost over 6 months, with a $5 fee added each month. Plan 2 requires you to make 12 equal payments of $50. Find the cost of each plan.

Suppose you have $250 in savings and you can afford a $50 monthly payment. Which payment plan should you choose? Explain your choice.

A practice test for Chapter 1 is provided on page 595.

Completing the CHAPTER Project

Use the following checklist to complete your project

☑ The items and their costs are listed on the order form.

☑ The table of shipping charges is included.

☑ Pictures of the items are included.

☑ A list of advantages and disadvantages of mail-order shopping is included.

PORTFOLIO   Select one of the assignments from this chapter and place it in your portfolio. Attach a note to it explaining why you selected it.

Section One: Multiple Choice

There are twelve multiple-choice questions in this section. Choose the best answer. If a correct answer is *not here*, mark the letter for Not Here.

1. What is the value of 2^5?

  A 10

  B 32

  C 25

  D 64

2. There will be 120 people at the annual orchestra awards banquet. Each table seats 8 people. Which procedure can be used to find the number of tables needed?

  F Add 8 to 120.

  G Multiply 8 and 120.

  H Divide 120 by 8.

  J Subtract 8 from 120.

3. Find the value of the expression $121 \div 11 + 4 \cdot 8$.

  A 120

  B 43

  C 12

  D 60

4. By scoring 2,491 points during the 1995-1996 regular season, Michael Jordan scored more points than any other NBA player. What is this number rounded to the nearest hundred?

  F 2,490

  G 2,400

  H 2,590

  J 2,500

5. Evaluate mn if $m = 120$ and $n = 50$.

  A 500

  B 600

  C 5,000

  D 6,000

Please note that Questions 6–12 have five answer choices.

6. Kenyatta is planning to buy a car that costs $9,786. To have a sunroof installed, the cost is $548 more. What is the price of the car with a sunroof?

  F $9,334

  G $10,334

  H $10,024

  J $11,334

  K Not Here

7. A sign at the school bookstore read:

Item	Price
Calculator	$5
Notebook	$1
Combination Lock	$3
3-Ring Binder	$2

Donavan purchased 3 notebooks, 2 binders, and a lock. Find the total cost of the items.

  A $8

  B $7

  C $12

  D $9

  E $10

8. The students at Grant Middle School are selling candy bars as a fundraiser. Each class that sells 400 candy bars will earn a field trip to a water park. Lori's class has sold 207 candy bars. How many more candy bars must her class sell to earn a field trip?

  F 193

  G 184

  H 173

  J 194

  K Not Here

9. Cord Photography is having a sale on camera equipment.

Camera Package	
Camera	$178
Lens	$ 96
Camera Strap	$ 23

Which is the best estimate for the total amount of the camera package?

A $100

B $300

C $400

D $200

E $600

10. The school cafeteria prepared 1,987 lunches last week. About how many lunches were prepared each day?

F 600

G 700

H 500

J 400

K Not Here

11. Which is a reasonable estimate when 298 is subtracted from 702?

A 400

B 500

C 900

D 1,000

E Not Here

12. Javiar drove his car for 4 hours at 45 miles per hour and for 3 hours at 50 miles per hour. Which expression could be used to find the total number of miles that Javier drove his car?

F $45 \div 4 \times 50 \div 3$

G $45 + 50 \times 4$

H $45 + 4 \times 50 + 3$

J $45 \times 4 + 50 \times 3$

K Not Here

Test-Taking Tip THE PRINCETON REVIEW

Many problems can be solved without much calculating if you understand the basic mathematical concepts. Always look carefully at what is asked, and think of possible shortcuts for solving the problem.

Section Two: Free Response

This section contains five questions for which you will provide short answers. Write your answers on your paper.

13. While installing gas pipes, Hector used pieces of pipe that measured 1 ft, 2 ft, 3 ft, and 4 ft. If he cut the pieces from a 20-ft pipe, how much pipe was left?

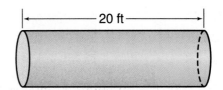

|———— 20 ft ————|

14. Write $2 \times 3 \times 5 \times 5 \times 7$ in exponential notation.

15. Sue recycled 41 aluminum cans. Her friend Tammy gave her some more aluminum cans, and then she had recycled 62 cans. To find how many cans she was given, Sue wrote $c + 41 = 62$. What is the value of c?

16. Find the value of $10 \times 10 + 10 \div 2$.

17. Solve $t - 16 = 11$.

interNET CONNECTION Test Practice **For additional test practice questions, visit:**

www.glencoe.com/sec/math/mac/mathnet

Statistics: Graphing Data

What you'll learn in Chapter 2

- to interpret and make frequency tables, bar graphs, line graphs, circle graphs, and stem-and-leaf plots,
- to use graphs to solve problems and make predictions,
- to find the mean, median, mode, and range of a set of data,
- to recognize when statistics and graphs are misleading, and
- to graph points in a coordinate system.

CHAPTER Project

PETS ARE (HU)MANS' BEST FRIENDS

This project involves conducting a survey about pets. You will make tables and graphs from your survey and present them to the class. You should then compare the results of your survey to other students' surveys.

Getting Started

- Look at the class list below. How many students have no pets? How many pets are there altogether?
- Survey your classmates to determine the number and kind of pets they have.

Name	Number and Type of Pet	Total Pets
Kevin	2 dogs, 2 turtles	4
Carmen	2 dogs, 2 horses, 1 cat, 4 fish	9
Lawanda	1 dog, 3 cats	4
Heather	1 dog, 1 cat	2
Tiffany	2 dogs, 1 cat	3
Javier	1 cat, 1 hamster	2
Sonia	2 dogs, 2 cats, 20 fish	24
Matthew	1 dog	1
Denny	1 dog, 1 cat	2
Hisano	2 dogs, 1 turtle, 2 hamsters	5
Benito	1 dog, 1 cat, 6 fish, 1 hamster	9
Codi	4 cats, 3 dogs, 2 hamsters, 2 horses, 1 rabbit	12
Paquita	6 dogs, 3 cats, 3 birds, 2 snakes, 3 rats	17
Alvin	2 birds, 1 rabbit	3
Nikki	1 dog, 1 bird	2
Kyle	1 cat, 1 turtle	2
Malik	1 cat	1
Diego	1 dog, 1 snake, 1 fish	3
Cassi	5 rabbits, 2 dogs, 3 cats	10
Stephanie	1 dog, 2 cats, 1 horse	4
Lindsay	4 dogs, 5 horses, 5 cats	14
Tyrone	0	0
Michael	0	0
Judi	1 dog	1

Technology Tips

- Use **computer software** to make graphs of your data.
- Use an **electronic encyclopedia** to do your research.
- Use a **word processor.**

 **Research** For up-to-date information on pets, visit:

www.glencoe.com/sec/math/mac/mathnet

Working on the Project

You can use what you'll learn in Chapter 2 to help you make your presentation.

Page	Exercise
49	10
57	17
81	15
89	Alternative Assessment

2-1 Frequency Tables

What you'll learn
You'll learn to make and interpret frequency tables.

When am I ever going to use this?
You can use frequency tables to keep track of votes in a class survey.

Word Wise
data
statistics
frequency table

Have you ever purchased something and the cashier asked you for your zip code? Zip codes are **data** that tell the store where its customers live. Data are pieces of information that are usually numerical. **Statistics** involves collecting, organizing, analyzing, and presenting data. In statistics, data are often organized in tables.

One type of table is a **frequency table**. A frequency table shows how many times each piece of data occurs. The frequency table below shows the results of a survey.

Favorite TV Show	Tally	Frequency
Show A	卌 卌 卌 卌 卌 III	28
Show B	卌 卌 卌 卌 卌 I	26
Show C	卌 卌 卌 卌 卌 卌 卌	35
Show D	卌 卌 卌 卌 卌 卌	30
Show E	卌 卌 卌 卌 卌 卌 卌 卌 IIII	44
Show F	卌 卌 卌	15

There is a tally mark for each response in the corresponding row. The total of each row of marks is given in the last column.

Example ① Real World APPLICATION

Food Marian's Diner surveyed its customers to determine the favorite lunch item—hamburger (HB), hot-dog (HD), pizza (P), or sub (S). Make a frequency table of the responses shown at the right.

HD	HB	S	P
P	S	HB	HD
S	HB	HD	P
P	S	HB	P
P	S	HD	P
S	P	P	P

Explore What do you know?

You know how each person responded.

You know how many people responded.

Plan Draw a table with three columns and tally the responses.

Solve In the first column, list each lunch item. In the second column, mark the tallies. In the third column, write the frequency or number of tallies.

Lunch Choice	Tally	Frequency
Hamburger	IIII	4
Hot Dog	IIII	4
Pizza	卌 卌	10
Sub	卌 I	6

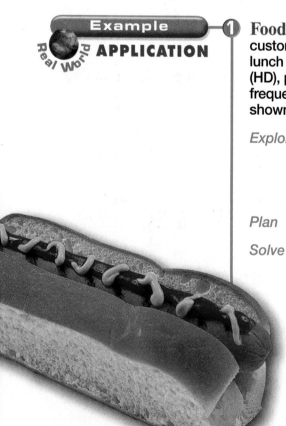

Examine   Check to see if the total number of scores in the frequency column matches the number of responses in the original list. If it doesn't, you need to check your work.

You can use the information in a completed frequency table in many problem-solving situations.

Example
CONNECTION

② Life Science Mr. Martinez' life science class is taking a field trip to the local zoo to do research for their class project. He decides to ask the students which behind-the-scenes tour they would like to take. They will take the tour receiving the most votes. Mr. Martinez made the frequency table below. Which tour should he schedule for the students?

Tour	Tally	Frequency
reptiles	⊮ ⊮ I	11
large cats	⊮ III	8
pachyderms	III	3
birds	⊮ I	6

Look for the tour that has the most tally marks. Mr. Martinez should schedule a tour about reptiles.

HANDS-ON   MINI-LAB

Work in groups of five.   📰 newspaper

Ten of the most commonly used words in written English are *the, and, of, to, in, is, you, that, it,* and *for*.

Try This

1. Select a portion of text from a newspaper article that is at least 4 inches long and 1 column wide.

2. Make a frequency table. Tally each time one of these ten words appears.

Talk About It

3. Which word appeared most often? Do you think it is the most commonly used word in written English?

4. Do you think the results would be the same if you selected a different section of the newspaper? Explain why or why not.

5. Do you think the results would be a lot different if you selected a portion of a novel instead of the newspaper? Explain why or why not.

Study Hint

Problem Solving When taking a survey, it is often helpful to have several responses for people to choose from. Otherwise, you may have many different responses with the same frequency—one.

Communicating Mathematics

Read and study the lesson to answer each question.

1. *Tell* an advantage of organizing data into a frequency table.

2. *Write* a sentence to tell what you can conclude when you analyze the favorite TV show data in the frequency table on page 46.

HANDS-ON MATH

3. Use the newspaper article from the Mini-Lab on page 47 to make a frequency table of the letters used in the article.

   a. Do you think the letter that appeared most often is used most often in the English language? Explain why or why not.

   b. Were any letters in the alphabet not in the article?

   c. What vowel appears most often?

   d. Compare your results with other classmates.

Guided Practice

4. *Physical Education* Coach Estes took a survey of the students in gym class to see which sport she should offer during the spring. She recorded the responses in the frequency table below.

Sport	Tally	Frequency
volleyball	卌 卌	
baseball	IIII	
softball	卌 III	
basketball	卌 II	

   a. Copy the table and complete the frequency column.

   b. Which sport should Coach Estes offer?

5. The ages of the guests at Linn's birthday party are shown below.

14	12	13	12	12	14
13	10	11	12	13	12

   a. Make a frequency table of the data.

   b. What was the most common age of the guests?

Applications and Problem Solving

Make a frequency table for each set of data.

6. To the nearest hour, how many hours of television did you watch last Saturday?

1	2	4	3	2	1	1	2	2	3

7. How many times were you absent in the last six weeks?

1	5	0	1	3	1	1	0	1
3	5	1	2	0	0	2	0	3

8. What were the daily high temperatures for September?

89	92	89	89	90	89	89	87	86	87
89	87	86	90	90	89	90	87	87	86
87	89	89	87	89	87	87	86	85	86

9. Business The frequency table shows the sizes of shoes sold in August at the Athlete's Locker.

Size	Tally	Frequency
9	~~JHT~~ ~~JHT~~ II	12
9½	~~JHT~~ ~~JHT~~ ~~JHT~~ I	16
10	~~JHT~~ ~~JHT~~ ~~JHT~~ ~~JHT~~ III	23
10½	~~JHT~~ ~~JHT~~ I	11
11	~~JHT~~ ~~JHT~~ ~~JHT~~	15

    a. Describe the data shown in the table.

    b. How many pairs of shoes were sold in August?

    c. Which size should the Athlete's Locker stock the most?

10. Working on the CHAPTER Project

    a. Make a frequency table that shows the number of pets owned by each student on the class list on page 45.

    b. Make a frequency table that shows the number of pets owned by each of your classmates.

11. Business Refer to the frequency table below.

Favorite Snack Food of Food Mart Customers		
Snack Food	**Tally**	**Frequency**
potato chips	~~JHT~~ I	6
tortilla chips	III	3
corn chips	~~JHT~~	5
pretzels	II	2
popcorn	III	3

    a. Which snack(s) should Food Mart keep well stocked?

    b. Are there any snacks that Food Mart should discontinue stocking? Explain.

12. Critical Thinking Refer to the beginning of the lesson. What are some reasons that a store would like to know the zip codes of its customers?

Mixed Review

13. Standardized Test Practice Name the number that is the solution of the equation $102 \div q = 17$. *(Lesson 1–7)*

    A 6   **B** 7

    C 8   **D** 9

14. Money Matters Before buying a computer, Jocelyn called several computer stores to compare prices. The highest price was $2,349, and the lowest price was $1,788. Estimate the difference between the highest and lowest prices.

(Lesson 1-3)

For **Extra Practice,** see page 559.

Scales and Intervals

What you'll learn

You'll learn to choose appropriate scales and intervals for frequency tables.

When am I ever going to use this?

Knowing how to choose scales and intervals can help you make a bar graph or line graph.

Word Wise

interval
scale

For her history project, Margaret is investigating the early state legislature of Massachusetts. She uses an atlas to find how far each county seat is from the capital, Boston. The distances in miles are 65, 134, 37, 85, 22, 101, 90, 102, 7, 105, 2, 41, 28, and 42. How can Margaret organize this data so it is easier to study the distances? *This problem will be solved in Example 1.*

Margaret can group the data in **intervals**. To do this, she needs to determine the **scale** for her frequency table. The scale must include all of the data.

Example 1

CONNECTION

History Refer to the beginning of the lesson. Organize Margaret's distance data into a frequency table.

Step 1 Determine the scale. Since the data includes distances from 2 to 134, begin the scale with 0 and end with 140.

Step 2 Determine the interval. To keep the number of intervals manageable, choose an interval of 20.

Step 3 Make the frequency table. Now Margaret can easily see the number of county seats in each interval.

All intervals need to be equal and not overlap.

Distances from County Seats to Boston		
Distance	**Tally**	**Frequency**
121–140	I	1
101–120	III	3
81–100	II	2
61–80	I	1
41–60	II	2
21–40	III	3
0–20	II	2

Example 2

APPLICATION

Sports The scores of the Glencoe Scholarship Golf tournament are shown below.

74, 87, 88, 97, 75, 78, 76, 69, 84, 83, 86, 88,
92, 93, 72, 89, 85, 67, 77, 82, 84, 85, 99, 81

a. Determine an appropriate scale for the data.

b. Make a frequency table for the data.

a. A scale from 66 to 100 is appropriate because it includes all of the scores.

b. Since the scale is small, an interval of 5 is used in the table at the right.

Golf Scores		
Score	Tally	Frequency
96–100	II	2
91–95	II	2
86–90	IIII	5
81–85	IIII II	7
76–80	III	3
71–75	III	3
66–70	II	2

You can use a frequency table to organize data about a group of people, such as your classmates.

HANDS-ON MINI-LAB

Work with a partner.　　　📄 paper　✏️ pencil

Study Hint

Problem Solving When the scale of a set of data is small, intervals can be 1, 2, 3, 4, or 5. When the scale is large, intervals are usually multiples of 10.

Try This

1. Each person should ask ten classmates how many hours a week each of them spend on the telephone. Record each response. Combine your results with your partner.
2. Choose a scale and appropriate intervals for a frequency table.
3. Make a frequency table.

Talk About It

4. What does your frequency table tell you?
5. Compare your frequency table to a frequency table from another pair of classmates. Are the scale and intervals the same? Explain why or why not.

CHECK FOR UNDERSTANDING

Communicating Mathematics

Read and study the lesson to answer each question.

1. *Explain* how to find an appropriate scale for the data in the chart.

2. *Write* a sentence or two explaining how to determine an interval for the data in the chart.

Distances Traveled to School by Students in Mr. Alonzo's Class					
18	14	30	62	40	47
34	12	56	57	63	54
32	27	33	46	17	13

3. Using the data you collected for the Mini-Lab on page 51, make another frequency table to show how the scale and interval would change if you had asked your classmates how many *minutes* they spend on the phone.

Guided Practice

Use the set of data below to answer Exercises 4–6.

$12, $8, $13, $6, $17, $7, $9, $13, $11, $10, $15, $20

4. What scale would you use in making a frequency table for this set of data?

5. Would an interval of 2 or 10 be better to use in making a frequency table for this set of data? Explain.

6. Make a frequency table of the data.

7. *Entertainment* Shelby's last eighteen video game scores were: 3,500; 10,200; 30,000; 65,300; 56,000; 28,200; 17,000; 43,200; 23,300; 41,900; 32,600; 29,100; 46,400; 39,500; 25,000; 37,800; 49,700; and 34,300.

   a. Make a frequency table for this set of data.

   b. In what interval are most of Shelby's scores?

EXERCISES

Practice

Choose the better scale for a frequency table for each set of data.

8. 3, 17, 9, 6, 2, 5, 12	**a.** 0 to 10	**b.** 0 to 20
9. 53, 88, 25, 47, 50, 18, 15, 43	**a.** 0 to 99	**b.** 20 to 80
10. 245, 144, 489, 348, 36, 284, 150, 94, 220	**a.** 0 to 499	**b.** 0 to 999

Choose the best interval for a frequency table for each set of data.

11. the data in Exercise 8	**a.** 2	**b.** 10	**c.** 20
12. the data in Exercise 9	**a.** 4	**b.** 5	**c.** 10
13. the data in Exercise 10	**a.** 10	**b.** 100	**c.** 1,000

Make a frequency table for each set of data.

14. the data in Exercise 8

15. the data in Exercise 9

16. the data in Exercise 10

17. *Geography* Refer to the beginning of the lesson. Suppose Margaret decided to do her project on the state of Texas instead of Massachusetts.

   a. Could she use the same scale she used for Massachusetts? Explain why or why not.

   b. Could she use the same interval she used for Massachusetts? Explain why or why not.

18. *Health* Read the nutrition label on at least 15 different food items and record the number of milligrams (mg) of sodium per serving.

   a. Make a frequency table of your data.

   b. What conclusions can you make from your table?

19. *Geography* The highest elevation in states with mountains over 5,000 feet high are shown below.

State	Elevation (ft)	State	Elevation (ft)
Alaska	20,320	New Mexico	13,161
Arizona	12,633	New York	5,344
California	14,494	North Carolina	6,684
Colorado	14,433	Oregon	11,239
Hawaii	13,796	South Dakota	7,242
Idaho	12,662	Tennessee	6,643
Maine	5,268	Texas	8,749
Montana	12,799	Utah	13,528
Nebraska	5,424	Virginia	5,729
Nevada	13,140	Washington	14,410
New Hampshire	6,288	Wyoming	13,804

   a. What scale would you use in making a frequency table for this set of data?

   b. Make a frequency table of the data.

20. *Critical Thinking* Create a set of at least 15 pieces of data so that a frequency table of the data would need a scale from 20 to 40.

Mixed Review

21. *Sports* The number of hits each player on the little league baseball team had in the last game of the season were 1, 3, 2, 4, 2, 3, 1, 0, and 2. Make a frequency table of the data. *(Lesson 2-1)*

22. *Algebra* Evaluate a^3 if $a = 2$. *(Lesson 1-6)*

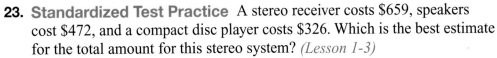

23. *Standardized Test Practice* A stereo receiver costs $659, speakers cost $472, and a compact disc player costs $326. Which is the best estimate for the total amount for this stereo system? *(Lesson 1-3)*

   A $900   **B** $1,100

   C $1,200   **D** $1,500

   E $1,900

For **Extra Practice,** see page 560.

Bar Graphs and Line Graphs

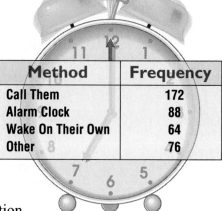

What you'll learn

You'll learn to construct bar graphs and line graphs.

When am I ever going to use this?

You can use a bar graph to compare the attendance at sporting events and a line graph to show population data.

Word Wise

bar graph
line graph

How do you wake up in the morning? Does someone wake you or do you use an alarm clock? The results of a survey of how 400 parents with children under 18 wake their children is shown in the frequency table.

Method	Frequency
Call Them	172
Alarm Clock	88
Wake On Their Own	64
Other	76

A graph is a more visual representation of data. A **bar graph** is used to compare the frequency of the amount in a category or class.

Example

1 Use the data from the survey to draw a vertical bar graph.

Step 1 Draw and label both scales.

Step 2 Determine the scale for the vertical axis. Choose an interval that best represents the data and mark off equal spaces on the vertical scale.

> **Scale** The data includes numbers from 76 to 172, so begin the scale with 0 and end with 180.

> **Interval** Since the vertical scale needs to go from 0 to 180, use an interval of 20.

Step 3 Mark off equal spaces on the horizontal axis and label each with a method.

Step 4 Draw a bar for each method. The height of each bar represents the number of parents that use that method.

Step 5 Label the graph with a title.

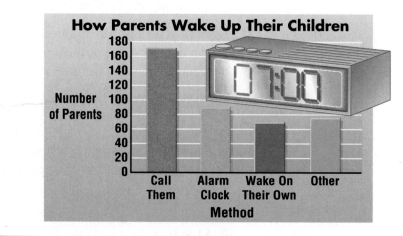

How Parents Wake Up Their Children

To show how data changes over a period of time, you could use a **line graph**. A line graph is usually used to show the change and direction of change over time.

Food Christopher takes a poll to find out how many soft drinks his classmates drink each day. The results are shown at the right. Draw a line graph using the data he collected.

Day	Frequency
Monday	27
Tuesday	22
Wednesday	18
Thursday	26
Friday	57
Saturday	61
Sunday	42

Step 1 Draw and label both scales.

Step 2 Determine the scale for the vertical axis. Choose an interval that best represents the data and mark off equal spaces on the vertical scale.

Scale The data includes numbers from 18 to 61, so begin the scale with 0 and end with 70.

Interval Since the vertical scale needs to go from 0 to 70, use an interval of 10.

Step 3 Mark off equal spaces on the horizontal scale and label them with the days of the week in order.

Step 4 Draw a dot to show the frequency for each day. Draw line segments to connect the dots.

Step 5 Label the graph with a title.

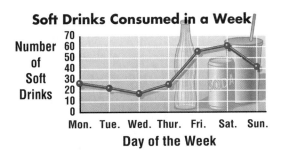

The graph shows that Christopher's classmates drink many more soft drinks on the weekend.

> **Study Hint**
>
> **Technology** You can use a graphing calculator to make a line graph without a title or labels. There are also many computer software programs available that will construct graphs as you enter the data.

CHECK FOR UNDERSTANDING

Communicating Mathematics

Read and study the lesson to answer each question.

1. *Examine* the bar graph in Example 1. Tell which method for waking up children was used the most.

2. *Examine* the line graph in Example 2. Tell which day the students drank the least number of soft drinks.

3. Write a sentence or two explaining how the vertical scale and the interval affect the look of a bar graph or line graph.

Guided Practice

Choose a scale and an interval for the vertical axis of a line graph for each set of data.

4. 28, 13, 9, 26, 19

5. 18, 32, 52, 68, 56, 31

6. Transportation The number of annual passengers arriving at and departing from airports are listed in the table.

   a. Choose an interval for a bar graph of the passenger data.

   b. Make a horizontal bar graph for the data.

World's Busiest Airports	
Airport	Passengers (thousands)
Atlanta	73,500
Chicago-O'Hare	72,500
Los Angeles	61,200
London Heathrow	60,700
Dallas-Ft. Worth	60,500
Tokyo	51,200
Frankfurt	42,700
San Francisco	40,000
Paris	38,600

Source: Airports Council International

EXERCISES

Practice

Make a bar graph for each set of data.

7. a vertical bar graph

Favorite Soft Drink	
Soft Drink	Frequency
Brand A	4
Brand B	15
Brand C	9
Brand D	19
Brand E	11
Brand F	8

8. a horizontal bar graph

Average Event Day Attendance (thousands)	
Sport	Attendance
NASCAR Races	83
NFL (football)	62
MLB (baseball)	31
NBA (basketball)	16
NHL (hockey)	14

Source: USA TODAY research

Make a line graph for each set of data.

9.

Average Snowfall for Brighton, Utah	
Month	Snow (in.)
Oct.	21
Nov.	50
Dec.	66
Jan.	69
Feb.	63
March	70
April	51
May	16

Source: The USA TODAY Weather Almanac

10.

Sales of Sing-Along Players in the U.S. (thousands)	
Year	Amount
1990	$77,900
1991	$94,800
1992	$119,000
1993	$116,000
1994	$86,900
1995	$94,300
1996	$82,100

Source: National Association of Music Merchants

Refer to the following tables for Exercises 11–15.

A.

Super Bowl Ticket Prices	
Year	Price
1967	$12
1972	15
1977	20
1982	40
1987	75
1992	150
1997	275

Source: The NFL

B.

Entertainment Expenses per Person	
Entertainment	Amount
Basic Cable	$110
Books	79
Home Video	73
Recorded Music	56
Daily Newspapers	49
Magazines	36
On-line/Internet	7

Source: *Statistical Abstract of the United States*

11. Determine a scale for the prices in Table A.

12. What would be the best interval for the amounts in Table B? Explain why you chose it.

13. Which table could be best represented with a line graph? Explain.

14. Make a bar graph for the data in Table B.

15. Make a line graph for the data in Table A.

Applications and Problem Solving

16. *Geography* The line graph shows Australia's population growth.

   a. Estimate the population of Australia in 1991.

   b. Estimate the number of years it took for the population to double.

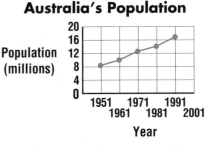

Australia's Population

17. *Working on the* CHAPTER Project Refer to your frequency tables from Exercise 10 on page 49.

   a. Make a bar graph using your frequency table from Exercise 10a.

   b. Make a bar graph using your frequency table from Exercise 10b.

18. *Critical Thinking* The bar graph shows the allowance of your five best friends. How might you use the information in this graph to argue for a raise in your allowance if you are currently getting $5 a week?

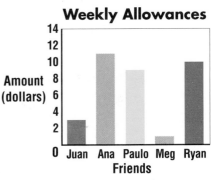

Weekly Allowances

Mixed Review

19. **Standardized Test Practice** Brianna's scores on her first six history quizzes were 76, 72, 83, 96, 81, and 85. Choose the best interval for a frequency table for this data. *(Lesson 2-2)*

   A 10   **B** 5   **C** 2   **D** 1

For **Extra Practice,** see page 560.

20. *Algebra* Evaluate $ab - c$ if $a = 3$, $b = 16$ and $c = 5$. *(Lesson 1-5)*

21. Evaluate $14 - 2 \times 5$. *(Lesson 1-4)*

PROBLEM SOLVING

2-3B Use a Graph

A Follow-Up of Lesson 2-3

The sixth grade class is planning a field trip to the science center. Their science teacher, Ms. Tai, has asked Steve and Maria to help her decide where they will stop for lunch. The choices are Burger Barn, Dairy Dine, Mardee's, Mickey's, and Sally's. Let's listen in!

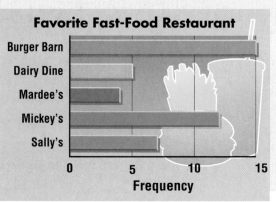

Favorite Fast-Food Restaurant

Burger Barn
Dairy Dine
Mardee's
Mickey's
Sally's

0   5   10   15
Frequency

I made a bar graph using our survey results. It looks like we should pick Burger Barn.

Yeah, but Ms. Tai told us it had to be healthy. I'm not sure Burger Barn is our best choice based on the pictograph I made.

Steve

What's a pictograph?

A pictograph is another type of graph used to compare numbers.

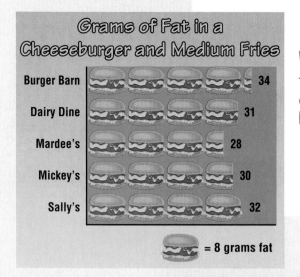

Grams of Fat in a Cheeseburger and Medium Fries

Burger Barn	34
Dairy Dine	31
Mardee's	28
Mickey's	30
Sally's	32

🍔 = 8 grams fat

Well, Mardee's has the least amount of fat, but it's the least favorite. How about Mickey's? It was the second most popular choice and the second lowest in fat!

Maria

THINK ABOUT IT

1. **Compare** the two graphs to see if you agree with Maria's suggestion that they select Mickey's. Explain your answer.

2. **Analyze** the data in the two graphs to see what your second choice would be. Explain your answer.

For **Extra Practice,** see page 560.

3. **Apply** what you have learned about the **use a graph** strategy to solve the following problem.

Ms. Tai has suggested that students bring a sack lunch. Then they could stop to have dessert on the way home. The bar graph shows the fat grams in a chocolate shake at each of the restaurants. Which restaurant would you recommend and why?

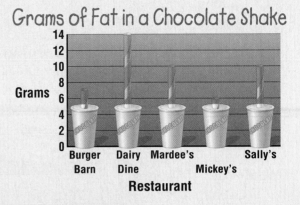

Grams of Fat in a Chocolate Shake

ON YOUR OWN

4. The second step of the 4-step plan for problem solving is *plan*. **Explain** how the type of graph helps you to make a plan for solving the problem.

5. **Write a Problem** that you could solve more easily if you could use a graph to examine the data in the problem.

6. **Examine** the line graph in Example 2 on page 55. Suppose Christopher's mother is going to buy 2-liter bottles of soft drinks. What day of the week would you recommend that she purchase based on the results shown in the graph?

MIXED PROBLEM SOLVING

STRATEGIES

Look for a pattern.
Solve a simpler problem.
Act it out.
Guess and check.
Draw a diagram.
Make a chart.
Work backward.

Solve. Use any strategy.

7. **Money Matters** Mr. Rivera bought 5 packages of batteries for $5.89 each. About how much did he spend?

8. **Algebra** Find x if the product of x and 28 is 252.

9. **Travel** Ogima travels 345 miles per week. How far does he drive in four weeks?

10. **Money Matters** Laura collected $2.00 from each student to buy a gift for their teacher. If 27 people contributed, how much money was collected?

11. **Standardized Test Practice** The line graph shows the number of people in Lubbock and Garland from 1960 to 1990.

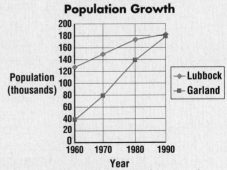

Population Growth

About how many more people were in Lubbock than in Garland in 1980?

A 90,000   B 70,000

C 35,000   D 5,000

2-4 Reading Circle Graphs

What you'll learn

You'll learn to interpret circle graphs.

When am I ever going to use this?

Knowing how to interpret circle graphs can help you analyze a budget.

Word Wise

circle graph

In Lesson 2-3B, you learned how to interpret sets of data displayed in bar graphs and line graphs. In the Mini-Lab below, you will make a **circle graph**.

HANDS-ON MINI-LAB

Work with a partner. 🔲 tape measure ✂ markers

🎞 tape ◇ adding machine tape 〰 string

The table shows the age groups of all the people who listen to professional football radio broadcasts. You can use this data to make a circle graph.

Age Group	Percent
18–24	11%
25–34	27%
35–44	27%
45–54	17%
55–64	9%
65+	9%

Try This

1. Cut a piece of adding machine tape that is one meter long.

2. For each age group, mark the length in centimeters that corresponds to the percent. For example, the 18–24 age group is 11% so mark off 11 centimeters and label it accordingly.

3. Tape the ends of the adding machine tape together to form a circle. The labels should be facing the inside of the circle.

4. Stand your adding machine tape circle on a flat surface and locate its center. Tape one end of a piece of string to the center. Tape the other end to the point where the adding machine tape is joined together. Repeat with five more pieces of string joining the center to each mark on the tape.

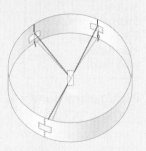

Talk About It

5. What is the total of the percents?

6. Write a short paragraph describing the circle graph. Describe the sizes of the sections in relation to each other.

7. The circle graph you made represents the same information as the table. Discuss the advantages and disadvantages of each.

A circle graph is used to compare parts of a whole. The interior of the circle represents a set of data, and the pie-shaped sections are the parts. All the percents given in a circle graph must add up to 100%.

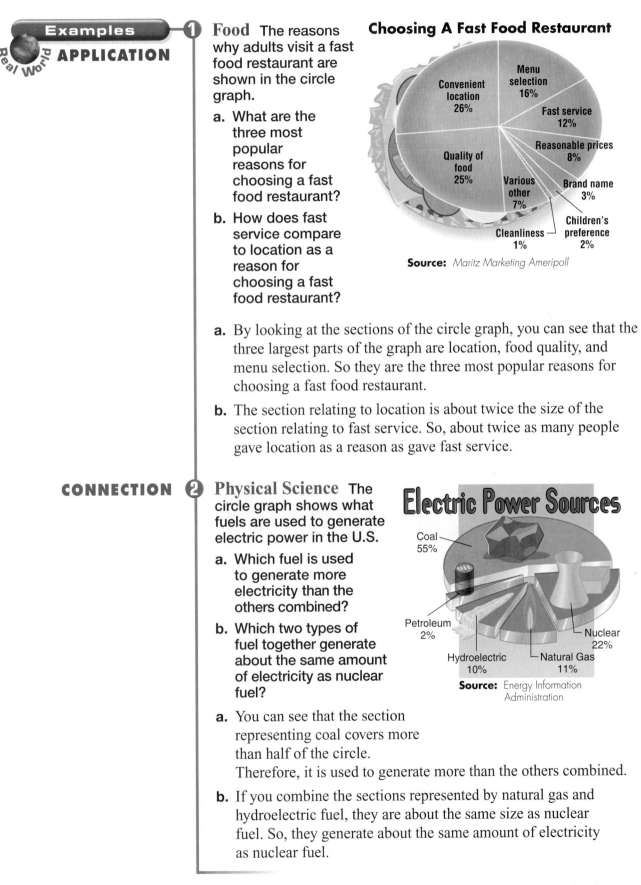

Examples

Real World APPLICATION

1 **Food** The reasons why adults visit a fast food restaurant are shown in the circle graph.

a. What are the three most popular reasons for choosing a fast food restaurant?

b. How does fast service compare to location as a reason for choosing a fast food restaurant?

Choosing A Fast Food Restaurant

Convenient location 26%
Menu selection 16%
Fast service 12%
Reasonable prices 8%
Quality of food 25%
Various other 7%
Brand name 3%
Children's preference 2%
Cleanliness 1%

Source: *Maritz Marketing Ameripoll*

a. By looking at the sections of the circle graph, you can see that the three largest parts of the graph are location, food quality, and menu selection. So they are the three most popular reasons for choosing a fast food restaurant.

b. The section relating to location is about twice the size of the section relating to fast service. So, about twice as many people gave location as a reason as gave fast service.

CONNECTION **2** **Physical Science** The circle graph shows what fuels are used to generate electric power in the U.S.

a. Which fuel is used to generate more electricity than the others combined?

b. Which two types of fuel together generate about the same amount of electricity as nuclear fuel?

Electric Power Sources

Coal 55%
Petroleum 2%
Hydroelectric 10%
Natural Gas 11%
Nuclear 22%

Source: Energy Information Administration

a. You can see that the section representing coal covers more than half of the circle. Therefore, it is used to generate more than the others combined.

b. If you combine the sections represented by natural gas and hydroelectric fuel, they are about the same size as nuclear fuel. So, they generate about the same amount of electricity as nuclear fuel.

Communicating Mathematics

Read and study the lesson to answer each question.

1. **Explain** when it is best to use a circle graph.

2. **Write** what the total of the percents in a circle graph should be.

HANDS-ON MATH

3. The table shows the number of hours per week that students ages 12–18 use a computer at school. Use the same procedure you used in the Mini-Lab to make a circle graph of the data in the table.

Time	Percent
Less than 1 hour	22%
1–2 hours	28%
2–4 hours	16%
4–6 hours	19%
More than 6 hours	11%
Don't know	4%

Source: *Consumer Electronics Manufacturers Association Survey*

Guided Practice

4. **Food** The circle graph shows the favorite sandwiches of children.

   a. The percents are 5%, 11%, 14%, 21% and 28%. Match each percent with the appropriate section of the graph.

   b. Suppose hamburgers and ham and cheese were combined. Would the combination be preferred by $\frac{1}{4}$ of children? Explain.

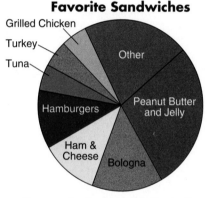

Favorite Sandwiches

Source: *Are You Normal?* by Bernice Kanner, St. Martin's, 1995

EXERCISES

Practice

5. **Food** The circle graph shows the retail sales by shape at the Pasta Factory Outlet.

   a. The percents are 13%, 15%, 31% and 41%. Match each percent with the appropriate section of the graph.

   b. Which two types of pasta have about the same amount of sales?

   c. What percent of sales comes from short and long pasta?

   d. Which two types together have about the same amount of sales as short pasta?

   e. Which two types together account for about half of the sales?

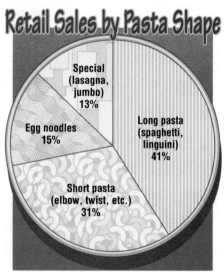

Retail Sales by Pasta Shape

6. _Money Matters_ The circle graph
shows the percent of people who
purchase athletic shoes by age group.

   a. What two age groups together make
about half of all athletic shoe purchases?

   b. What two age groups together make
about one fourth of all athletic shoe
purchases?

   c. What is the difference between the
percent of people 65 and older and
the 18–24 age group?

Who Buys Athletic Shoes?

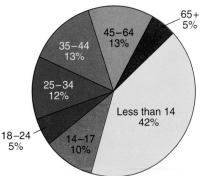

Source: _The Sporting Goods Market_

7. _Entertainment_ The circle graph
shows the format of recent
pre-recorded music sales.

   a. The two formats with the lowest
percent of sales each had 1%.
What are they?

   b. Cassette album sales were five
times cassette single sales and
cassette single sales were five
times as much as music video
sales. The rest are CDs. What
are the percents for each of
these formats?

Music Sales

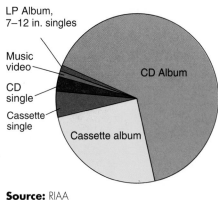

Source: RIAA

8. _Critical Thinking_ Which would _not_ be appropriate data to show in
a circle graph? Explain your reasoning.

   a. data showing what teens say would 'most' encourage them to volunteer

   b. data showing how many hours parents help their children with
homework each week

   c. data showing the five most wasted foods in the United States

   d. data showing why people dine out

Mixed Review

9. _Life Science_ Janet has a leaf collection that includes 15 birch, 8 willow,
5 oak, 10 maple, and 8 miscellaneous leaves. Make a bar graph for this
data. _(Lesson 2-3)_

10. Standardized Test Practice In March, an average of 280 videos were
rented per day at Brian's Video Barn. Which is the best estimate for the
number of videos rented for the whole month of March? _(Lesson 1-3)_

   A 3,000

   B 6,000

   C 8,000

   D 9,000

   E 10,000

For **Extra Practice,**
see page 561.

Making Predictions

What you'll learn

You'll learn to make predictions from line graphs.

When am I ever going to use this?

Making predictions can be used to plan housing and highway projects.

The line graph shows the population of the world since 1900. In 1992, the United Nations predicted that world population could increase to over 6 billion by the year 2000.

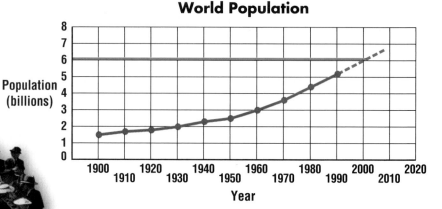

World Population

Source: United Nations Department for Economic and Social Information and Policy Analysis, Population Division.

Line graphs can be used to assist in making predictions as follows.

Step 1 Extend the graph as shown by the dashed line.

Step 2 Draw a horizontal line from the point where the extension intersects the 2000 gridline.

Step 3 You can see that the population of the world could be over 6 billion by the year 2000.

Example

CONNECTION

① **Geography** Refer to the graph above.

a. **Predict the world population in the year 2020. Identify any assumptions you use to make your prediction.**

Extend the dashed line so it intersects the 2020 gridline. If the extension aligns with the segment between 1980 and 1990, world population could be about 8 billion by 2020.

b. **Suppose the population growth between 1990 and 2000 matches the rate of growth between 1940 and 1950 instead of between 1980 and 1990. What would the population be in the year 2000?**

The line segment between 1940 and 1950 is not as steep as the one between 1980 and 1990. The population would probably be less than 6 billion.

Communicating Mathematics

Read and study the lesson to answer each question.

1. *Predict* the world population in 2010. Identify any assumptions you use to make your prediction.

2. *You Decide* Jamonte predicts that world population will be over 8 billion in 2010. Tashima predicts it will be less than 7 billion. Whose prediction do you agree with? Explain your reasoning.

Guided Practice

3. *Sports* The line graph shows the winning times in the 3,000 meter steeplechase at the last thirteen Olympics.

Olympic 3,000-Meter Steeplechase, 1948–1996

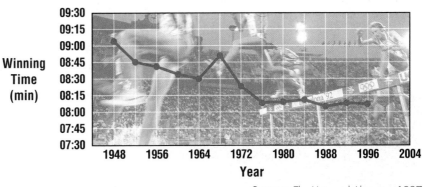

Source: The Universal Almanac, 1997

a. Would you have predicted the winning time in 1968? What might have happened?

b. Would you predict the winning time in 2004 to be under 8 minutes? Explain why or why not.

EXERCISES

Practice

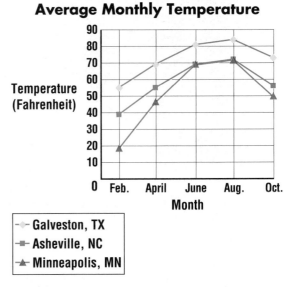

interNET
CONNECTION
For the latest weather statistics, visit:
www.glencoe.com/sec/math/mac/mathnet

4. *Earth Science* The line graph shows the average monthly temperature for three U.S cities.

a. Predict the temperature for Galveston in December, Asheville in January, and Minneapolis in November.

Average Monthly Temperature

b. How much colder would you expect it to be in Asheville than in Galveston in November?

c. How much colder would you expect it to be in Minneapolis than in Asheville in December?

d. How much colder would you expect it to be in Minneapolis than in Galveston in January?

Legend:
- Galveston, TX
- Asheville, NC
- Minneapolis, MN

Applications and Problem Solving

5. *Physical Science* Angelina and Bart each dropped a different ball from several distances and recorded the highest point of the first bounce. The graph shows their results.

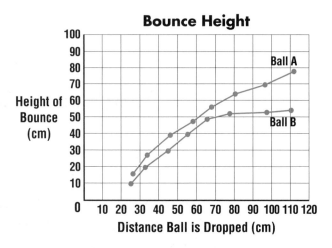

Bounce Height

Height of Bounce (cm)

Distance Ball is Dropped (cm)

Ball A

Ball B

a. What would you predict about Ball A if the distance it is dropped continues to increase?

b. What would you predict about Ball B if the distance it is dropped continues to increase?

c. How might you explain these differences?

d. Do you think that each ball will always give a higher bounce with a higher drop height? Explain.

6. *Sports* Tom and Jim have been running each afternoon preparing for the track team try-outs. They hope to qualify for the 1,600-meter run. The graph shows their best times over the last ten weeks.

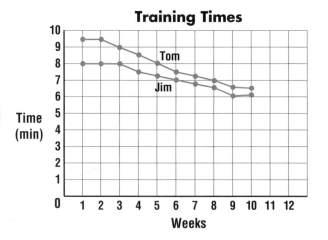

Training Times

Time (min)

Weeks

Tom

Jim

a. Who has shown the most improvement?

b. Who has the better times?

c. Who do you predict will be the faster runner in 15 weeks? What will be his time? Explain.

For **Extra Practice,** see page 561.

7. *Critical Thinking* Suppose the lines in Exercise 6 cross after week 12. What does the point where they cross represent?

8. *Food* Refer to the circle graph in Example 1 on page 61. Which two reasons together account for half of the responses? *(Lesson 2-4)*

9. **Standardized Test Practice** If $m = 6$ and $n = 4$, find the value of the expression $2m - n$. *(Lesson 1-5)*

   A 2   **B** 4   **C** 6   **D** 8

CHAPTER 2 — Mid-Chapter Self Test

Make a frequency table for the set of data. *(Lesson 2-1)*

1. To the nearest inch, how tall are you?

   57   54   58   59   60   57   63   57   55   56   60   57

2. Use the set of data 58, 92, 23, 14, 62, 79, 43, 85, 91, 18, 50, 47, 25, 88. *(Lesson 2-2)*
   a. Choose the better scale for a frequency table for the data: 0 to 99 or 0 to 999.
   b. Choose the best interval for a frequency table for the data: 1, 10, or 100
   c. Make a frequency table for this set of data.

3. Make a bar graph for the data in the table at the right. *(Lesson 2-3)*

Favorite Ice Cream Flavor	
Flavor	**Frequency**
Chocolate	7
Strawberry	9
Vanilla	5

4. The circle graph shows the average family size in the U.S. *(Lesson 2-4)*
   a. Which two sizes of families together have about the same percent as the size with the greatest percent?
   b. What percent of all families have five or more members?

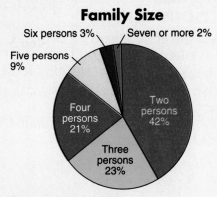

Family Size

Six persons 3%   Seven or more 2%
Five persons 9%
Four persons 21%
Two persons 42%
Three persons 23%

Source: U.S. Census Bureau

5. *Economics* Keshia is saving her money to purchase a video game. She starts with $40 and adds $10 the first week, $11 the second week, $12 the third week, and so on. The graph at the right shows a record of her savings for the first 10 weeks. *(Lesson 2-5)*
   a. Estimate when Keshia will double her money (have at least $100).
   b. Estimate when she will double her money again (have at least $200).
   c. Can you predict when she might double her money again (have at least $400)? Explain your reasoning.

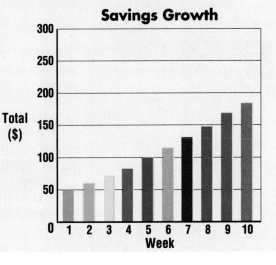

Savings Growth

Total ($)

Week

Stem-and-Leaf Plots

What you'll learn

You'll learn to construct stem-and-leaf plots.

When am I ever going to use this?

You can use a stem-and-leaf plot to record a large set of data collected for a science project.

Word Wise

stem-and-leaf plot
stem
leaf

D'Ante's brother told him that if he wants to improve his grades, he should watch less television and study more. D'Ante took a survey to find how many hours of television other kids in his grade watch in a week. The results are shown in the chart. What could D'Ante conclude from his data? *This problem will be solved in Example 2.*

2	18	23	14	13
4	5	9	11	13
21	3	19	8	16
9	10	3	3	10
12	15	4	16	19

One way to display a large data set to make it easy to read is to construct a **stem-and-leaf plot**. To display the data 25, 8, 14, 25, 12, and 21 in a stem-and-leaf plot, follow these steps.

Step 1 Find the least and the greatest number. Identify the tens digit in each. The least number, 8, has 0 in the tens place. The greatest number, 25, has 2 in the tens place.

Step 2 Draw a vertical line and write the tens digits from least to greatest to the left of the line. These digits form the **stems**.

```
0 |
1 |
2 |
```

Step 3 Write the units digits to the right of the line, with the corresponding stem. The units digits form the **leaves**.

```
0 | 8
1 | 4 2
2 | 5 5 1
```

Step 4 Order the leaves in each row from least to greatest.

Stem	Leaf
0	8
1	2 4
2	1 5 5

Step 5 Include a key or an explanation.

$1|2 = 12$

Example 1
APPLICATION

Racing The Iditarod Dog Sled Race is more than 1,000 miles across Alaska. The distances between the checkpoints are shown in the chart.

a. Make a stem-and-leaf plot of the data.

b. In which interval do most of the distances fall? Why would this information be useful to competitors?

20	29	14	52	34
45	30	48	93	48
23	38	90	65	25
18	60	70	90	40
58	48	28	18	55
22				

Source: *USA TODAY* research

a. The least number is 14, and the greatest number is 93. So the stems start with 1 and end with 9.

Stem	Leaf
1	4 8 8
2	0 9 3 5 8 2
3	4 0 8
4	5 8 8 0 8
5	2 8 5
6	5 0
7	0
8	
9	3 0 0

Order the leaves from least to greatest.

➡

1|4 = 14 miles

Stem	Leaf
1	4 8 8
2	0 2 3 5 8 9
3	0 4 8
4	0 5 8 8 8
5	2 5 8
6	0 5
7	0
8	
9	0 0 3

Include 8 as a stem even though none of the numbers have an 8 in the tens place.

b. Most of the distances are between 20 and 48 miles. This information would be helpful in calculating the amount of supplies needed between checkpoints.

Example 2

Real World APPLICATION

Surveys Refer to the beginning of the lesson. What could D'Ante conclude from his survey?

Make a stem-and-leaf plot of the survey data to make it easy to summarize. The stem 1 has the greatest number of leaves. D'Ante could conclude that most of the students watch 10-19 hours of television per week.

Stem	Leaf
0	2 3 3 3 4 4 5 8 9 9
1	0 0 1 2 3 3 4 5 6 6 8 9 9
2	1 3

2|1 = 21 hours

CHECK FOR UNDERSTANDING

Communicating Mathematics

Read and study the lesson to answer each question.

1. ***Tell*** when a stem-and-leaf plot would be the best way to represent a set of data.

2. ***Explain*** how a stem-and-leaf plot is like a bar graph.

3. ***Find*** a list of data in a magazine and make a stem-and-leaf plot. Describe how the plot summarizes the data.

Guided Practice

4. Determine the stems for the data set 9, 42, 33, 21, 11, 6, 40, 5, 5, 9.

5. Make a stem-and-leaf plot for the following set of data.

   27, 53, 58, 34, 24, 36, 20, 38, 43, 45, 35, 54, 78, 47, 58, 36

6. ***Traffic*** Ms. Sangita recorded the number of cars that passed through the intersection in her neighborhood every morning for 3 weeks.

46	53	61	41	70	38	50
33	67	52	69	32	45	62
39	51	66	72	52	59	61

   a. Make a stem-and-leaf plot of her data.

   b. In what intervals do most of the data fall?

Practice

Determine the stems for each set of data.

7. 41, 44, 28, 22, 39, 26, 33, 17, 14, 56, 22

8. 135, 106, 100, 132, 92, 136, 89, 128, 112

9. 6, 15, 4, 3, 11, 22, 2, 9, 18, 20, 41, 45, 18

Make a stem-and-leaf plot for each set of data.

10. 35, 29, 27, 28, 27, 30, 44, 33, 32, 45, 38, 28, 22, 21, 31, 31, 38, 24

11. 124, 99, 140, 133, 94, 130, 124, 167, 162, 92, 99, 132

12. 92¢, 83¢, 94¢, 54¢, 85¢, 54¢, 96¢, 89¢, 75¢

13. Make a stem-and-leaf plot of the temperatures (°F) 57°, 34°, 57°, 56°, 27°, 58°, 46°, 53°, 28°, 34°, 9°, 56°, 57°, 45°, 34°, and 29°.

Applications and Problem Solving

14. *Agriculture* Cotton crops are typically harvested when all of the pods on the plants are open. Mr. Smotherman wants to rent harvesting equipment when all of the pods are open. The chart shows the dates in October during the past 19 years when the cotton pods were fully open.

10	19	12	21
18	7	11	16
12	18	20	14
15	16	8	22
9	12	13	

   a. Make a stem-and-leaf plot of the data.

   b. Estimate what day Mr. Smotherman will need to rent the harvesting equipment.

15. *Life Science* As part of a science project, Lai'sung recorded the number of Monarch butterflies he saw every afternoon for one month during migrating season.

6	9	26	11	5
7	9	27	10	8
5	9	41	9	11
13	16	37	18	9
19	18	22	22	14
17	23	19	27	29

   a. Make a stem-and-leaf plot of the data.

   b. On how many days did Lai'sung see more than 20 butterflies?

16. *Critical Thinking* Refer to Exercise 6. The City Council told Ms. Sangita that they would place a traffic light at the intersection in her neighborhood if, on average, there were more than 50 cars that pass through the intersection each morning. Use the stem-and-leaf plot and one other type of graph of the same data to prepare an argument for the City Council meeting.

Mixed Review

17. *Business* Use the line graph to estimate sales in July at First Motors. *(Lesson 2-5)*

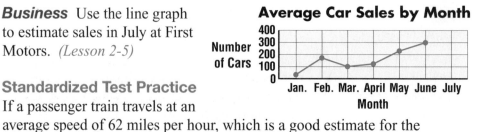

Average Car Sales by Month

18. **Standardized Test Practice** If a passenger train travels at an average speed of 62 miles per hour, which is a good estimate for the number of hours it will take to travel 293 miles? *(Lesson 1-3)*

   A 3 h   **B** 4 h   **C** 5 h   **D** 7 h   **E** 10 h

For **Extra Practice,** see page 561.

2-7 Mean, Median, and Mode

What you'll learn

You'll learn to find the mean, median, mode, and range to describe a set of data.

When am I ever going to use this?

You can use mean, median, and mode to describe your classmates' allowances.

Word Wise

measures of
  central
  tendency
mean
median
mode
average
range

The table shows the prices of ten compact 35mm cameras.

It is often helpful to find one number to represent one aspect of a set of data. Some of these numbers are known as **measures of central tendency**. Three common measures of central tendency are **mean**, **median**, and **mode**.

Brand	Price
A	$160
B	$80
C	$230
D	$215
E	$180
F	$220
G	$170
H	$220
I	$300
J	$185

Each measure is a different type of **average**. When people use the word *average,* they are usually referring to the mean.

Mean	The mean of a set of data is the sum of the data divided by the number of pieces of data.

The mean of the camera prices is

$$\frac{160 + 80 + 230 + 215 + 180 + 220 + 170 + 220 + 300 + 185}{10} \text{ or } 196.$$

The median is another measure of central tendency.

Median	The median is the middle number when the data are arranged in numerical order.

To find the median camera price, first order the prices. Then find the middle number.

$$\underline{80 \quad 160 \quad 170 \quad 180 \quad 185} \quad \underline{215 \quad 220 \quad 220 \quad 230 \quad 300}$$

5 numbers   *5 numbers*

There are two middle numbers, 185 and 215. To find the median, you need to find the mean of these two numbers.

$$\frac{185 + 215}{2} \text{ or } 200$$

The median is 200, so the median price is $200.

Study Hint

Reading Math The median value means that there are just as many values above the median as below it.

A third measure of central tendency is the mode.

Mode	The mode of a set of data is the number(s) or item(s) that appear most often.

To find the mode of the camera price data, look for the dollar amount that appears most often. The amount $220 appears twice and is the mode. If Brand E would have cost $185, then $185 would have been listed twice also, and the data would have two modes. *A set of data may have no mode, one mode, or multiple modes.*

Example ①

APPLICATION

Real World

Entertainment The graph shows the number of channels available from different cable companies. Find the mean, median, and mode of the data.

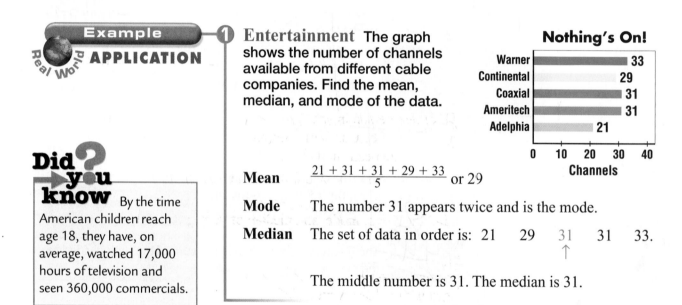

Mean $\dfrac{21 + 31 + 31 + 29 + 33}{5}$ or 29

Mode The number 31 appears twice and is the mode.

Median The set of data in order is: 21   29   31   31   33.

The middle number is 31. The median is 31.

The stem-and-leaf plots show the test scores of two sixth grade science classes. Both classes have the same mean score, the same median score, and the same mode score. However, the individual scores are not all the same. How do the two sets of scores differ?

Third Period			Fourth Period	
2			2	0 8
3			3	5 8
4	3 5		4	5
5	2 4 5 6		5	0
6	0 0 0 8		6	0 0
7	2 5 8		7	2 5
8	1 5		8	8 9
9		8|1 = 81	9	1 6 8

Another aspect of data is how it is spread out. One way to describe the spread of a set of data is to use the **range**. The range is the difference between the greatest number and the least number in a set of data.

The range of the science test scores are as follows.

Third Period: 85 − 43 or 42   Fourth Period: 98 − 20 or 78

The scores of the fourth period class are more spread out, or dispersed, than those of the third period class.

2 Refer to the beginning of the lesson. What is the range of the camera prices?

First order the prices.

80   160   170   180   185   215   220   220   230   300

The range is 300 − 80 or 220.

The range of the camera prices is $220.

In this Mini-Lab, you'll use beans to learn more about measures of central tendency.

HANDS-ON MINI-LAB

Work in groups of five.   bowl of beans   calculator

Try This

- Have one person in your group reach into the bowl and grab a handful of beans. Count and record the number of beans. Replace the beans in the bowl.
- Repeat the previous step for the other four members of your group.
- Find the mean, mode, and median of your group's results.

Talk About It

1. Suppose you add the number 4 to the set of data so that your group now has six pieces of data.
   a. Find the mean, mode, and median of the new set of data.
   b. Which measure of central tendency is affected the most?
   c. Which measure of central tendency is most representative of the data before adding the 4? after adding the 4?

2. Suppose you add the number 500 to your original set of data.
   a. Find the mean, mode, and median of the new set of data.
   b. Which measure of central tendency is affected the most?
   c. Which measure of central tendency is most representative of the data before adding the 500? after adding the 500?

3. In general, which measure of central tendency is most affected by adding an extreme value? Explain.

CHECK FOR UNDERSTANDING

Communicating Mathematics

Read and study the lesson to answer each question.

1. *State* which measure—mean, median, or mode—best represents the original set of data in the Mini-Lab. Explain your reasoning.

2. *Explain* how to find the median of a set of data that has an even number of data in the set.

3. Add a number to your original set of data in the Mini-Lab so that the mean and median stay the same.

Guided Practice

List the data in each set from least to greatest. Then find the median, mode, mean, and range.

4. 3, 7, 1, 5, 5, 9, 2

5. 36, 26, 10, 57, 29, 83, 27, 88, 37

6. *Food* If 18 snacks in a vending machine cost 60¢ each, 9 snacks cost 50¢ each, and 8 snacks cost 80¢ each, what is the mean cost for all 35 snacks?

EXERCISES

Practice

Find the mean, median, mode, and range for each set of data.

7. 52, 48, 32, 40, 54, 38

8. 23, 27, 33, 31, 30, 27, 25

9. 13, 17, 16, 16, 14, 14, 16, 14

10. 9, 15, 9, 12, 7, 9, 15, 4

11. 31, 25, 18, 40, 31, 18, 26, 32, 39, 44, 37

12. 475, 377, 273, 379, 477, 573, 475, 377, 385, 484

Refer to the beginning of the lesson for Exercises 13–15.

13. Suppose the Brand I camera was removed from the list. How would this affect the mean?

For **Extra Practice,** see page 562.

14. Suppose both the Brand I and the Brand B cameras were removed from the list. How would this affect the median?

15. If the Brand C camera cost as much as the Brand I camera, would this change the mean? Explain why or why not.

Applications and Problem Solving

Family
Activity

Find a bar graph in a newspaper or magazine in your home. Work with a family member to find the mean, median, and mode of the data in the graph.

16. *Earth Science* The Channel 6 meteorologist made the graph at the right.

a. What was the mean high temperature for June?

b. What was the range of the high temperatures?

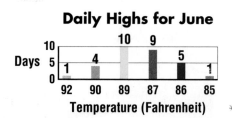

Daily Highs for June

17. *Business Management* The manager of Dairy Land keeps a record of the sizes of the shakes sold. Would the mean, median, or mode be more useful to her? Explain.

18. *Critical Thinking* List six numbers such that the mean is 14, the median is 14, the modes are 12 and 14, and the range is 5.

Mixed Review

19. *Statistics* Construct a stem-and-leaf plot for 12, 7, 23, 9, 10, 20, 0, 4, 19, 13, 5, 7, 2, 13, 18, and 2. *(Lesson 2-6)*

20. **Standardized Test Practice** Shamrock, Texas, had a population of 2,206 in 1996. What was the population rounded to the nearest thousand? *(Lesson 1-3)*

A 2,000   **B** 2,200   **C** 2,020   **D** 2,210

What's the Average?

Get Ready This game is for four players.

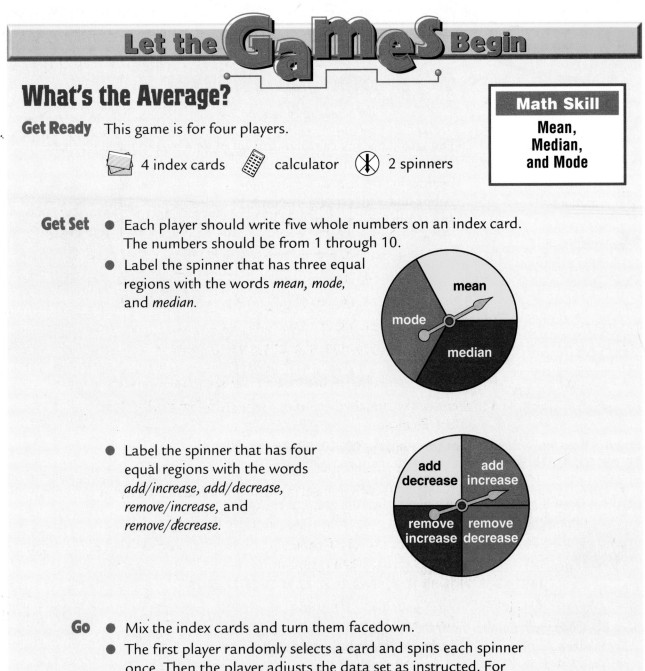

4 index cards   calculator   2 spinners

Math Skill
Mean,
Median,
and Mode

Get Set
- Each player should write five whole numbers on an index card. The numbers should be from 1 through 10.
- Label the spinner that has three equal regions with the words *mean, mode,* and *median.*

- Label the spinner that has four equal regions with the words *add/increase, add/decrease, remove/increase,* and *remove/decrease.*

Go
- Mix the index cards and turn them facedown.
- The first player randomly selects a card and spins each spinner once. Then the player adjusts the data set as instructed. For example, if the player gets *mode* and *add/decrease,* the player must add a piece of data to the data set so the mode decreases. If the player gets *median* and *remove/increase,* the player must remove a piece of data from the data set so the median of the set increases.
- The other members then check.
- A player scores two points for each correct solution and loses one point for each incorrect solution.
- The first player to get 10 points wins.

interNET
CONNECTION Visit www.glencoe.com/sec/math/mac/mathnet for more games.

Lesson 2-7 Mean, Median, and Mode   **75**

COOPERATIVE LEARNING

2-7B Box-and-Whisker Plots

A Follow-Up of Lesson 2-7

index cards

cardboard

markers

push pins

string

Did you know you can analyze a set of data by looking at its shape? In this lab, you will show the shape of a set of data by drawing a *box-and-whisker plot*. A box-and-whisker plot displays and summarizes a set of data.

TRY THIS

Work in groups of four.

Make a box-and-whisker plot for the data 20, 10, 22, 25, 18, 22, 16, 26, 23, 27, 14, and 19.

Step 1 Draw a number line on a piece of cardboard. Label it with an appropriate scale. Since the data goes from 10 to 27, begin the scale with 10 and end with 29.

10 11 12 13 14 15 16 17 18 19 20 21 22 23 24 25 26 27 28 29

Step 2 Copy each number from the data set on an index card. Place each number in its appropriate place above the number line.

22

10   14 16 18 19 20 22 23 25 26 27

10 11 12 13 14 15 16 17 18 19 20 21 22 23 24 25 26 27 28 29

Step 3 Mark the median and the *extreme* (least and greatest) *values* in the data by placing push pins on the number line. Label each number. The data is now separated into two halves.

22

10   14 16 18 19 20 22 23 25 26 27

10 11 12 13 14 15 16 17 18 19 20 21 22 23 24 25 26 27 28 29

lower extreme   median   upper extreme

Step 4 Now find the median of each half of the data. Mark each by placing a push pin on the number line. The median of the lower half of the data is called the *lower quartile*. The median of the upper half of the data is called the *upper quartile*. Label these on the number line. The data is now separated into fourths or *quartiles,* each having the same number of data.

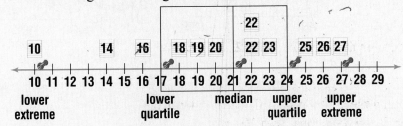

Step 5 Draw a box using the quartiles as the left and right edges. Then draw a line segment through the median.

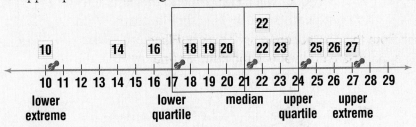

Step 6 Construct the whiskers. Tie a piece of string between the lower quartile and the least number. Tie another piece of string between the upper quartile and the greatest number.

The box-and-whisker plot shows that the bottom fourth of the data is fairly spread out while the top fourth has a small range.

ON YOUR OWN

1. ***Draw*** a box-and-whisker plot for 37, 12, 25, 9, 33, 17, 51, 45, 35, and 39.
2. If the whiskers in a box-and-whisker plot are longer than the box, what does this tell you about the data?
3. ***Reflect Back*** Make a box-and-whisker plot for the data in the stem-and-leaf plot in Example 2 on page 69.

Misleading Statistics

What you'll learn

You'll learn to recognize when statistics and graphs are misleading.

When am I ever going to use this?

Knowing how to recognize misleading statistics will help you make informed decisions.

During political campaigns, politicians often use statistics to make a point. In a recent election, opposing candidates presented the two graphs shown below.

- Do both graphs show the same information?

- Which graph suggests a drastic cut in government spending? Which graph suggests fairly steady spending?

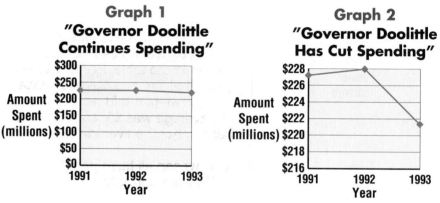

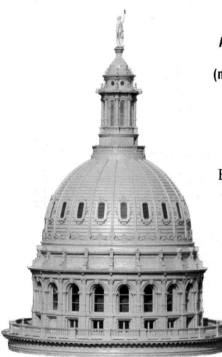

Both graphs represent the same data. By using a different scale, each candidate was able to take the same information and tell a very different story.

Graph 2 is a result of an expanded vertical scale. For this reason, Graph 2 is visually misleading when compared to the complete scale in Graph 1. This should be indicated using a small "break" in the vertical axis as shown at the right. This shows that the axis is not to scale between $0 and $220.

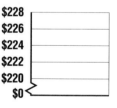

Example

CONNECTION

1 **Geography** The graphs display the same data.

a. **Is Graph A misleading? Explain.**

Graph A is misleading because there is no title and there are no labels on either scale. It should also include a break in the vertical axis between 0 and 30.

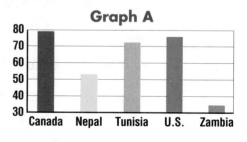

b. Is Graph B misleading? Explain.

Graph B is not misleading. The scales are labeled, and the graph includes a title.

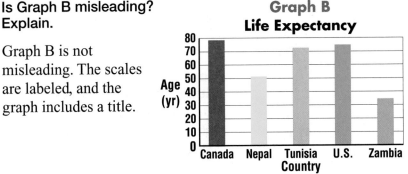

Graph B
Life Expectancy

Another way statistics can be misleading is by not using the best measure of central tendency.

APPLICATION

Allowances The results of a class survey on weekly allowances are shown in the table. Felipe told his sister that the average allowance was $7, Megan told her parents the average was $5, and Alba told her brother the average was $4.45.

Allowance	Frequency
$7	6
$6	0
$5	5
$4	1
$3	4
$2	2
$1	2

a. How can all three people think they are correct?

b. Which do you think is the best measure and why?

a. Each person used a different measure of central tendency to describe the set of data. Felipe used the mode, Megan used the median, and Alba used the mean.

b. The mode is misleading because $7 is also the highest allowance. The median is misleading because $5 is the second highest allowance. In this case, the best measure is the mean.

CHECK FOR UNDERSTANDING

Communicating Mathematics

Read and study the lesson to answer each question.

1. *Describe* at least two ways that a graph can be misleading.

2. *Write* a set of eight pieces of data that would best be represented by the median.

3. *Write* a few sentences explaining how the mean of a set of data might be used to mislead a consumer.

Guided Practice

Tell whether the *mean*, *median*, or *mode* would be best to describe each set of data. Explain each answer.

4. 500, 400, 390, 405, 390

5. car buyer's favorite color

6. populations of California, Texas, New York, Florida, Pennsylvania, and Idaho

7. **Stock Market** The graphs below display the same information.

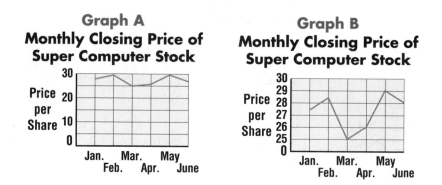

a. Explain why these graphs look different.

b. If a stockbroker wanted to convince a customer to invest in the stock market, which graph would she probably show the customer? Explain.

EXERCISES

Practice

8. **Money Matters** A local telephone company employee made two graphs to show the cost of calling Port Arthur.

a. Which graph makes a call to Port Arthur look more economical? Why?

b. Which graph would you show to a telephone customer? Explain.

c. Which graph would you show to a company stockholder? Explain.

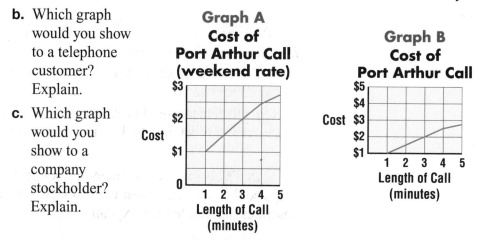

Determine the measure of central tendency that you think is best for each category. Explain your choice.

Home	Condo	Cape Cod	Ranch	Split Level	2–Story
Bathrooms	1	1	$1\frac{1}{2}$	2	$2\frac{1}{2}$
Bedrooms	2	2	2	4	5
Taxes	$1,200	$2,080	$1,800	$1,740	$2,990
Price	$62,900	$99,450	$124,800	$114,900	$239,400

9. number of bathrooms

10. number of bedrooms

11. taxes

12. price

Applications and Problem Solving

13. *Money Matters* An advertisement for *CD Spins* states that their average price for a CD is $11. How may the owner of *CD Spins* be using average to give the impression of lower prices?

14. *Language Arts* Dora is writing a book report on a book about Henry Aaron entitled *I Had A Hammer*. To show that Aaron is the lifetime home run leader, Dora draws the graphs below.

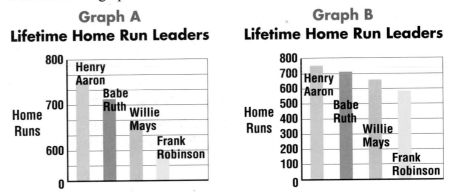

Henry Aaron

a. Which graph makes it look like Willie Mays hit about twice as many home runs as Frank Robinson?

b. Which graph should Dora use for a presentation? Explain.

c. How could Graph A be changed to be less misleading?

15. *Working on the* **CHAPTER Project** Refer to your frequency tables from Exercise 10 on page 49.

a. Find the mean, median, and mode for the data in your frequency table from Exercise 10a.

b. Tell which measure of central tendency you think would be best to describe this set of data. Explain your answer.

16. *Critical Thinking* Find a graph in a newspaper or magazine. Redraw the graph so that the data appear to show different results. Which graph describes the data better? Explain.

Mixed Review

17. **Standardized Test Practice** The Chill hockey team played 32 games. They scored a total of 128 goals. What was the mean (average) number of goals scored per game? *(Lesson 2-7)*

A 4

B 5

C 96

D 160

18. *Sports* Erica's bowling scores last weekend were 119, 134, 135, 125, 143, and 130. What is an appropriate scale for her scores? *(Lesson 2-2)*

For **Extra Practice,** see page 562.

19. *Patterns* Find the next three numbers in the pattern 1, 4, 9, 16, ... *(Lesson 1-2)*

Integration: Geometry
Graphing Ordered Pairs

What you'll learn

You'll learn to use ordered pairs to locate points and organize data.

When am I ever going to use this?

In geography, latitude and longitude are used to separate Earth into a coordinate grid.

Word Wise

coordinate system
origin
x-axis
y-axis
ordered pair
x-coordinate
y-coordinate

In the cartoon, how do the other students know who the machine has caught daydreaming?

Notice how the chart on the front of the machine forms a grid pattern. The grid is used to identify the daydreamer by using a code similar to what is used on road maps. Obviously, everyone knows how to locate the seat designated by 4-B.

In mathematics, we can locate a point by using a **coordinate system**.

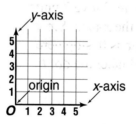

The coordinate system is formed when two number lines intersect at their zero points. This point is called the **origin**. The horizontal number line is called the **x-axis**, and the vertical number line is called the **y-axis**.

You can name any point on a coordinate system by using an **ordered pair** of numbers. The first number in an ordered pair is called the **x-coordinate**, and the second number is called the **y-coordinate**. An ordered pair is written in this form:

$$(4, 2)$$

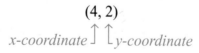

x-coordinate ↑ ↑ *y-coordinate*

The *x*-coordinate represents the number of horizontal units the point is from 0, and the *y*-coordinate represents the number of vertical units it is from 0.

Example ① Name the ordered pair for point *K*.

Start at the origin. Move right along the *x*-axis until you are under point *K*. Since you moved two units to the right, the *x*-coordinate of the ordered pair is 2.

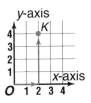

Now move up until you reach point *K*. Since you moved up 4 units, the *y*-coordinate is 4.

The ordered pair for point *K* is (2, 4).

You can also graph a point on a coordinate system. To graph a point means to place a dot at the point named by an ordered pair.

Example ② Graph each point.

a. *A*(6, 4)
Start at the origin. Move 6 units to the right on the *x*-axis. Then move 4 units up to locate the point. Draw a dot and label the dot *A*.

b. *B*(7, 0)
Start at the origin. Move 7 units to the right on the *x*-axis. Since the *y*-coordinate is 0, stop here. Draw a dot and label the dot *B*.

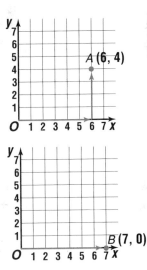

Ordered pairs can also be used to show how data are related.

Example ③
INTEGRATION

Algebra Pazi went bowling at The Bowling Castle. Games were $2 each. If *x* represents the number of games she bowled, the expression 2*x* gives the total cost.

a. Make a list of ordered pairs in this form: (number of games bowled, cost of games).

b. Then graph the ordered pairs.

(continued on the next page)

Lesson 2-9 Integration: Geometry   Graphing Ordered Pairs   **83**

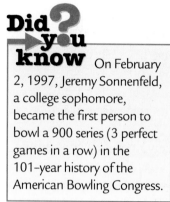
a. Evaluate the expression $2x$ for 1, 2, and 3 games. A table of ordered pairs that result is shown below.

b.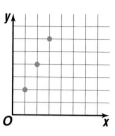

x (games)	2x	y (cost)	(x, y)
1	2 (1)	2	(1, 2)
2	2 (2)	4	(2, 4)
3	2 (3)	6	(3, 6)

(3, 6) means 3 games cost $6.

CHECK FOR UNDERSTANDING

Communicating Mathematics

Read and study the lesson to answer each question.

1. **Tell** how to graph the ordered pair (7, 9).

2. **Draw** a coordinate system and label the origin, x-axis, and y-axis.

3. **You Decide** Lina says that the ordered pairs (8, 3) and (3, 8) name the same point. Raul disagrees. Who is correct and why?

Guided Practice

Use the grid at the right to name the point for each ordered pair.

4. (3, 7)   5. (9, 5)   6. (0, 8)

Use the grid at the right to name the ordered pair for each point.

7. *E*   8. *D*   9. *C*

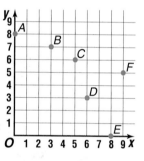

10. **Physical Science** Odell and Elise conducted an experiment to see how the mass of an object affected the distance a spring stretched. On graph paper, draw a coordinate grid. Then graph the ordered pairs (mass, distance). Use the x-axis for mass and the y-axis for distance. What did they observe?

Stretching of a Spring	
Mass	**Distance**
100 g	3 cm
200 g	6 cm
300 g	9 cm
400 g	12 cm
500 g	15 cm

EXERCISES

Practice

Use the grid at the right to name the point for each ordered pair.

11. (9, 3)   12. (5, 9)

13. (2, 0)   14. (5, 1)

15. (1, 7)   16. (8, 2)

17. (5, 6)   18. (0, 2)

19. (7, 5)   20. (3, 4)

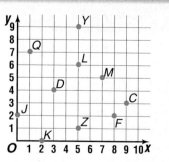

Use the grid at the right to name the ordered pair for each point.

21. G 22. R

23. A 24. P

25. H 26. I

27. C 28. M

29. B 30. S

Applications and
Problem Solving

31. **Art** Draw a coordinate grid. Number each axis from 0 to 20. Draw a picture on your grid. List the ordered pairs of points where sides of the drawing meet. Have a classmate draw your picture by graphing your ordered pairs and connecting the dots.

32. **Cartography** The grid is a simplified version of the map of present-day Washington, D.C.

    a. Which ordered pair indicates the location of the Supreme Court?

    b. Which location does (6, 3) indicate?

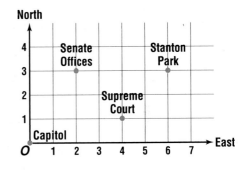

33. **Critical Thinking** Write a short paragraph that tells how graphing ordered pairs compares to graphing statistical data in bar graphs or line graphs.

Mixed Review

34. **Statistics** Find the mean for the set of Tim's history test scores: 88, 90, 87, 91, 49. Why is this a misleading statistic? *(Lesson 2-8)*

35. **Standardized Test Practice** Four girls have heights of 152, 158, 164, and 168 centimeters. Which height is a reasonable average height of the girls? *(Lesson 2-7)*

    A 120 cm   **B** 140 cm   **C** 160 cm

    D 180 cm   **E** 200 cm

36. **Life Science** Make a horizontal bar graph for the set of data in the chart. *(Lesson 2-3)*

For **Extra Practice,**
see page 562.

37. **Algebra** Solve the equation $56 \div n = 8$. *(Lesson 1-7)*

Top Animal Speeds (mph)	
Antelope	61
Cheetah	70
Coyote	43
Elk	45
Gazelle	50
Gray Fox	42
Hyena	40
Lion	50
Quarterhorse	47.5
Wildebeest	50
Zebra	40

inter NET
CONNECTION **Chapter Review** For additional review, visit: **www.glencoe.com/sec/math/mac/mathnet**

Vocabulary

After completing this chapter, you should be able to define each term, concept, or phrase and give an example or two of each.

Algebra

coordinate system (p. 82)
ordered pair (p. 82)
origin (p. 82)
x-axis (p. 82)
x-coordinate (p. 82)
y-axis (p. 82)
y-coordinate (p. 82)

Statistics and Probability

average (p. 71)
bar graph (p. 54)
box-and-whisker plot (p. 76)
circle graph (p. 60)
data (p. 46)
extreme value (p. 76)
frequency table (p. 46)

interval (p. 50)
leaf (p. 68)
line graph (p. 54)
lower quartile (p. 77)
mean (p. 71)
measures of central tendency (p. 71)
median (p. 71)
mode (p. 71)
range (p. 72)
scale (p. 50)
statistics (p. 46)
stem (p. 68)
stem-and-leaf plot (p. 68)
upper quartile (p. 77)

Problem Solving

use a graph (p. 58)

Understanding and Using the Vocabulary

Choose the letter of the term that best matches each phrase.

1. separates the scale into equal parts
2. the sum of a set of data divided by the number of pieces of data
3. a plot that uses place value to display a large data set
4. the number that appears most often in a set of data
5. the middle number when a set of data is arranged in numerical order
6. the horizontal number line of a coordinate system
7. the vertical number line of a coordinate system
8. the first number in an ordered pair
9. the second number in an ordered pair
10. the name of a point on a coordinate system in the form (x-coordinate, y-coordinate)

a. stem-and-leaf plot
b. mean
c. median
d. mode
e. ordered pair
f. interval
g. x-axis
h. x-coordinate
i. y-axis
j. y-coordinate
k. box-and-whisker plot

In Your Own Words

11. *Explain* the difference between a bar graph and a line graph.

Objectives & Examples

Upon completing this chapter, you should be able to:

● make and interpret frequency tables
(*Lesson 2-1*)

Make a frequency table for this list of favorite fruits.

| banana | grapes | apple | banana |
| apple | apple | grapes | apple |

Fruit	Tally	Frequency
apples	IIII	4
bananas	II	2
grapes	II	2

● choose appropriate scales and intervals for frequency tables (*Lesson 2-2*)

Find an appropriate scale and an interval for 7, 11, 24, 9, 16, 19, and 18.

scale = 0 to 25, interval of 5

● construct bar graphs and line graphs
(*Lesson 2-3*)

Number of Pets in Animal Shelter	
Season	Pets
Winter	65
Spring	75
Summer	85
Autumn	70

Line Graph

Number of Pets
100
75
50
25
0
W S S A
Season

● interpret circle graphs (*Lesson 2-4*)

Magazine Sales by Grade

Sixth graders sold the most subscriptions, earning $906.80.

Review Exercises

Use these exercises to review and prepare for the chapter test.

Make a frequency table for each set of data.

12. How many brothers and sisters do you have?

3	1	2	0	4	1
2	1	0	1	3	1
0	2	1	3	1	0

13. What is your favorite color?

blue	red	pink	blue	orange
red	blue	red	red	yellow
green	blue	brown	pink	green

Choose an appropriate scale and an interval for each set of data.

14. 4, 7, 6, 9, 2, 3, 6

15. 13, 38, 9, 23, 17, 43, 23

16. 16, 8, 9, 15, 10, 11, 17, 4, 7, 15

17. 138, 152, 112, 127, 173, 136, 98, 125, 145

18. Make a line graph from the data below.

Alma's Grades During Junior High	
Grade	Frequency
A	3
B	9
C	5
D	1
F	0

19. Make a vertical bar graph of the data in Exercise 18.

Use the graph.

20. What is the most common race time?

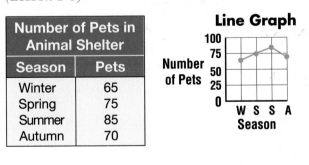
10K Race Times (Minutes)

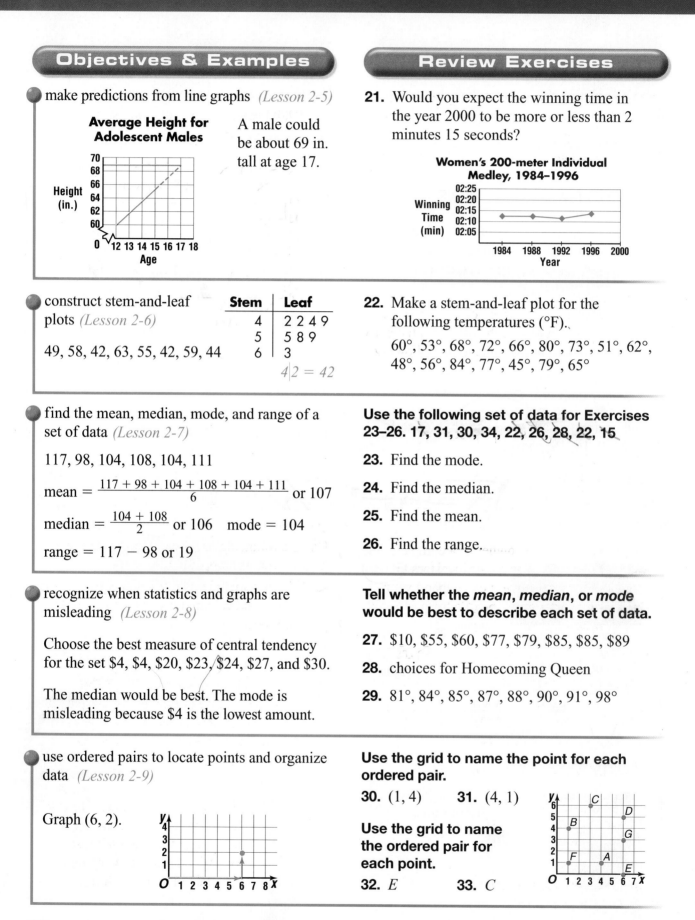

Objectives & Examples

make predictions from line graphs *(Lesson 2-5)*

Average Height for Adolescent Males

Height (in.)

70 68 66 64 62 60

0 12 13 14 15 16 17 18
Age

A male could be about 69 in. tall at age 17.

construct stem-and-leaf plots *(Lesson 2-6)*

49, 58, 42, 63, 55, 42, 59, 44

Stem	Leaf
4	2 2 4 9
5	5 8 9
6	3

4|2 = 42

find the mean, median, mode, and range of a set of data *(Lesson 2-7)*

117, 98, 104, 108, 104, 111

mean = $\frac{117 + 98 + 104 + 108 + 104 + 111}{6}$ or 107

median = $\frac{104 + 108}{2}$ or 106   mode = 104

range = 117 − 98 or 19

recognize when statistics and graphs are misleading *(Lesson 2-8)*

Choose the best measure of central tendency for the set $4, $4, $20, $23, $24, $27, and $30.

The median would be best. The mode is misleading because $4 is the lowest amount.

use ordered pairs to locate points and organize data *(Lesson 2-9)*

Graph (6, 2).

Review Exercises

21. Would you expect the winning time in the year 2000 to be more or less than 2 minutes 15 seconds?

Women's 200-meter Individual Medley, 1984–1996

Winning Time (min)

02:25 02:20 02:15 02:10 02:05

1984 1988 1992 1996 2000
Year

22. Make a stem-and-leaf plot for the following temperatures (°F).

60°, 53°, 68°, 72°, 66°, 80°, 73°, 51°, 62°, 48°, 56°, 84°, 77°, 45°, 79°, 65°

Use the following set of data for Exercises 23–26. 17, 31, 30, 34, 22, 26, 28, 22, 15

23. Find the mode.

24. Find the median.

25. Find the mean.

26. Find the range.

Tell whether the *mean*, *median*, or *mode* would be best to describe each set of data.

27. $10, $55, $60, $77, $79, $85, $85, $89

28. choices for Homecoming Queen

29. 81°, 84°, 85°, 87°, 88°, 90°, 91°, 98°

Use the grid to name the point for each ordered pair.

30. (1, 4)   **31.** (4, 1)

Use the grid to name the ordered pair for each point.

32. *E*   **33.** *C*

Applications & Problem Solving

34. Entertainment The frequency table shows the number of hours per week sixth graders play video games. *(Lesson 2-1)*

a. In what interval did most of the responses fall?

Hours Playing Video Games per Week	
0–2	10
3–4	11
5–6	5
7–8	3

b. How many people play video games less than 5 hours per week?

35. Television A survey of the number of minutes a junior high student watches TV on a school day produced the following data.

95, 85, 69, 75, 90, 45,
92, 65, 50, 40, 75

Make a stem-and-leaf plot for the data. *(Lesson 2-6)*

36. Use a Graph Was the marriage rate in 1980 higher or lower than in 1990? *(Lesson 2-3B)*

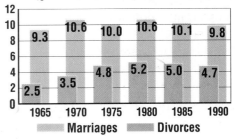

U.S. Marriage and Divorce Rates (per 1,000 of population)

Alternative Assessment

● Open Ended

Suppose you are the owner of a company. You want to convince a group of investors to contribute money so you can expand your business. Your monthly profits the past year are shown in the table.

January	$2,645	July	$3,200
February	$2,200	August	$2,900
March	$2,205	September	$2,050
April	$2,900	October	$1,950
May	$2,750	November	$2,400
June	$3,050	December	$3,250

Find the mean, median, and mode.

Explain how you would determine whether to present the mean, median, or mode as your "average" monthly profit.

A practice test for Chapter 2 is provided on page 596.

● Completing the CHAPTER Project

Use the following checklist to make sure your presentation is complete.

☑ You have included the frequency tables, bar graphs, and calculations of mean, median, and mode.

☑ The two new bar graphs show the types of pets in both classes.

☑ Your fact sheet about your favorite pet includes a picture or drawing of the pet, the cost of taking care of the pet for one year, and any other interesting facts about the pet.

PORTFOLIO Choose a topic of interest to you. Take a survey to collect data about it. Make a frequency table and a bar or line graph. Place all of these materials in your portfolio.

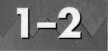
Section One: Multiple Choice

There are twelve multiple-choice questions in this section. Choose the best answer. If a correct answer is *not here,* mark the letter for Not Here.

1. How is the product $3 \times 3 \times 5 \times 5 \times 7$ expressed in exponential notation?

   A $3 \times 3 \times 5 \times 5 \times 7$

   B $3 \times 3 \times 5^2 \times 7$

   C $3^2 \times 5 \times 5 \times 7$

   D $3^2 \times 5^2 \times 7$

2. What is the value of $8 + 6 \times 10$?

   F 68

   G 120

   H 140

   J 480

3. If $n = 4$, $m = 3$, and $x = 12$, what is the value of $x \cdot m + n$?

   A 19

   B 40

   C 24

   D 84

4. Which average would be best to describe this data?

   195, 188, 70, 185, 190

   F range

   G mean

   H median

   J mode

5. Evaluate the expression $c - b$ if $c = 9$ and $b = 3$.

   A 3

   B 6

   C 9

   D 12

6. Minato found 5 seashells. His friend Soto gave him some more, and then he had 10. To find out how many seashells he was given, Minato wrote $x + 5 = 10$. What is the value of x?

   F 2

   G 5

   H 15

   J 35

Please note that Questions 7–12 have five answer choices.

7. Three chimpanzees weigh 62, 75, and 72 pounds. Which weight is a reasonable average weight of the chimpanzees?

   A 80 lb

   B 75 lb

   C 70 lb

   D 65 lb

   E 60 lb

8. The graph shows the number of visitors who toured Gund Arena on Saturday.

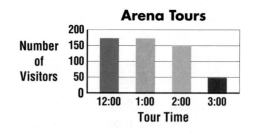

Arena Tours

How many people altogether toured the arena before 3:00 P.M.?

   F 275

   G 350

   H 400

   J 500

   K 550

9. Jason bought 6 hockey tickets that ranged in price from $7 to $10. Which is a reasonable cost for all 6 tickets?

   A $25

   B $40

   C $54

   D $70

   E $85

10. Serena's scout troop collected 4,230 aluminum cans for a service project. Scott's class collected 3,348 cans. How many more cans did Serena's scout troop collect than Scott's class?

    F 756

    G 880

    H 882

    J 892

    K Not Here

11. The base price of a mini-van is $18,876. With an air conditioner, the mini-van costs $678 more. What is the price of the mini-van with an air conditioner?

    A $19,454

    B $19,444

    C $18,954

    D $18,198

    E Not Here

12. Every class that sells 400 tickets to the school carnival earns a bowling party. Kylie's class sold 209 tickets. How many more tickets must her class sell to earn the bowling party?

    F 101

    G 181

    H 191

    J 211

    K Not Here

Test-Taking Tip THE PRINCETON REVIEW

You may be able to eliminate all or most of the answer choices by estimating. Also, look to see which answer choices are not reasonable for the information given in the problem.

Section Two: Free Response

This section contains four questions for which you will provide short answers. Write your answers on your paper.

13. Translate the sentence into an equation and solve.
    The sum of 7 and r is equal to 22.

14. Find the mode, median, and mean for the set of data. 72, 68, 59, 83, 74, 52

15. If $14 + p = 39$, what is the value of p?

16. The line graph shows the number of houses sold in Mentor and Medina from 1960 to 1990.

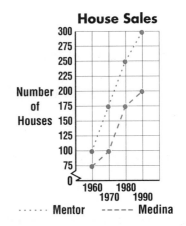

House Sales

····· Mentor   ---- Medina

About how many more houses were sold in Mentor than in Medina in 1980?

www.glencoe.com/sec/math/mac/mathnet

CHAPTER 3

Adding and Subtracting Decimals

What you'll learn in Chapter 3

- to model, read, write, compare, order, and round decimals,

- to show relationships between metric units of length and estimate and measure line segments,

- to determine whether answers are reasonable, and

- to estimate and find sums and differences of decimals.

CHAPTER Project

VACATION DESTINATION

In this project, you will add and subtract decimals to help you plan a one- or two-week vacation for your family. You will need to research the cost of transportation, lodging, food, and admission to any tourist attractions that you want to visit.

Getting Started

- Look at the table. Estimate the cost for two adults and two children to attend each of the three tours. Make a bar graph to compare your estimates.

- How could you estimate the cost for 4 people to attend the Hawaiian Luau?

Costs for Selected Activities on Maui, Hawaii		
Activity	**Costs**	
Whale Watching Tour	$27.50 (adult, over 12)	$15.00 (children, 12 and under)
Tropical Plantation Tour	$8.50 (adult, over 12)	$3.50 (children, 12 and under)
Bike Tour from Haleakala Crater	$59.00 (basic, all ages)	$99.00 (deluxe, all ages)
Snorkeling Cruise to Molokini	$33.00 (afternoon, all ages)	$39.95 (morning, all ages)
Hawaiian Luau	$44.95 (all ages)	
Parasailing	$29.95 (morning, all ages)	$39.95 (anytime, all ages)

Technology Tips

- Use a **spreadsheet** to find the cost of your vacation.
- Use **computer software** to help you make graphs.
- Use a **word processor**.

 **Research** For up-to-date information on family vacations, visit:

www.glencoe.com/sec/math/mac/mathnet

Working on the Project

You can use what you'll learn in Chapter 3 to help you make your plans.

Page	Exercise
115	35
120	34
125	Alternative Assessment

COOPERATIVE LEARNING

3-1A Decimals Through Hundredths

A Preview of Lesson 3-1

▦. base-ten blocks

You know that base-ten blocks can be used to model whole numbers. Did you know that they can also be used to model decimals? For decimals, the blocks have the meanings shown at the right.

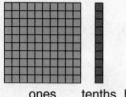

ones   tenths   hundredths

TRY THIS

Work in groups of four.

1 Model three tenths by using base-ten blocks.

- You can also model three tenths by trading the tenths for hundredths. How many hundredths are there?

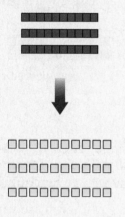

2 Show four tenths and seven hundredths with base-ten blocks. What decimal have you modeled?

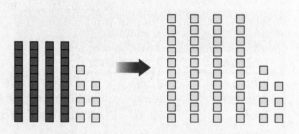

- Trade the tenths for hundredths and count the hundredths blocks.
- There are 47 blocks, so 47 hundredths is the decimal modeled.

ON YOUR OWN

1. Three tenths is the same as how many hundredths?

2. Show six tenths and four hundredths with base-ten blocks. Trade the tenths for hundredths. How many hundredths do you have now?

3. How many tenths are the same as ninety hundredths? Model using base-ten blocks.

4. **Look Ahead** If you separated a hundredth block into 100 equal parts, what decimal would be modeled by seventeen of the new parts?

3-1

Decimals Through Ten-Thousandths

What you'll learn

You'll learn to model, read, and write decimals.

When am I ever going to use this?

You can use decimals to record winning times at a track meet.

Word Wise

place value

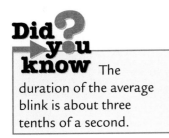

Did you know The duration of the average blink is about three tenths of a second.

Close your eyes and then open and close them as fast as you can. You have just modeled what happens when a camera takes a picture. The time a 35-millimeter camera's shutter stays open can range from a second to less than a millisecond.

A millisecond is $\frac{1}{1,000}$ of a second. What are some other ways to express $\frac{1}{1,000}$?

Decimals are another way to write fractions when the denominators are 10, 100, 1,000, and so on. We use **place-value** positions to name decimals.

In a decimal, the decimal point separates the ones place and the tenths place. The fraction $\frac{1}{1,000}$ can be written as the decimal 0.001 or in words as *one-thousandth*.

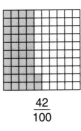

ones	tenths	hundredths	thousandths
0 .	0	0	1

Example

INTEGRATION

Statistics Charmaine asked 100 students at Western Hills Middle School to identify their favorite animal. 42 out of 100 or $\frac{42}{100}$ said their favorite animal is a dog.

a. Model the decimal for $\frac{42}{100}$.

Display forty-two hundredths blocks. Trade forty blocks for four tenths blocks.

$$\frac{42}{100}$$

b. Write the decimal in a place-value chart.

ones	tenths	hundredths
0 .	4	2

There are four tenths and two hundredths in the model. Write 4 in the tenths place and 2 in the hundredths place.

c. Write the decimal in words.

Notice that the last digit, 2, is in the hundredths place. The decimal in words is *forty-two hundredths*.

You can also write mixed numbers, such as $3\frac{7}{10}$, as decimals.

Example ② a. Model the decimal for $3\frac{7}{10}$.

b. Write the decimal in a place-value chart.

c. Write the decimal in words.

a. Display three ones and seven tenths blocks.

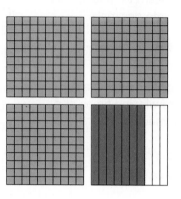

b. Write 3 in the ones place and 7 in the tenths place.

c. The decimal is *three and seven tenths*.

To write very small decimals, you can extend the place-value chart farther to the right.

Example ③ **Life Science** The egg of the Vervain hummingbird weighs

CONNECTION about $\frac{128}{10,000}$ ounce.

a. Write this fraction as a decimal in a place-value chart.

b. Write the decimal in words.

a.

b. The last digit, 8, is in the ten-thousandths place. So the decimal is *one hundred twenty-eight ten-thousandths*.

Communicating Mathematics

Read and study the lesson to answer each question.

1. *Write* the fraction, decimal, and word name for the decimal shown by the model.

2. *Draw* a model for 2.25.

3. *You Decide* Elizabeth says that thirty-eight thousandths is written as 0.0038. Mercedes says that it is written as 0.038. Who is correct? Explain.

Guided Practice

Write each fraction or mixed number as a decimal.

4. $\frac{9}{10}$

5. $\frac{43}{100}$

6. $\frac{563}{1,000}$

7. $4\frac{29}{10,000}$

8. Write four and three tenths as a decimal.

9. Write seven hundred two ten-thousandths as a decimal.

10. *Auto Racing* In 1996, Buddy Lazier of Vail, Colorado, won the Indianapolis 500 with an average speed of one hundred forty-seven and nine hundred fifty-six thousandths miles per hour. Write the speed as a decimal.

Practice

Write each fraction or mixed number as a decimal.

11. $\frac{3}{10}$

12. $\frac{99}{100}$

13. $\frac{7}{100}$

14. $10\frac{1}{10}$

15. $19\frac{53}{100}$

16. $\frac{375}{1,000}$

17. $\frac{9}{1,000}$

18. $\frac{5,561}{10,000}$

19. $47\frac{47}{1,000}$

20. $3\frac{27}{10,000}$

21. $172\frac{1}{10,000}$

22. $\frac{2,384}{1,000}$

Write each expression as a decimal.

23. twenty and nine tenths

24. eleven hundredths

25. three and three thousandths

26. one hundred nine and fifteen thousandths

27. eight hundred one ten-thousandths

28. six hundred ten and three hundred six ten-thousandths

Applications and Problem Solving

29. *Industrial Technology* A micrometer caliper is a device used to measure the thickness of an object. Many metric micrometer calipers can measure to one hundredth of a millimeter. Write one hundredth as a decimal.

30. Banking As a safeguard against error, the dollar amount on a check is written in both standard form and in words. Write $455.98 in words.

31. Critical Thinking Use the digits 0, 0, 5, 7 to make the greatest possible decimal and the least possible decimal.

Mixed Review

32. Geometry *(Lesson 2-9)*

    a. Which of the points on the graph has the coordinates (0, 4)?

    b. Write the coordinates of point *N*.

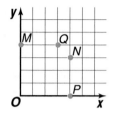

33. Standardized Test Practice Seven salespeople at Kip's Keyboards sold 13, 14, 10, 6, 3, 25, and 13 keyboards during the month of February. What is the mean for this set of data? *(Lesson 2-7)*

    A 3   **B** 12   **C** 13   **D** 25

34. Statistics Tammy took a survey of ten of her classmates to find the number of hours each person reads for pleasure each week. The results were 3, 0, 6, 10, 2, 20, 19, 8, 2, and 4. Would an interval of 1, 5, or 100 be best to form a frequency table of the data? Explain. *(Lesson 2-2)*

35. Money Martin collected $38 for the holiday children's fund. Phyllis collected half as much. Use the four-step plan to determine how much money Phyllis collected. *(Lesson 1-1)*

> For **Extra Practice**, see page 563.

Tiger *by Bud Blake*

1. Write one tenth as a decimal.

2. Suppose Hugo had said he learned ten thousandths of what he was supposed to. Would he be talking about the same amount? Explain.

Aviation

Mayte Greco

PILOT

Mayte Greco took flying lessons as a teenager. Now, the mother of five children, she is a pilot and owns her own air charter company in Florida. She also volunteers one day a week to search for Cuban refugees on rafts. She then radios the Coast Guard and waits until they come to the rescue.

Most companies that employ pilots require at least two years of college. To become a pilot for an airline company, you will need a college degree. Courses in engineering, meteorology, physics and mathematics are helpful in preparing for a pilot's career.

For more information:
Air Line Pilots Association
535 Herndon Parkway
P.O. Box 1169
Herndon, VA 22070

interNET
CONNECTION
www.glencoe.com/sec/
math/mac/mathnet

I'd like to fly people to their favorite vacation spots. I could stop there for a day or two myself!

Your Turn
Research the future of careers in flying. Will more pilots be needed in the next 10 to 20 years? What are the salaries for pilots?

3-2A Measurement

A Preview of Lesson 3-2

tape measure

The basic unit of length in the metric system is the *meter*. All other metric units of length are defined in terms of the meter.

The chart below summarizes the most commonly used metric units of length.

Unit	Symbol	Meaning
millimeter	mm	thousandth
centimeter	cm	hundredth
meter	m	one
kilometer	km	thousand

A metric ruler or tape measure is easy to read. The ruler below is labeled using *centimeters*.

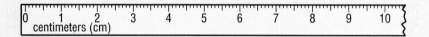

The pencil below is about 12 centimeters long.

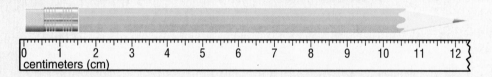

To read *millimeters*, you count each individual unit or mark on the metric ruler. There are ten millimeter marks for each centimeter mark. The pencil is about 124 millimeters long.

$$124 \text{ mm} = 12.4 \text{ cm}$$

There are 100 centimeters in a meter. Since there are 10 millimeters in one centimeter, there are 10×100 or 1,000 millimeters in a meter. The pencil is $\frac{124}{1,000}$ of a meter or 0.124 meters long.

$$124 \text{ mm} = 0.124 \text{ m}$$

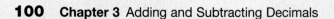

Work in groups of four.

Copy the table. Using a tape measure, measure the objects and complete the table.

Object	Measure		
	m	cm	mm
height of door			
width of door			
length of classroom			
length of math book			
length of pencil			
length of table or desk			
width of hallway			
length of sheet of paper			
length of chalkboard eraser			
length of your hand			
width of your little finger			

ON YOUR OWN

1. On your table, mark the unit of measure that is most appropriate for each item.

2. What pattern do you notice in the relationship between the measure you marked and the size of the object?

3. What patterns do you notice in the relationship between the numbers in the columns?

4. Look around your classroom. Select three objects that you think would be best measured in meters, three objects you think would be best measured in millimeters, and three objects you think would be best measured in centimeters. Explain your choices.

5. *Look Ahead* Copy the table below. Write the name of a common object that you think has a length that corresponds to each length in the first column.

Length	Object
5 centimeters	
15 centimeters	
3 meters	
1 meter	
75 centimeters	

Integration: Measurement
Length in the Metric System

In the 1996 Summer Olympics, Michael Johnson became the first man to win both the 200-meter and 400-meter races.

Most Olympic events use the **metric system** to measure lengths. The basic unit of length in the metric system is the **meter**. All other metric units of length are defined in terms of the meter.

Decimals are used in the metric system. The chart summarizes the most commonly used metric units of length.

Unit	Symbol	Meaning	Size	Model
millimeter	mm	thousandth	0.001 m	thickness of a dime
centimeter	cm	hundredth	0.01 m	width of large paperclip
meter	m	one	1.0 m	width of doorway
kilometer	km	thousand	1,000 m	six city blocks

Examples

Write the unit of length: millimeter, centimeter, meter or kilometer, that you would use to measure each of the following. Then estimate each measure.

1 thickness of a nickel

Since a nickel is thin, the *millimeter* is the appropriate unit. The thickness of a nickel is a little more than the thickness of a dime, so the thickness of a nickel is about 2 millimeters.

2 length of a baseball bat

Since the length of a baseball bat is close to the width of a doorway, the *meter* is the appropriate unit. The length of a baseball bat is about 1 meter.

3 On the map, what is the distance between San Antonio and Houston in centimeters? in millimeters?

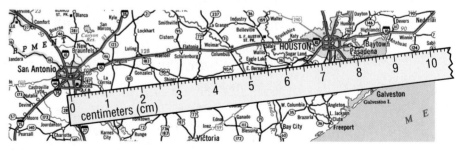

The distance is about 7.3 centimeters, or 73 millimeters.

④ Life Science Use a centimeter ruler to measure the length of the body of the chameleon as shown in the photo below.

In the photo, the length of the body of the chameleon is about 9 centimeters.

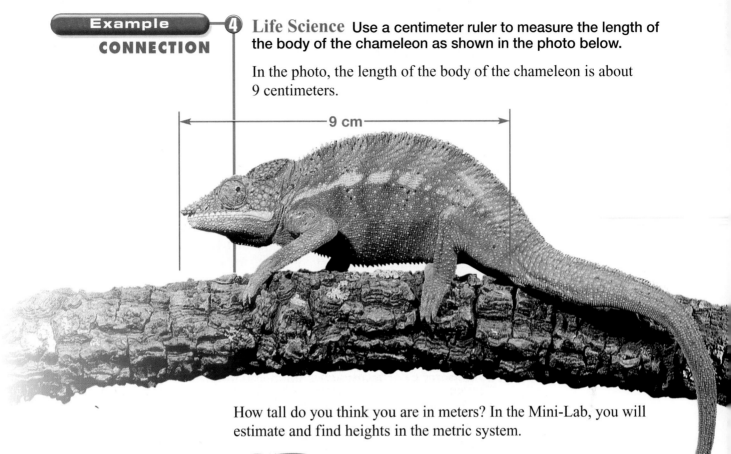

—— 9 cm ——

How tall do you think you are in meters? In the Mini-Lab, you will estimate and find heights in the metric system.

MINI-LAB

Work with a partner.   ∿ string   ▭ tape measure

Try This
- Cut a length of string that is as long as your partner is tall.
- Estimate each other's height in centimeters and in meters. Then measure your string to find your height.

Talk About It
1. What is your height in centimeters? Answers will vary.
2. What is your height in meters? Answers will vary.
3. Write a paragraph explaining how you decide what metric unit to use to measure length. Use specific examples in your explanation.

CHECK FOR UNDERSTANDING

Communicating Mathematics

Read and study the lesson to answer each question.

1. ***Draw*** a line that is 10.5 centimeters long.

2. ***Tell*** whether you would use centimeters or meters to measure the length of this book. Explain your answer.

HANDS-ON MATH

3. Collect everyone's height data from the Mini-Lab. Organize the data and draw a bar graph to display the data.

Write the unit of length: millimeter, centimeter, meter or kilometer, that you would use to measure each of the following. Then estimate each measure.

4. the length of a football field   5. the width of a quarter

Use a centimeter ruler to measure each line segment.

6. ━━━━━━━━━━━   7. ━━━━━━━━

8. *Sports* Runners often participate in races that are 10 kilometers long. How many meters are in 10 kilometers?

EXERCISES

Write the unit of length: millimeter, centimeter, meter or kilometer, that you would use to measure each of the following. Then estimate each measure.

9. the length of a skateboard   10. the thickness of a pencil

11. the height of a giraffe   12. the length of a swimming pool

Use a centimeter ruler to measure each line segment.

13. ━━━   14. ━━━━━

15. ━━━━━━━   16. ━━━━━━━━━━

Use a centimeter ruler to measure one side of each figure.

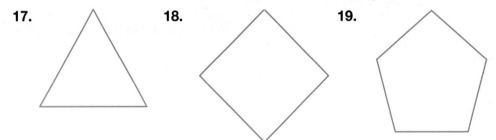

17.   18.   19.

20. How many millimeters are in 6.8 centimeters?

21. How many centimeters are in 3 meters?

22. *Language Arts* The metric prefix *cent* means *hundred*. Make a list of words that contain this prefix.

23. *Geometry* On a sheet of centimeter grid paper, draw a square such that the perimeter (distance around the square) is 20 centimeters.

24. *Critical Thinking* Order from least to greatest.
0.0037 km   3.9 m   55 cm   0.49 m   999 mm

25. **Standardized Test Practice**   Choose the decimal that represents *twelve and sixty-three thousandths. (Lesson 3-1)*
A 1206.3   **B** 120.63   **C** 12.063   **D** 0.12063

26. *Statistics* Construct a horizontal bar graph using the guinea pig weights of 1.8, 1.754, 2.09, 1.91, and 2.1 pounds. *(Lesson 2-3)*

For **Extra Practice**, see page 563.

27. *Algebra* Find the value of $3x - 2y$ if $x = 6$ and $y = 4$. *(Lesson 1-5)*

Comparing and Ordering Decimals

What you'll learn

You'll learn to compare decimals and order a set of decimals.

When am I ever going to use this?

Knowing how to compare and order decimals can help you find library books.

Mrs. Lee's science class is studying acids and bases. Acidity is expressed using the pH scale, which ranges from 0 to 14.

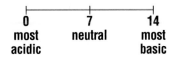

Substance	pH
apple	2.9
baking soda	9.0
carrot	5.1
drinking water	7.0
household ammonia	13.0
lemon	2.1
lye	14.0
rainwater	5.8
tomato	4.2

The results of Carmen's group are shown in the table.

Carmen's group needs to compare the pH of the carrot and of rainwater, 5.1 and 5.8, respectively. There are two ways to compare decimals. You can compare the digits in each place-value position, or you can use a number line.

Method 1 Use place value.

Line up the decimal points of the two numbers. Then start at the left, comparing the digits in the same place-value position that are not equal. The decimal with the greater digit is the greater decimal.

carrot: 5.**1**
rainwater: 5.**8**

1 and 8 are not equal. 1 tenth < 8 tenths, so 5.1 < 5.8.

Method 2 Use a number line.

On a number line, numbers to the right are greater than numbers to the left.

5.1 is left of 5.8, 5.1 < 5.8.

Since 5.1 < 5.8, the carrot is more acidic than the rainwater.

Study Hint

Reading Math
Remember that the symbol < is read as *is less than* and the symbol > is read as *is greater than*. The symbol always points toward the lesser number.

Example

1 **Which is greater, 2.037 or 2.033?**

Method 1 Use place value.

2.03**7**   *Line up decimal points.*
2.03**3**   *Starting at the left, compare each place-value.*

7 and 3 are not equal.

7 thousandths > 3 thousandths, so 2.037 > 2.033.

(continued on the next page)

Method 2 Use a number line.

Compare the decimals on a number line.

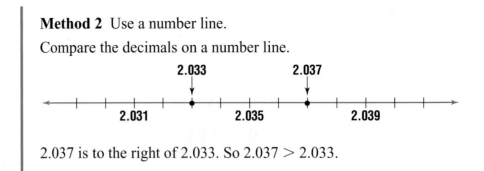

2.037 is to the right of 2.033. So 2.037 > 2.033.

You can also use grid paper to compare decimals.

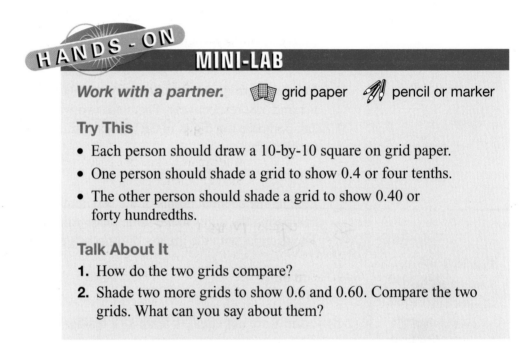

MINI-LAB

Work with a partner.   grid paper   pencil or marker

Try This

- Each person should draw a 10-by-10 square on grid paper.
- One person should shade a grid to show 0.4 or four tenths.
- The other person should shade a grid to show 0.40 or forty hundredths.

Talk About It

1. How do the two grids compare?
2. Shade two more grids to show 0.6 and 0.60. Compare the two grids. What can you say about them?

You can annex, or place zeros to the right of a decimal so that each decimal has the same number of decimal places.

Example ② **Which is the greatest number, 19.48, 19.481, or 19.4?**

Line up the decimal points.

Annex a zero so that each has the same number of decimal places.

19.48	19.480
19.481	19.481
19.4	19.400

Since 8 hundredths is greater than 0 hundredths, 19.480 > 19.400.

Since 1 thousandth is greater than 0 thousandths, 19.481 > 19.480.

The greatest number is 19.481.

Sports The table shows the top rebounder for each team in the Midwest Division of the NBA for the 1997 regular season. Order the rebounding averages from least to greatest.

Player	Team	Rebounds/Game
Barkley	Houston	13.5
Green	Dallas	7.9
Malone	Utah	9.9
Gugliotta	Minnesota	8.7
Er. Johnson	Denver	11.1
Perdue	San Antonio	9.8
Reeves	Vancouver	8.1

Use a number line to order the average number of rebounds.

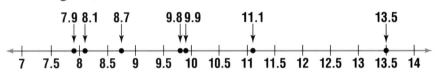

The rebounding averages in order from least to greatest are 7.9, 8.1, 8.7, 9.8, 9.9, 11.1, and 13.5.

CHECK FOR UNDERSTANDING

Communicating Mathematics

Read and study the lesson to answer each question.

1. **Draw** a number line that compares 3.89 and 3.91.

2. **Tell** how the decimals 1.018 and 1.01, graphed on the number line, compare.

HANDS-ON MATH

3. Use base-ten blocks to compare 1.63 and 1.54.

Guided Practice

State the greatest number in each group.

4. 6.02 or 6.20
5. 0.042, 0.06 or 0.051
6. 198.6, 198.06, 189.5, or 198.067

Order each set of decimals from least to greatest.

7. 13.507; 13.05; 13.9; 13.84

8. 0.2; 0.09; 0.19; 0.21; 2.1; 21.9; 0.002

9. **Sports** The table shows the scores for the top five teams in women's gymnastics at the 1996 Summer Olympics. Order the scores from greatest to least.

Team	Score
China	385.867
Romania	388.246
Russia	388.404
Ukraine	385.841
USA	389.225

Practice

State the greatest number in each group.

10. 16.099 or 160.98

11. 0.331 or 0.303

12. 18.607 or 18.06

13. 1.018 or 1.01

14. 0.03, 0.31, or 0.039

15. 47.553, 47.5, or 47.053

16. 547.484 or 547.4843

17. 0.068, 0.07, or 0.7

Order each set of decimals from least to greatest.

18. 94.7
101.1
99.7
98.5

19. 15
15.8
14.95
15.01

20. 37.5
35.7
35.849
36.06

Order each set of decimals from greatest to least.

21. 0.025
0.0316
0.0306
0.0249
0.0208

22. 43.8
42.998
43.16
42.022
43.6789

23. 379.8778
378.87
397.877
379.9
379.88

24. Which is the least, 10.59 or 10.599?

25. Which is the greatest, 0.0621, 0.603, or 0.06?

Applications and Problem Solving

26. *Library Science* Shannon has to return the stack of books shown at the right to the library shelves. Order the call numbers on the books from least to greatest.

27. *Statistics* Refer to the beginning of the lesson.
   a. Order the substances from the most acidic to the most basic.
   b. What is the median pH reading?

28. *Critical Thinking* Antonio has more money than Beatriz. Antonio has less money than César. Dexter has $0.10 more than Eric. Use the decimals at the right to determine how much money each person has.

$1.70
$0.79
$1.07
$1.18
$0.89

Mixed Review

29. *Measurement* Measure the line segment. *(Lesson 3-2)*

_____

30. Write thirty-seven thousandths as a decimal. *(Lesson 3-1)*

31. *Statistics* Make up a set of data that has no mode and state why it has no mode. *(Lesson 2-7)*

32. **Standardized Test Practice** Florida has 67 counties. What is this number rounded to the nearest ten? *(Lesson 1-3)*
   A 60   **B** 65   **C** 70   **D** 75

3-4

Rounding Decimals

What you'll learn

You'll learn to round decimals.

When am I ever going to use this?

Knowing how to round decimals is helpful in estimating with money.

A Mexican restaurant created the world's largest burrito in Anaheim, California, on July 31, 1995. It weighed 4,217 pounds and was 3,112.99 feet long. Do you think the weight and length were reported this way in the local papers?

Newspapers often round numbers so it is easier to read. The length of the burrito may have been reported as 3,113 feet.

Decimals may be rounded to any place-value position.

Example 1

Round 1.63 to the nearest tenth.

To round 1.63, look at the number line below.

1.63

1.60   1.65   1.70

On the number line, 1.63 is closer to 1.6 than it is to 1.7.

1.63 rounded to the nearest tenth is 1.6.

LOOK BACK

You can refer to Lesson 1-3 for information on rounding whole numbers.

You can also round decimals without using a number line.

Rounding Decimals	• Look at digit to the right of the place being rounded. • The digit remains the same if the digit to the right is 0, 1, 2, 3, or 4. • Round up if the digit to the right is 5, 6, 7, 8, or 9.

Example 2

APPLICATION

Money Matters Ayashe purchased a carbon monoxide detector for $27.44. To the nearest dollar, how much did she spend?

Look at the digit to the right of the ones place.

27.44   *Since 4 < 5, the digit in the*
 ↑   *ones place stays the same.*
ones place

To the nearest dollar, Ayashe spent $27.

Example ——③ Round 3.4672 to the nearest hundredth.

Look at the digit to the right of the hundredths place.

$$3.4672 \qquad \textit{Round up since 7 > 5.}$$
$$\uparrow$$
$$\textit{hundredths place}$$

3.4672 rounded to the nearest hundredth is 3.47.

CHECK FOR UNDERSTANDING

Communicating Mathematics

Read and study the lesson to answer each question.

1. **Explain** how to round $10.79 to the nearest dollar.

2. **Draw** a number line to show why 2.983 rounded to the nearest tenth is 3.0.

3. **You Decide** Sharon says that 456.789 rounded to the nearest hundredth is 456.79. Carlos says it is 500. Who is correct? Explain.

Guided Practice

Round each number to the underlined place-value position.

4. 0.2<u>8</u>   5. 8.2<u>0</u>2   6. 0.<u>1</u>487   7. 1<u>9</u>.59

8. **Money Matters** The unit price of a twenty-ounce bottle of a soft drink is $0.0445 per ounce. How much is this to the nearest cent?

EXERCISES

Practice

Round each number to the underlined place-value position.

9. 1<u>8</u>.44   10. 0.8<u>4</u>9   11. 20.<u>4</u>5   12. 2.4<u>8</u>5

13. 49.<u>7</u>75   14. 68.<u>8</u>8   15. 19.7<u>7</u>5   16. 48.<u>8</u>02

17. 99.<u>9</u>8   18. 42.7<u>8</u>96   19. 6.99<u>9</u>8   20. 4.0<u>0</u>4

21. Round $1.69 to the nearest dollar.

22. Round 49,237.1589499 to the nearest ten-thousandth.

Applications and Problem Solving

23. **Entertainment** KTXQ in Dallas, Texas, can be found by tuning to 102.1 on the radio. The DJs round the call number to the nearest whole number. What number do the DJs use to refer to KTXQ?

24. **Money Matters** Gasoline prices are usually calculated to the thousandth place. If you purchased gas that cost $1.199 per gallon, what price would you say you paid?

25. **Critical Thinking** The table shows the average density of the nine planets. If you had to determine the mean, median, and mode of this set of data, would you round to the nearest tenth, nearest whole number, or use the exact numbers? Explain.

Planet	Grams per cubic cm
Mercury	5.42
Venus	5.25
Earth	5.52
Mars	3.94
Jupiter	1.33
Saturn	0.69
Uranus	1.27
Neptune	1.71
Pluto	2.03

Mixed Review

26. Order the decimals 2.9, 2.38, 2.474, 2.91, and 2.88 from greatest to least. *(Lesson 3-3)*

27. **Statistics** Is the mean a misleading measure of central tendency for this set of data: 17, 21, 20, 19, 17, 21, 18, 22, and 21? Explain. *(Lesson 2-8)*

28. **Standardized Test Practice** The graph shows the number of boys and girls who signed up to play in the Northtown Youth Soccer League. Which is a reasonable conclusion that can be drawn from the information in the graph? *(Lesson 2-3)*

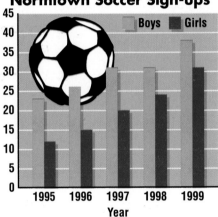

Northtown Soccer Sign-ups

  A More girls than boys signed up for soccer in 1998.

  B The number of girls playing soccer has been catching up with the number of boys since 1995.

  C The number of boys who joined soccer decreased every year.

  D Girls did not want to play soccer before 1995.

  E There were more soccer sign-ups in 1997 than any other year.

For **Extra Practice,** see page 564.

CHAPTER 3

Mid-Chapter Self Test

Express each fraction as a decimal. *(Lesson 3-1)*

1. $\frac{7}{10}$

2. $\frac{43}{100}$

3. $\frac{681}{1,000}$

4. $\frac{409}{10,000}$

Measure each line segment to the nearest centimeter. *(Lesson 3-2)*

5. ———————

6. ———————

7. **Library** Miguel Aguilar went to the library to get some information about bees for a project. He is looking for books with the call numbers 638.178 and 638.186. Which number comes first? *(Lesson 3-3)*

Round each number to the underlined place-value position. *(Lesson 3-4)*

8. 6.8

9. 3.401

10. 181.98

Estimating Sums and Differences

What you'll learn

You'll learn to estimate decimal sums and differences.

When am I ever going to use this?

You'll estimate sums and differences when you need to determine if you have enough money to pay for the groceries in your cart.

Word Wise

clustering

Do you participate in a sport? The graph shows the cost of treating sports injuries in emergency rooms in a recent year. About how much did treating these injuries cost?

Since you only need to know about how much the cost was, you can estimate. One way to estimate is to round the amounts to the same place-value position and then add.

Where Does It Hurt?

Sport	Injuries	Cost (billions)
Cycling	599,874	$4.29
Basketball	693,933	$3.60
Snow skiing	330,289	$2.40
Football	390,180	$2.20
Skating, all types	322,311	$1.96

Source: Consumer Product Safety Commission, American Academy of Orthopedic Surgeons

Round each number to the nearest billion.

$$\begin{array}{rcl} \$4.29 & \to & \$4 \\ \$3.60 & \to & \$4 \\ \$2.40 & \to & \$2 \\ \$2.20 & \to & \$2 \\ +\$1.96 & \to & +\$2 \\ \hline & & \$14 \end{array}$$

The cost of sports injuries was about $14 billion.

Example ① **APPLICATION**

Real World

Advertising The table shows the household exposure of the top ten advertisers on prime-time TV between August 26 and September 1. About how many households were exposed to Advertiser 1's and Advertiser 2's ads?

To estimate, round each number to the nearest hundred million.

Advertiser	Household Exposure (millions)
1	390.4
2	216.5
3	183.4
4	172.7
5	172.6
6	136.6
7	131.4
8	126.3
9	106.7
10	106.6

$$\begin{array}{rcl} 390.4 & \to & 400 \quad \textit{9 > 4, so round up.} \\ +216.5 & \to & +200 \quad \textit{1 < 5, so the digit stays the same.} \\ \hline & & 600 \end{array}$$

About 600 million households were exposed to Advertiser 1's and Advertiser 2's ads.

CONNECTION

② Earth Science The largest cut diamond is the 545.67-carat gem known as the Golden Jubilee. Before it was cut, it weighed 775.50 carats. About how many carats of the uncut diamond were not used?

$$
\begin{array}{ccc}
775.50 & \rightarrow & 776 \\
-545.67 & \rightarrow & -546 \\
\hline
& & 230
\end{array}
$$

Round to the nearest carat. Then subtract.

About 230 carats of the uncut diamond were not used.

APPLICATION

③ Money Matters Several video games are on sale. The sale price for Video game A is $42.99, and the sale price for Video game B is $55.88.

a. About how much less is Video game A than Video game B?

$$
\begin{array}{ccc}
\$55.88 & \rightarrow & \$56 \\
- \ 42.99 & \rightarrow & - \ 43 \\
\hline
& & 13
\end{array}
$$

Round to the nearest dollar. Then subtract.

Video game A is about $13 less than Video game B.

b. If you bought both games while they were on sale, about how much would they cost?

$$
\begin{array}{ccc}
\$55.88 & \rightarrow & \$56 \\
+42.99 & \rightarrow & +43 \\
\hline
& & 99
\end{array}
$$

Round to the nearest dollar. Then add.

The games would cost about $99.

Clustering is another way you can estimate sums and differences. Clustering is used when numbers are close to the same number.

Example

APPLICATION

④ Business The table shows the charges for business related calls and faxes made by Mrs. Moore last week. About how much were her business related phone charges last week?

Minutes	Amount ($)
1.0	0.31
1.0	0.28
1.0	0.28
1.0	0.26
1.0	0.30

Since each amount is about $0.30, add this amount 5 times.

$$0.30 + 0.30 + 0.30 + 0.30 + 0.30 = 1.50$$

Mrs. Moore spent about $1.50 last week.

Communicating
Mathematics

Read and study the lesson to answer each question.

1. *Explain* how to use rounding to estimate the difference between $16.98 and $4.29.

2. *Describe* a situation where it makes sense to use the clustering method to estimate a sum.

3. *Write* a paragraph explaining how estimation would help you buy items at a store.

Guided Practice

Estimate using rounding.

4. 7.75 + 8.95   5. 18.52 + 31.3   6. $20 − $1.82

Estimate using clustering.

7. 11.36 + 10.84 + 11 + 10.5

8. $3.44 + $3.40 + $3.50 + $3.49

9. Estimate the sum of 20.09, 20.58, and 19.98.

10. About how much is $53.38 minus $32.68?

11. *Physical Science* Enrique has two samples of the same chemical. He wants to store both samples in a 0.5-liter container. One sample is 0.38 liter. The other sample is 0.21 liter. Can he store both samples in the container? Explain.

EXERCISES

Practice

Estimate using rounding.

12. 0.43 + 0.94   13. 8.78 − 5.09   14. 68.69 − 7.43

15. 31.556 + 17.405   16. 0.612 + 0.185   17. $31.30 − $18.52

18. 0.8 − 0.7383   19. $57.98 − $26.95   20. 5.34 + 6.33 + 1.9

Estimate using clustering.

21. 4.38 + 3.68 + 4.42   22. $11.46 + $10.57 + $10.88 + $11

23. 0.805 + 1.006 + 0.64 + 0.9   24. 6.72 + 5.9 + 6.143 + 6.5037

25. $54.45 + $54.07 + $53.99   26. 95.98 + 98.15 + 104.5 + 104.95

27. About how much more is $64.50 than $39.95?

28. Estimate the sum 3.456 + 2.888 + 3.393 + 3.483.

29. About how much above $109.99 is an amount of $199.98?

30. Estimate the difference between 1.685 and 0.454.

31. About how much above 98.6°F is a body temperature of 102.4°F?

32. Estimate the sum of $7.25, $6.88, $6.75, $7.02, and $6.97.

33. *Measurement* The largest commercial building in terms of floor area is the flower auction building of the Cooperative VBA in Aalsmer, Netherlands. The original floor surface of 3.7 million square feet has now been extended to 5.27 million square feet. By about how many square feet was the floor surface increased?

34. *Recycling* The table shows the amount received by Mrs. Barsch's class for turning in aluminum cans. About how much money did they receive?

Week	Money
1	$6.80
2	$6.60
3	$7.20
4	$7.00

35. *Working on the* **CHAPTER Project** Refer to the table on page 93.
   a. In Mandy's family, three people are over 12, and one is under 12. Can her whole family go on the Tropical Plantation Tour for less than $35?
   b. Using any strategy, estimate the cost of eating in restaurants for one day for your family or your group. Show the numbers you used to make your estimate.

36. *Critical Thinking* Four same-priced items are purchased. Based on rounding, the estimate of the total was $16.
   a. What is the maximum price each item could have cost?
   b. What is the minimum price each item could have cost?

Mixed Review

37. *Sports* Batting averages are rounded to the nearest thousandth. In 1996, Mike Stanley, catcher for the Boston Red Sox, had a batting average of 0.2695 to the nearest ten-thousandth. How will his 1996 batting average be listed on his baseball card? *(Lesson 3-4)*

38. *Sports* The chart shows the times for four runners in a 100-meter race. In what order did the participants cross the finish line? *(Lesson 3-3)*

Runner	Time
Sarah	14.31 s
Camellia	13.84 s
Fala	13.97 s
Debbie	13.79 s

39. *Statistics* Find the median of the data: 23, 19, 22, 22, 20, 20, 19, 22. *(Lesson 2-7)*

40. **Standardized Test Practice** The circle graph shows pie sales at a local bakery. What part of the total sales is peanut butter and strawberry? *(Lesson 2-4)*
   A 0.20
   B 0.17
   C 0.12
   D 0.08

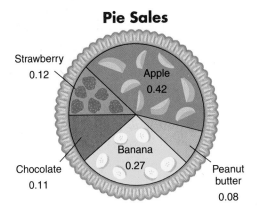

Pie Sales

Strawberry 0.12
Apple 0.42
Banana 0.27
Chocolate 0.11
Peanut butter 0.08

41. *Algebra* Evaluate the expression $a + b \div c$ if $a = 5$, $b = 12$, and $c = 4$. *(Lesson 1-5)*

For **Extra Practice**, see page 564.

PROBLEM SOLVING

3-6A Reasonable Answers

A Preview of Lesson 3-6

Akira and Brian were shopping for CDs and noticed there were hundreds of CDs. Let's listen in!

Wow! Look at the selection they have here. Do you think over a billion CDs have been sold yet?

I know where we can find the answer. I have an almanac in my backpack. We were using it today in social studies class. Let's look.

There it is. The table shows the number of CDs that were shipped for sale from 1984 to 1994.

It looks like almost a billion CDs were shipped in 1994 alone!

Year	CDs Shipped (millions)
1984	5.8
1985	22.6
1986	53.0
1987	102.1
1988	149.7
1989	207.2
1990	286.5
1991	333.3
1992	407.5
1993	495.4
1994	662.1

Akira

How did you figure that? There were less than 700 million CDs shipped that year.

Brian

THINK ABOUT IT

1. **Compare and contrast** Akira's and Brian's thinking. Whose estimate do you think is correct? Explain why.

2. **Explain** why both boys might be considered correct.

3. **Choose** two years when a total of about 200 million CDs were shipped. Explain your reasoning.

4. **Apply** what you have learned from Akira and Brian's situation to solve the following problem.

Kelsey buys a small beanbag animal for $5.54. She pays for the purchase with a $20 bill. Which is a more reasonable estimate for the amount of change she should receive: $12 or $15? Explain why.

For **Extra Practice,** see page 564.

ON YOUR OWN

5. The last step of the 4-step plan for problem solving asks you to *examine* your solution. *Explain* how the place value to which you round affects how you examine a solution.

6. *Write* a definition of a reasonable answer in your own words.

7. *Explain* how the strategies you used in estimating with large decimal numbers can be applied to estimating with small decimal numbers.

MIXED PROBLEM SOLVING

STRATEGIES

Look for a pattern.
Solve a simpler problem.
Act it out.
Guess and check.
Draw a diagram.
Make a chart.
Work backward.

Solve. Use any strategy.

8. *Write a Problem* using the following numbers and phrases: $4.59; 3 centimeters; 75° F; 9:00 A.M., and 6 hours.

9. *Statistics* During the fall fund-raising project, Roberta's class sold magazine subscriptions. The pictograph shows how many magazines they sold during a 6-week period.

Subscriptions Sold Per Week

Week 1–6

4 subscriptions = ▢

a. How many subscriptions were sold during week 3?

b. What does the half-magazine shown during week 4 mean?

c. During what week did the class sell 16 subscriptions?

10. *Sports* Suppose 235,532 people attended a 4-game home stand to see the Texas Rangers during the 1997 season. Which is a more reasonable estimate for the number of people that attended each game: 50,000 or 60,000? Explain.

11. *Statistics* Based on the data in the table on page 116, what would you predict about the shipment of CDs from 1995 to 2000? Explain how you made your prediction.

12. *Education* Robert E. Lee High School will graduate 678 seniors on June 8. The ceremony will be held in the gymnasium. The gymnasium holds 2,100 people in addition to the graduates. Is it reasonable to offer each graduate three tickets for family and friends? Explain.

13. **Standardized Test Practice** Which diagram does *not* show 0.7?

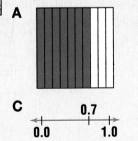
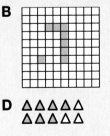

A

B

C
0.7
0.0   1.0

D △△△△△
  △△△△△

Adding and Subtracting Decimals

What you'll learn

You'll learn to add and subtract decimals.

When am I ever going to use this?

You'll add and subtract decimals when you balance your checking account.

The table shows the top ten movies based on money earned on a recent weekend. What was the total earned by the top four movies?

In order to find the total, you need to add the four numbers.

Adding decimals is like adding whole numbers. Make sure that you line up the decimal points before you add or subtract.

Movie	Money Earned (millions)
A	$45.1
B	25.5
C	17.4
D	17.3
E	17.25
F	8.1
G	5.6
H	4.4
I	3.4
J	1.4

$$
\begin{array}{r}
\overset{2\ 1}{45.1} \\
25.5 \\
17.4 \\
+17.3 \\
\hline
105.3
\end{array}
$$

Estimate:
50 + 30 + 20 + 20 = 120

The top four movies earned a total of $105.3 million. The estimate shows that the answer is reasonable.

Example

APPLICATION **1**

Entertainment Refer to the table above. How much more money was earned by Movie A than by Movie J?

Explore What do you know?
You know how much money each movie earned.

What do you need to know?
You need to know how much more money Movie A earned than Movie J.

Plan You need to subtract the smaller amount earned from the larger amount earned.

Estimate first. 45 − 1 = 44

Solve
$$
\begin{array}{r}
\overset{4\ 11}{4\cancel{5}.\cancel{1}} \\
-\ 1.4 \\
\hline
43.7
\end{array}
$$
Movie A earned $43.7 million more than Movie J.

Examine The estimate shows that the answer is reasonable.

Sometimes it is necessary to annex zeros in order to subtract decimals.

2 Find the difference of 7 and 2.35.

Estimate: 7 − 2 = 5

$$\begin{array}{ll} 7.00 & \textit{Annex two zeros.} \\ -2.35 & \textit{Rename and subtract.} \end{array}$$

$$\begin{array}{r} \overset{6\ 9\ 10}{7.\cancel{0}\cancel{0}} \\ -2.35 \\ \hline 4.65 \end{array}$$

The estimate shows that the answer is reasonable.

3 Find the sum of 47.68 and 7.8.

Estimate: 50 + 10 = 60

47.68 $\boxed{+}$ 7.8 $\boxed{=}$ 55.48

The estimate shows that the answer is reasonable.

INTEGRATION

4 **Algebra** **Evaluate x + y if x = 4.56 and y = 19.367.**

$x + y = 4.56 + 19.367$ *Replace x with 4.56 and y with 19.367.*

$$\begin{array}{ll} 4.560 & \textit{Annex a zero and line up the decimal points.} \\ +19.367 & \textit{Add.} \\ \hline 23.927 \end{array}$$

The value is 23.927.

Study Hint

Estimation When you use a calculator to add or subtract decimals, estimate to check that the result is reasonable. If it isn't, check to see if each decimal point was entered correctly.

CHECK FOR UNDERSTANDING

Communicating Mathematics

Read and study the lesson to answer each question.

1. *Write* directions explaining how to subtract 2.67 from 3.

2. *Explain* how you would find the sum of 3.701, 0.49, and 2.4 using paper and pencil.

Math Journal

3. *Write* a paragraph explaining how adding and subtracting decimals compares to adding and subtracting whole numbers.

Guided Practice

Add or subtract.

4.	6.4	5.	1.34	6.	41.39	7.	3.7
	+3.3		+0.9		−23.17		−2.95

8. $67.38 − 37.46$

9. $3.702 + 0.49 + 2.4$

10. *Algebra* Solve the equation $c = 0.085 + 2.487$.

Practice

Add or subtract.

11.　2.3
　　+4.1

12.　0.37
　　+0.55

13.　0.67
　　−0.43

14.　42.76
　　−31.59

15.　$6.78
　　+ 4.99

16.　8
　　+6.76

17.　8.267
　　−6.52

18.　17.6
　　−4.73

19. 6.6 − 4.58

20. 5.77 − 2.374

21. 0.563 + 5.8 + 6.89

22. 23.4 + 9.865 + 18.26

23. Find the sum of 84.34 and 67.235.

24. How much is 46 minus 23.78?

25. How much more than $102.90 is $115?

26. *Statistics* Find the average (mean) of 8.12, 7.6, and 8. Round to the nearest hundredth.

Solve each equation.

27. $29.2 − 2.78 = b$

28. $e = 5.162 + 0.6099$

29. $c = 4 − 1.9$

30. $478.98 − 46 = k$

31. *Algebra* What is the value of $a − b$ if $a = 34.6$ and $b = 23.88$?

32. *Algebra* Evaluate $r + s + t$ if $r = 45.1$, $s = 16$, and $t = 8.091$.

Applications and Problem Solving

33. *Food* The graph shows how many pounds of turkey the average person eats during the year.

　a. How many more pounds of turkey does the average person eat from October through December than from January through March?

　b. How many pounds of turkey does the average person eat in one year?

Source: National Turkey Federation

34. *Working on the* CHAPTER Project Refer to the table on page 93. In Dawn's family, two people are over 12, and three are under 12. Find the exact cost for her family to take the Whale-Watching Tour and attend the Hawaiian Luau.

35. *Critical Thinking* Arrange the digits 1, 2, 3, 4, 5, 6, 7, and 8 into two decimals so that their difference is as close to 0 as possible. Use each digit only once.

36. *Money Matters* Marta plans to buy a baseball for $6.50, a baseball glove for $37.99, and a baseball cap for $13.79. Estimate the cost of these items before tax is added. *(Lesson 3-5)*

37. Standardized Test Practice How long is the pencil in centimeters? *(Lesson 3-2)*

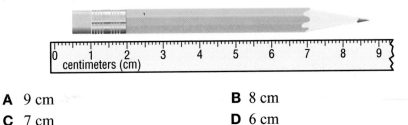

A 9 cm	**B** 8 cm
C 7 cm	**D** 6 cm

For **Extra Practice**, see page 565.

38. *Statistics* Is the mode a misleading measure of central tendency for this set of data: 21, 20, 19, 13, 21, 18, 12, 21? Explain. *(Lesson 2-8)*

39. *Algebra* Identify the number that is the solution of the equation $42 \div h = 14$; 3, 4, or 5. *(Lesson 1-7)*

Let the GameS Begin

The 1-Meter Dash

Math Skill

Adding and Subtracting Decimals

Get Ready This game is for two players.

⚀ 2 number cubes   ⊛ 2 spinners   ◑ 2 counters   🪜 meterstick

Get Set ● Make a spinner that has two equal sections. Mark the sections with the words *add* and *subtract*.

● Each person starts with a counter at opposite ends of a meterstick.

Go ● The first player rolls the number cubes and spins the spinner.

● The player forms a decimal using the numbers on the number cube and then rolls the number cubes again.

● The player forms another decimal, performs the operation shown on the spinner, and then moves the counter that many centimeters along the meterstick. For example, suppose a player rolls a 2 and a 4, spins *subtract*, and then rolls a 5 and 6. That player could move $6.5 - 2.4$ or 4.1 centimeters.

● The winner is the first player to have their counter go beyond the other end of the meterstick.

interNET CONNECTION Visit www.glencoe.com/sec/math/mac/mathnet for more games.

🖳 *inter*NET
CONNECTION
Chapter Review For additional lesson-by-lesson review, visit:
www.glencoe.com/sec/math/mac/mathnet

Vocabulary

After completing this chapter, you should be able to define each
term, concept, or phrase and give an example or two of each.

Measurement
centimeter (p. 100)
meter (p. 102)
metric system (p. 102)
millimeter (p. 100)

Number and Operations
clustering (p. 113)
place value (p. 95)

Problem Solving
reasonable answers (pp. 116–117)

Understanding and Using the Vocabulary

**Determine whether each statement is *true* or *false*. If the statement is
false, replace the underlined word or number to make it true.**

1. The number 0.07 is <u>greater</u> than 0.071.

2. A millimeter equals <u>one thousandth</u> of a meter.

3. A centimeter equals <u>one tenth</u> of a meter.

4. When rounding decimals, the digit in the place being rounded should be
   rounded up if the digit to its right is <u>6</u>.

5. The length of the cassette tape below is about 10 <u>millimeters</u>.

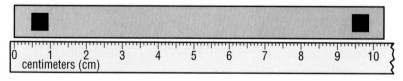

6. In 643.082 the digit 2 names the number two <u>hundredths</u>.

7. Six hundred and twelve thousandths written as a decimal is <u>0.612</u>.

8. Estimating 3.3 + 2.9 + 3.4 + 3.09 by computing 3 + 3 + 3 + 3 is an
   example of <u>clustering</u>.

9. If the amount of your purchase is $6.74, then a reasonable amount of
   change from <u>$20</u> is $13.26.

In Your Own Words

10. ***Explain*** when and why the clustering strategy is used to estimate sums.

Objectives & Examples

Upon completing this chapter, you should be able to:

● model, read, and write decimals *(Lesson 3-1)*

Model $\frac{3}{100}$ using base-ten blocks.

Write $\frac{73}{100}$ as a decimal.

Write 7 in the tenths place and 3 in the hundredths place.

$$0 \bullet 7 \mid 3$$
ones tenths hundredths

Write 0.024 in words.

The last digit, 4, is in the thousandths place. The decimal in words is *twenty-four thousandths.*

● show relationships among metric units of length and measure line segments *(Lesson 3-2)*

0 1 2 3 4
centimeters (cm)

The line segment measures 4.1 centimeters or 41 millimeters.

Review Exercises

Use these exercises to review and prepare for the chapter test.

Write each fraction or mixed number as a decimal.

11. $\frac{8}{100}$

12. $8\frac{9}{10}$

13. $14\frac{17}{1,000}$

14. $\frac{643}{10,000}$

Write each expression as a decimal.

15. two tenths

16. thirty-four hundredths

17. fifty-three thousandths

18. thirty and twelve ten-thousandths

19. *Life Science* An amoeba is one of the larger creatures of the microscopic world. The length of a typical amoeba is 0.0008 meter. Write 0.0008 in words.

Use a centimeter ruler to measure each line segment.

20. ——————

21. ————————

22. ————

Use a centimeter ruler to measure one side of the figure.

23.

Objectives & Examples

compare decimals and order a set of decimals
(Lesson 3-3)

Is 5.43 greater than 5.427?

Since 3 hundredths is greater than 2 hundredths, 5.43 is greater than 5.427.

Write 45.93, 46.4, 45.89, and 45.311 in order from least to greatest.

The decimals in order from least to greatest are 45.311, 45.89, 45.93, and 46.4.

round decimals *(Lesson 3-4)*

Round 4.739 to the nearest tenth.

The digit to the right of the tenths place is 3. So, 7 remains the same.

4.739 rounded to the nearest tenth is 4.7.

estimate decimal sums and differences
(Lesson 3-5)

a. rounding

$$\begin{array}{rcr} 7.79 & \to & 7.8 \\ -2.32 & \to & -2.3 \\ \hline & & 5.5 \end{array}$$

b. clustering

$7.96 + 8.1 + 8.23 + 7.7 \to 8 + 8 + 8 + 8$
$\to 8 \times 4 = 32$

add and subtract decimals *(Lesson 3-6)*

Find the difference of 7.3 and 2.89.

$$\begin{array}{l} \overset{6\ \ 1210}{7.3\,\cancel{0}} \quad \textit{Line up the decimal points.} \\ -2.89 \quad \textit{Annex a zero.} \\ \hline 4.41 \quad \textit{Rename and subtract.} \end{array}$$

Review Exercises

State the greatest number in each group.

24. 5.218 or 5.207　　**25.** 11.6 or 11.13

26. 13.02, 13.022, or 13.21

Order each set of decimals from least to greatest.

27. 0.0319, 0.31, 0.032, 0.0289

28. 75.3, 7.598, 7.8, 75.6, 75.09

Order each set of decimals from greatest to least.

29. 6.32, 6.75, 6.39, 6.02

30. 17.023, 17.0201, 17.0463, 17.045, 17.002

Round each number to the underlined place-value position.

31. 7̲.29　　　　　　**32.** 76.8̲02

33. 13.5̲81　　　　　**34.** 69.9̲99

Estimate.

35. 4.86 − 1.131　　**36.** $34.29 + $17.58

37. 6.19 + 5.98 + 5.7 + 6 + 6.3

38. *Money Matters* Inali worked three days last week. He earned $19.85 on Monday, $17.75 on Thursday, and $21.30 on Saturday. About how much did he earn in all?

Add or subtract.

39. $\begin{array}{r} 15.63 \\ -\ 2.718 \\ \hline \end{array}$　　**40.** $\begin{array}{r} 4.63 \\ +4.72 \\ \hline \end{array}$

41. 25.6 + 47.92 + 3.1 + 0.48

42. Solve $x = 12 - 3.45$.

Applications & Problem Solving

43. Inventory Control Cheryl works in a warehouse. She must place items on shelves according to their stock number. Arrange the set of stock numbers in order from least to greatest. *(Lesson 3-3)*

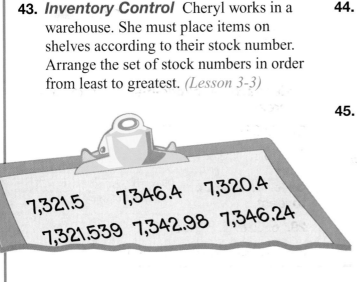

7,321.5   7,346.4   7,320.4

7,321.539   7,342.98   7,346.24

44. Money Matters Lamont wrote a check for groceries. On the check, he wrote the amount of purchase as "Seventy-six and $\frac{32}{100}$ dollars." Write the amount as a decimal. *(Lesson 3-1)*

45. Money Matters Mr. Marlin bought three pairs of athletic shoes for his children. The shoes cost $39.99, $75.50, and $89.90. Estimate the amount of money Mr. Marlin spent on shoes. *(Lesson 3-5)*

46. Reasonable Answers Jackson bought a pennant at a souvenir shop. He paid for the $7.59 pennant with a $10 bill. Should he expect about $4 or $2 in change? *(Lesson 3-6A)*

Alternative Assessment

● **Open Ended**

Suppose you are planning a family picnic. The prices of the items you want are shown in the table.

Item	Cost
chips	2 bags for $4.48
bread	3 loaves for $3.29
cookies	12 for $3.50
apples	4 for $1.25

Estimate the cost to feed your family.

Suppose you bought the quantities listed in the cost column. Write the total dollar amount in words. If you have a $20 bill, would you expect about $10 or $8 in change?

A practice test for Chapter 3 is provided on page 597.

● **Completing the CHAPTER Project**

Use the following checklist to make sure your plan is complete.

☑ A schedule of travel times and activities.

☑ All travel costs including transportation, lodging, food, souvenirs, and entertainment. You may have to estimate the cost of some items.

☑ A paragraph describing why you chose your vacation spot.

● PORTFOLIO   Write an example to illustrate using each type of estimation (rounding and clustering), and include them in your portfolio.

Section One: Multiple Choice

There are twelve multiple-choice questions in this section. Choose the best answer. If a correct answer is *not here*, mark the letter for Not Here.

1. The class scores on a history quiz were: 8, 9, 8, 8, 7, 9, 10, 7, 10, 8, 8, 9, 9, 10, 9, and 8. What was the mode?

   A 8

   B 8.5

   C 8.56

   D 8 and 9

2. Jodi keeps her fish in a tank that has a base 60 centimeters wide and 100 centimeters long. If she gets a tank that is 20 centimeters wider but with the same length as the older tank, what will be the dimensions of the new tank?

   F 40 cm by 100 cm

   G 60 cm by 120 cm

   H 80 cm by 100 cm

   J 80 cm by 120 cm

3. Write $2 \times 3^3 \times 5 \times 7^2$ as the product of factors.

   A $2 \times 3 \times 5 \times 7$

   B $2 \times 3 \times 3 \times 5 \times 7^2$

   C $2 \times 3 \times 3 \times 3 \times 5 \times 7 \times 7$

   D $3^3 \times 7^2$

4. What is the solution of $6a = 42$?

   F 4

   G 5

   H 6

   J 7

5. Which shows the number *five and eighty-six thousandths?*

   A 5.086

   B 5.0086

   C 5.860

   D 5.86

6. The graph shows the number of donuts sold at two stores.

DONUT SALES

About how many more donuts were sold at The Superstore than at Max's Market on Wednesday?

   F 0

   G 20

   H 40

   J 60

Please note that Questions 7–12 have five answer choices.

7. A TV costs $975, a home theater system costs $395, and a VCR costs $169. Which is the best estimate for the total amount for this entertainment package?

   A $1,000

   B $1,200

   C $1,300

   D $1,600

   E $2,000

8. Four boxes of cereal have weights of 22, 11, 29, and 21 ounces. Which weight is a reasonable average weight of the 4 boxes of cereal?

   F 50 ounces

   G 40 ounces

   H 30 ounces

   J 20 ounces

   K 10 ounces

9. A tree farm has 438 seedlings to be planted in 6 rows. How many seedlings need to be planted in each row?

    A 89

    B 73

    C 66

    D 63

    E Not Here

10. Lin paid $275.59 for a 25-inch color TV. The tax was $15.60. What was the price of the TV?

    F $291.19

    G $269.99

    H $260.99

    J $259.99

    K Not Here

11. If an airplane travels 435 miles per hour, how many miles will it travel in 5 hours?

    A 2,055 mi

    B 2,075 mi

    C 2,155 mi

    D 2,175 mi

    E Not Here

12. Each day Toru drives his delivery truck 2.75 kilometers to his first stop, 0.5 kilometer to his second stop, 7.8 kilometers to his third stop, and 5.42 kilometers back to the distribution center. How many kilometers does Toru drive his truck each day?

    F 15.47 km

    G 13.47 km

    H 13.22 km

    J 4.4 km

    K Not Here

interNET CONNECTION Test Practice **For additional test practice questions, visit:**

www.glencoe.com/sec/math/mac/mathnet

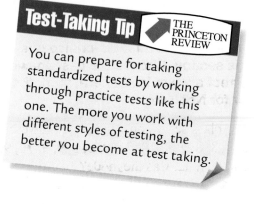

Test-Taking Tip THE PRINCETON REVIEW

You can prepare for taking standardized tests by working through practice tests like this one. The more you work with different styles of testing, the better you become at test taking.

Section Two: Free Response

This section contains eight questions for which you will provide short answers. Write your answers on your paper.

13. There will be 120 people at the music awards banquet. Each table seats 8 people. How can you find the number of tables needed?

14. Any number that makes an equation true is a(n) _____.

15. The lengths in miles of the Great Lakes are Lake Superior, 350; Lake Michigan, 307; Lake Huron, 206; Lake Erie, 241; and Lake Ontario, 193. What is the median of this data?

16. Round 5.99<u>9</u>8 to the underlined place-value position.

17. What is the value of $8 + 6 \times 10$?

18. State the greater of the numbers, 1.089 or 1.09.

19. Which measure of central tendency would be best to describe the colors of cars at a used car lot?

20. Evaluate the expression $a - b$ if $a = 42.2$ and $b = 35.9$.

Interdisciplinary Investigation

"O" IS FOR OLYMPICS

What do Atlanta, Georgia, and Athens, Greece, have in common? Both cities were sites for the modern Olympic Games. In 1896, Athens hosted the first modern Olympic Games with 13 nations participating. Now almost 200 nations compete every four years.

Are athletes today faster? How do the times of runners today compare to runners in 1896? Are swimmers getting faster every year? How do the times for men and women compare?

What You'll Do

In this investigation, you will collect data about an Olympic event. You will graph the data and predict future records in the event.

Materials 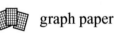 calculator

graph paper

Procedure

1. Work in groups of four. Find a source of winning Olympic times for each year that the games were held. Select one timed event in which both women and men participate. The event needs to have been held at least ten times. Two examples are the 100-meter dash and the 100-meter freestyle (swimming).

2. Separate your group into pairs. One pair will work with the men's times, and the other pair with the women's times for the event you selected. Make a table to display the times. Write all times in seconds. For example, 1 m 22.2 s is 60 s + 22.2 s or 82.2 s.

3. Graph your results.

4. Predict the winning times for men and women for the year 2012. Does your prediction seem reasonable?

Technology Tips

- Use a **calculator** to determine the mean and median.

- Use a **graphing calculator** or **graphing software** to display your graphs.

Making the Connection

Use the Olympic data you collected as needed to help in these investigations.

Language Arts

Your friend says, "Men are better than women in sports and always will be." Write a letter to your friend stating whether you agree or disagree with this statement. Use the data from the investigation to support your opinion.

Foreign Language

Select a winner in an Olympic event from a country that speaks a foreign language that you know or are interested in. Write an article about the athlete's victory using that language.

Social Studies

Select a popular sport. Research the history of this sport and write a short report.

Go Further

- Predict when you think men and women will have the same time in the event you used in the investigation. What do you think the fastest time will ever be for that event?

- If possible, time everyone in your class in the 100-meter dash. How do the times compare to the Olympic times?

 **Research** For more information on the history of the Olympics, visit the website below.

Data Collection and Comparison To share and compare your data with other students in the U.S., visit:

www.glencoe.com/sec/math/mac/mathnet

 You may want to place your work on this investigation in your portfolio.

CHAPTER 4

Multiplying and Dividing Decimals

CHAPTER Project

BE YOUR OWN BOSS!

Owning your own business can be a lot of work. But, it can be very rewarding. It also requires creativity. In this project, you will learn just a little of what goes into making a small business successful. After choosing a product you'd like to sell, you'll explain why you chose it and then determine its selling price. Then you should develop a pricing strategy for large orders. Finally, you need to design an advertisement for your product.

Getting Started

- Decide on a product to sell.
- Determine how much it will cost you to make this product.
- Decide on the price that you want to ask for your product.
- Decide whether you want to do any special promotions such as offering a discount on early orders.

Technology Tips

- Use a **spreadsheet** to organize your price chart and make calculations.
- Use **desktop publishing software** to design your advertisement.

*inter*NET CONNECTION Research **For up-to-date information on starting a small business, visit:**

www.glencoe.com/sec/math/mac/mathnet

Working on the Project

You can use what you'll learn in Chapter 4 to help you sell your product.

Page	Exercise
135	32
143	30
173	Alternative Assessment

COOPERATIVE LEARNING

4-1A Multiplying Decimals by Whole Numbers

A Preview of Lesson 4-1

grid paper

markers

scissors

In this lab, you will use grid paper to draw decimal models. You can use decimal models to multiply a decimal by a whole number. Remember, □ represents one hundredth, ⬚ represents one tenth, and 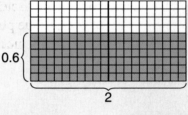 represents one.

In the following activity, the number of rows will represent the first factor of the multiplication, and the number of columns will represent the second factor. The number of shaded rows or columns will represent the product.

TRY THIS

Work in groups of three.

To model 0.6 × 2, follow these steps.

Step 1 Draw two 10-by-10 squares on grid paper to represent the factor 2.

Step 2 Shade six rows to represent 0.6.

Step 3 Cut off the shaded rows and arrange them to form as many 10-by-10 squares as possible.

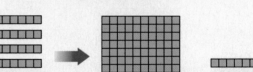

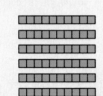

There is one 10-by-10 square, which is 1, and two rows, which is 0.2.

Therefore, 0.6 × 2 = 1.2.

ON YOUR OWN

Draw decimal models to show each product.

1. 4 × 0.6

2. 3 × 0.7

3. 0.5 × 5

4. Write a multiplication problem using decimals for the model shown.

5. *Look Ahead* Find the product of 8 and 0.4 without using models.

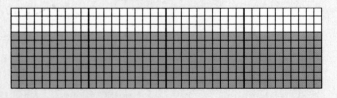

4-1

Multiplying Decimals by Whole Numbers

What you'll learn

You'll learn to estimate and find the products of decimals and whole numbers.

When am I ever going to use this?

Knowing how to multiply a decimal by a whole number can help you find the amount of money you earn.

Word Wise

compatible numbers

LOOK BACK

You can refer to Lesson 1-3 to review rounding.

Mrs. Sopher spends $2.25 for the Saturday and Sunday edition of her local newspaper. What does this cost per year? *This problem will be solved in Example 2.*

When multiplying a decimal by a whole number, multiply as with whole numbers. One way to determine where to place the decimal point in the product is to use estimation.

There are two methods to estimate products. You can use rounding or **compatible numbers**.

Method 1   Use rounding.

First, round each factor to its greatest place-value position. Then multiply. Do not round 1-digit factors.

Method 2   Use compatible numbers.

It is easy to find the product of compatible numbers mentally.

Examples

1 Find 9 × 78.42.

Estimate using rounding. Round 78.42 to 80. 9 × 80 = 720.

$$
\begin{array}{r}
\overset{7\,3\;\;1}{78.42} \\
\times \quad\; 9 \\
\hline
705.78
\end{array}
$$

Multiply as with whole numbers.

Since the estimate is 720, place the decimal point after 5.

Check with a calculator. 78.42 ⊠ 9 ⊟ 705.78 ✓

APPLICATION

2 **Money Matters** Refer to the beginning of the lesson. How much does Mrs. Sopher spend per year for newspapers?

Explore You know how much she spends each weekend. You want to know how much she spends in a year.

(continued on the next page)

Plan   There are 52 weeks in a year. Find $2.25 × 52.
Estimate using compatible numbers.

$2.25 → $2 *Round to the greatest place value.*
× 52 → ×50 *Round to 50 since it is easy to*
                     multiply 2 and 5 mentally.

Since $2 × 5 = 10$, then $2 × 50 = 100$.

Solve
$$
\begin{array}{r}
\$2.25 \\
\times\ 52 \\
\hline
450 \\
1125 \\
\hline
117.00
\end{array}
$$
*Since the estimate is 100, place
the decimal point after 7.*

Mrs. Sopher spends $117.00 a year for newspapers.

Examine   Compared to the estimate, the answer is reasonable.

You can also determine where to place the decimal point in the product by counting the number of decimal places in the decimal factor. The product must have the same number of decimal places. If more decimal places are needed, annex zeros.

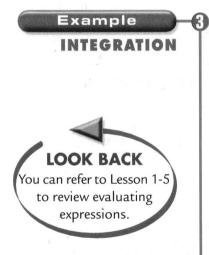

Example

INTEGRATION

LOOK BACK
You can refer to Lesson 1-5 to review evaluating expressions.

③ Algebra Evaluate the expression $3a$ if $a = 0.032$.

$3a = 3 × 0.032$ *Replace a with 0.032.*

$$
\begin{array}{r}
0.032 \\
\times\ \ \ 3 \\
\hline
0.096
\end{array}
$$
← *three decimal places*

← *Annex a zero on the left to make three decimal places.*

Check your answer by adding.
$$
\begin{array}{r}
0.032 \\
0.032 \\
+\ 0.032 \\
\hline
0.096
\end{array}
$$ ✓

The product is 0.096.

CHECK FOR UNDERSTANDING

**Communicating
Mathematics**

Read and study the lesson to answer each question.

1. *Explain* how you could use compatible numbers to estimate the product of 40.23 and 251.

2. *Write* a multiplication problem for the model shown.

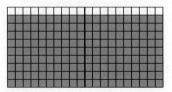

3. You Decide Montega uses a calculator to find the product of 34.78 and 452. He gets 15,720.56 for an answer. Is the answer reasonable? Explain why or why not.

Guided Practice

Use estimation to place the decimal point in each product.

4. $0.88 \times 3 = 264$ **5.** $12.6 \times 19 = 2394$ **6.** $254 \times 3.82 = 97028$

Multiply.

7. $\begin{array}{r} 0.2 \\ \times 8 \\ \hline \end{array}$ **8.** $\begin{array}{r} 5.02 \\ \times\ 3 \\ \hline \end{array}$ **9.** 64×0.005

10. Algebra Evaluate the expression $12n$ if $n = 5.6$.

11. Astronomy Pluto, normally the farthest planet from the Sun, is also the slowest. Its average speed around the Sun is 10,604 miles per hour. Earth, by contrast, travels 6.28 times faster. What is the average speed of Earth?

EXERCISES

Practice

Multiply.

12. $\begin{array}{r} 0.6 \\ \times 5 \\ \hline \end{array}$ **13.** $\begin{array}{r} 0.28 \\ \times\ 4 \\ \hline \end{array}$ **14.** $\begin{array}{r} 2.03 \\ \times\ 7 \\ \hline \end{array}$ **15.** $\begin{array}{r} 3.42 \\ \times\ 9 \\ \hline \end{array}$

16. $\begin{array}{r} 0.007 \\ \times\ \ 8 \\ \hline \end{array}$ **17.** $\begin{array}{r} 10.7 \\ \times\ 6 \\ \hline \end{array}$ **18.** $\begin{array}{r} 0.67 \\ \times 33 \\ \hline \end{array}$ **19.** $\begin{array}{r} 3.25 \\ \times 802 \\ \hline \end{array}$

20. $1{,}250 \times 2.5$ **21.** 0.0125×754 **22.** $2{,}967 \times 0.071$

Solve each equation.

23. $x = 36 \times 0.007$ **24.** $4.8 \times 235 = y$ **25.** $p = 2{,}388 \times 1.65$

26. Algebra Evaluate $112d$ if $d = 0.98$.

27. What is the solution of $n = 58.002 \cdot 367$?

28. Will 11.7×5 be closer to 55 or 60? Explain.

29. How does knowing that $5^2 = 25$ help you find the answer to 5×0.5?

Applications and Problem Solving

30. Statistics The graph shows the broadcast television ratings for baseball. In which year were the ratings twice as high as the 1996 ratings?

FOUL BALL!

'93	Network A	3.8
'94	Network B	6.2
'95	Network B	5.8
'96	Network C	2.9

Sources: Major League Baseball and *USA TODAY* research

31. Money Exchange If the Japanese yen (¥) is worth $0.0078, what is the value of ¥ 3,750?

32. Working on the CHAPTER Project Determine how much it would cost someone to buy 1 item, 2 items, 3 items, . . . , through 10 items of your product. Organize this data in a table.

33. *Critical Thinking* Create a multiplication problem where the product is 45.89 and one of the factors is a whole number.

Mixed Review

34. Standardized Test Practice Jaden plans to buy a soccer ball for $21.99, soccer cleats for $45.50, and a soccer shirt for $14.79. What is the cost of these items before tax is added? *(Lesson 3-6)*

A $82.28

B $81.18

C $80.28

D $80.18

E Not Here

Andy's Sit-Ups

35. *Fitness* The line graph shows Andy's progress doing sit-ups. How many sit-ups do you predict he will be doing in October? *(Lesson 2-5)*

36. *School* At Gillian School, there are 36 student council representatives as shown in the circle graph. Which grade (both boys and girls) has the most representatives? *(Lesson 2-4)*

Student Council Representatives

Girls 6th grade — 8
Boys 6th grade — 4
Girls 8th grade — 5
Boys 7th grade — 4
Girls 7th grade — 9
Boys 8th grade — 6

37. *Sports* The table shows the total points scored in the eight games in the second round of the midwest region during the 1997 NCAA Division I Women's Basketball Tournament. *(Lesson 1-3)*

a. Estimate the total points scored over the two days.

b. Explain how you could use a combination of methods to estimate the two-day total.

c. Suppose you wanted to make a bar graph comparing the total points scored in each game. Would you round to the nearest hundred or to the nearest ten? Explain.

Total Points Scored	
March 14	**March 15**
118	106
126	138
141	145
145	148

For **Extra Practice,** see page 565.

Using the Distributive Property

What you'll learn

You'll learn to compute products mentally using the distributive property.

When am I ever going to use this?

Knowing how to use the distributive property can help you multiply mentally.

Word Wise

distributive property

Christine's drum teacher charges $15 for a lesson. If Christine had 4 lessons in November and 3 lessons in December, how much did she pay her drum teacher for those two months?

Before you solve this problem, let's look at another way to show multiplication, using grouping symbols such as parentheses. For example, you can write 3×5 as $3(5)$ or $(3)5$. Let's include parentheses with the rules for order of operations.

Order of Operations	1. Do all operations within grouping symbols first.
	2. Do all powers before other operations.
	3. Multiply and divide in order from left to right.
	4. Add and subtract in order from left to right.

Now go back to Christine's problem.

Method 1 To find the total cost of the lessons, you can multiply the charge per lesson by the total number of lessons.

charge per lesson   *number of lessons*

$$15(4 + 3) = 15(7)$$

Use the order of operations. Add inside the parentheses first.

$$= 105$$

Study Hint

Reading Math

Read the expression $15(4 + 3)$ as *fifteen times the quantity four plus three.*

Method 2 You can find the cost for each month first, and then add to find the total cost of the lessons.

money paid in November   *money paid in December*

$$15 \times 4 + 15 \times 3 = 60 + 45$$

Use the order of operations. Do the multiplication first.

$$= 105$$

The solution is the same using either method. Christine paid the drum teacher $105 for lessons in November and December. So, the following sentence is true.

$$15(4 + 3) = 15 \times 4 + 15 \times 3$$

This is an example of the **distributive property**.

Distributive Property	Symbols:	Arithmetic	$5(3 + 6) = 5 \cdot 3 + 5 \cdot 6$
		Algebra	For any numbers a, b, and c, $a(b + c) = ab + ac$.

The distributive property allows us to solve problems in parts. This makes it easy to solve some multiplication problems mentally.

Examples

1 Find 6 × 45 mentally using the distributive property.

Estimate: 6 × 50 = 300

$$6 \times 45 = 6(40 + 5) \qquad \textit{Use 40 + 5 for 45.}$$
$$= 6 \times 40 + 6 \times 5$$
$$= 240 + 30$$
$$= 270 \qquad \textit{Compare to the estimate.}$$

APPLICATION

2 **Money Matters** Adria has a paper route to earn some extra money. She earns \$0.18 per customer per week. If she has 80 customers, how much will she make weekly?

Estimate: 80 × \$0.20 = \$16

$$80 \times 0.18 = 80(0.1 + 0.08)$$
$$= 80 \times 0.1 + 80 \times 0.08$$
$$= 8 + 6.4$$
$$= 14.4$$

Adria will make \$14.40. *Compare to the estimate.*

CHECK FOR UNDERSTANDING

Communicating Mathematics

Read and study the lesson to answer each question.

1. ***Tell*** the order of operations you would use to find $9(8^2 + 6)$.

2. ***Explain*** how to use the distributive property to solve 5.5 × 8.

3. ***Write*** a paragraph explaining how the distributive property can help you solve a problem mentally.

Guided Practice

4. Rewrite 12 × 5 + 12 × 8 using the distributive property.

Find each product mentally. Use the distributive property.

5. 8 × 18   6. 52 × 3   7. 6.4 × 5

8. Find the product of 7 and 30.9 mentally.

9. ***Money Matters*** Emilio is taking Karate lessons. The instructor charges \$25.00 for each lesson. If Emilio has 5 lessons in one month, how much does he owe the instructor?

Practice

Rewrite each expression using the distributive property.

10. $4(30 + 6)$ **11.** $3(20 + 7)$ **12.** $15 \times 20 + 15 \times 0.4$

Find each product mentally. Use the distributive property.

13. 6×14 **14.** 103×7 **15.** 3×72

16. 2.4×11 **17.** 8.1×6 **18.** 20×4.3

19. 20.7×6 **20.** 14×110 **21.** 30×3.09

22. Find the product of 12 and 15 mentally.

23. Mentally multiply 60 and 10.5.

24. *Algebra* What is the value of $8(x + 0.7)$ if x is 30?

Applications and Problem Solving

25. *Money Matters* The company that employs Mrs. Leshnock pays her 31.5¢ for each mile she drives her personal car for business purposes. How much money will she receive if she drives her car 50 miles on company business?

26. *Travel* The Garcia family drove from Houston to San Antonio. They filled their car's tank with 12.8 gallons of gas before they left Houston. They filled the tank with 10.2 gallons when they arrived in San Antonio. If gas cost $1.29 per gallon at both gas stations, how much did they spend on gas?

The Alamo in San Antonio, Texas

27. *Critical Thinking* Evaluate each expression.

     a. $0.2(2 - 0.4)$ **b.** $0.1(1 - 0.5)$

     c. $0.15(5 - 1.4)$ **d.** $0.75(10 - 0.25)$

Mixed Review

28. **Standardized Test Practice** The sign below was on a sale rack. If Jhan bought 3 items from this sale rack, a reasonable total cost, without tax is — *(Lesson 4-1)*

     A $13.00.

     B $26.00.

     C $75.00.

     D $100.00.

     E $110.00.

SALE ITEMS! $12.99 to $29.99

29. *Weather* Washington, D.C., has an average annual precipitation of 35.86 inches. Round this amount to the nearest tenth. *(Lesson 3-4)*

30. *Statistics* The number of students in Mrs. Jing's history class during the first two weeks of the new semester were: 27, 31, 25, 19, 31, 32, 24, 26, 33, and 31. Construct a stem-and-leaf plot for the data. *(Lesson 2-6)*

For **Extra Practice,** see page 565.

COOPERATIVE LEARNING

4-3A Multiplying Decimals

A Preview of Lesson 4-3

grid paper

markers

In the Hands-On Lab on page 132, you used decimal models to multiply a whole number and a decimal. In this lab, you will use decimal models to multiply two decimals.

TRY THIS

Work with a partner.

1 To model 0.7 × 0.5, follow these steps.

- Draw a 10-by-10 square. Recall that each small square represents 0.01.
- Color seven rows of the model blue to represent 0.7.
- Color five columns of the model yellow to represent 0.5.

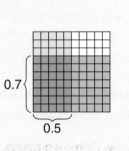

There are 35 small squares that are shaded green.

Therefore, 0.7 × 0.5 = 0.35.

2 To model 0.9 × 2.3, you need to draw three 10-by-10 squares.

- Line up the squares as shown.
- Color 9 rows blue to represent 0.9.
- Color 23 columns yellow to represent 2.3.

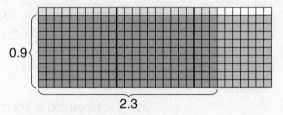

There are 207 small squares that are shaded green. This would cover two 10-by-10 squares, which is 2, and seven small squares, which is 0.07.

Therefore, 0.9 × 2.3 = 2.07.

ON YOUR OWN

Write a multiplication problem using decimals for each model.

1.

2.

Use decimal models to show each product.

3. 0.4 × 0.6

4. 1.3 × 0.2

5. 2.5 × 1.1

6. *Look Ahead* Find the product of 1.7 and 2.2 without using models.

Multiplying Decimals

What you'll learn

You'll learn to multiply decimals.

When am I ever going to use this?

Knowing how to multiply decimals can help you find the amount of tip to leave at a restaurant.

On September 1, 1997, the minimum wage increased to $5.15 an hour. If Domingo works 17.5 hours a week and earns minimum wage, how much does he make each week before taxes?

To answer this question, you need to multiply 5.15 by 17.5. When you multiply decimals, multiply as with whole numbers. One way to place the decimal point is to use estimation.

Estimate: 5 · 20 = 100 Round 5.15 to 5 and 17.5 to 20.

$$
\begin{array}{r}
5.15 \\
\times 17.5 \\
\hline
2\,575 \\
36\,05 \\
51\,5 \\
\hline
90.125
\end{array}
$$

Multiply as with whole numbers.

Since the estimate is 100, place the decimal point after 0.

Rounded to the nearest penny, Domingo makes $90.13 before taxes.

Another way to place the decimal point in the product is by counting the decimal places in each factor. The product will have the same number of decimal places as the sum of the number of decimal places in the factors.

Examples

1 **Find 2.4 · 5.9.**   *Estimate: 2 · 6 = 12*

$$
\begin{array}{r}
5.9 \\
\times 2.4 \\
\hline
23\,6 \\
118 \\
\hline
14.16
\end{array}
$$

← *one decimal place*
← *one decimal place*

← *two decimal places*

The product is 14.16.   *Compared to the estimate, the answer is reasonable.*

2 **Find 1.45 × 0.7.**   *Estimate: 1 × 1 = 1*

$$
\begin{array}{r}
1.45 \\
\times 0.7 \\
\hline
1.015
\end{array}
$$

← *two decimal places*
← *one decimal place*
← *three decimal places*

The product is 1.015.

3 Multiply 0.02 and 1.36.   *Estimate: 0 · 1 = 0*

  1.36 ← *two decimal places*
  ×0.02 ← *two decimal places*
  0.0272 ← *To make four decimal places,*
      annex a zero on the left.

The product is 0.0272.

INTEGRATION **4** **Algebra** Evaluate 4.3n if n = 10.89.

4.3n = 4.3 · 10.89 *Replace n with 10.89.*

Estimate: 4 · 11 = 44

  10.89
  × 4.3
  3267
  3356
  46.827   *Compare to the estimate.*

CHECK FOR UNDERSTANDING

Communicating Mathematics

Read and study the lesson to answer each question.

1. *Explain* how you know where the decimal point would go in the product 4.507 × 0.09 = 040563.

2. *Write* the multiplication sentence represented by the decimal model.

Guided Practice

3. Use estimation to place the decimal point in the product 3.4 · 1.2 = 408.

Multiply.

4. 0.4 × 8.3   5. 8.54 · 3.27   6. 39.6 × 2.417

7. Evaluate 0.002y if y = 3.9.

8. *Money Matters* Helaku wants to buy a new video game. The game he wants costs $41.99. The sales tax is calculated by multiplying the total of the merchandise by 0.0575. If the video game is the only thing Helaku purchases, how much sales tax will he pay?

EXERCISES

Practice **Use estimation to place the decimal point in each product.**

9. 5.6 × 12.43 = 69608   10. 0.03 × 1.24 = 00372

Multiply.

11. 0.35 · 1.4   12. 5.2 × 0.065   13. 3.06 · 4.28

14. 0.9 × 0.15   15. 18.37 · 908.44   16. 0.003 × 0.012

Solve each equation.

17. $p = 1.3 \cdot 7.3$

18. $q = 0.3 \cdot 0.012$

19. $0.6(0.031) = m$

20. $28.2(4.4) = n$

Evaluate each expression if $a = 1.06$, $b = 0.002$, and $c = 5.5$.

21. ab

22. $a(b + c)$

23. $a(c - b)$

24. abc

25. Find the product of 24.8 and 4.389.

26. What is 2.15 multiplied by 3.84?

27. *Algebra* What is the product of x and y if $x = 1.16$ and $y = 0.006$?

28. *Algebra* Find the value of prt if $p = \$250$, $r = 0.03$, and $t = 7.5$.

29. *Life Science* The graph shows a breakdown by age group of the 26.2 million Americans who were hearing impaired as of January, 1996. The number of people between the ages of 65 and 74 is 3.95 times the number of people in the 0-17 age group. Complete the graph by finding the number of people in the 65-74 age group.

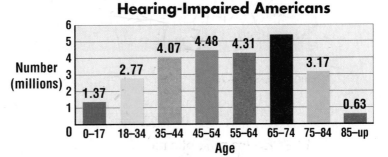

Hearing-Impaired Americans

Source: American Speech–Language–Hearing Association

30. *Working on the* **CHAPTER Project** Refer to the prices you determined in Exercise 32 on page 135.

   a. Suppose your after-tax prices are determined by multiplying the prices by 1.065. Find the after-tax price for your product.

   b. Suppose your senior citizen discount is determined by multiplying the after-tax prices by 0.90. Find the discount prices.

31. *Critical Thinking* Write a multiplication problem where the product of two decimals is between 0.05 and 0.75.

Mixed Review

32. Find 8.5×4 mentally. Use the distributive property. *(Lesson 4-2)*

33. *Standardized Test Practice* Shanee recorded the number of miles she rode her bicycle each day for four days. Which list shows that data in correct order from least to greatest? *(Lesson 3-3)*

   A 7.6, 21.4, 13.7, 9.3

   B 21.4, 13.7, 21.4

   C 13.7, 7.6, 21.4, 9.3

   D 7.6, 9.3, 13.7, 21.4

For **Extra Practice**, see page 566.

34. Express $\frac{8}{1,000}$ as a decimal. *(Lesson 3-1)*

35. *Algebra* Evaluate y^3 if $y = 6$. *(Lesson 1-6)*

SPREADSHEETS

4-3B Reading Spreadsheets

A Follow-Up of Lesson 4-3

computer

spreadsheet software

A computer *spreadsheet* arranges data and formulas in a column and row format. A spreadsheet is used for organizing and analyzing data.

A spreadsheet is made up of *cells*. A cell can contain data, labels, or formulas. Each cell has an address. A cell's address is a combination of the letter from the top of a particular column and the number from the left of a particular row. The cell B4 is the cell in column B and row 4. *Cell B4 in the spreadsheet below contains the amount $29.90.*

The spreadsheet below shows a price list for a retail store.

TRY THIS

Work with a partner.

The formulas in the cells in column D find the product of the numbers that are entered in columns B and C.

Use the spreadsheet to determine the discounts by making the following substitutions.

C2 = 0.2, C3 = 0.25, C4 = 0.15, C5 = 0.2, C6 = 0.35, C7 = 0.1

	A	B	C	D	E
1	Item	Regular Price	Rate of Discount	Discount	Sale Price
2	Blouse	$35.99		= B2*C2	= B2-D2
3	Skirt	$49.99		= B3*C3	= B3-D3
4	Jeans	$29.90		= B4*C4	= B4-D4
5	Shorts	$15.00		= B5*C5	= B5-D5
6	Sweatshirt	$18.99		= B6*C6	= B6-D6
7	T-Shirts	$19.95		= B7*C7	= B7-D7

ON YOUR OWN

1. Name the cell that holds the data $25.41.

2. What is stored in cell D5?

3. The formulas in the cells in column E find the difference of the numbers in column B and column D. Use the results from the activity above to determine each sale price.

4. How could you use the spreadsheet to determine the amount of sales tax on each item?

5. Suppose column C was labeled *Quantity* and column D was labeled *Total*. Would the formulas in column D change?

Integration: Geometry
Perimeter and Area

Alicia is redecorating her bedroom and wants to add a wallpaper border around the entire room. How much will she need?

Alicia needs to find the **perimeter** of the bedroom so she can know how much border to buy. The perimeter (P) of any closed figure is the distance around the figure. You can find the perimeter by adding the measures of the **sides** of the figure.

$P = 15 + 9 + 10 + 3 + 5 + 12$

$P = 54$

The perimeter of Alicia's room is 54 feet. So, Alicia needs 54 feet of wallpaper border.

To find the perimeter of a rectangle, add the measures of its four sides. Let P represent the measure of the perimeter, ℓ its length, and w its width.

$P = \ell + w + \ell + w$

$P = \ell + \ell + w + w$

$P = 2\ell + 2w$

Perimeter of a Rectangle	Words:	The perimeter of a rectangle is two times the length (ℓ) plus two times the width (w).
	Symbols: $P = 2\ell + 2w$ Model:	w [rectangle] w with ℓ on top and bottom

Example

1 Find the perimeter of a rectangle with a length of 18.3 meters and a width of 7.5 meters.

$P = 2\ell + 2w$

$P = 2(18.3) + 2(7.5)$ *Replace ℓ with 18.3 and w with 7.5.*

$P = 36.6 + 15$

$P = 51.6$

The perimeter is 51.6 meters.

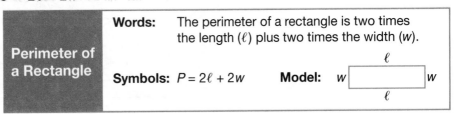

Since each side of a square has the same length, you can multiply the measure of any of its sides (*s*) by 4 to find its perimeter. Thus, the formula for the perimeter of a square can be written as $P = 4s$.

Example **2** Find the perimeter of a square cement block whose sides each measure 17.875 inches.

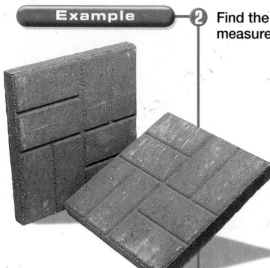

17.875 in.

$P = 4s$

$P = 4 \times 17.875$ *Replace s with 17.875.*

4 $\boxed{\times}$ 17.875 $\boxed{=}$ 71.5

The perimeter of the cement block is 71.5 inches.

In addition to the perimeter, we often solve problems by using the **area** of a closed figure. Area is the number of square units needed to cover a surface.

By counting, you can see that the rectangle at the right has an area of 30 square units. You can also find its area by multiplying its length, 6 units, by its width, 5 units.

You can find the area of any rectangle by multiplying its length and its width.

Area of a Rectangle	**Words:** The area of a rectangle is the product of its length (ℓ) and width (*w*).
	Symbols: $A = \ell \cdot w$ **Model:** 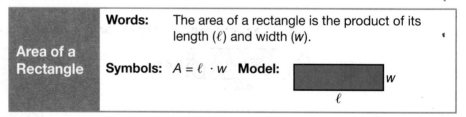

Area is expressed in square units. For example, if the length and width are measured in feet, the area will be written in square feet, or ft^2.

Example **3**

Real World **APPLICATION**

Sports International soccer fields are rectangular and measure 100 meters by 73 meters. A soccer field needs to be recovered with sod. How much sod do the groundskeepers need for the field?

$A = \ell \cdot w$

$A = 100 \cdot 73$ *Replace ℓ with 100 and w with 73.*

$A = 7,300$

The groundskeepers need 7,300 square meters of sod.

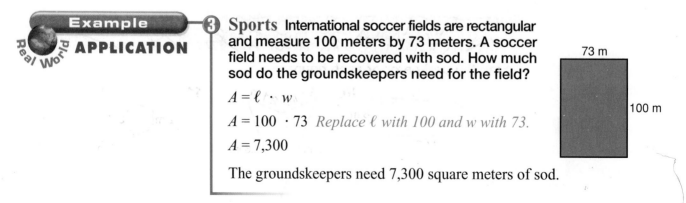

73 m

100 m

Integration: Geometry
Perimeter and Area

What you'll learn

You'll learn to find the perimeters and the areas of rectangles and squares.

When am I ever going to use this?

Knowing how to find perimeter can help you buy the right amount of framing. Knowing how to find area can help you buy the right amount of paint or carpet.

Word Wise

perimeter
side
area

Alicia is redecorating her bedroom and wants to add a wallpaper border around the entire room. How much will she need?

Alicia needs to find the **perimeter** of the bedroom so she can know how much border to buy. The perimeter (P) of any closed figure is the distance around the figure. You can find the perimeter by adding the measures of the **sides** of the figure.

$$P = 15 + 9 + 10 + 3 + 5 + 12$$

$$P = 54$$

The perimeter of Alicia's room is 54 feet. So, Alicia needs 54 feet of wallpaper border.

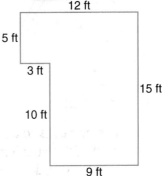

To find the perimeter of a rectangle, add the measures of its four sides. Let P represent the measure of the perimeter, ℓ its length, and w its width.

$$P = \ell + w + \ell + w$$
$$P = \ell + \ell + w + w$$
$$P = 2\ell + 2w$$

Perimeter of a Rectangle	**Words:** The perimeter of a rectangle is two times the length (ℓ) plus two times the width (w).
	Symbols: $P = 2\ell + 2w$ **Model:**

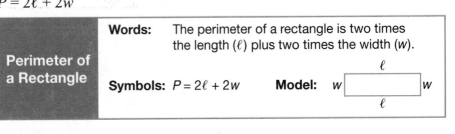

Example

Study Hint

Mental Math You can rewrite the formula for perimeter using the distributive property:
$2\ell + 2w = 2(\ell + w)$.

1. Find the perimeter of a rectangle with a length of 18.3 meters and a width of 7.5 meters.

$$P = 2\ell + 2w$$
$$P = 2(18.3) + 2(7.5) \quad \textit{Replace } \ell \textit{ with 18.3 and w with 7.5.}$$
$$P = 36.6 + 15$$
$$P = 51.6$$

The perimeter is 51.6 meters.

Since each side of a square has the same length, you can multiply the measure of any of its sides (*s*) by 4 to find its perimeter. Thus, the formula for the perimeter of a square can be written as $P = 4s$.

Example 2 Find the perimeter of a square cement block whose sides each measure 17.875 inches.

17.875 in.

$P = 4s$

$P = 4 \times 17.875$ *Replace s with 17.875.*

4 ☒ 17.875 ☐ 71.5

The perimeter of the cement block is 71.5 inches.

In addition to the perimeter, we often solve problems by using the **area** of a closed figure. Area is the number of square units needed to cover a surface.

By counting, you can see that the rectangle at the right has an area of 30 square units. You can also find its area by multiplying its length, 6 units, by its width, 5 units.

You can find the area of any rectangle by multiplying its length and its width.

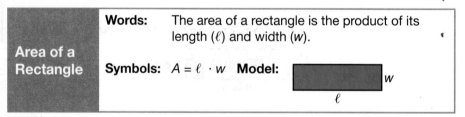

| **Area of a Rectangle** | **Words:** The area of a rectangle is the product of its length (ℓ) and width (*w*). |
| | **Symbols:** $A = \ell \cdot w$ **Model:** |

w

ℓ

Area is expressed in square units. For example, if the length and width are measured in feet, the area will be written in square feet, or ft².

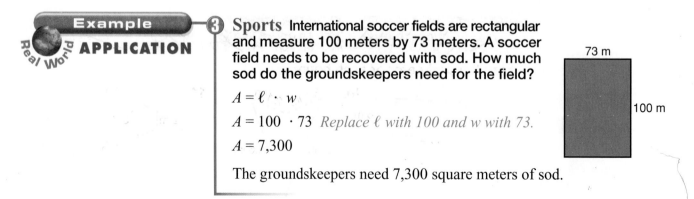

Example 3

Real World APPLICATION

Sports International soccer fields are rectangular and measure 100 meters by 73 meters. A soccer field needs to be recovered with sod. How much sod do the groundskeepers need for the field?

73 m

100 m

$A = \ell \cdot w$

$A = 100 \cdot 73$ *Replace ℓ with 100 and w with 73.*

$A = 7,300$

The groundskeepers need 7,300 square meters of sod.

You can write the formula for the area of a square as $A = s^2$, because each side of a square has the same length.

Example **4** Find the area of a square whose sides measure 4.5 meters.

$A = s^2$

$A = 4.5^2$ *Replace s with 4.5.*

$A = 20.25$

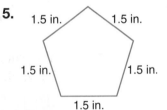

4.5 m

The area of the square is 20.25 square meters.

CHECK FOR UNDERSTANDING

Communicating Mathematics

Read and study the lesson to answer each question.

1. **Draw** and label a rectangle that has a perimeter of 36 units.

2. **Explain** how to find the area of a square.

3. Use grid paper to draw as many different rectangles as you can with a perimeter of 12 units. **Make a table** listing the length, width, perimeter, and area of each rectangle. Use the relationships between the lengths and widths to create formulas that could be used to find the perimeter and area of any rectangle.

Guided Practice

Find the perimeter of each figure.

4.
13 cm

9 cm

13 cm

5.
1.5 in.   1.5 in.

1.5 in.   1.5 in.

1.5 in.

Find the perimeter and the area of each figure.

6.
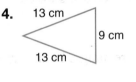
rectangle   2.6 m

8.8 m

7.

square   30.4 cm

8. What is the area of a rectangle with a length of 10.25 inches and a width of 8.5 inches?

EXERCISES

Practice

Find the perimeter of each figure.

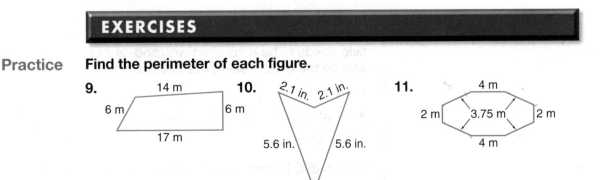

9.
14 m

6 m

17 m

10.
2.1 in.   2.1 in.

5.6 in.   5.6 in.

11.
4 m

2 m   3.75 m   2 m

4 m

Find the perimeter and the area of each figure.

12.

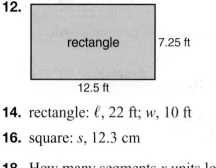

rectangle   7.25 ft

12.5 ft

13.

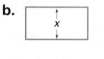

square   1.5 in.

14. rectangle: ℓ, 22 ft; w, 10 ft

15. rectangle: ℓ, 4.8 m; w, 1.6 m

16. square: s, 12.3 cm

17. square: s, 11.6 in.

18. How many segments x units long will fit on the perimeter of each figure?

a.

b.

Applications and Problem Solving

19. Find the area and perimeter of the figure made up of rectangles.

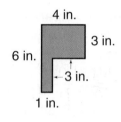

4 in.

3 in.

6 in.

3 in.

1 in.

20. *Gardening* Mrs. Collins has a rectangular strawberry garden that is 6.1 meters long and 2.5 meters wide.

a. She wants to put a fence around her garden to keep the rabbits out. How much fencing does she need?

b. She needs to cover the garden with a net to keep birds from eating the strawberries. How much netting does she need?

21. *Sports* Badminton was added for the Summer Olympic games in Barcelona, Spain. A regulation badminton court has the measurements shown in the diagram.

a. Tape is to be applied to the floor to form a badminton court's boundaries. Do you need to find the area or the perimeter to determine how much tape you need?

b. What is the area of the court?

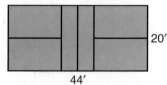

20'

44'

22. *Geometry* Suppose you increase a pair of opposite sides of a square by two units. Will the area of the rectangle be four more square units than the area of the square? Use a model in your explanation.

23. *Critical Thinking* Using the figure shown, draw another figure that has twice the area and still has the same perimeter.

Mixed Review

24. *Algebra* Evaluate the expression zwx if $w = 0.2$, $x = 20.7$, and $z = 3.01$. *(Lesson 4-3)*

25. **Standardized Test Practice** Ernesto bought 7 spiral notebooks for his classes. Each notebook cost $2.29, including tax. What was the total cost of the notebooks? *(Lesson 4-1)*

    A $8.93   **B** $16.93   **C** $16.03   **D** $17.03   **E** Not Here

For **Extra Practice,** see page 566.

COOPERATIVE LEARNING

4-4B Area and Perimeter

A Follow-up of Lesson 4-4

grid paper

In this lab, you will use grid paper to explore how perimeter and area are related.

TRY THIS

Work in groups of three.

Step 1 Copy the table at the right.

Step 2 On centimeter grid paper, draw a rectangle with a length of 6 and a width of 1.

Step 3 Find the perimeter and the area of the rectangle and record them in your table.

Step 4 Repeat Steps 2 and 3 for the remaining dimensions in your table.

Length	Width	Perimeter	Area
6	1		
5	2		
4	3		

ON YOUR OWN

1. Each person in your group should copy the table at the right.

   a. One person should complete the table for all rectangles whose perimeter is 16 with whole number sides.

   b. Another person should complete the table for all rectangles whose perimeter is 18 with whole number sides.

   c. The third person should complete the table for all rectangles whose perimeter is 20 with whole number sides.

Length	Width	Area

2. Suppose you want to enclose a rectangular garden with the greatest area.

   a. What would be the dimensions of the garden if you have 50 feet of fence?

   b. What would be the dimensions of the garden if you have 48 feet of fence?

3. What can you conclude about the relationship between area and perimeter?

4. *Reflect Back* Find the dimensions of a rectangle with the least perimeter possible if its area is 24 square units.

4-5A Solve a Simpler Problem

A Preview of Lesson 4-5

Jemeka and Macy volunteered to plant some flowering bushes along the side of a neighborhood senior center. The wall of the building is 50 feet long. A gardener told them to plant the bushes 2 feet apart to give them room to grow. They are trying to determine how many plants to buy. Let's listen in!

50 divided by 2 is 25. I guess we need 25 plants.

I'm not sure that's right. Let's make it easier. Suppose the wall was 4 feet long. We'd need 3 bushes, not 2. Right?

Let's see. There would be one at the front corner, one in the middle 2 feet from the corner, and one at the back corner. When 3 bushes are planted 2 feet apart, the bushes on the end are 4 feet apart. You're right!

Macy

And if the wall was 6 feet long, we'd need one more bush for a total of 4 bushes. Is there a pattern here?

Yes! The number of plants needed is half the total distance plus one. So for this 50-foot wall, we'll need 26 plants, not 25. It was hard to see that until we worked with a shorter wall.

Jemeka

THINK ABOUT IT

Work with a partner.

1. **Explain** why Macy's first method of solving the problem seemed reasonable.

2. **Think** of another way the girls could have solved the problem.

3. **Apply** what you have learned about the **solve a simpler problem** strategy to solve the following problem.

Mike's Subs has made a submarine sandwich for eighteen people to share. How many cuts must be made to divide the sandwich equally among the eighteen people?

For **Extra Practice,** see page 566.

ON YOUR OWN

4. The last step of the 4-step plan for problem solving asks you to *examine* your solution. *Explain* how you can use estimation with decimals to help you examine a solution.

5. *Tell* why solving a simpler problem can sometimes help you solve a more difficult problem.

6. *Explain* how using the distributive property is similar to the solving a simpler problem strategy.

MIXED PROBLEM SOLVING

STRATEGIES

Look for a pattern.
Solve a simpler problem.
Act it out.
Guess and check.
Draw a diagram.
Make a chart.
Work backward.

Solve. Use any strategy.

7. Gardening Marco wants to fence in his tomato garden that is 8 feet by 6 feet. How much fencing does he need?

8. Decorating How many 1-foot square tiles are needed to cover a kitchen floor that is 14 feet by 10 feet?

9. Statistics The bar graph shows revenues for several popular magazines.

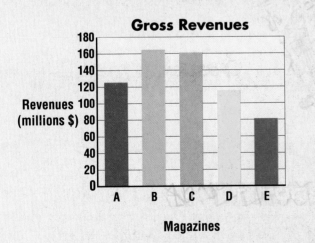

Gross Revenues

Revenues (millions $)

Magazines

a. About how much higher were the revenues of magazine A than magazine E?

b. About how much higher were magazine B's revenues than magazine D's?

c. About how much more revenue did the magazine with the highest revenue have than magazine C?

10. Decorating The diagram shows Paula's dining room and living room area.

a. If she wants to carpet this area, how much carpet will she need?

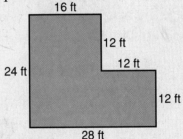

16 ft
12 ft
12 ft
24 ft
12 ft
28 ft

b. Draw another room that will use the same amount of carpet.

11. Sports Luis bowled a 125, a 148, and a 162 during league competition. Before this competition, his average was 128 pins. Compare his previous average to his competition average.

12. Entertainment The Strand Theater seats 536 people. At last night's movie, there was one empty seat for every three people. How many people were in the movie theater?

13. Standardized Test Practice Nick bought 5 packages of strawberries. Each package weighed 1.15 pounds. What was the weight of the 5 packages?

A 4.85 lb

B 5.15 lb

C 5.45 lb

D 5.75 lb

E Not Here

Dividing Decimals by Whole Numbers

What you'll learn

You'll learn to divide decimals by whole numbers.

When am I ever going to use this?

Knowing how to divide a decimal by a whole number can help you find your share of a group purchase.

A package that included in-line skates, a helmet, and pads was on sale for $119.50. Josh didn't have enough money to pay for them so, in order to get the sale price, he put them on layaway. The store's layaway policy requires one payment now and a payment equal to that each week for 4 weeks. How much will he pay each time?

You need to divide $119.50 by 5.

First estimate. $100 \div 5 = 20$

When dividing a decimal by a whole number, place the decimal point in the quotient directly above the decimal point in the dividend. Then divide as you do with whole numbers.

$$
\begin{array}{r}
23.90 \\
5\overline{)119.50} \\
-10 \\
\hline
19 \\
-15 \\
\hline
4\,5 \\
-4\,5 \\
\hline
00
\end{array}
$$

Place the decimal point.
Then divide as with whole numbers.

He needs to pay $23.90 each week. Checking this answer against the estimate, the answer is reasonable.

Examples

Find each quotient.

1 $29.8 \div 2$

Estimate: $30 \div 2 = 15$

$$
\begin{array}{r}
14.9 \\
2\overline{)29.8} \\
-2 \\
\hline
9 \\
-8 \\
\hline
1\,8 \\
-1\,8 \\
\hline
0
\end{array}
$$

Place the decimal point.
Divide as with whole numbers.

$29.8 \div 2 = 14.9$

The estimate shows that the answer is reasonable.

2 $8.58 \div 12$

Estimate: $10 \div 10 = 1$

$$
\begin{array}{r}
0.715 \\
12\overline{)8.580} \\
-8\,4 \\
\hline
18 \\
-12 \\
\hline
60 \\
-60 \\
\hline
0
\end{array}
$$

Place the decimal point.
Divide as with whole numbers.
Annex a zero to continue dividing.

$8.58 \div 12 = 0.715$

The estimate shows that the answer is reasonable.

Usually when you divide with decimals, the answer does not come out even. You need to round the quotient to a specified place-value position. Always divide to one more place-value position than the place to which you are rounding.

 **Example**

APPLICATION

③ Money Matters Katie and 5 of her friends bought a six-pack of fruit juice after their soccer game. Each friend wants to pay for her share. If the six-pack costs $3.29, how much does each friend owe to the nearest cent?

Study Hint

Estimation When dealing with money, always round the quotient up.

$$
\begin{array}{r}
0.548 \\
6\overline{)3.290} \\
-3\ 0 \\
\hline
29 \\
-24 \\
\hline
50 \\
-48 \\
\hline
2
\end{array}
$$

Estimate: $3.00 \div 6 = \$0.50$
Place the decimal point.
Annex a zero.
Divide to the thousandths place.

Katie and her five friends cannot pay $0.548 each. $0.548 rounded to the nearest cent is $0.55.

Each person owes $0.55. The estimate shows that the answer is reasonable.

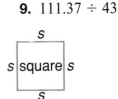

CHECK FOR UNDERSTANDING

Communicating Mathematics

Read and study the lesson to answer each question.

1. *Explain* why you can use an estimate to place the decimal point in the quotient for $34.56 \div 25$.

2. *Tell* how to round a quotient to a specific place-value position.

3. *You Decide* Payat uses a calculator to find $75.89 \div 38$. He gets 1.997105263 for an answer. Is the answer reasonable? Explain.

Guided Practice

Find each quotient.

4. $13\overline{)15.6}$

5. $6\overline{)2.49}$

6. $55.6 \div 8$

Round each quotient to the nearest tenth.

7. $55.39 \div 7$

8. $78.75 \div 35$

9. $111.37 \div 43$

10. **Geometry** The perimeter of the square is 30.24 meters. What is the length of one side?

s
s square s
s

Practice   **Find each quotient.**

11. $25\overline{)6.25}$

12. $16\overline{)117.44}$

13. $37\overline{)784.4}$

14. $475.2 \div 32$

15. $479.96 \div 52$

16. $256.36 \div 34$

Round each quotient to the nearest tenth.

17. $29.48 \div 11$

18. $28.56 \div 42$

19. $263.25 \div 81$

Round each quotient to the nearest hundredth.

20. $344.736 \div 76$

21. $650.23 \div 29$

22. $4.567 \div 21$

Solve each equation.

23. $x = 29.37 \div 3$

24. $8.956 \div 10 = y$

25. $z = 302.5 \div 55$

26. Find 72.89 divided by 39 to the nearest thousandth.

27. What is the solution of $m = 276.33 \div 61$?

28. *Algebra* Evaluate $b \div c$ if $b = 34.56$ and $c = 32$.

Applications and Problem Solving

29. *Money Matters* On Monday, Amber was paid $20 for watering her neighbor's plants the previous week. On Tuesday, her mother paid her $14 for raking 14 bags of leaves. She also made $10 baby-sitting on Friday. What was Amber's average pay for these three jobs?

30. *Sports* The Pickerington girls track team ran the 4-by-100 meter relay in 48.9 seconds. What was the average time of each runner?

31. *Spreadsheets* The spreadsheet shows the unit price for a jar of jelly. To find the unit price, divide the cost of the item by its size. Find the unit price for the next three items. Round to the nearest cent.

	A	B	C	D
1	Item	Cost	Size	Unit Price
2	Jelly	$1.59	12 oz	0.1325
3	Cereal	$3.35	18 oz	
4	Bread	$1.19	16 oz	
5	Ketchup	$0.89	14 oz	

32. *Money Matters* Suppose you wanted to fill a car's 12-gallon gas tank. If you can afford to spend $14.25 on gasoline, what price per gallon should you look for? Round to the nearest cent.

33. *Critical Thinking* Create a division problem that meets all of the following conditions.
- the divisor is a whole number
- the dividend is a decimal
- the quotient is 0.125 when rounded to the nearest thousandth
- the quotient is 0.12 when rounded to the nearest hundredth

Mixed Review

34. Geometry Find the perimeter of the rectangle. *(Lesson 4-4)*

11.75 m
3.4 m

35. Weather During two weeks of torrential rains, Antwon measured the rainfall. He recorded 11.33 centimeters in the first week and 15.75 centimeters in the second week. About how much rain fell in the two weeks? *(Lesson 3-5)*

36. Careers Prospective employees were told that the current employees' earnings averaged $475 a week. In the past, 6 people have earned weekly amounts of $200, $400, $410, $260, $320, and $1,260. Were the prospects being misled? Explain. *(Lesson 2-8)*

37. Standardized Test Practice
The line graph shows the number of people in Marysville and Plain City from 1960 to 1990. About how many more people were in Marysville than in Plain City in 1980? *(Lesson 2-3)*

For **Extra Practice,** see page 567.

    A 2,000   **B** 3,500
    C 6,000   **D** 7,500

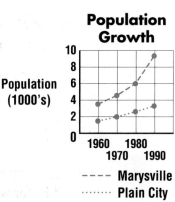
Population Growth
Population (1000's)
10
8
6
4
2
0
1960 1980
  1970 1990
---- Marysville
········ Plain City

CHAPTER 4

Mid-Chapter Self Test

1. Money Matters Mrs. Ortiz has bought a new car. Her payments will be $385.55 a month for 48 months. About how much will she pay in all? *(Lesson 4-1)*

Find each product mentally. Use the distributive property. *(Lesson 4-2)*

2. 6×17

3. 10.5×40

Solve each equation. *(Lesson 4-3)*

4. $a = 28.5 \cdot 0.61$

5. $b = 0.006(3.4)$

6. Geometry Find the perimeter of a square if each side is 7.2 inches long. *(Lesson 4-4)*

Find the perimeter and the area of each figure. *(Lesson 4-4)*

7.
rectangle 0.2 cm
4.7 cm

8.
square
3.1 m

Round each quotient to the nearest hundredth. *(Lesson 4-5)*

9. $597.3 \div 62$

10. $69.597 \div 45$

COOPERATIVE LEARNING

4-6A Dividing by Decimals

A Preview of Lesson 4-6

base-ten blocks

In this lab, you will use base-ten blocks to model dividing a decimal by a decimal. Recall that each 10-by-10 block represents 1, each row or column represents 0.1, and each small square represents 0.01.

TRY THIS

Work with a partner.

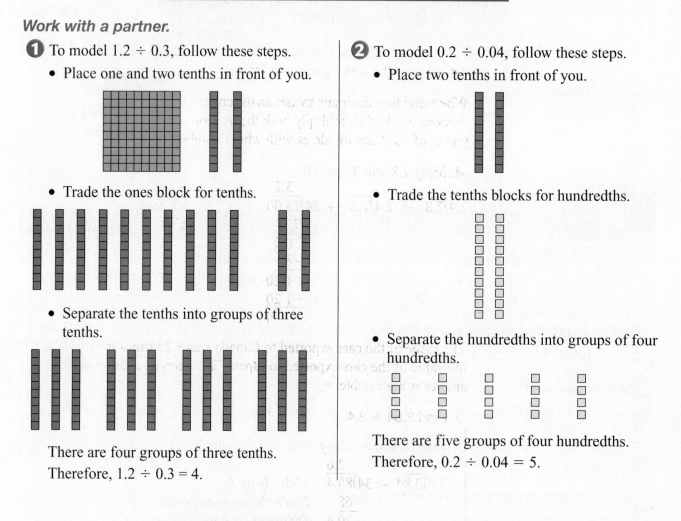

1 To model 1.2 ÷ 0.3, follow these steps.

- Place one and two tenths in front of you.

- Trade the ones block for tenths.

- Separate the tenths into groups of three tenths.

There are four groups of three tenths.
Therefore, 1.2 ÷ 0.3 = 4.

2 To model 0.2 ÷ 0.04, follow these steps.

- Place two tenths in front of you.

- Trade the tenths blocks for hundredths.

- Separate the hundredths into groups of four hundredths.

There are five groups of four hundredths.
Therefore, 0.2 ÷ 0.04 = 5.

ON YOUR OWN

Use base-ten blocks to show each quotient.

1. 1.4 ÷ 0.7   **2.** 2.6 ÷ 0.2   **3.** 0.9 ÷ 0.03   **4.** 1.25 ÷ 0.25

5. Look Ahead Find the quotient for 2.4 ÷ 0.8 without using models.

4-6

Dividing by Decimals

What you'll learn

You'll learn to divide decimals by decimals.

When am I ever going to use this?

Decimal division can be used to compare the rates of speed over different distances.

The circle graph shows where the United States exported $16.9 billion worth of cars in a recent year.

How many times greater was the value of the cars exported to Canada than to Japan?

You need to divide 7.8 by 2.4.

First, estimate the answer.

$$8 \div 2 = 4$$

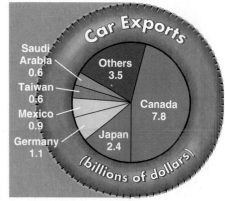

Source: U.S. Census Bureau

The value of the cars exported to Canada was about 4 times greater than the value of the cars exported to Japan.

When dividing decimals by decimals, change the divisor to a whole number. To do this, multiply both the divisor and dividend by the same power of 10. Then divide as with whole numbers.

Multiply 7.8 and 2.4 by 10.

$$2.4\overline{)7.8} \;\rightarrow\; 2.4\overline{)7.8} \;\rightarrow\; 24\overline{)78.00}$$

```
        3.25
  24)78.00
    -72
      6 0
     -4 8
      1 20
     -1 20
         0
```

Multiply to check.

```
     3.25
    ×2.4
   1 300
   6 50
   7.800  ✓
```

The value of the cars exported to Canada was 3.25 times greater than the value of the cars exported to Japan. The estimate shows that the answer is reasonable.

Study Hint

Mental Math When multiplying by powers of ten, move the decimal to the right as many places as the number of zeros in the power of ten.

Example

1 Find $8.84 \div 3.4$.

Estimate: $9 \div 3 = 3$

$$3.4\overline{)8.84} \;\rightarrow\; 34\overline{)88.4}$$

```
        2.6
   34)88.4
    - 68
      20 4
    - 20 4
         0
```

Multiply by 10.
Place the decimal point.
Divide as with whole numbers.

$8.84 \div 3.4 = 2.6$. The estimate shows that the answer is reasonable.

Check: $2.6 \times 3.4 = 8.84$ ✓

2 Find $15.176 \div 0.028$.

$$
\begin{array}{r}
542. \\
0.028\overline{)15.176} \rightarrow 28\overline{)15176.} \\
-140 \\
\hline
117 \\
-112 \\
\hline
56 \\
-56 \\
\hline
0
\end{array}
$$

Multiply by 1,000.
Place the decimal point.
Divide.

$15.176 \div 0.028 = 542$ **Check:** $0.028 \times 542 = 15.176$ ✓

INTEGRATION **3** **Measurement** Mr. Henderson's flower garden is 11.25 meters long. He wants to make a border along one side using bricks that are 0.25 meter long. How many bricks does he need?

$$
\begin{array}{r}
45 \\
0.25\overline{)11.25} \rightarrow 25\overline{)1125} \\
-100 \\
\hline
125 \\
-125 \\
\hline
0
\end{array}
$$

Multiply by 100.
Place the decimal point.
Divide.

Mr. Henderson needs 45 bricks.

CHECK FOR UNDERSTANDING

Communicating Mathematics

Read and study the lesson to answer each question.

1. *Write* a decimal division problem that would require you to multiply the divisor and dividend by 10 before you divide.

2. *Explain* why the quotient for $2.35 \div 0.58$ should be about 4.

Guided Practice

Find each quotient.

3. $2.7\overline{)51.3}$
4. $2.5\overline{)43.25}$
5. $0.18\overline{)124.92}$

6. $302.5 \div 5.5$
7. $70.59 \div 1.3$
8. $3.965 \div 0.065$

9. *Algebra* Evaluate $a \div b$ if $a = 7.502$ and $b = 3.41$.

EXERCISES

Practice

Find each quotient.

10. $4.9\overline{)160.72}$
11. $0.98\overline{)109.76}$
12. $3.25\overline{)2,606.5}$

13. $3.6\overline{)1.44}$
14. $89.6\overline{)206.08}$
15. $0.008\overline{)0.0072}$

16. $0.32 \div 0.005$
17. $210.657 \div 0.071$
18. $9.425 \div 0.0125$

19. $0.0186 \div 0.031$
20. $40.99524 \div 14.3$
21. $5,885.9514 \div 703.22$

Solve each equation.

22. $b = 17.01 \div 0.81$

23. $p = 94.16 \div 0.88$

24. $127.6625 \div 102.13 = x$

25. $0.0078 \div 0.002 = y$

26. What is 48.355 divided by 0.095?

27. Find 7.502 divided by 2.2.

28. *Algebra* Evaluate $s \div t$ if $s = 1.43$ and $t = 1.3$.

29. *Algebra* Evaluate $(p + q) \div r$ if $p = 0.7$, $q = 1.4$, and $r = 2.5$.

Applications and Problem Solving

30. *Sports* At the 1996 Olympics, American sprinter Michael Johnson set a world record of 19.32 seconds for the 200-meter dash. A honeybee can fly the same distance in 40.572 seconds. How many times faster than a honeybee is Michael Johnson?

31. *Transportation* The Aero Spacelines Super Guppy, a converted Boeing C-97, can carry 87.5 tons. Tanks that weigh 4.5 tons each are to be loaded onto the Super Guppy. What is the maximum number of tanks it can transport?

32. *Food* The table shows how many pounds of breakfast cereal are consumed each year by the average person in selected countries.

a. To the nearest tenth, how many times greater was the amount consumed in Great Britain than in France?

b. To the nearest tenth, how many times greater was the amount consumed in the United States than in Canada?

Country	Amount (lb)
Canada	6.02
France	1.78
Great Britain	7.34
South Korea	0.07
United States	11.9

33. *Critical Thinking* Replace each ■ to make a true sentence.

■.8 ■ 6 ÷ 0.35 = 2.5 ■

Mixed Review

34. Find the quotient when 68.52 is divided by 12. *(Lesson 4-5)*

35. Write the number 204.2398 in words. *(Lesson 3-1)*

36. **Standardized Test Practice**
The stem-and-leaf plot shows the high temperatures for 20 cities on February 23. Which interval had the fewest high temperatures? *(Lesson 2-6)*

A $20° - 29°$ **B** $30° - 39°$

C $50° - 59°$ **D** $70° - 79°$

Stem	Leaf
2	5
3	0 2 4 5
4	1 2 6 9
5	
6	0 3 3 4 4 5 5 6
7	0 0 1

$4|2 = 42$

For **Extra Practice,** see page 567.

37. *Algebra* Solve $b \times 6 = 42$ mentally. *(Lesson 1-7)*

BUSINESS

Cedric Walker

Small business owner

Cedric Walker is the co-owner, along with Cal Dupree, of the nation's only African-American-owned, operated, and performed circus. After spending three years doing research, Walker opened the circus in 1994.

A person who is interested in owning a small business should take classes in economics, mathematical analysis, business ethics, communications, mathematics of finance, and marketing. A college degree in communications, marketing, business law, business management, or accounting should be considered.

For more information:

The Young Entrepreneurs' Organization
10101 North Glebe Road
Arlington, VA 22207
(703) 519-6700

inter*NET* CONNECTION

www.glencoe.com/sec/
math/mac/mathnet

I can't wait to design my own product and start my own business!

Your Turn

Part of what makes a small business owner successful is offering a product or service that people need. Survey your class or your school and find out what people need that a small business could provide. Write a paragraph describing what you discovered and how you might be able to fill the need with your own small business.

Zeros in the Quotient

What you'll learn

You'll learn to divide decimals involving special cases of zero in the quotient.

When am I ever going to use this?

Knowing how to divide all kinds of decimals can help you find the unit cost of a jar of peanut butter.

Did you know?

The men's gymnastic team from Japan won the gold medal in the team combined exercises for five consecutive Olympics, from 1960 through 1976.

On March 30, 1980, Shigeru Iwasaki somersaulted backward 54.68 yards in 10.8 seconds. To the nearest hundredth, how many yards per second did he somersault?

First, estimate the answer.

$$50 \div 10 = 5$$

He went about 5 yards per second.

Method 1 Use a calculator.

54.68 ÷ 10.8 = 5.062962963

To the nearest hundredth, he went 5.06 yards per second. Compared to the estimate, the answer is reasonable.

Method 2 Use paper and pencil.

You can see why a zero appears in this quotient.

$$10.8)\overline{54.68} \rightarrow$$

```
        5.062
108)546.800
   −540
      6 8
     −0
      6 80
     −6 48
        320
       −216
        104
```

Divide to the thousandths place to round to the nearest hundredth. 68 < 108 so a zero is written in the quotient.

Example 1 **Find 9.03 ÷ 0.301.**

```
          30.0
0.301)9.030
     −9 03
        00
```
Write a zero in the ones place of the quotient.

$$9.03 \div 0.301 = 30.0$$

Check: $0.301 \times 30 = 9.03$ ✓

Real World APPLICATION

2 **Money Matters** Mr. Alvarez put 11.5 gallons of gas in his van. If he paid $13.90, what was the price per gallon?

$$11.5\overline{)13.90} \rightarrow \begin{array}{r} 1.208 \\ 115\overline{)139.000} \\ \underline{-115} \\ 24\ 0 \\ \underline{-23\ 0} \\ 1\ 00 \\ \underline{-0} \\ 1\ 000 \\ \underline{-920} \\ 80 \end{array}$$ *Estimate: 10 ÷ 10 = 1*

The price of the gasoline was about $1.21 per gallon.

CHECK FOR UNDERSTANDING

Communicating Mathematics

Read and study the lesson to answer each question.

1. *Write* a decimal division problem that would require a zero in the quotient.

2. *Explain* how you could estimate the quotient for 17.5 ÷ 0.02.

3. *Write* a letter to a student explaining why 156.14 ÷ 14.8 is not 1.55.

Guided Practice

Find each quotient to the nearest hundredth.

4. $13\overline{)27}$

5. $36\overline{)145}$

6. $61\overline{)2,441}$

7. $75.3 \div 15$

8. $0.68 \div 6$

9. $3.41 \div 0.2$

10. Solve $y = 15.25 \div 0.5$.

11. *Algebra* Evaluate $d \div r$ if $d = 51.714$ and $r = 5.07$.

EXERCISES

Practice

Find each quotient to the nearest hundredth.

12. $12\overline{)24.32}$

13. $0.25\overline{)75.25}$

14. $0.45\overline{)0.117}$

15. $24\overline{)125}$

16. $3.98\overline{)4.02}$

17. $7.89\overline{)86.3}$

18. $1.43 \div 13$

19. $83.25 \div 4.11$

20. $234.4 \div 11.2$

21. $2.5 \div 0.24$

22. $0.0125 \div 1.7$

23. $95.23 \div 47.6$

Solve each equation.

24. $g = 6.018 \div 5.9$

25. $d = 59.59 \div 0.59$

26. $7,576.4 \div 37.6 = f$

27. $3.3894 \div 4.2 = m$

28. What is 0.62524 divided by 30.8?

29. *Algebra* Evaluate $(a + b) \div c$ if $a = 0.016$, $b = 0.008$, and $c = 7.5$.

30. *Music* A compact disc has 11 tracks. If the entire CD takes 34.8 minutes to play, what is the average time for each track to the nearest hundredth of a minute?

31. *Geometry* The area of the patio in the diagram is 27.36 square meters. The width of the patio is 4.5 meters. Find the length of the patio.

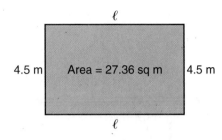

32. *Physical Science* Sound travels through air at 330 meters per second. How long will it take a bat's cry to reach its prey and echo back if the prey is 1 meter away? Round to the nearest thousandth of a second.

33. *Critical Thinking* If a decimal greater than 0 and less than 1 is divided by a lesser decimal, would the quotient be always less, sometimes less, or never less than 1?

Mixed Review

34. Find 5.06 divided by 2.3. *(Lesson 4-6)*

35. *Standardized Test Practice* Mrs. Zacharias asked her mathematics students to decide whether 95 ceramic tiles, each measuring 1.5 feet-by-1.5 feet, would be enough to completely cover the floor of her classroom. What other piece of information would allow the students to answer the question? *(Lesson 4-4)*

  A the weight of each tile
  B the number of classrooms in the school
  C the number of students in Mrs. Zacharias' classroom
  D the area of the classroom floor
  E the number of tiles in a package

36. *Write a Problem* about a real-life situation involving the photo at the left that can be represented by a mean of 20° F. *(Lesson 2-7)*

37. *Statistics* Choose an appropriate scale for a frequency table for the following set of data: 24, 67, 11, 9, 52, 38, 114, and 98. *(Lesson 2-2)*

For **Extra Practice,**
see page 567.

Integration: Measurement
Mass and Capacity in the Metric System

What you'll learn

You'll learn to use metric units of mass and capacity.

When am I ever going to use this?

Knowing how to use metric units of mass and capacity can help you solve problems in science class.

Word Wise

gram (g)
kilogram (kg)
milligram (mg)
liter (L)
milliliter (mL)

Pablo Lara is a weightlifter from Cuba who won the gold medal in the Olympics in Atlanta, Georgia. Pablo competed in the 76-kilogram division. This means his mass could be no more than 76 kilograms.

In the metric system, all units are defined in terms of a basic unit. The basic unit of mass in the metric system is the **kilogram (kg)**. A kilogram is 1,000 **grams (g)**, and a **milligram (mg)** is 0.001 gram.

Many items you use every day are measured in grams, kilograms, or milligrams.

- A small paper clip has a mass of about 1 gram.
- Your math textbook has a mass of about 1 kilogram.
- A grain of salt has a mass of about 1 milligram.

Examples

Write the unit of mass: milligram, gram, or kilogram, that you would use to measure each of the following. Then estimate the mass.

1 a compact disc

Since a compact disc has a small mass, the *gram* is the appropriate unit.

The mass of a compact disc is about 50 grams.

2 a laptop computer

Since a laptop computer has a mass close to several copies of your textbook, the *kilogram* is the appropriate unit.

The mass of a laptop computer is about 4 kilograms.

Did you know Pablo Lara lifted 451.75 pounds (204.9 kg) in the clean and jerk and 358 pounds (162.4 kg) in the snatch, yet he weighs only 167.5 pounds.

The basic unit of capacity in the metric system is the **liter (L)**. A liter is a little more than a quart. The **milliliter (mL)** is 0.001 liter. It takes 1,000 milliliters to make a liter.

- A small pitcher has a capacity of about 1 liter.
- An eyedropper has a capacity of about 1 milliliter.

Write the unit of capacity: milliliter or liter, that you would use to measure each of the following. Then estimate the capacity.

❸ a bathtub

Since a bathtub holds a large amount, the *liter* is the appropriate unit.

An average bathtub has a capacity of about 80 liters.

❹ 10 drops of food coloring

Since 10 drops of food coloring is a small amount, the *milliliter* is the appropriate unit.

The capacity is about 1 milliliter.

CHECK FOR UNDERSTANDING

Communicating Mathematics

Read and study the lesson to answer each question.

1. *Tell* which unit represents a greater mass, 1 gram or 1 kilogram.

2. *Explain* what a milliliter and a milligram have in common.

Guided Practice

Write the unit that you would use to measure each of the following. Then estimate the mass or capacity.

3. the mass of a horse
4. an aspirin
5. a large bottle of soft drink
6. a tennis ball

7. *Food* A bread company makes loaves of bread that have a mass of 1 kilogram. A loaf of honey whole wheat is much larger than a loaf of blueberry apricot. Explain why this happens.

EXERCISES

Practice

Write the unit that you would use to measure each of the following. Then estimate the mass or capacity.

8. gas in the tank of a car
9. a bottle of cough syrup
10. vanilla used in a cookie recipe
11. the mass of an Olympic boxer
12. a can of soup
13. a bag of sugar
14. a glass of iced tea
15. a dollar bill
16. a pitcher of water
17. a canary
18. the liquid in a thermometer
19. a table tennis ball

Name an item that you think has the given measure.

20. about 250 mL
21. about 15 kg

Applications and Problem Solving

22. *Life Science* The human brain averages about 0.025 of a human's total body mass. Find your mass in kilograms and use a calculator to determine the approximate mass of your brain.

23. *Medicine* A pharmacist has 84.73 grams of a prescription medicine. She wants to separate it into 48 capsules. To the nearest thousandth, how many grams will go into each capsule?

24. *Money Matters* A can of soft drink contains 355 milliliters.

    a. How many milliliters are in a 6-pack of soft drink?

    b. Two liters of soft drink costs the same as a 6-pack of the same soft drink. Which is the better buy? Explain.

25. *Life Science* The table shows part of the recommended daily diet for a gibbon.

    a. Determine how many grams of carrots, bananas, and celery are needed in one week.

    b. Is the amount of carrots, bananas, and celery in part a more or less than 4 kilograms?

    c. Visit a grocery store and select two or three of the items to weigh. About how many of these items would be needed for one week?

380 g lettuce
270 g orange
150 g spinach
143 g sweet potato
270 g banana
147 g carrot
210 g celery
210 g green beans

26. *Critical Thinking* Marva filled a 250-milliliter beaker with sand. She said its mass was 250 milligrams. Is she right? Explain.

Mixed Review

27. Divide 1,269.45 by 6.3. *(Lesson 4-7)*

28. *Physical Science* One cubic meter of carbon dioxide has a mass of 1.977 kilograms. Round this number to the nearest tenth. *(Lesson 3-4)*

29. *Standardized Test Practice* If a passenger train travels an average of 58 miles per hour, which is a good estimate for the number of hours it will take to travel 286 miles? *(Lesson 1-3)*

    A 3 h   **B** 4 h   **C** 5 h   **D** 6 h   **E** 18 h

For **Extra Practice**, see page 568.

Let the Games Begin

Scavenger Hunt

Math Skill
Metric Measurement

Get Ready   This game is for three to four players.

     5 index cards per player   tape

Get Set   Each player is assigned a unit of measure. For example, one player may have centimeters, another player may have liters, and the third player may have kilograms.

Go
- Each player will write their unit on each of the index cards and find items in the class that would best be measured using that unit.
- Players should pass their cards to the person on their left.
- The winner is the first player to tape the cards to the appropriate items.

interNET CONNECTION Visit www.glencoe.com/sec/math/mac/mathnet for more games.

What you'll learn

You'll learn to change units within the metric system.

When am I ever going to use this?

You'll often need to change metric units when you are simplifying measurements.

Integration: Measurement
Changing Metric Units

A chemistry experiment requires 3.05 grams of 3% hydrogen peroxide to produce one liter of oxygen. If a scientist wanted 24 liters of oxygen, how many kilograms of 3% hydrogen peroxide will she need? *This problem will be solved in Example 3.*

To change from one unit to another within the metric system, you either multiply or divide by powers of 10. The chart below shows the relationship between the units in the metric system and the powers of 10.

Each place value is ten times the place value to its right.

Study Hint

Mental Math
Remember you can multiply or divide by a power of ten by moving the decimal point.

To change from a larger unit to a smaller unit, you need to multiply. To change from a smaller unit to a larger unit, you need to divide.

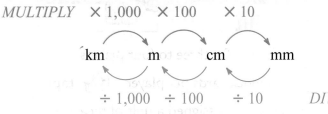

Examples

1 345 mL = _?_ L

To change from milliliters to liters, divide by 1,000 since 1 L = 1,000 mL.

$345 \div 1,000 = 0.345$
345 mL = 0.345 L

2 0.9 cm = _?_ mm

To change from centimeters to millimeters, multiply by 10 since 1 cm = 10 mm.

$0.9 \times 10 = 9$
0.9 cm = 9 mm

CONNECTION

③ Physical Science Refer to the beginning of the lesson. How many kilograms of 3% hydrogen peroxide will the scientist need?

Explore You know it took 3.05 grams of 3% hydrogen peroxide to produce 1 liter of oxygen. You want to know how many kilograms are needed for 24 liters.

Plan Multiply 3.05 times 24 to find the number of grams needed. Then change this number from grams to kilograms.

Solve $3.05 \times 24 = 73.2$

Divide to change from smaller units to larger units.

$73.2 \div 1,000 = 0.0732$

She will need 0.0732 kilogram of 3% hydrogen peroxide.

Examine Since 1 g = 0.001 kg, the answer is reasonable.

> **Study Hint**
>
> **Reading Math** When a measurement is less than 1, the unit of measure stays singular. For example, 0.45 g is read as forty-five hundredths (of a) gram.

④ 0.45 g = _?_ mg

To change from grams to milligrams, multiply by 1,000 since 1 g = 1,000 mg.

$0.45 \times 1,000 = 450$

0.45 g = 450 mg

⑤ 8,960 m = _?_ km

To change from meters to kilometers, divide by 1,000 since 1 km = 1,000 m.

$8,960 \div 1,000 = 8.96$

8,960 m = 8.96 km

CHECK FOR UNDERSTANDING

Communicating Mathematics

Read and study the lesson to answer each question.

1. *Explain* how to change milliliters to liters.

2. *Tell* whether you would use grams or kilograms to weigh a large screen television.

3. *You Decide* Parker and Dinh each changed 3.45 kilograms to grams. Parker's answer was 34.5 grams, and Dinh's answer was 0.345 grams. Who is correct? Explain your reasoning.

Guided Practice

Write whether you multiply or divide to change each measurement. Then complete.

4. 328 mL = _?_ L

5. _?_ mm = 0.7 cm

6. _?_ g = 150 mg

7. 5.02 kg = _?_ g

8. _?_ mL = 1.5 L

9. 150 m = _?_ km

10. *Geography* The circumference of Earth is about 40,000 kilometers. How many meters is it around Earth?

Practice

Complete.

11. 210 mm = _?_ cm   **12.** _?_ L = 253 mL   **13.** 2.5 L = _?_ mL

14. 0.0068 kg = _?_ g   **15.** _?_ m = 524 cm   **16.** _?_ m = 2,400 mm

17. _?_ mL = 0.0817 L   **18.** 149 mg = _?_ g   **19.** _?_ g = 1,953 mg

20. 0.593 km = _?_ m   **21.** 3.29 L = _?_ mL   **22.** 975 cm = _?_ mm

23. 5.25 kg = _?_ g   **24.** _?_ g = 427 mg   **25.** _?_ mL = 0.01 L

26. _?_ cm = 6.908 km   **27.** 6,700 mg = _?_ kg   **28.** 1.2 km = _?_ mm

29. How many centimeters are in 0.58 meter?

30. Change 0.6 milligram to grams.

31. How many liters are in 213 milliliters?

Applications and Problem Solving

32. *Earth Science* Seismic waves are waves generated by an earthquake. The focus of an earthquake is the point in Earth's interior where seismic waves originate. The focus can be between 5 and 700 kilometers below the surface of Earth. How many meters can the focus be below the surface?

33. *Food* A can holds 355 milliliters of soft drink. How many cans would equal 2 liters?

34. *Critical Thinking* A liter of water at 4°C has a mass of 1 kilogram. What is the mass of 1 milliliter of water at 4°C?

Mixed Review

35. *Measurement* To measure the water in a washing machine, would you use liters or milliliters as your units? *(Lesson 4-8)*

36. Find 4,507.72 divided by 8.5. *(Lesson 4-6)*

37. *Algebra* Evaluate the expression $z + w$ if $z = 6.45$ and $w = 71.2$. *(Lesson 3-6)*

38. **Standardized Test Practice** The graph shows the average number of greeting cards purchased by Americans recently. Which is a reasonable conclusion that can be drawn from the information in the graph? *(Lesson 2-3)*

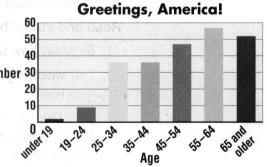

Greetings, America!

Source: American Greetings Corporation

A People over 65 buy the most greeting cards.

B People 19 to 24 bought twenty times as many cards as people under 19.

C The number of cards bought increases in every age group.

D People 65 and older bought about five times as many cards as people 19 to 24.

E People under 19 don't buy greeting cards.

For **Extra Practice**, see page 568.

CHAPTER 4

Study Guide and Assessment

interNET CONNECTION Chapter Review For additional lesson-by-lesson review, visit:
www.glencoe.com/sec/math/mac/mathnet

Vocabulary

After completing this chapter, you should be able to define each term, concept, or phrase and give an example or two of each.

Algebra
distributive property (p. 137)

Geometry
area (p. 146)
perimeter (p. 145)
sides (p. 145)

Number and Operations
compatible numbers (p. 133)

Measurement
gram (p. 164)
kilogram (p. 164)
liter (p. 164)
milligram (p. 164)
milliliter (p. 164)

Problem Solving
solve a simpler problem (pp. 150–151)

Understanding and Using the Vocabulary

State whether each sentence is *true* or *false*. If false, replace the underlined word or number to make a true sentence.

1. You can use compatible numbers to <u>estimate</u> products.
2. Using the distributive property, $4(3 + 2) = 4 \cdot 3 + \underline{3} \cdot 2$.
3. The <u>area</u> of a rectangle is two times the length plus two times the width.
4. The perimeter of a rectangle with a length of 4 meters and a width of 1 meter is <u>5</u> meters.
5. The area of a rectangle is the <u>product</u> of its length and width.
6. If each side of a square is 4.2 feet, then its <u>area</u> is 17.64 square feet.
7. The basic unit of mass in the metric system is the <u>kilogram</u>.
8. A <u>milligram</u> is 1,000 grams.
9. The basic unit of capacity in the metric system is the <u>liter</u>.
10. It takes <u>0.001 milliliter</u> to make a liter.
11. To change from a larger unit to a smaller unit, you need to <u>divide</u>.
12. To change from a smaller unit to a larger unit, you need to <u>multiply</u>.

In Your Own Words

13. *Explain* how to estimate the product of 6.9 and 88.

Objectives & Examples

Upon completing this chapter, you should be able to:

● estimate and find the products of decimals and whole numbers *(Lesson 4-1)*

Find 12.32×12.

$$\begin{array}{r} 12.32 \\ \times\ 12 \\ \hline 147.84 \end{array}$$
Estimate: $12 \times 12 = 144$
Multiply as with whole numbers.
Since the estimate is 144, place the decimal point after 7.

● compute products mentally using the distributive property *(Lesson 4-2)*

Find 90×3.8.
Estimate: $90 \times 4 = 360$
$$\begin{aligned} 90 \times 3.8 &= 90(3 + 0.8) \\ &= 90 \times 3 + 90 \times 0.8 \\ &= 270 + 72 \\ &= 342 \end{aligned}$$

● multiply decimals *(Lesson 4-3)*

Find $38.76 \cdot 4.2$.
$$\begin{array}{r} 38.76 \\ \times\ 4.2 \\ \hline 7752 \\ 15504 \\ \hline 162.792 \end{array}$$
← two decimal places
← one decimal place

← three decimal places

● find the perimeters and the areas of rectangles and squares *(Lesson 4-4)*

Find the perimeter and the area of the figure.

3.2 m

6.1 m

$$\begin{aligned} P &= 2\ell + 2w \\ &= 2(6.1) + 2(3.2) \\ &= 12.2 + 6.4 \\ &= 18.6 \text{ m} \end{aligned}$$

$$\begin{aligned} A &= \ell \cdot w \\ &= 6.1 \cdot 3.2 \\ &= 19.52 \text{ m}^2 \end{aligned}$$

Review Exercises

Use these exercises to review and prepare for the chapter test.

Multiply.

14. $\begin{array}{r} 17.31 \\ \times\ 40 \end{array}$ **15.** $\begin{array}{r} 2.15 \\ \times 227 \end{array}$

16. 38.5×791 **17.** 6×7.91

Solve each equation.

18. $m = 201 \times 3.94$
19. $51.2 \times 1{,}891 = r$

Find each product mentally. Use the distributive property.

20. 40×8.9
21. 30×10.76
22. 9×16
23. 5×6.8

Multiply.

24. $8.74 \cdot 2.23$ **25.** 0.04×5.1
26. 0.04×0.0063 **27.** $11.089 \cdot 5.6$

Solve each equation.

28. $d = 2.6 \cdot 3.9$ **29.** $112.45(4.8) = n$

Find the perimeter of each figure.

30.   **31.**

4.3 km

2 km   4.9 km

8.3 m

4.2 m   4.7 m

10.1 m

Find the perimeter and the area of each figure.

32. rectangle: $\ell = 25$ in.; $w = 8.1$ in.
33. rectangle: $\ell = 11.2$ mi; $w = 9.3$ mi
34. square: $s = 100$ m
35. square: $s = 7.6$ ft

Applications & Problem Solving

59. Fitness Tyra exercises every morning by running once around a rectangular neighborhood. The length of the rectangle is 1.7 kilometers, and the width is 0.9 kilometer. How far does she run? *(Lesson 4-4)*

60. Solve a Simpler Problem The figure below is to be painted on a wall as part of a mural. Find the area of the figure. *(Lesson 4-5A)*

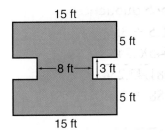

61. Money Matters Drew bought a new car and financed it over a 48-month period. The total of all the payments he will make is $15,336.96. How much is each car payment? *(Lesson 4-5)*

62. Money Matters Mariano wants to collect $131.25 to replace his friend's CD player that was lost in a fire. He is asking everyone to contribute $1.25. How many people will he need to collect from? *(Lesson 4-7)*

Alternative Assessment

Open Ended

Suppose you run a landscape company. The label on a bag of fertilizer states that it will cover 3 square meters. How can you determine how many bags to bring with you to fertilize your client's yard?

Suppose one client has a rectangular yard with dimensions of 7.3 meters by 5.2 meters. For how many bags of fertilizer will you charge the client?

Completing the CHAPTER Project

Use the following checklist to make sure your project is complete.

☑ You have included your price chart for multiple items, your prices with tax included, and your senior citizen discount prices.

☑ Your paragraph describes why you chose the product that you did and how you determined its price.

☑ You have designed an advertisement for your product. Be sure to include your business name, a picture of the product, the price of the item, and anything else that you think would help sell the product.

PORTFOLIO   Select an item from this chapter that you feel shows your best work. Place it in your portfolio. Explain why you selected it.

A practice test for Chapter 4 is provided on page 598.

Section One: Multiple Choice

There are twelve multiple-choice questions in this section. Choose the best answer. If a correct answer is *not here*, mark the letter for Not Here.

1. Kenda found 18 crayons. Her classmate Marissa gave her some more, and then she had 42. To find out how many crayons she was given, Kenda wrote $18 + c = 42$. What is the value of c?

   A 60

   B 52

   C 34

   D 24

2. The exponential form of $3 \times 3 \times 7 \times 7 \times 7 \times 11$ is $3^2 \times 7^? \times 11$. What should replace the question mark?

   F 1

   G 2

   H 3

   J 7

3. Julian is 1.54 meters tall. How many centimeters tall is Julian?

   A 0.0154 cm

   B 15.4 cm

   C 154 cm

   D 1,540 cm

4. Fairfield has an average precipitation of 35.22 inches in June and July. Round this amount to the nearest tenth.

   F 40

   G 35.2

   H 35.22

   J 35.3

Please note that Questions 5–12 have five answer choices.

5. A sign in a store window reads as follows:

   ALL ITEMS
   $5.99
   to
   $13.99

   If Manny bought 3 items in the store, what would be a reasonable total cost?

   A $6.00

   B $12.00

   C $24.00

   D $45.00

   E $60.00

6. 8×3.21 is about—

   F 24.

   G 32.

   H 240.

   J 320.

   K 2,568.

7. Jesse wants to buy ice-cream cones for himself and 3 friends. Each cone costs $1.19. Which is the best estimate of the amount of money Jesse will need in order to buy the cones?

   A $1

   B $3

   C $5

   D $7

   E $9

8. Aubrey charges $1.75 per hour for walking her neighbor's dogs. If she works for 8 hours, how much will she earn?
   F $17.50
   G $16.00
   H $14.00
   J $7.50
   K Not Here

9. Three watermelons have weights of 18, 17, and 22 pounds. Which weight is a reasonable average weight of the three watermelons?
   A 10 lb
   B 20 lb
   C 25 lb
   D 45 lb
   E 60 lb

10. Bill's Boy Scout troop collected $439 for the Muscular Dystrophy Association. Kelly's Girl Scout troop collected $672. How much more did Kelly's troop collect than Bill's?
    F $133
    G $233
    H $243
    J $247
    K Not Here

11. Rashid cut a ribbon into 5 equal pieces. If each piece was 2.4 meters long, how long was the ribbon before he cut it?
    A 0.48 m
    B 1.2 m
    C 12 m
    D 120 m
    E Not Here

12. Solve the equation $y = 43.1 + 7.256$.
    F 51.456
    G 50.356
    H 11.566
    J 7.687
    K Not Here

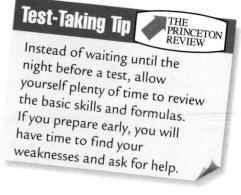

Test-Taking Tip   THE PRINCETON REVIEW

Instead of waiting until the night before a test, allow yourself plenty of time to review the basic skills and formulas. If you prepare early, you will have time to find your weaknesses and ask for help.

Section Two: Free Response

This section contains seven questions for which you will provide short answers. Write your answers on your paper.

13. Five students took Mr. Castillo's history test. The scores were 93, 82, 95, 95, and 75. Find the mean, median and mode of the scores.

14. Maria plans to buy dog food for $15.99, a bone for $5.79, and a dog toy for $2.99. What is the cost of these items before tax is added?

15. To measure the water in a swimming pool, would you use liters or milliliters?

16. Evaluate $(t + w) \div z$ if $t = 0.007$ $w = 0.014$, and $z = 2.5$.

17. 23.5 mL = _?_ L

18. Without computing, explain why 312.28 divided by 29.6 is not 105.5.

19. Which is a better buy, a 306-gram can of soup for $0.42 or a 539-gram can for $0.73? Explain.

interNET **CONNECTION**   **Test Practice** For additional test practice questions, visit:

www.glencoe.com/sec/math/mac/mathnet

CHAPTER 5

Using Number Patterns, Fractions, and Ratios

**What** you'll learn in Chapter 5

- to use divisibility patterns to find the prime factorization and greatest common factor,
- to solve problems by making an organized list,
- to simplify fractions and to express mixed numbers as improper fractions and vice versa,
- to measure length in the customary system,
- to find the least common multiple of two or more numbers and compare and order fractions, and
- to write decimals as fractions and vice versa.

CHAPTER Project

HOT OFF THE PRESS

Imagine that you have been hired as a reviewer for a magazine publisher. Your job is to investigate competitors' magazines to determine what parts of the magazines have articles and what parts have advertisements. In this project, you will use fractions, ratios, and decimals to help you prepare a report that summarizes your findings. You can choose any four magazines for your project.

Getting Started

- Look at the table below. Find the total number of pages in each of the given magazines.
- Use the table to determine which magazine has the least number of pages of articles and advertisements. Which has the greatest number of pages of articles and advertisements?

Magazine	Pages		
	Articles	Ads	Table of Contents
A	58	20	1
B	47	19	1
C	48	2	1
D	80	29	2

Technology Tips

- Use a **spreadsheet** to express fractions as decimals.
- Use computer **software** to make graphs.

*inter*NET CONNECTION For up-to-date information on magazines, visit: www.glencoe.com/sec/math/mac/mathnet

Working on the Project

You can use what you'll learn in Chapter 5 to help you investigate magazines.

Page	Exercise
196	36
219	27
223	Alternative Assessment

Divisibility Patterns

What you'll learn

You'll learn to use divisibility rules for 2, 3, 5, 6, 9, and 10.

When am I ever going to use this?

Knowing how to use divisibility rules can help you form groups when going on a field trip.

green

A popular brand of crayons were first introduced in 1903. Each box contained eight crayons. Today, crayons are sold in boxes of 8, 16, 24, 32, 48, 64, and 96 crayons. Suppose you have a box of 24 crayons. Can you divide the crayons evenly among three children?

To solve this problem, you can divide 24 by 3. $24 \div 3$ means to separate 24 into 3 equal groups.

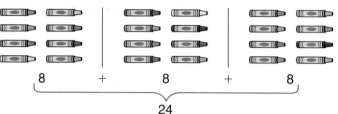

Since $8 + 8 + 8 = 24$, $24 \div 3 = 8$. The quotient, 8, is a whole number, so you can say that 24 is *divisible* by 3. The 24 crayons can be divided evenly among 3 children.

You can test for divisibility mentally by using divisibility rules. The divisibility rules for 2, 3, 5, 6, 9, and 10 are as follows.

A number is divisible by:

- 2 if the ones digit is divisible by 2.
- 3 if the sum of the digits is divisible by 3.
- 5 if the ones digit is 0 or 5.
- 6 if the number is divisible by both 2 and 3.
- 9 if the sum of the digits is divisible by 9.
- 10 if the ones digit is 0.

Examples

1 **Is 46 divisible by 2?**

The ones digit is 6. Since $6 \div 2 = 3$, 6 is divisible by 2. So, 46 is divisible by 2.

2 **Is 428 divisible by 3?**

The sum of the digits is $4 + 2 + 8$, or 14. Since 14 is not divisible by 3, 428 is not divisible by 3.

Check by using a calculator.

428 ÷ 3 = *142.66667* *Since the result is not a whole number, the remainder is not zero.*

428 is not divisible by 3.

3 Is 2,736 divisible by 2, 3, 5, 6, 9, or 10?

2: Yes. The ones digit, 6, is divisible by 2.

3: Yes. The sum of the digits, 18, is divisible by 3.

5: No. The ones digit is not 0 or 5.

6: Yes. The number is divisible by both 2 and 3.

9: Yes. The sum of the digits, 18, is divisible by 9.

10: No. The ones digit, 6, is not 0.

So, of the given numbers, 2,736 is divisible by 2, 3, 6, and 9.

APPLICATION

Real World

4 **Gardening** Lyn has 4 dozen marigolds. She wants to plant the flowers in rows so that each row has the same number of flowers. Can she plant the flowers in 3 equal rows?

Explore   You know that Lyn has 4 dozen marigolds. You need to find out whether she can plant the flowers in 3 equal rows.

Plan   There are 12 in a dozen, so 4 dozen equals 48. Use divisibility rules to see whether 48 is divisible by 3.

Solve   The sum of the digits, 12, is divisible by 3. Thus, she can plant the flowers in 3 equal rows.

Examine   By using a model, you find that 48 flowers can be planted in 3 rows of 16.

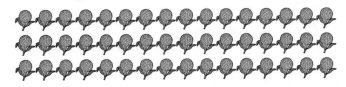

CHECK FOR UNDERSTANDING

Communicating Mathematics

Read and study the lesson to answer each question.

1. *Draw* a picture showing how a box of 24 crayons can be equally divided among 2, 6, or 12 children.

2. *Explain* how to determine whether 129 is divisible by 3.

Math Journal

3. *Write* the divisibility rules for 2, 3, 5, 6, 9, and 10.

Guided Practice

Determine whether the first number is divisible by the second number.

4. 243; 3   5. 2,081; 2   6. 963; 9

State whether each number is divisible by 2, 3, 5, 6, 9, or 10.

7. 17   8. 104   9. 1,032

10. **World Records** On February 28, 1898, in Milton, Massachusetts, Henry Helm Clayton and A.E. Sweetland set a record when their kite reached an altitude of 12,471 feet. Determine whether 12,471 is divisible by 2, 3, 5, 6, 9, or 10.

EXERCISES

Practice

Determine whether the first number is divisible by the second number.

11. 435; 5   12. 240; 10   13. 624; 6

14. 333; 9   15. 1,330; 3   16. 1,752; 2

17. 5,514; 3   18. 507; 5   19. 11,112; 6

State whether each number is divisible by 2, 3, 5, 6, 9, or 10.

20. 208   21. 576   22. 175

23. 397   24. 403   25. 4,077

26. 918   27. 2,860   28. 9,910

29. Is 747 divisible by 9?

30. Is 8,760 divisible by 5?

31. Determine whether 3,517 is divisible by 6.

32. Find a number that is divisible by both 3 and 5.

33. Find a number that is divisible by 2, 9, and 10.

Applications and Problem Solving

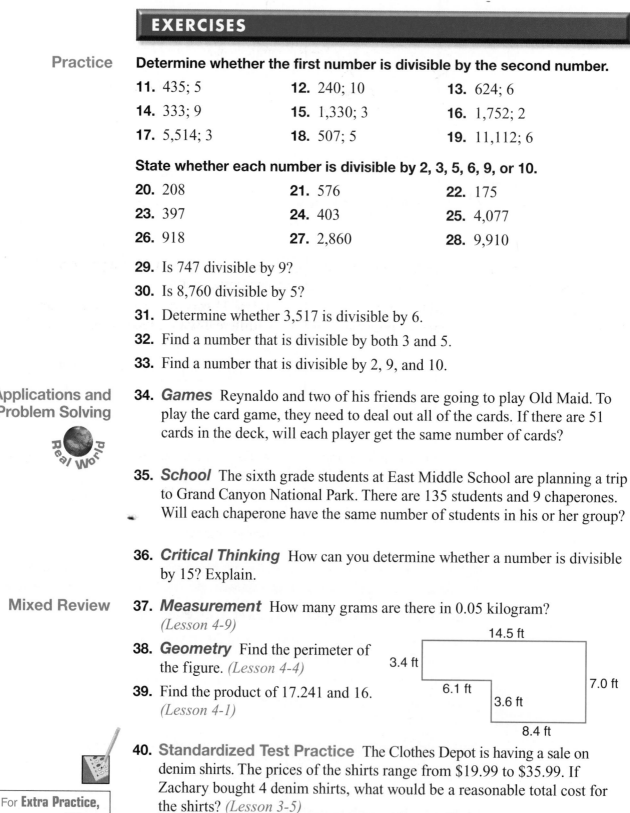

34. **Games** Reynaldo and two of his friends are going to play Old Maid. To play the card game, they need to deal out all of the cards. If there are 51 cards in the deck, will each player get the same number of cards?

35. **School** The sixth grade students at East Middle School are planning a trip to Grand Canyon National Park. There are 135 students and 9 chaperones. Will each chaperone have the same number of students in his or her group?

36. **Critical Thinking** How can you determine whether a number is divisible by 15? Explain.

Mixed Review

37. **Measurement** How many grams are there in 0.05 kilogram? *(Lesson 4-9)*

38. **Geometry** Find the perimeter of the figure. *(Lesson 4-4)*

14.5 ft
3.4 ft
6.1 ft
3.6 ft
7.0 ft
8.4 ft

39. Find the product of 17.241 and 16. *(Lesson 4-1)*

For **Extra Practice**, see page 568.

40. **Standardized Test Practice** The Clothes Depot is having a sale on denim shirts. The prices of the shirts range from $19.99 to $35.99. If Zachary bought 4 denim shirts, what would be a reasonable total cost for the shirts? *(Lesson 3-5)*

　　A $60　　　B $70　　　C $120　　　D $160　　　E $180

5-2A Rectangular Arrays

A Preview of Lesson 5-2

square tiles

A *composite number* is a number that has more than two factors. A *prime number* is a number that has exactly two factors. You can use rectangular arrays to find out whether a number is composite or prime. *You can refer to Lesson 1-6 to review factors.*

TRY THIS

Work with a partner.

1 Determine whether 6 is a prime number or a composite number.

- Use 6 square tiles to build as many different-shaped rectangles as possible.

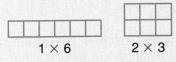

1 × 6   2 × 3

- There is a rectangle that is 1 unit by 6 units and one that is 2 units by 3 units. So, the factors of 6 are 1, 2, 3, and 6.

Since 6 has four factors, it is a composite number.

2 Determine whether 11 is a prime number or a composite number.

- Use 11 square tiles to build as many different-shaped rectangles as possible.

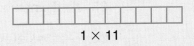

1 × 11

- There is one rectangle whose dimensions are 1 unit by 11 units. So, the factors of 11 are 1 and 11.

Since 11 has exactly two factors, it is a prime number.

ON YOUR OWN

1. Repeat the process with areas of 2 square units through 12 square units.
2. Which numbers of square tiles had only one arrangement?
3. Which numbers of square tiles had more than one arrangement?
4. Is there a relationship between a multiplication table and the number of arrangements you have? If so, describe it.
5. Make a guess about which numbers between 12 and 25 square units can have more than one rectangular shape. Explain why you selected those numbers.
6. *Look Ahead* Write a sentence describing the characteristics of prime and composite numbers.

Prime Factorization

Each rectangle shown below has an area of 12 square units. The dimensions of the rectangles are 1 by 12, 2 by 6, and 3 by 4. So, the factors of 12 are 1, 2, 3, 4, 6, and 12.

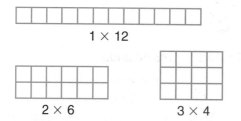

1×12

2×6 3×4

Since 12 has six factors, it is a **composite number**. A composite number is any whole number greater than one with more than two factors.

In Lesson 5-2A, you found that there was only one rectangle you could make with 2, 3, 5, 7, and 11 square tiles. Numbers such as these are **prime numbers**. A prime number has exactly two factors, 1 and the number itself.

The numbers 0 and 1 are neither prime nor composite. Zero has an endless number of factors. The number 1 has only one factor, itself.

Example **1** Determine whether 18 is a prime or composite number. Draw rectangles to explain your answer.

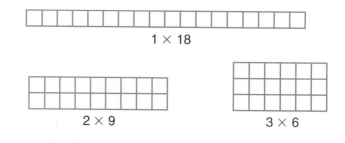

1×18

2×9 3×6

The factors of 18 are 1, 2, 3, 6, 9, and 18. Since 18 has more than two factors, it is a composite number.

Every composite number can be expressed as a product of prime numbers. This is called the **prime factorization** of a number. To find the prime factorization of a number, you can use a **factor tree** or division.

Examples

2 Find the prime factorization of 36.

Method 1 Use a factor tree.

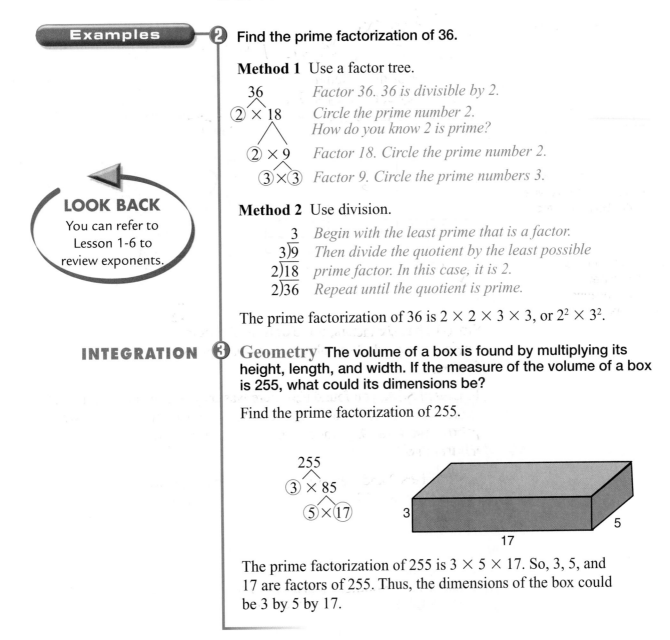

36   *Factor 36. 36 is divisible by 2.*

② × 18   *Circle the prime number 2. How do you know 2 is prime?*

② × 9   *Factor 18. Circle the prime number 2.*

③×③   *Factor 9. Circle the prime numbers 3.*

Method 2 Use division.

$$\begin{array}{r} 3 \\ 3\overline{)9} \\ 2\overline{)18} \\ 2\overline{)36} \end{array}$$

Begin with the least prime that is a factor. Then divide the quotient by the least possible prime factor. In this case, it is 2. Repeat until the quotient is prime.

The prime factorization of 36 is $2 \times 2 \times 3 \times 3$, or $2^2 \times 3^2$.

LOOK BACK

You can refer to Lesson 1-6 to review exponents.

INTEGRATION **3** **Geometry** The volume of a box is found by multiplying its height, length, and width. If the measure of the volume of a box is 255, what could its dimensions be?

Find the prime factorization of 255.

255

③ × 85

⑤×⑰

The prime factorization of 255 is $3 \times 5 \times 17$. So, 3, 5, and 17 are factors of 255. Thus, the dimensions of the box could be 3 by 5 by 17.

CHECK FOR UNDERSTANDING

Communicating Mathematics

Read and study the lesson to answer each question.

1. *Draw* a picture to determine whether 10 is a prime or composite number.

2. *Explain* the difference between a prime number and a composite number.

3. *You Decide* Michelle says that the prime factorization of 120 is $2 \times 2 \times 2 \times 3 \times 5$. Lorenzo says that it is $2 \times 5 \times 3 \times 2 \times 2$. Who is correct? Explain your reasoning.

Tell whether each number is *prime*, *composite*, or *neither*.

4. 17   **5.** 0   **6.** 23   **7.** 57

Find the prime factorization of each number.

8. 49   **9.** 75   **10.** 32   **11.** 104

12. *Geography* The state of North Carolina is made up of 100 counties. Express 100 as a product of primes.

EXERCISES

Practice **Tell whether each number is *prime*, *composite*, or *neither*.**

13. 1	**14.** 15	**15.** 29	**16.** 44
17. 45	**18.** 53	**19.** 56	**20.** 31
21. 87	**22.** 110	**23.** 93	**24.** 114

Find the prime factorization of each number.

25. 42	**26.** 81	**27.** 65	**28.** 38
29. 17	**30.** 24	**31.** 18	**32.** 40
33. 102	**34.** 97	**35.** 120	**36.** 48

37. What is the least prime number that is greater than 80?

38. Write 24, 90, and 98 as products of primes using exponents.

39. What is the greatest square of a whole number that divides 72 with a remainder of 0? What is the greatest cube of a whole number that divides 72 with a remainder of 0?

Applications and Problem Solving

Real World

40. *Math History* In 1742, Christian Goldbach of Russia suggested that any even number greater than two could be expressed as the sum of two prime numbers. For example, $28 = 11 + 17$ and $30 = 13 + 17$. Find two prime numbers whose sum is the given number.

     a. 32   **b.** 40   **c.** 58

41. *Number Patterns* Twin primes are two prime numbers that are consecutive odd integers such as 3 and 5, 5 and 7, and 11 and 13. Find all of the twin primes that are less than 100.

42. *Critical Thinking* Every odd number greater than 7 can be expressed as the sum of three prime numbers. Which three prime numbers have a sum of 57?

Mixed Review **43.** Determine whether or not 462 is divisible by 6. *(Lesson 5-1)*

44. **Standardized Test Practice** Seth purchased 3 CDs for $51.12. If each CD sold for the same amount, what was the price of each CD? *(Lesson 4-5)*

     A $15.06   **B** $13.10   **C** $10.98   **D** $17.04   **E** Not Here

45. *Algebra* Evaluate $a - b$ if $a = 7.3$ and $b = 2.94$. *(Lesson 3-6)*

46. *Statistics* Find the mean of the numbers 120, 112, 88, 100, 141, and 147. *(Lesson 2-7)*

47. Write 17^7 as a product. *(Lesson 1-6)*

For **Extra Practice,** see page 569.

GRAPHIC DESIGN

CYBER GRAFIX

I'd like to design graphics for a website.

Kathryn Sharar Prusinski
GRAPHIC DESIGNER

Kathryn Sharar Prusinski is co-owner of CyberGraphix, a company that designs online magazines and other computer-related products. She holds a degree in advertising design and a certificate in advanced computer graphics. Ms. Prusinski designed several online materials for the Smithsonian and Sesame Street.

Most graphic designers have a bachelor's degree in graphic design, art, or art history. Graphic designers should have an understanding of computer technology, especially painting and graphic design tools. Computer and mathematics courses are an essential part of preparing for a career in graphics design.

For more information:
Graphic Arts Technical Foundation
200 Deer Run Road
Sewickley, PA 15143-2600

interNET
CONNECTION
www.glencoe.com/sec/
math/mac/mathnet

Your Turn
Write and design a brochure that describes the advantages of a career in the graphic design industry.

5-3A Make a List

A Preview of Lesson 5-3

Harvi and Kendra are making a banner for spirit week using three sheets of paper. Harvi is trying to figure out how many different banners they can make using two colors of paper. Let's listen in!

Harvi

Our school colors are red and blue so we should use red and blue paper. How many different banners can we make using three sheets of paper?

Well, let's arrange the sheets of paper in different ways and see.

Arrangements

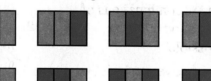

It's going to be hard to remember each arrangement. While you're arranging the paper, I'll list the possibilities.

That's a good idea.

Well, there are eight different possibilities.

RRR   RRB   RBR   RBB
BRR   BBR   BRB   BBB

Kendra

THINK ABOUT IT

Work with a partner.

1. **Analyze** the eight different possibilities. Do you agree or disagree with the possibilities? Explain your reasoning.

2. **Think** of another way that Harvi and Kendra could have solved the problem.

3. **Apply** the **make a list** strategy to solve the following problem.

   Molly's Muffin Shop sells apple, cranberry, blueberry, and strawberry muffins in three sizes: mini, regular, and jumbo. How many different options does a customer have to buy a muffin?

For **Extra Practice,** see page 569.

ON YOUR OWN

4. The third step of the 4-step plan for problem solving asks you to *solve* the problem. ***Explain*** the advantages of making an organized list over a random list when solving the problem.

5. ***Write a Problem*** that can be solved by making a list. Then ask a classmate to solve the problem.

6. ***Look Back*** Explain how you can use the make a list strategy to answer Exercise 41 on page 184.

MIXED PROBLEM SOLVING

Solve. Use any strategy.

STRATEGIES

Look for a pattern.
Solve a simpler problem.
Act it out.
Guess and check.
Draw a diagram.
Make a chart.
Work backward.

7. ***Manufacturing*** A sweater company offers 6 different styles of sweaters in 5 different colors. How many combinations of style and color are possible?

8. ***Money Matters*** Luisa saved $125 in May and $175 in June. How much does she need to save in July to average $150 per month in savings?

9. ***Number Patterns*** Colleen is planning to travel to Dallas, San Antonio, and Houston. She cannot decide in which order to schedule her visits. How many choices does she have?

10. ***Food*** The graph shows that strawberry and grape jelly account for more than half of the yearly jelly sales in the United States. *About* how much more money is spent on strawberry and grape jelly than the other types of jelly?

Yearly Jelly Sales

$366.2 million

$291.5 million

Strawberry and Grape

All others

Source: Nielsen Marketing Research

11. ***Travel*** The Guerrero's vacation lasted 12 days. They budgeted $75 per night for lodging and $65 per day for food. How much did they budget for these two items for the entire trip?

12. ***Number Sense*** How many different 2-digit numbers can you make using the digits 1, 3, 5, and 7 if no number is repeated?

13. **Standardized Test Practice** Samantha is going on vacation. She has asked her neighbor to feed her 2 dogs while she is gone. Each dog eats 8.5 cans of dog food per week. What else do you need to know to find how many cans of food Samantha should buy before she goes on vacation?

    A the number of ounces in each can of dog food

    B the price of one can of dog food

    C the number of weeks Samantha will be on vacation

    D the breed of each dog

    E the weight of each dog

14. ***Manufacturing*** How many different license plates can you make with the letters M, N, and P and the numbers 1, 2, and 3 if each license plate starts with three letters and ends with three numbers?

Greatest Common Factor

What you'll learn

You'll learn to find the greatest common factor of two or more numbers.

When am I ever going to use this?

Knowing how to find the greatest common factor of two or more numbers can help you coordinate a parade.

Word Wise

greatest common factor (GCF)

Suppose that 27 flag line and 45 drill team members were selected to participate in the Flag Day parade. The parade director wants all members to line up in rows that have the same number of people. Flag line and drill team members will be in different rows. What is the greatest number of people that can be in each row?

To answer this question, you can use the factors of 27 and 45.

factors of 27: 1, 3, 9, 27
factors of 45: 1, 3, 5, 9, 15, 45

Notice that 1, 3, and 9 are common factors of both 27 and 45. The greatest of the common factors of two or more numbers is called the **greatest common factor (GCF)** of the numbers. So, the greatest common factor of 27 and 45 is 9.

The greatest number of people that can be in each row is 9.

To find the GCF of two or more numbers, you can make a list or use prime factorization.

Method 1 Make a list.
- List all of the factors of each number.
- Identify the common factors.
- The greatest of the common factors is the GCF.

Method 2 Use prime factorization.
- Write the prime factorization of each number.
- Identify all of the common prime factors.
- The product of the common prime factors is the GCF.

Example **1** Find the GCF of 24 and 32 by making a list.

factors of 24: 1, 2, 3, 4, 6, 8, 12, 24   *List all of the factors*
factors of 32: 1, 2, 4, 8, 16, 32   *of each number.*

The common factors are 1, 2, 4, and 8.
The GCF of 24 and 32 is 8.

---

2 Find the GCF of 36 and 54 by using prime factorization.

36
②× 18
②× 9
③×③

54
②× 27
③× 9
③×③

Write the prime factorization of each number.

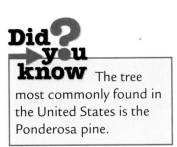

The common prime factors are 2, 3, and 3.

The GCF of 36 and 54 is 2 × 3 × 3, or 18.

CONNECTION

3 **Life Science** Kaja has 32 fir, 48 cedar, and 80 maple tree seedlings to sell at the annual Agriculture Fair. He wants to display them in rows with the same number of seedlings in each row with no mixing of rows.

a. What is the greatest number of seedlings that he could put in each row?

b. How many rows of each tree seedlings will there be?

a. factors of 32: 1, 2, 4, 8, 16, 32
factors of 48: 1, 2, 3, 4, 6, 8, 12, 16, 24, 48
factors of 80: 1, 2, 4, 5, 8, 10, 16, 20, 40, 80

List all of the factors of each number.

The common factors are 1, 2, 4, 8, and 16.
The GCF of 32, 48, and 80 is 16.
So, Kaja could put 16 tree seedlings in a row.

b. Since 2 × 16 = 32, there would be 2 rows of fir seedlings.
Since 3 × 16 = 48, there would be 3 rows of cedar seedlings.
Since 5 × 16 = 80, there would be 5 rows of maple seedlings.

CHECK FOR UNDERSTANDING

Communicating Mathematics

Read and study the lesson to answer each question.

1. **Explain** how to find the GCF of 12 and 18.

2. **Tell** how the GCF of two numbers could be one of the numbers. Give two examples.

Guided Practice

Find the GCF of each set of numbers by making a list.

3. 10, 15   **4.** 8, 30   **5.** 14, 35, 84

Find the GCF of each set of numbers by using prime factorization.

6. 15, 45   **7.** 8, 88   **8.** 16, 24, 72

9. **School** Carena has two rolls of streamers to use in decorating the school gym for a pep rally. One roll is 64 feet long, and the other roll is 72 feet long. If she wants to make all of the streamers the same length, what is the greatest length each streamer can be?

Practice

Find the GCF of each set of numbers by making a list.

10. 8, 32   **11.** 6, 44   **12.** 42, 56

13. 75, 30   **14.** 9, 18, 42   **15.** 22, 55, 88

Find the GCF of each set of numbers by using prime factorization.

16. 16, 56   **17.** 17, 34   **18.** 35, 65

19. 96, 108   **20.** 16, 52, 76   **21.** 12, 18, 60

Find the GCF of each set of numbers using either method.

22. 15, 36   **23.** 21, 45   **24.** 42, 90

25. 34, 85   **26.** 6, 57, 99   **27.** 19, 95, 152

28. List all of the factors of 54 and 90. Then list the common factors.

29. Find the GCF of 39, 48, and 51.

30. Write two numbers whose GCF is 18.

Real World

Applications and Problem Solving

31. *Remodeling* Shawn is covering a portion of his bathroom wall with equal-sized ceramic square tiles. The portion of the wall to be tiled measures 16 inches wide and 72 inches long.

    a. What is the largest square tile that can be used so that no tiles will need to be cut?

    b. What is the total number of tiles Shawn will need?

32. *Business* Marsha has 45 apples, 75 pears, and 105 oranges to sell at her family's farm market. She wants to put the fruit in bags so that there are the same number of pieces of each fruit in each bag without mixing the fruit.

    a. What is the greatest number of pieces of fruit that can be put in each bag?

    b. How many bags of each kind of fruit will there be?

33. *Critical Thinking* Two composite numbers that have 1 as their only common factor are said to be *relatively prime*. Find two composite numbers less than 25 that are relatively prime.

Mixed Review

34. *Standardized Test Practice* Which expression represents the prime factorization of 378? *(Lesson 5-2)*

    A $2 \times 3 \times 7^3$ **B** $2^3 \times 3 \times 7$ **C** $2 \times 3^3 \times 7$ **D** $2^3 \times 3^3 \times 7^3$

35. *Decorating* Antonieta is buying new carpet for her bedroom. If her bedroom measures 5.3 yards by 4.7 yards, how many square yards of carpeting will she need? *(Lesson 4-4)*

36. Estimate $4.231 + 3.98 + 4 + 4.197 + 3.76$. *(Lesson 3-5)*

37. Write *nine and sixteen thousandths* as a decimal. *(Lesson 3-1)*

For **Extra Practice,** see page 569.

COOPERATIVE LEARNING

5-4A Fractions and Ratios

A Preview of Lesson 5-4

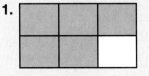 ruler

You have used models to help you understand decimals. You can also use models to help you understand ratios and fractions. A *ratio* is a comparison of two numbers by division. The ratio comparing 3 to 4 can be stated as 3 out of 4, 3 to 4, 3:4, or $\frac{3}{4}$.

TRY THIS

Work with a partner.

1 Model the ratio 1 *out of* 4.
- Copy the rectangle shown.

- Separate the rectangle into four equal parts.

- Shade 1 part.

One out of four parts is shaded.

Therefore, the model represents the ratio *1 out of 4*. Note that the model also represents the fraction $\frac{1}{4}$.

2 Model the fraction $\frac{3}{5}$.
- Copy the rectangle shown.

- Separate the rectangle into five equal parts.

- Shade 3 parts.

Three out of five of the parts are shaded.

Therefore, the model represents the fraction $\frac{3}{5}$ as well as the ratio *3 out of 5*.

ON YOUR OWN

Write a ratio and a fraction for each model shown.

1.

2.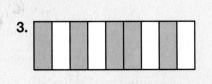

3.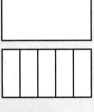

Work with a partner.

3 Compare $\frac{4}{6}$ and $\frac{2}{3}$.

- Draw two identical rectangles as shown.

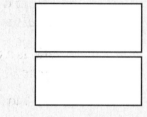

- Separate the upper rectangle into 6 equal parts. Separate the lower rectangle into 3 equal parts.

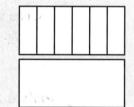

- Shade 4 parts of the upper rectangle and 2 parts of the lower rectangle.

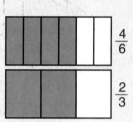

$\frac{4}{6}$

$\frac{2}{3}$

Since the areas shaded are the same, the fractions are equivalent.

Write the pair of fractions represented by each model.

4.

5.

Draw models for each pair of fractions. Then tell whether the pair are equivalent.

6. $\frac{1}{3}, \frac{2}{6}$

7. $\frac{2}{8}, \frac{1}{4}$

8. *Look Ahead* What fraction is represented by the model shown? Name another fraction that is represented by the model.

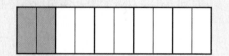

Simplifying Fractions and Ratios

What you'll learn

You'll learn to express fractions and ratios in simplest form.

When am I ever going to use this?

Knowing how to write fractions and ratios in simplest form can help you analyze baseball statistics.

Word Wise

ratio
equivalent fractions
simplest form

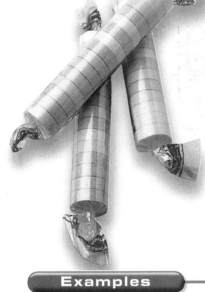

A roll of candy contains about 16 candies. Tyra has a roll in which 4 of the 16 candies are pink. You can compare the number of pink candies to the total number of candies using a **ratio**.

A ratio is a comparison of two numbers by division. The ratio that compares 4 to 16 can be written in several ways.

$$4 \text{ to } 16 \qquad\qquad 4 \text{ out of } 16 \qquad\qquad 4:16$$

Ratios can be expressed as fractions. In this case, the fraction is $\frac{4}{16}$.

You can write the fraction $\frac{4}{16}$ as $\frac{2}{8}$ and also as $\frac{1}{4}$. Fractions that name the same number are called **equivalent fractions**.

Method 1 Use a model.

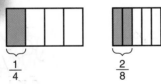

$$\frac{1}{4} \qquad\qquad \frac{2}{8}$$

The rectangles are the same size and the same part or fraction of each rectangle is shaded. So, the fractions are equivalent. That is, $\frac{1}{4} = \frac{2}{8}$.

Method 2 Use paper and pencil.

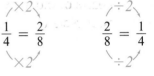

$$\overset{\times 2}{\frac{1}{4} = \frac{2}{8}}_{\times 2} \qquad\qquad \overset{\div 2}{\frac{2}{8} = \frac{1}{4}}_{\div 2}$$

Multiply or divide the numerator and the denominator of a fraction by the same nonzero number.

Examples

Replace each ■ with a number so that the fractions are equivalent.

1 $\frac{3}{8} = \frac{■}{24}$

Since 8 × 3 = 24, multiply the numerator and denominator by 3.

$$\overset{\times 3}{\frac{3}{8} = \frac{■}{24}}_{\times 3}, \text{ so } \frac{3}{8} = \frac{9}{24}.$$

2 $\frac{15}{25} = \frac{3}{■}$

Since 15 ÷ 3 = 5, divide the numerator and denominator by 5.

$$\overset{\div 5}{\frac{15}{25} = \frac{3}{■}}_{\div 5}, \text{ so } \frac{15}{25} = \frac{3}{5}.$$

A fraction is in **simplest form** when the GCF of the numerator and denominator is 1. To write a fraction in simplest form, find the GCF of the numerator and the denominator. Then divide the numerator and denominator by the GCF.

3 Write $\frac{4}{10}$ in simplest form.

factors of 4: 1, 2, 4   *Find the GCF of the numerator*
factors of 10: 1, 2, 5, 10   *and the denominator.*

The GCF of 4 and 10 is 2.

$$\overset{\div 2}{\underset{\div 2}{\frac{4}{10} = \frac{2}{5}}}$$ *Divide the numerator and denominator by the GCF, 2.*

Since the GCF of 2 and 5 is 1, the fraction $\frac{2}{5}$ is in simplest form.

> **Study Hint**
>
> **Technology** You can use a fraction calculator to simplify fractions. For example, to simplify $\frac{4}{16}$, enter
>
> 4 ☐ / ☐ 16 ☐ SIMP ☐ ☐ = ☐ .
>
> Repeat ☐ SIMP ☐ ☐ = ☐
> until the N/D → n/d does not appear on the screen.

4 Write $\frac{18}{21}$ in simplest form.

factors of 18: 1, 2, 3, 6, 9, 18   *Find the GCF of the numerator*
factors of 21: 1, 3, 7, 21   *and the denominator.*

The GCF of 18 and 21 is 3.

$$\overset{\div 3}{\underset{\div 3}{\frac{18}{21} = \frac{6}{7}}}$$ *Divide the numerator and denominator by the GCF, 3.*

Since the GCF of 6 and 7 is 1, the fraction $\frac{6}{7}$ is in simplest form.

INTEGRATION

5 Statistics Approximately 26 out of 100 American households have two or more VCR's. Express the ratio 26:100 in simplest form.

factors of 26: 1, 2, 13, 26   *Find the GCF.*
factors of 100: 1, 2, 4, 5, 10, 20, 25, 50, 100

The GCF of 26 and 100 is 2.

$$\overset{13}{\underset{50}{\frac{26}{100}}} = \frac{13}{50}$$ *Simplify. Divide both the numerator and the denominator by the GCF, 2.*

In simplest form, the ratio is 13:50.

Communicating Mathematics

Read and study the lesson to answer each question.

1. *Examine* the figure shown.
   a. Write the fraction that represents the shaded part of the figure.
   b. Find the GCF of the numerator and denominator of the fraction.
   c. Write the fraction in simplest form.
   d. Draw a model that represents the fraction in simplest form.

2. *Explain* how you can tell whether a fraction is in simplest form.

Guided Practice

Replace each ▪ with a number so that the fractions are equivalent.

3. $\frac{1}{3} = \frac{▪}{27}$ 4. $\frac{12}{16} = \frac{3}{▪}$ 5. $\frac{▪}{5} = \frac{9}{15}$

State whether each fraction or ratio is in simplest form. If not, write each fraction or ratio in simplest form.

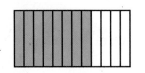

6. $\frac{5}{8}$ 7. 4 to 36

8. $\frac{42}{49}$ 9. 79:100

10. *Statistics* Approximately 36 of every 100 people in the United States listen to compact discs on portable CD players. Express the ratio 36:100 as a fraction in simplest form.

Practice

Replace each ▪ with a number so that the fractions are equivalent.

11. $\frac{1}{2} = \frac{▪}{8}$ 12. $\frac{10}{15} = \frac{2}{▪}$ 13. $\frac{3}{4} = \frac{▪}{12}$

14. $\frac{14}{18} = \frac{▪}{9}$ 15. $\frac{8}{9} = \frac{▪}{27}$ 16. $\frac{30}{35} = \frac{6}{▪}$

17. $\frac{13}{78} = \frac{1}{▪}$ 18. $\frac{4}{5} = \frac{▪}{40}$ 19. $\frac{57}{60} = \frac{19}{▪}$

State whether each fraction or ratio is in simplest form. If not, write each fraction or ratio in simplest form.

20. $\frac{4}{20}$ 21. $\frac{10}{38}$ 22. 8:25   23. 18 out of 24

24. 28 to 77   25. $\frac{15}{100}$ 26. $\frac{27}{54}$ 27. 21:35

28. 41:85   29. $\frac{42}{50}$ 30. $\frac{11}{67}$ 31. 12 out of 48

32. Express the ratio 12 out of 144 in simplest form.

33. Write a fraction that can be expressed as $\frac{2}{3}$ in simplest form.

34. *Games* A popular crossword board game contains 100 tiles, most of which are labeled with a letter. The table shows that the letter O appears on 8 of the tiles.

A-9	J-1	S-4
B-2	K-1	T-6
C-2	L-4	U-4
D-4	M-2	V-2
E-12	N-6	W-2
F-2	O-8	X-1
G-3	P-2	Y-2
H-2	Q-1	Z-1
I-9	R-6	
BLANK-2		

  a. Write a ratio comparing the number of times the letter O appears to the total number of wooden tiles.

  b. Write this ratio in simplest form.

*inter*NET
C O N N E C T I O N
For the latest major league statistics, visit:
www.glencoe.com/sec/math/mac/mathnet

35. *Sports* The chart shows the number of at-bats and the number of hits for certain baseball players during the first few games of the season.

Player	At-Bats	Hits
Gwynn	40	16
Ripken	45	15
Gonzalez	36	11
Williams	42	16
Bonds	36	12
Rodriguez	28	10

  a. For each player, write a fraction in simplest form that shows the number of at-bats compared to the total number of hits.

  b. Which two players had the same "batting average"?

36. *Working on the* **CHAPTER Project** Refer to page 177.

  a. Count the number of pages of articles, advertisements, and table of contents in each issue of your four chosen magazines. Record the data in a table.

  b. For each issue, write a ratio in simplest form that compares the number of pages of articles to the total number of pages, the number of pages of ads to the total number of pages, and the number of pages of table of contents to the total number of pages.

37. *Critical Thinking* A fraction is equivalent to $\frac{3}{4}$ and the sum of the numerator and denominator is 84. What is the fraction?

Mixed Review

38. Find the greatest common factor of 40 and 36. *(Lesson 5-3)*

39. *Standardized Test Practice* Which expression is equivalent to $(2.3 \times 4) + (6.7 \times 4)$? *(Lesson 4-2)*

  A $4(2.3 \times 6.7)$ **B** $6.7(2.3 + 4)$
  C $4(2.3 + 6.7)$ **D** $2.3(4 + 6.7)$

40. *Money Matters* If Andreina works Monday through Friday baby-sitting and makes $10.75 each day she works, how much will she make in three weeks? *(Lesson 4-1)*

41. Order the decimals 27.025, 26.98, 27.13, 27.9, and 27.131 from least to greatest. *(Lesson 3-3)*

For **Extra Practice,** see page 570.

42. Find $45 \div 3 \times 3 - 7 + 12$. *(Lesson 1-4)*

COOPERATIVE LEARNING

5-4B Experimental Probability

A Follow-Up of Lesson 5-4

number cube

A number cube is marked with 1, 2, 3, 4, 5, and 6. If you roll it 60 times, what is the *probability*, or chance, it will show a 2? a 4? a 6? *Experimental probability* is a ratio that compares the number of ways a certain outcome occurs to the total number of outcomes. You can conduct an experiment to find the experimental probability of rolling a 2, a 4, and a 6.

TRY THIS

Work with a partner.

To find the probability that the number cube will show a 2, 4, or 6, follow these steps.

- Copy the chart shown.
- Estimate the number of times the number cube will show a 2, 4, and 6 if it is rolled 60 times.
- Roll the number cube 60 times. Make a tally of each roll.
- Suppose the number cube shows a 4 eleven times out of 60 rolls. The ratio, or experimental probability, of showing a 4 is 11 out of 60 or $\frac{11}{60}$. Based on your results, what is the experimental probability of showing a 2? a 4? a 6? Write a ratio to represent the experimental probability.

	2	4	6
estimate			
actual			

ON YOUR OWN

Six index cards are labeled L, O, C, K, E, and R. Without looking, Jackie chooses a card, records its letter, and replaces it. She repeats the activity 48 times. The chart shows the results of her experiment.

L	O	C	K	E	R
ЖЖ III	ЖЖ II	ЖЖ ЖЖ	ЖЖ ЖЖ	ЖЖ ЖЖ	ЖЖ III

1. What is the experimental probability of choosing an O?
2. What is the experimental probability of choosing an E?
3. What is the experimental probability of choosing an O or a C?
4. **Reflect Back** Find the experimental probability of choosing a vowel. Write the answer in simplest form.

Mixed Numbers and Improper Fractions

Tennis anyone? Tennis racquets come in a variety of grip sizes. Three of these are $4\frac{3}{8}$ inches, $4\frac{1}{4}$ inches, and $3\frac{7}{8}$ inches.

The numbers $4\frac{3}{8}$, $4\frac{1}{4}$, and $3\frac{7}{8}$ are examples of **mixed numbers**.

A mixed number shows the sum of a whole number and a fraction.

In the following Mini-Lab, you'll learn that mixed numbers can be written as fractions.

HANDS-ON MINI-LAB

Work with a partner.

ruler

Try This

- Draw a rectangle like the one shown. Shade the rectangle to represent the whole number 1.
- Draw an identical rectangle beside the first one. Separate this rectangle into four equal parts to show fourths. Shade one part to represent $\frac{1}{4}$.
- Separate the whole number portion into $\frac{1}{4}$s.

Talk About It

1. How many shaded $\frac{1}{4}$s are there?
2. What fraction is equivalent to $1\frac{1}{4}$?

A fraction, like $\frac{5}{4}$ or $\frac{6}{5}$, with a numerator that is greater than or equal to the denominator is called an **improper fraction**. To express a mixed number as an improper fraction, you can use a model.

Example ❶ Express $3\frac{1}{2}$ as an improper fraction.

Change the whole number into halves. Then count the total number of halves.

There are seven $\frac{1}{2}$s in $3\frac{1}{2}$. So, $3\frac{1}{2}$ can be expressed as $\frac{7}{2}$.

Study Hint

Reading Math

A mixed number like $3\frac{1}{2}$ is read as *three and one half*. $3\frac{1}{2}$ means $3 + \frac{1}{2}$.

A shortcut to writing a mixed number as an improper fraction is to first multiply the whole number by the denominator and add the numerator. Then write this sum over the denominator.

$$1\frac{1}{4} = \frac{(1 \times 4) + 1}{4} = \frac{5}{4}$$

Example

CONNECTION ❷ **Life Science** The body of the vampire bat measures $2\frac{3}{4}$ inches. Express the body length of a vampire bat as an improper fraction.

$2\frac{3}{4} = \frac{(2 \times 4) + 3}{4}$ *Multiply 2 by 4 and add 3.*

$= \frac{11}{4}$ *Then write the result over 4.*

The body length of a vampire bat can be expressed as $\frac{11}{4}$ inches.

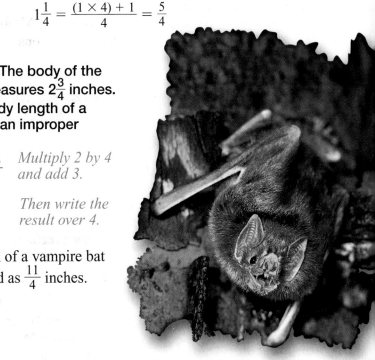

You can also express an improper fraction as a mixed number. To do this, divide the numerator by the denominator.

Example ❸ Express $\frac{7}{3}$ as a mixed number.

Divide 7 by 3.

$$\begin{array}{r} 2 \\ 3\overline{)7} \\ -6 \\ \hline 1 \end{array}$$ *Write the remainder in the numerator of a fraction that has the divisor as the denominator.*

$\frac{7}{3} = 2\frac{1}{3}$

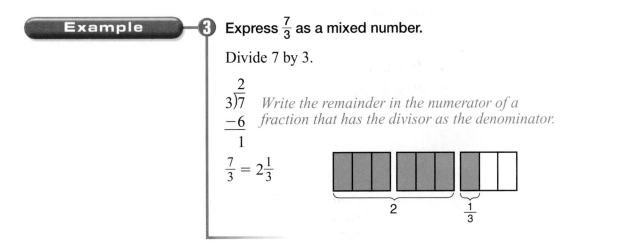

Communicating Mathematics

Read and study the lesson to answer each question.

1. *Define* improper fraction.

2. *Write* a mixed number and an improper fraction for the model shown.

HANDS-ON MATH

3. *Fold* each of three paper plates into four equal parts to show fourths. Shade nine parts. What improper fraction do the shaded parts represent? What mixed number is equivalent to this improper fraction?

Guided Practice

Express each mixed number as an improper fraction.

4. $2\frac{1}{2}$ 5. $1\frac{2}{5}$ 6. $4\frac{3}{4}$

Express each improper fraction as a mixed number.

7. $\frac{11}{4}$ 8. $\frac{13}{2}$ 9. $\frac{24}{6}$

10. Express *nine-fourths* as a mixed number.

EXERCISES

Practice

Express each mixed number as an improper fraction.

11. $3\frac{1}{5}$ 12. $2\frac{3}{8}$ 13. $1\frac{1}{8}$ 14. $4\frac{3}{4}$ 15. $1\frac{7}{9}$

16. $4\frac{5}{6}$ 17. $8\frac{2}{3}$ 18. $5\frac{8}{9}$ 19. $7\frac{3}{5}$ 20. $12\frac{4}{7}$

Express each improper fraction as a mixed number.

21. $\frac{19}{3}$ 22. $\frac{13}{8}$ 23. $\frac{15}{8}$ 24. $\frac{17}{5}$ 25. $\frac{29}{4}$

26. $\frac{23}{6}$ 27. $\frac{19}{9}$ 28. $\frac{32}{8}$ 29. $\frac{25}{7}$ 30. $\frac{48}{11}$

31. Express *six and seven-eighths* as an improper fraction.

32. Find the mixed number that is equivalent to *thirty-eight ninths*.

Applications and Problem Solving

33. *Sports* To win the U.S. Triple Crown, a horse must win the Kentucky Derby, the Preakness Stakes, and the Belmont Stakes. The table shows the distance of each of these races. Express each distance as an improper fraction.

Race	Distance
Kentucky Derby	$1\frac{1}{4}$ mi
Preakness Stakes	$1\frac{3}{16}$ mi
Belmont Stakes	$1\frac{1}{2}$ mi

34. *Entertainment* The table shows the running time of some movies.

Movie	Running Time (min)
A	88
B	76
C	84
D	69

  a. For each movie, write a fraction that compares the running time to the number of minutes in an hour.

  b. Express each fraction as a mixed number in simplest form.

35. *Critical Thinking* Explain, in your own words, how you can determine whether a fraction is less than, equal to, or greater than 1.

Mixed Review **36.** Express $\frac{35}{42}$ in simplest form. *(Lesson 5-4)*

37. Find the prime factorization of 204. *(Lesson 5-2)*

38. **Standardized Test Practice** Jalisa is planning to buy baseball cards that cost $2.27 per pack including tax. How many packs of baseball cards can she buy with $32? *(Lesson 4-6)*

  A 12 **B** 13 **C** 14 **D** 15 **E** Not Here

For **Extra Practice**, see page 570.

39. Write *thirteen hundredths* as a decimal. *(Lesson 3-1)*

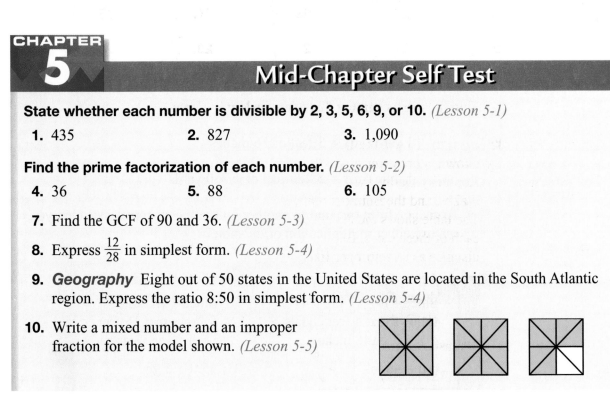

CHAPTER 5

Mid-Chapter Self Test

State whether each number is divisible by 2, 3, 5, 6, 9, or 10. *(Lesson 5-1)*

1. 435 **2.** 827 **3.** 1,090

Find the prime factorization of each number. *(Lesson 5-2)*

4. 36 **5.** 88 **6.** 105

7. Find the GCF of 90 and 36. *(Lesson 5-3)*

8. Express $\frac{12}{28}$ in simplest form. *(Lesson 5-4)*

9. *Geography* Eight out of 50 states in the United States are located in the South Atlantic region. Express the ratio 8:50 in simplest form. *(Lesson 5-4)*

10. Write a mixed number and an improper fraction for the model shown. *(Lesson 5-5)*

Integration: Measurement
Length in the Customary System

What you'll learn

You'll learn to measure line segments and objects with a ruler divided in halves, fourths, and eighths.

When am I ever going to use this?

Knowing how to use the customary system can help you measure everyday objects.

Word Wise

inch
foot
yard
mile

Did you know that, for many people, the length of their arm is about 8 times the length of their index finger?

HANDS-ON MINI-LAB

Work with a partner. 📏 tape measure 〰 string

Try This
- On a long piece of string, mark off eight segments using the length of your index finger. Measure this length to the nearest inch. Record.
- Measure the length of your arm from your shoulder to the end of your index finger to the nearest inch. Record.

Talk About It
1. How does the total length of the eight segments on the string compare to the actual length of your arm?
2. Divide the actual length of your arm by 8 to get an estimate of the length of your index finger. How does this estimate compare to the actual length of your index finger?

One of the most commonly used customary units of length is the **inch**. Some others are the **foot**, **yard**, and **mile**.

$$1 \text{ foot (ft)} = 12 \text{ inches (in.)}$$

$$1 \text{ yard (yd)} = 3 \text{ feet or } 36 \text{ inches}$$

$$1 \text{ mile (mi)} = 5,280 \text{ feet or } 1,760 \text{ yards}$$

To change from one unit to another unit in the customary system, you can use either multiplication or division.

Examples

1 4 ft = _?_ in.

Since 1 ft = 12 in., it follows that 4 feet equals 4 × 12, or 48 inches.

4 ft = 48 in.

2 15 ft = _?_ yd

Since 1 yd = 3 ft, it follows that 15 feet equals 15 ÷ 3, or 5 yards.

15 ft = 5 yd

Sometimes you need to measure objects using units less than an inch. Most rulers are separated into eighths.

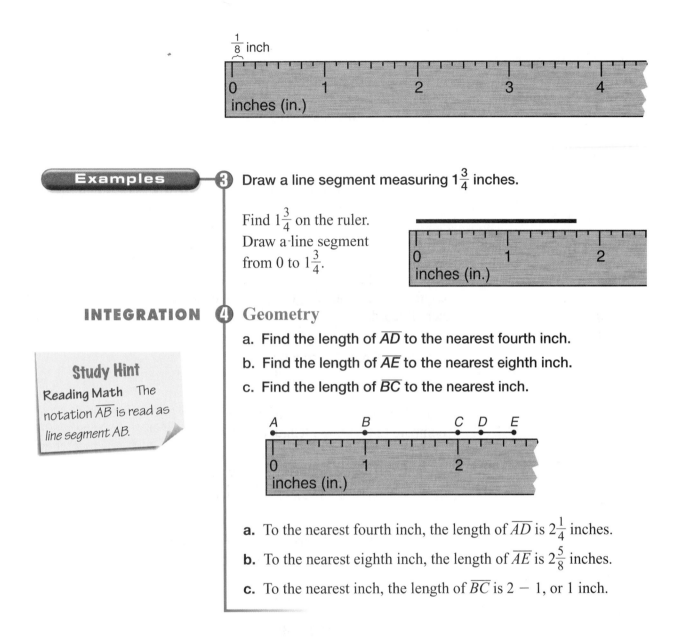

$\frac{1}{8}$ inch

Examples

3 Draw a line segment measuring $1\frac{3}{4}$ inches.

Find $1\frac{3}{4}$ on the ruler. Draw a line segment from 0 to $1\frac{3}{4}$.

INTEGRATION **4** Geometry

a. Find the length of $\overline{AD}$ to the nearest fourth inch.

b. Find the length of $\overline{AE}$ to the nearest eighth inch.

c. Find the length of $\overline{BC}$ to the nearest inch.

Study Hint

Reading Math   The notation $\overline{AB}$ is read as line segment AB.

a. To the nearest fourth inch, the length of $\overline{AD}$ is $2\frac{1}{4}$ inches.

b. To the nearest eighth inch, the length of $\overline{AE}$ is $2\frac{5}{8}$ inches.

c. To the nearest inch, the length of $\overline{BC}$ is $2 - 1$, or 1 inch.

CHECK FOR UNDERSTANDING

Communicating Mathematics

Read and study the lesson to answer each question.

1. *Tell* how you would change 24 inches to feet.

2. *Explain* why $\frac{4}{16}$ inch is the same measure as $\frac{1}{4}$ inch.

HANDS-ON

3. A *cubit* and a *span* are examples of *nonstandard units* of measurement. A cubit is the measure from a person's elbow to the end of the middle finger. A span is the measure from the end of the thumb to the end of the little finger as the hand is outstretched. ***Use*** a ruler or tape measure to find the approximate length of your cubit and your span in the customary system.

Complete.

4. 7 ft = _?_ in.

5. 36 in. = _?_ ft

Draw a line segment of each length.

6. $1\frac{1}{8}$ inches

7. $2\frac{1}{4}$ inches

Find the length of each line segment or object to the nearest half, fourth, or eighth inch.

8.

9.

10. *Auto Mechanics* Mrs. Huang uses wrenches of different sizes when working on cars. She needs to choose a wrench that will tighten the bolt shown. What size wrench will she need: $\frac{1}{2}$ inch, $\frac{5}{8}$ inch, or $\frac{3}{4}$ inch?

EXERCISES

Complete.

11. 60 in. = _?_ ft

12. 3 ft = _?_ yd

13. 3 yd = _?_ in.

14. 78 in. = _?_ ft

15. 2 mi = _?_ ft

16. 225 ft = _?_ yd

Draw a line segment of each length.

17. $1\frac{1}{4}$ inches

18. $1\frac{1}{2}$ inches

19. $\frac{3}{8}$ inch

20. 2 inches

21. $1\frac{5}{8}$ inches

22. $2\frac{1}{2}$ inches

With a family member, gather several objects such as a pencil, a book, and a spoon. Then take turns estimating the length of the objects. Record your estimates. Then measure the lengths to see whose estimates were closer to the measured length.

Find the length of each line segment or object to the nearest half, fourth, or eighth inch.

23. ▬

24.

25.

26.

27. ▬▬

28. ▬▬▬▬▬▬▬

29. Which is greater: 16 inches or $1\frac{1}{2}$ feet?

30. Order the measurements 1 yard, 24 inches, 4 feet, and 40 inches from least to greatest.

31. Order the measurements $\frac{1}{2}$ inch, $\frac{1}{8}$ inch, 1 inch, and $\frac{1}{4}$ inch from greatest to least.

32. *Postal Service* The first animated character to be featured on a 32-cent stamp was issued May 22, 1997, by the United States Postal Service. What is the measure of the width and height of the stamp to the nearest eighth inch?

33. *Geometry* Find the measure of each line segment.

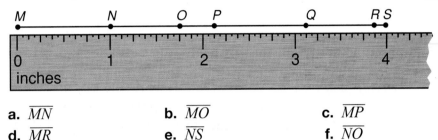

a. $\overline{MN}$ b. $\overline{MO}$ c. $\overline{MP}$
d. $\overline{MR}$ e. $\overline{NS}$ f. $\overline{NO}$

34. *Critical Thinking* How many eighth inches are in a foot? How many fourth inches are in a yard?

Mixed Review

35. Express $5\frac{3}{8}$ as an improper fraction. *(Lesson 5-5)*

36. Find the greatest common factor of 45 and 75. *(Lesson 5-3)*

37. *Standardized Test Practice* The sides of a regulation football field measure 120 yards and about 53.3 yards. About how far does a person have to walk to go all the way around a regulation football field? *(Lesson 4-4)*

A 173.3 yd   **B** 246.6 yd   **C** 346.6 yd   **D** 6,396 yd

38. Solve the equation $n = 3.569 + 781.2$. *(Lesson 3-6)*

For **Extra Practice**,
see page 570.

MATH IN THE MEDIA

Peanuts

1. Do you agree or disagree with the method Sally used to measure Snoopy's mouth? Explain.

2. Tell how Sally got a measurement of 3 inches.

5-7 Least Common Multiple

What you'll learn

You'll learn to find the least common multiple of two or more numbers.

When am I ever going to use this?

You can use the least common multiple to help you lay tile on a floor or wall.

Word Wise

multiple
common multiples
least common multiple
(LCM)

José is shopping for party supplies. He finds a package of 10 plates, a package of 16 napkins, and a package of 8 cups. What is the least number of packages of plates, napkins, and cups José can buy so that he has the same number of plates, napkins, and cups? *This problem will be solved in Example 3.*

To answer this question, you can use multiples. A **multiple** of a number is the product of the number and any whole number.

HANDS-ON
MINI-LAB

Work with a partner. ✂ scissors 📏 ruler

Try This

- Cut six strips of paper 2 inches long by 1 inch wide.
- Cut six strips of paper 3 inches long by 1 inch wide.
- Place the 2-inch strips of paper end to end to make a train.
- Repeat the process with the 3-inch strips of paper. Place this train below the 2-inch strip train.

←2 in.→
←— 3 in.—→

Talk About It

1. Sketch a diagram of the trains. What are the first six multiples of 2 and of 3?
2. Describe where the ends of the strips of paper are lined up. What are the measures of these lengths?
3. At what measurement, other than zero, do the strips of paper line up for the first time?

Notice that 0, 6, and 12 are multiples of both 2 and 3. They are called **common multiples**. The least of the common multiples of two or more numbers, other than zero, is called the **least common multiple (LCM)**. The least common multiple of 2 and 3 is 6.

Example

1 **Determine whether 65 is a multiple of 13.**

multiples of 13: 0, 13, 26, 39, 52, 65, 78,...   *13 × 0 = 0*
                                                    13 × 1 = 13
So, 65 is a multiple of 13.   *13 × 2 = 26*
                                                       ⋮

To find the LCM, you can use either one of the following methods.

Method 1 Make a list.

- List several multiples of each number.
- Identify the common multiples.
- The least of the common multiples is the LCM.

Method 2 Use a calculator.

- List the multiples of the greater number.
- Divide the multiples of the greater number by the lesser number until you get a whole number quotient.

Method 3 Use prime factorization.

- Write the prime factorization of each number.
- Identify all common prime factors. Then find the product of the prime factors using each common prime factor only once and any remaining factors. This product is the LCM.

Examples

2 Find the LCM of 4 and 7.

Method 1 Make a list.

multiples of 4:
0, 4, 8, 12, 16, 20, 24, 28,...

multiples of 7:
0, 7, 14, 21, 28, 35, 42,...

Method 2 Division

multiples of 7:
0, 7, 14, 21, 28, 35, 42,...

$7 \div 4 = 1.75$

$14 \div 4 = 3.5$

$21 \div 4 = 5.25$

$28 \div 4 = 7$ ✓

The LCM of 4 and 7 is 28.

Study Hint

Problem Solving You can make a list to find the multiples.

APPLICATION

3 **Shopping** Refer to the beginning of the lesson. What is the least number of packages of plates, napkins, and cups José can buy so that he has the same number of plates, napkins, and cups?

Explore   You need to find the LCM of 10, 16, and 8.

Plan   Use prime factorization.

Solve $10 = 2 \times 5$
$16 = 2 \times 2 \times 2 \times 2$
$8 = 2 \times 2 \times 2$

The LCM of 10, 16, and 8 is $2 \times 2 \times 2 \times 2 \times 5$, or 80.

Since $10 \times 8 = 80$, he will need 8 packages of plates.
Since $16 \times 5 = 80$, he will need 5 packages of napkins.
Since $8 \times 10 = 80$, he will need 10 packages of cups.

(continued on the next page)

Examine Each package of plates contains 10 plates.
So, eight packages will contain 10 × 8, or 80 plates.

Each package of napkins contains 16 napkins.
So, five packages will contain 16 × 5, or 80 napkins.

Each package of cups contains 8 cups.
So, ten packages will contain 8 × 10, or 80 cups.

Since he will have exactly 80 of each item, the answer is correct.

CHECK FOR UNDERSTANDING

Communicating Mathematics

Read and study the lesson to answer each question.

1. *Define* least common multiple.

2. *Explain* to a classmate how to find the LCM of 8 and 12.

HANDS-ON MATH

3. Using paper strips, make a model to find the LCM of 5 and 6. What is the LCM of 5 and 6?

Guided Practice

Determine whether the first number is a multiple of the second number.

4. 40; 5   5. 32; 8   6. 133; 3

Find the LCM for each set of numbers.

7. 3, 9   8. 5, 12   9. 2, 11   10. 9, 12, 15

11. *Party Planning* Refer to the chart shown. What is the least number of packages of each item that should be purchased so that there are the same number of hot dogs, hot dog buns, and plates?

The Food Shop	
Item	**Quantity**
Hot Dogs	10/pkg
Hot Dog Buns	8/pkg
Plates	20/pkg

EXERCISES

Practice

Determine whether the first number is a multiple of the second number.

12. 15; 9   13. 30; 6   14. 14; 8

15. 32; 16   16. 27; 4   17. 35; 3

18. 84; 7   19. 115; 3   20. 142; 7

Find the LCM for each set of numbers.

21. 6, 12   22. 5, 8   23. 3, 7   24. 18, 24

25. 16, 20   26. 12, 21   27. 13, 16   28. 21, 28

29. 4, 6, 9   30. 15, 25, 75   31. 10, 35, 40   32. 14, 28, 32

33. What is the least common multiple of 3, 9, and 18?

34. Find the LCM of 7, 8, and 28.

35. *Mechanics* Two interlocking
gears have 48 teeth and 28 teeth,
respectively. The lead teeth of
each gear are opposite each
other. How many complete
rotations must the smaller
gear make before the lead
teeth are lined up again?

36. *Remodeling* Mrs. Guzman is replacing the tile on the wall in her
bathroom. She places 3-inch tiles in the first row, 4-inch tiles in the
second row, and 5-inch tiles in the third row. At what point will all three
tiles be lined up?

37. *Critical Thinking* The LCM of two numbers is $2^3 \times 3^2$. Find two pairs
of numbers that fit this description.

Mixed Review

38. **Measurement** Draw a line segment measuring $\frac{5}{8}$ inch. *(Lesson 5-6)*

39. Express the fraction $\frac{47}{5}$ as a mixed number. *(Lesson 5-5)*

40. **Standardized Test Practice** Dom is planning a trip to England and
wishes to exchange his U.S. currency for British pound. If one U.S.
dollar equals 0.623 pounds, about how many pounds will Dom get for
$126? *(Lesson 4-1)*

   A 86   **B** 79   **C** 75   **D** 57   **E** Not Here

41. Estimate $6.291 + 234.38$. *(Lesson 3-5)*

For **Extra Practice,**
see page 571.

42. *Statistics* Find the mean for the scores 19, 17, 28, 32, and 23.
(Lesson 2-7)

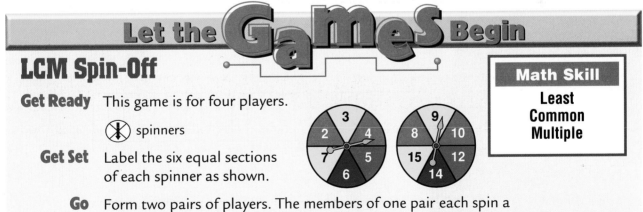

Let the Games Begin

LCM Spin-Off

Get Ready This game is for four players.

 ⊗ spinners

Get Set Label the six equal sections
of each spinner as shown.

Math Skill
Least
Common
Multiple

Go Form two pairs of players. The members of one pair each spin a
spinner. The members in the other pair compete to be the first to
name the LCM of the two numbers. The first member to correctly
name the LCM gets 1 point. Pairs take turns spinning the spinner and
guessing the LCM. The first pair to get 5 points wins.

inter NET
CONNECTION Visit www.glencoe.com/sec/math/mac/mathnet for more games.

Comparing and Ordering Fractions

What you'll learn

You'll learn to compare and order fractions.

When am I ever going to use this?

Knowing how to order fractions can help you determine better buys when grocery shopping.

Word Wise

least common
denominator (LCD)

The graph shows that $\frac{3}{5}$ of smog is caused by cars and $\frac{3}{20}$ is caused by power plant emissions. Is more smog caused by cars or by power plants?

Causes of Smog

Industry	Burning Waste	Cars	Heating Buildings	Power Plants
$\frac{3}{20}$	$\frac{1}{20}$	$\frac{3}{5}$	$\frac{1}{20}$	$\frac{3}{20}$

Source: Environmental Protection Agency

You can solve the problem by comparing the fractions $\frac{3}{5}$ and $\frac{3}{20}$.

One way to compare fractions is to express them as fractions with the same denominator. Any common denominator could be used. But the **least common denominator (LCD)** makes the computation easier.

The least common denominator is the LCM of the denominators. To find the LCD of $\frac{3}{5}$ and $\frac{3}{20}$, you need to find the LCM of 5 and 20.

multiples of 5: 0, 5, 10, 15, 20,…

multiples of 20: 0, 20, 40, 60, 80, …

The LCM of the denominators, 5 and 20, is 20.

$$\frac{3}{5} = \frac{\blacksquare}{20} \xrightarrow{\times 4} \frac{3}{5} = \frac{12}{20}$$

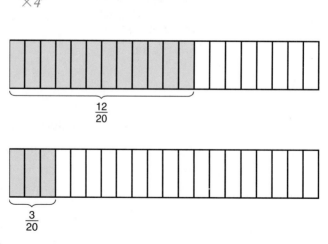

$\frac{12}{20}$

$\frac{3}{20}$

Since $12 > 3$, $\frac{12}{20} > \frac{3}{20}$. Therefore, $\frac{3}{5} > \frac{3}{20}$. So, more smog is caused by cars than by power plants.

Replace each ● with < , > , or = to make a true sentence.

1 $\frac{3}{5}$ ● $\frac{2}{3}$

The LCM of 5 and 3 is 15. Express $\frac{3}{5}$ and $\frac{2}{3}$ as fractions with a denominator of 15.

$$\frac{3}{5} = \frac{\blacksquare}{15}, \text{ so } \frac{3}{5} = \frac{9}{15}.$$ (×3)

$$\frac{2}{3} = \frac{\blacksquare}{15}, \text{ so } \frac{2}{3} = \frac{10}{15}.$$ (×5)

Since $9 < 10$, $\frac{9}{15} < \frac{10}{15}$. Therefore, $\frac{3}{5} < \frac{2}{3}$.

2 $\frac{9}{10}$ ● $\frac{4}{5}$

The LCM of 10 and 5 is 10. Express $\frac{4}{5}$ as a fraction with a denominator of 10.

$$\frac{4}{5} = \frac{\blacksquare}{10} \rightarrow \frac{4}{5} = \frac{8}{10}$$ (×2)

Since $9 > 8$, $\frac{9}{10} > \frac{8}{10}$. Therefore, $\frac{9}{10} > \frac{4}{5}$.

APPLICATION

Real World

3 **Travel** The graph shows how people prefer to pay for travel expenses. Do more people prefer to pay for travel expenses with credit cards or with cash?

Paying for Travel Expenses

Cash	$\frac{1}{2}$
Credit Cards	$\frac{7}{25}$
Traveler's Checks	$\frac{11}{50}$

Source: Cirrus System, Inc.

You need to compare the fractions $\frac{7}{25}$ and $\frac{1}{2}$. The LCM of 25 and 2 is 50. So, express each fraction with a denominator of 50.

$$\frac{7}{25} = \frac{\blacksquare}{50}, \text{ so } \frac{7}{25} = \frac{14}{50}.$$ (×2)

$$\frac{1}{2} = \frac{\blacksquare}{50}, \text{ so } \frac{1}{2} = \frac{25}{50}.$$ (×25)

Since $14 < 25$, $\frac{14}{50} < \frac{25}{50}$. Therefore, $\frac{7}{25} < \frac{1}{2}$.

So, more people prefer to pay with cash than with credit cards.

Communicating Mathematics

Read and study the lesson to answer each question.

1. *Explain* how the LCD can be used to compare fractions.

2. *State* the LCD of $\frac{3}{8}$ and $\frac{5}{6}$.

3. *Write* one or two sentences describing how to compare $\frac{2}{5}$ and $\frac{4}{9}$.

Guided Practice

Find the LCD for each pair of fractions.

4. $\frac{2}{3}, \frac{1}{6}$

5. $\frac{3}{5}, \frac{3}{4}$

Replace each ● with < , > , or = to make a true sentence.

6. $\frac{5}{6}$ ● $\frac{7}{8}$

7. $\frac{6}{9}$ ● $\frac{2}{3}$

8. $\frac{5}{12}$ ● $\frac{3}{4}$

9. *Food* The graph shows where people usually eat dessert in their homes. Do more people eat dessert in their kitchen or in their living room?

Where We Eat Dessert

Kitchen $\frac{3}{10}$

Dining Room $\frac{7}{50}$

Living Room $\frac{9}{50}$

Bedroom $\frac{7}{25}$

Den $\frac{1}{10}$

Source: The Alden Group

Practice

Find the LCD for each pair of fractions.

10. $\frac{1}{2}, \frac{3}{8}$

11. $\frac{3}{4}, \frac{5}{6}$

12. $\frac{7}{12}, \frac{5}{8}$

13. $\frac{3}{5}, \frac{4}{9}$

14. $\frac{1}{6}, \frac{5}{9}$

Replace each ● with < , > , or = to make a true sentence.

15. $\frac{1}{3}$ ● $\frac{2}{9}$

16. $\frac{3}{5}$ ● $\frac{3}{4}$

17. $\frac{10}{16}$ ● $\frac{5}{8}$

18. $\frac{1}{2}$ ● $\frac{3}{8}$

19. $\frac{2}{3}$ ● $\frac{1}{5}$

20. $\frac{9}{15}$ ● $\frac{11}{15}$

21. $\frac{9}{28}$ ● $\frac{5}{14}$

22. $\frac{3}{10}$ ● $\frac{1}{4}$

23. $\frac{5}{7}$ ● $\frac{15}{21}$

24. Which is greater, $\frac{2}{5}$ or $\frac{3}{7}$?

25. Which is less, $\frac{1}{3}$ or $\frac{2}{7}$?

26. Order the fractions $\frac{1}{2}, \frac{3}{5}, \frac{5}{6}, \frac{2}{3}$ from least to greatest.

27. Order the fractions $\frac{1}{6}, \frac{2}{5}, \frac{3}{7}, \frac{3}{5}$ from greatest to least.

28. *Entertainment* According to a research company, during an average week from 8:00 P.M. until 11:00 P.M., male teens spend $\frac{1}{3}$ of this time watching TV, and males over the age of 18 years of age spend $\frac{3}{7}$ of this time watching TV. Who spends more time watching TV in the evening, male teens or males over the age of 18?
Source: Nielson Media Research

29. *Money Matters* Hiroshi needs to purchase a can of garbanzo beans to make a salad for his family picnic. The table shows the cost of garbanzo beans at his neighborhood market. Write a fraction that compares the cost of each can to its weight. Which can is the better buy? Explain.

Isley's Market Garbanzo Beans	
Weight	**Cost**
8 oz	$0.45
16 oz	$0.90

30. *Critical Thinking* I am a fraction in simplest form. I am not improper. My numerator and denominator are twin primes. The sum of my numerator and denominator is equal to a dozen. Who am I? (*Hint*: Twin primes are prime numbers that have a difference of two.)

Mixed Review

31. Find the least common multiple of 15, 20, and 25. *(Lesson 5-7)*

32. *Measurement* How many yards are in 138 feet? *(Lesson 5-6)*

33. Write the fraction $\frac{100}{112}$ in simplest form. *(Lesson 5-4)*

34. **Standardized Test Practice** Kenny drives his dairy truck 7.5 miles to his first stop, 8.4 miles to his second stop, 9.2 miles to his third stop, and 10.5 miles back to the dairy company. How many miles does he travel in his dairy truck altogether? *(Lesson 3-6)*

A 36.5 mi
B 38.4 mi
C 35.6 mi
D 39.2 mi
E Not Here

35. *Algebra* Which of the numbers 21, 23, or 27 is the solution of $41 - m = 18$? *(Lesson 1-7)*

36. *Patterns* Draw the next three figures in the pattern. *(Lesson 1-2)*

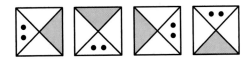

For **Extra Practice,** see page 571.

Writing Decimals as Fractions

What you'll learn

You'll learn to express terminating decimals as fractions in simplest form.

When am I ever going to use this?

Decimals and fractions are used interchangeably in situations where weight and length are measured.

Word Wise

terminating decimal

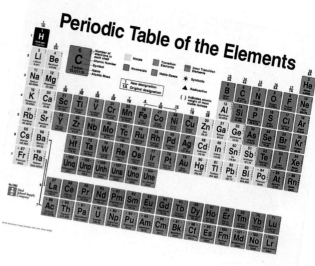

The periodic table gives the symbol, name, atomic number, and atomic mass of each element. The atomic mass of lead (Pb) is 207.2.

The number 207.2 is an example of a **terminating decimal**. Terminating decimals can be written as fractions with denominators of 10, 100, 1,000, and so on. For example, 207.2 can be written as the mixed number $207\frac{2}{10}$ or $207\frac{1}{5}$ in simplest form.

Examples

Express each decimal as a fraction or mixed number in simplest form.

1 0.8

$0.8 = \frac{8}{10}$ *Write the decimal as a fraction. 0.8 means eight tenths.*

$= \frac{\overset{4}{\cancel{8}}}{\underset{5}{\cancel{10}}}$ *Simplify. Divide the numerator and denominator each by the GCF, 2.*

$= \frac{4}{5}$

2 0.28

$0.28 = \frac{28}{100}$ *Write the decimal as a fraction. 0.28 means twenty-eight hundredths.*

$= \frac{\overset{7}{\cancel{28}}}{\underset{25}{\cancel{100}}}$ *Simplify. Divide by the GCF, 4.*

$= \frac{7}{25}$

3 15.125

$15.125 = 15\frac{125}{1,000}$ *Write the decimal as a mixed number. 15.125 means fifteen and one hundred twenty-five thousandths.*

$= 15\frac{\overset{1}{\cancel{125}}}{\underset{8}{\cancel{1,000}}}$ *Simplify. Divide by the GCF, 125.*

$= 15\frac{1}{8}$

Study Hint

Mental Math Here are some commonly used decimal-fraction equivalencies.

$0.5 = \frac{1}{2}$ | $0.25 = \frac{1}{4}$

$0.2 = \frac{1}{5}$ | $0.125 = \frac{1}{8}$

Transportation The F-16C Fighting Falcon can reach a maximum speed of mach 2.05 or 1,320 miles per hour. Express its mach speed as a mixed number in simplest form.

$$2.05 = 2\frac{5}{100} \quad \text{\textit{Write the decimal as a mixed number.}}$$
$$\text{\textit{2.05 means two and five hundredths}}$$

$$= 2\frac{\overset{1}{\cancel{5}}}{\underset{20}{\cancel{100}}} \quad \text{\textit{Simplify. Divide by the GCF, 5.}}$$

$$= 2\frac{1}{20}$$

The mach speed can be expressed as $2\frac{1}{20}$.

CHECK FOR UNDERSTANDING

Communicating Mathematics

Read and study the lesson to answer each question.

1. **Explain** how to express 0.36 as a fraction in simplest form.

2. **Tell** what decimal is represented by the model shown. Then write the decimal as a fraction in simplest form.

3. **You Decide** Ann says that to express 3.55 as a fraction, you use a denominator of 100. Aisha says to use a denominator of 1,000. Who is correct? Explain your reasoning.

Guided Practice

Express each decimal as a fraction or mixed number in simplest form.

4. 0.6 **5.** 0.45 **6.** 2.08 **7.** 4.375

8. **Food** The best-selling packaged cookie in the world has a diameter of 1.75 inches. Express the diameter of the cookie as a fraction in simplest form.

EXERCISES

Practice

Express each decimal as a fraction or mixed number in simplest form.

9. 3.2 **10.** 5.26 **11.** 8.65 **12.** 0.04

13. 5.64 **14.** 6.018 **15.** 2.4 **16.** 4.303

17. 13.009 **18.** 1.234 **19.** 7.89 **20.** 9.82

21. Write *thirty-eight hundredths* as a decimal and as a fraction in simplest form.

22. Express *twelve and sixteen thousandths* as a decimal and as a mixed number in simplest form.

Lesson 5-9 Writing Decimals as Fractions **215**

23. *Travel* The traffic sign shown tells
drivers the distance and direction in
which a landmark is located from an
exit ramp. What fraction of a mile is
each landmark located from the exit?

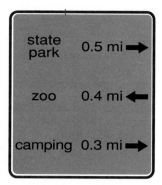

state park	0.5 mi ➡
zoo	0.4 mi ⬅
camping	0.3 mi ➡

24. *History* Greenbacks, or paper money, were
issued during the Civil War to help the North
and South pay for the war. Today, a dollar bill is
6.14 inches long, 2.61 inches wide, and weighs
0.033 ounce. Express the length, width, and
weight of a dollar bill as fractions or mixed
numbers in simplest form.

25. *Physical Science* The table shows
the atomic mass of certain elements.
Express the atomic mass of each
element as a mixed number in
simplest form.

Element	Atomic Mass
Krypton (Kr)	83.8
Selenium (Se)	78.96
Sulfur (S)	32.06
Carbon (C)	12.011

26. *Critical Thinking* *True* or *False*? Every terminating decimal can be
written as a fraction with a denominator that is divisible by 2 and 5.
Explain your reasoning.

Mixed Review

27. Which fraction is greater, $\frac{13}{40}$ or $\frac{3}{7}$? *(Lesson 5-8)*

28. *Geometry* Find the perimeter of the rectangle
if $h = 3.5$ and $g = 2.1$. *(Lesson 4-4)*

h feet

g feet

29. Standardized Test Practice Madison wants to buy banana splits for
herself and 3 friends. Each banana split costs $2.49. Which is the best
estimate of the amount of money Madison will need in order to buy the
banana splits? *(Lesson 1-3)*

    A $4   **B** $6   **C** $8   **D** $10   **E** $12

30. *Entertainment* For a popular dog film, animators drew a total of
6,469,952 spots on the animated dogs. Round this number to the nearest
thousand. *(Lesson 1-3)*

31. *Patterns* Draw the next figure in the pattern shown. *(Lesson 1-2)*

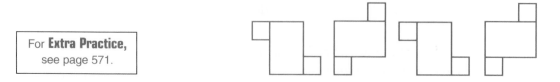

For **Extra Practice,**
see page 571.

5-10 Writing Fractions as Decimals

What you'll learn

You'll learn to express fractions as terminating and repeating decimals.

When am I ever going to use this?

Knowing how to express numbers in different forms can help you interpret weather statistics.

Word Wise

repeating decimal
bar notation

Home gardening has become a popular hobby. The top five vegetables grown in home gardens are tomatoes, peppers, onions, cucumbers, and beans. An estimated $\frac{2}{5}$ of home gardeners grow beans.

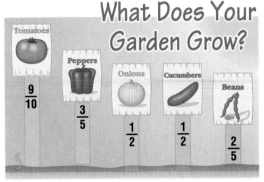

What Does Your Garden Grow?

Tomatoes $\frac{9}{10}$ Peppers $\frac{3}{5}$ Onions $\frac{1}{2}$ Cucumbers $\frac{1}{2}$ Beans $\frac{2}{5}$

Source: National Gardening Association, Gallup Organization

Any fraction can be written as a decimal using division.

$\frac{2}{5}$ means $2 \div 5$.

$$
\begin{array}{r}
0.4 \\
5\overline{)2.0} \\
-20 \\
\hline
0
\end{array}
$$

Write 2 as 2.0.
Place the decimal point in the quotient. Divide as with whole numbers.

The fraction $\frac{2}{5}$ can be written as 0.4. A decimal like 0.4 is called a *terminating decimal* because the division ends, or terminates, when the remainder is zero.

Example ① Express $\frac{1}{8}$ as a decimal.

$$
\begin{array}{r}
0.125 \\
8\overline{)1.000} \\
-8 \\
\hline
20 \\
-16 \\
\hline
40 \\
-40 \\
\hline
0
\end{array}
$$

Divide 1 by 8.

Therefore, $\frac{1}{8} = 0.125$.

Not all fractions are terminating decimals. Decimals like 0.5555555… are called **repeating decimals** because the digits repeat. The **bar notation** $0.\overline{5}$ can be used to indicate that the 5 digit repeats forever. Several repeating decimals are shown.

$$0.433333333… = 0.4\overline{3} \qquad \textit{The digit 3 repeats.}$$
$$2.121212121… = 2.\overline{12} \qquad \textit{The digits 12 repeat.}$$
$$13.567567567… = 13.\overline{567} \qquad \textit{The digits 567 repeat.}$$

Examples

2 Express $\frac{3}{11}$ as a decimal. Use bar notation to show a repeating decimal.

Method 1 Use paper and pencil.

$$\begin{array}{r} 0.2727… \\ 11\overline{)3.0000} \\ -22 \\ \hline 80 \\ -77 \\ \hline 30 \\ -22 \\ \hline 80 \\ -77 \\ \hline 3 \end{array}$$

Divide 3 by 11.

The pattern will continue.

Method 2 Use a calculator.

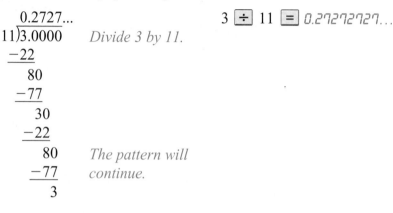

3 ÷ 11 = 0.27272727…

Therefore, $\frac{3}{11} = 0.2727…$ or $0.\overline{27}$.

CONNECTION 3 **Earth Science** The average annual precipitation in Georgia is about $48\frac{3}{5}$ inches. How many inches of precipitation does Georgia average each month?

Divide $48\frac{3}{5}$ by 12. First, express $48\frac{3}{5}$ as a decimal.

48 + 3 ÷ 5 = 48.6 $48\frac{3}{5} = 48 + \frac{3}{5}$

Then divide by 12.

48.6 ÷ 12 = 4.05

Georgia averages about 4.05 inches of precipitation each month.

CHECK FOR UNDERSTANDING

Communicating Mathematics

Read and study the lesson to answer each question.

1. *Explain* to a classmate how to express a fraction as a decimal.

2. *Give* an example of a terminating decimal and a repeating decimal.

Guided Practice

Write each repeating decimal using bar notation.

3. 0.4444444…

4. 10.34343434…

Express each fraction or mixed number as a decimal. Use bar notation to show a repeating decimal.

5. $\frac{3}{8}$ **6.** $2\frac{4}{11}$ **7.** $\frac{7}{12}$

8. Express the fraction $\frac{7}{8}$ as a decimal.

EXERCISES

Practice

Write each repeating decimal using bar notation.

9. 0.77777777…   **10.** 1.33333333…   **11.** 2.45454545…

12. 17.0909090…   **13.** 0.83183183…   **14.** 5.01289289…

Express each fraction or mixed number as a decimal. Use bar notation to show a repeating decimal.

15. $\frac{3}{4}$ **16.** $4\frac{1}{5}$ **17.** $\frac{2}{9}$ **18.** $2\frac{5}{8}$

19. $\frac{5}{11}$ **20.** $7\frac{1}{3}$ **21.** $\frac{11}{12}$ **22.** $\frac{7}{10}$

23. Express *four and one-eleventh* as a decimal.

24. Find the decimal equivalent of the fraction *four fifteenths*.

Applications and Problem Solving

25. *Food* A chocolate drop weighs $\frac{1}{6}$ ounce. Find the decimal equivalent of $\frac{1}{6}$ ounce.

26. *Geography* Japan is a group of islands in the Pacific Ocean. Tokyo is Japan's largest city. The city receives about $5\frac{1}{8}$ inches of rain each month. How many inches of rain does Tokyo receive in a year?

27. *Working on the* CHAPTER Project Refer to the table you made on page 196 in Exercise 36.

   a. Express each of your fractions as a decimal. If necessary, round to the nearest hundredth. Which of your magazines has the highest decimal portion devoted to articles? to advertisements? Which has the lowest decimal portion devoted to articles? to advertisements?

   b. Make a bar graph for the set of data.

28. *Critical Thinking* Tell how you can determine whether a fraction in simplest form will be expressed as a terminating or repeating decimal by looking at the denominator.

Mixed Review

29. *Earth Science* Mercury moves in its orbit at a speed of 29.75 miles per second. Write this speed as a mixed number in simplest form. *(Lesson 5-9)*

30. Find the LCM of 24 and 30. *(Lesson 5-7)*

31. **Standardized Test Practice** Express 24 as a product of primes. *(Lesson 5-2)*

   A $2 \times 4 \times 3$ **B** $2 \times 2 \times 2 \times 6$

   C $2 \times 2 \times 8$ **D** $2 \times 2 \times 2 \times 3$

32. Find $2 + 17 - 16 \div 4$. *(Lesson 1-4)*

For **Extra Practice,** see page 572.

Vocabulary

After completing this chapter, you should be able to define each term, concept, or phrase and give an example or two of each.

Number and Operations
bar notation (p. 218)
common multiples (p. 206)
composite number (pp. 181, 182)
equivalent fractions (p. 193)
factor tree (p. 183)
greatest common factor (GCF) (p. 188)
improper fraction (p. 198)

least common denominator (LCD) (p. 210)
least common multiple (LCM) (p. 206)
mixed number (p. 198)
multiple (p. 206)
prime factorization (p. 183)
prime number (pp. 181, 182)
ratio (pp. 191, 193)
repeating decimal (p. 218)
simplest form (p. 194)
terminating decimal (p. 214)

Measurement
foot (p. 202)
inch (p. 202)
mile (p. 202)
yard (p. 202)

Probability
experimental (p. 197)

Problem Solving
make a list (p. 186)

Understanding and Using the Vocabulary

Choose the letter of the term that best matches each phrase.

1. a comparison of two numbers by division
2. the product of a number and any whole number
3. a number having more than two factors
4. numbers showing the sum of a whole number and a fraction
5. a number having exactly two factors, 1 and itself
6. the LCM of 15 and 9
7. the greatest common factor of 56 and 70

a. multiple
b. prime number
c. 45
d. improper fraction
e. composite number
f. 14
g. ratio
h. mixed number

In Your Own Words

8. *Explain* the relationship between inch, foot, yard, and mile.

Objectives & Examples

Upon completing this chapter, you should be able to:

● use divisibility rules for 2, 3, 5, 6, 9, and 10 *(Lesson 5-1)*

Is 630 divisible by 2, 3, 5, 6, 9, or 10?
It is divisible by 2, 3, 5, 6, 9, and 10.

Review Exercises

Use these exercises to review and prepare for the chapter test.

State whether each number is divisible by 2, 3, 5, 6, 9, or 10.

9. 51
10. 300
11. 423
12. 1,250

Objectives & Examples

Review Exercises

● find the prime factorization of a composite number *(Lesson 5-2)*

16
②× 8
②× 4
②×②

The prime factorization of 16 is
2 × 2 × 2 × 2 or 2^4.

Tell whether each number is *prime*, *composite*, or *neither*.

13. 37   **14.** 78

15. 1   **16.** 47

Find the prime factorization of each number.

17. 54   **18.** 75

19. 124   **20.** 36

● find the greatest common factor of two or more numbers *(Lesson 5-3)*

Find the GCF of 12 and 18.

factors of 12: 1, 2, 3, 4, 6, 12

factors of 18: 1, 2, 3, 6, 9, 18

The GCF of 12 and 18 is 6.

Find the GCF of each set of numbers by making a list.

21. 30, 36   **22.** 39, 26

Find the GCF of each set of numbers by using prime factorization.

23. 18, 28   **24.** 12, 24, 30

● express fractions and ratios in simplest form *(Lesson 5-4)*

Write $\frac{12}{36}$ in simplest form.

factors of 12: 1, 2, 3, 4, 6, 12

factors of 36: 1, 2, 3, 4, 6, 9, 12, 18, 36

The GCF of 12 and 36 is 12.

$$\frac{12}{36} \overset{\div 12}{\underset{\div 12}{=}} \frac{1}{3}$$

In simplest form, the fraction is $\frac{1}{3}$.

Replace each ■ with a number so that the fractions are equivalent.

25. $\frac{5}{6} = \frac{■}{24}$ **26.** $\frac{15}{35} = \frac{3}{■}$

State whether each fraction or ratio is in simplest form. If not, write each fraction or ratio in simplest form.

27. $\frac{15}{18}$ **28.** 2 out of 9

29. 14 to 16   **30.** $\frac{24}{28}$

● express mixed numbers as improper fractions and vice versa *(Lesson 5-5)*

Express $1\frac{2}{3}$ as an improper fraction.

$$1\frac{2}{3} = \frac{(1 \times 3) + 2}{3}$$

$$= \frac{5}{3}$$

Express each mixed number as an improper fraction.

31. $3\frac{3}{5}$ **32.** $4\frac{2}{7}$

Express each improper fraction as a mixed number.

33. $\frac{19}{5}$ **34.** $\frac{36}{7}$

Objectives & Examples

measure line segments and objects with a ruler divided in halves, fourths, and eighths *(Lesson 5-6)*

Draw a line segment measuring $1\frac{3}{8}$ inches.

_____

find the least common multiple of two or more numbers *(Lesson 5-7)*

Find the LCM of 8 and 12.

multiples of 8: 0, 8, 16, 24, 32, …

multiples of 12: 0, 12, 24, 36, …

The LCM of 8 and 12 is 24.

compare and order fractions *(Lesson 5-8)*

$\frac{3}{7}$ ● $\frac{4}{9}$ *The LCM of 7 and 9 is 63.*

$$\frac{3}{7} = \frac{27}{63} \qquad \frac{4}{9} = \frac{28}{63}$$

with $\times 9$ and $\times 7$ shown

Since $\frac{27}{63} < \frac{28}{63}, \frac{3}{7} < \frac{4}{9}$.

express terminating decimals as fractions in simplest form *(Lesson 5-9)*

Express 1.72 as a mixed number in simplest form.

$1.72 = 1\frac{72}{100}$ or $1\frac{18}{25}$

express fractions as terminating and repeating decimals *(Lesson 5-10)*

Express $\frac{5}{6}$ as a decimal.

5 ÷ 6 = 0.833333…

$\frac{5}{6} = 0.8\overline{3}$

Review Exercises

Draw a line segment of each length.

35. $1\frac{3}{4}$ inches   **36.** $2\frac{7}{8}$ inches

37. Find the length of the line segment to the nearest half, fourth, or eighth inch.

_____

38. Determine whether 82 is a multiple of 12.

Find the LCM for each set of numbers.

39. 15, 25   **40.** 28, 35

41. 7, 12   **42.** 12, 15, 20

Find the LCD for each pair of fractions.

43. $\frac{3}{5}, \frac{1}{4}$ **44.** $\frac{5}{6}, \frac{7}{8}$

Replace each ● with $<$, $>$, or $=$ to make a true sentence.

45. $\frac{5}{9}$ ● $\frac{6}{11}$ **46.** $\frac{8}{12}$ ● $\frac{6}{9}$

Express each decimal as a fraction or mixed number in simplest form.

47. 0.8   **48.** 0.04

49. 3.65   **50.** 7.36

Express each fraction or mixed number as a decimal. Use bar notation to show a repeating decimal.

51. $\frac{5}{8}$ **52.** $1\frac{5}{11}$

Applications & Problem Solving

53. Make a List When school is cancelled because of snow, the superintendent has to call four radio stations to let them know. How many ways can the superintendent make the phone calls? *(Lesson 5-3A)*

54. Landscaping Brandon used $10\frac{1}{4}$ bags of mulch around the flower beds in his yard. Write $10\frac{1}{4}$ as an improper fraction. *(Lesson 5-5)*

55. Money Matters Zu-Wang worked $6\frac{2}{5}$ hours last week. If he earns \$4.75 per hour, how much did he earn last week? *(Lesson 5-10)*

56. School The graph shows the results of a survey on field trip preferences. Where do more students prefer to go on their field trip? *(Lesson 5-8)*

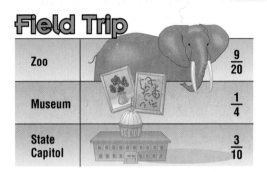

Field Trip

Zoo	$\frac{9}{20}$
Museum	$\frac{1}{4}$
State Capitol	$\frac{3}{10}$

Alternative Assessment

Open Ended

Suppose you are buying party supplies for a younger sibling's birthday party. You find a package of 6 decks of cards, a package of 8 candy bars, and a package of 4 cans of modeling dough. How can you determine the least number of packages of cards, candy, and modeling dough to buy so that you have the same number of each?

Find the least number of packages to buy so that you will have the same number of each item. If there will be 15 children at the party, what is the least number of packages of each item you would have to buy?

Completing the CHAPTER Project

Use the following checklist to make sure your report for your boss is complete.

☑ The table showing the number of pages of articles, ads, and table of contents in each issue of your magazines is included.

☑ The data comparing the number of pages of articles, ads, and table of contents to the total number of pages in each magazine is correct.

☑ The data showing what portion of each magazine is devoted to articles and ads is correct.

☑ The bar graphs of the data are included.

PORTFOLIO   Select one of the assignments from this chapter and place it in your portfolio. Attach a note to it explaining why you selected it.

A practice test for Chapter 5 is provided on page 599.

Section One: Multiple Choice

There are thirteen multiple-choice questions in this section. Choose the best answer. If a correct answer is *not here*, choose the letter for Not Here.

1. Which is a true sentence?

A $0.5 > 0.48$

B $0.48 > 0.5$

C $1.4 > 2.5$

D $5.01 < 4.08$

2. Find the area of the rectangle.

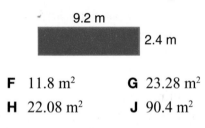

9.2 m

2.4 m

F 11.8 m^2

G 23.28 m^2

H 22.08 m^2

J 90.4 m^2

3. Shalena's birthday is 7 weeks and 5 days away. How many days away is her birthday?

A 49 days

B 54 days

C 35 days

D 12 days

4. Simplify the fraction $\frac{18}{144}$.

F $\frac{1}{6}$

G $\frac{1}{8}$

H $\frac{1}{14}$

J $\frac{1}{12}$

5. Xavier High School has a student enrollment of 2,155 students. What is this number rounded to the nearest thousand?

A 2,000

B 2,100

C 2,150

D 2,200

Please note that Questions 6–13 have five answer choices.

6. Greg purchased 4 paperback books. Each book costs between $4.99 and $8.99. What is a reasonable total cost for the books?

F $10   **G** $30

H $15   **J** $45

K $50

7. Mitena is buying 3 binders that cost $2.89 each and 4 pens that cost $0.49 each. Which expression can be used to find the total cost of the items?

A $3 \times 2.89 \times 4 \times 0.49$

B $3 \times 2.89 + 4 \times 0.49$

C $3 + 2.89 + 4 + 0.49$

D $3 + 2.89 \times 4 + 0.49$

E $3 \times 2.89 \div 4 \times 0.49$

8. The graph shows the number of teams who entered the annual 3-on-3-basketball tournament from 1994 to 1998.

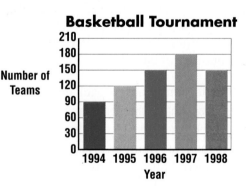

Basketball Tournament

Number of Teams

1994 1995 1996 1997 1998
Year

How many more teams entered the tournament in 1997 than in 1995?

F 60   **G** 30

H 90   **J** 120

K 150

9. If $n = 4.35 \div 1.5$, what is the value of n to the nearest tenth?

   A 2.9

   B 0.09

   C 0.29

   D 1.2

   E Not Here

10. Nashala wants to buy movie tickets for herself and three friends. Each ticket costs $2.95. Which is the best estimate for the total cost of the tickets?

   F $3

   G $6

   H $9

   J $12

   K Not Here

11. Find the value of $23 - 6 + 2 \times 5$.

   A 195

   B 7

   C 27

   D 29

   E Not Here

12. The Framing Center charges $12.95 to frame one 3-inch by 5-inch picture. How much would it cost to frame three 3-inch by 5-inch pictures?

   F $35.95

   G $42.99

   H $38.85

   J $45.79

   K Not Here

13. Solve $11.05 \div 0.65$.

   A 9   **B** 15

   C 14   **D** 17

   E Not Here

inter NET
CONNECTION Test Practice **For additional test practice questions, visit:**

www.glencoe.com/sec/math/mac/mathnet

Test-Taking Tip THE PRINCETON REVIEW

It is a good idea to review formulas before taking a standardized test. Usually the formulas are given in the test booklet, but reviewing the formulas before taking a test may give you an advantage.

Section Two: Free Response

This section contains six questions for which you will provide short answers. Write your answers on your paper.

14. Erika's bowling scores were 119, 134, 135, 125, 143, and 135. What is the mode of her scores?

15. What is the value of $y + 42$ if $y = 87$?

16. Express the decimal 4.456 as a mixed number in simplest form.

17. Levon is shopping for groceries. He buys a 12-pack of cola for $3.98, a bottle of laundry detergent for $4.79, and 2 frozen pizzas for $5.86 each. What is the total cost of the items before tax is added?

18. What is the perimeter of a rectangular bulletin board that is 4 feet wide and 6 feet long?

19. Angel purchased a red, a black, and a blue T-shirt. The total cost for the shirts was $14.37. How much change should he receive from a $20 bill?

CHAPTER 6

Adding and Subtracting Fractions

What you'll
learn in Chapter 6

- to round fractions and mixed numbers,
- to estimate and find sums and differences of fractions and mixed numbers,
- to solve problems by eliminating possibilities, and
- to add and subtract measures of time.

CHAPTER Project

TRAIL TO THE
TREASURE TROVE

Countless stories and movies have been written about people in search of hidden or lost treasure. Often, these people found or bought maps describing the location of the treasure. In this project, you will create and draw your own map for locating a hidden treasure. You will then provide instructions so that someone can find your treasure by using your map.

Getting Started

- On your map, the distance between the starting point and the treasure, in a straight line, must be $5\frac{3}{4}$ inches.
- You will use the landmarks and distances shown in the table when you draw your map.

Landmarks	Distance
starting point to first landmark	$\frac{7}{8}$ in.
center of fountain to tallest oak tree	$1\frac{3}{8}$ in.
tallest oak tree to bench near pond	$2\frac{1}{2}$ in.
bench near pond to seal-shaped rock	$1\frac{5}{8}$ in.
last landmark to treasure	$1\frac{1}{8}$ in.

Technology Tips

- Use a **word processor** to list your instructions.
- Use **computer software** to draw your map.
- Use an **electronic encyclopedia** to learn more about maps and map making.

interNET CONNECTION Research **For up-to-date information on maps and map making, visit:**

www.glencoe.com/sec/math/mac/mathnet

Working on the Project

You can use what you'll learn in Chapter 6 to help you make your treasure map.

Page	Exercise
231	34
252	37
261	Alternative Assessment

Rounding Fractions and Mixed Numbers

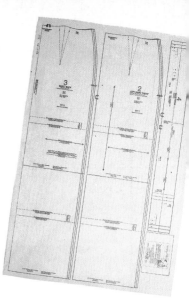

What you'll learn

You'll learn to round fractions and mixed numbers.

When am I ever going to use this?

Knowing how to round fractions and mixed numbers can help you order enough food for a party.

Sophia is making a skirt to wear in her class play. The pattern for the skirt calls for $2\frac{7}{8}$ yards of fabric. She rounds the amount of fabric needed to the nearest whole number because she can use the extra material to make a matching headband.

To round fractions and mixed numbers to the nearest unit, you can use the following guidelines. A number line can help you decide how to round.

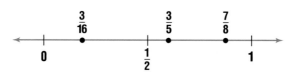

LOOK BACK

You can refer to Lesson 1-3 to review rounding.

- If the numerator is almost as large as the denominator, round the number up to the next whole number.

  $\frac{7}{8}$ rounds to 1.   *7 is almost as large as 8.*

- If the numerator is about half of the denominator, round the fraction to $\frac{1}{2}$.

  $\frac{3}{5}$ rounds to $\frac{1}{2}$.   *3 is about half of 5.*

- If the numerator is much smaller than the denominator, round the number down to the next whole number.

  $2\frac{3}{16}$ rounds to 2.   *3 is much smaller than 16.*

Example ① Round $3\frac{4}{5}$ to the nearest half.

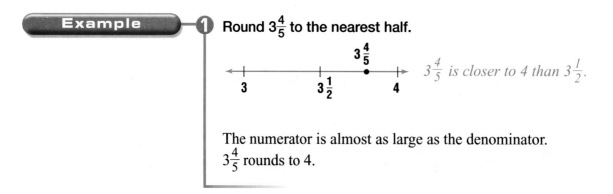

$3\frac{4}{5}$ is closer to 4 than $3\frac{1}{2}$.

The numerator is almost as large as the denominator. $3\frac{4}{5}$ rounds to 4.

2 Round $\frac{3}{8}$ to the nearest half.

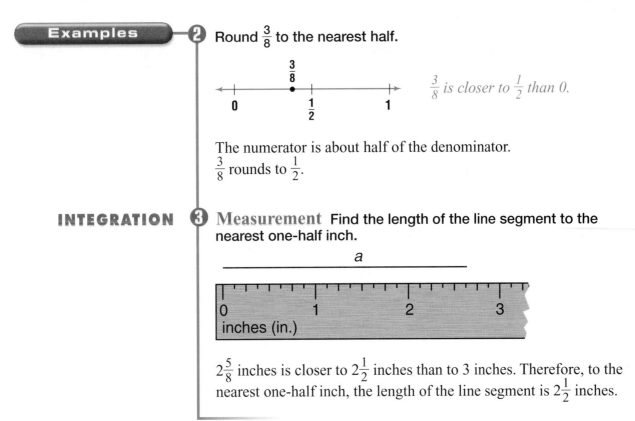

$\frac{3}{8}$ *is closer to* $\frac{1}{2}$ *than 0.*

The numerator is about half of the denominator. $\frac{3}{8}$ rounds to $\frac{1}{2}$.

INTEGRATION

3 **Measurement** Find the length of the line segment to the nearest one-half inch.

$2\frac{5}{8}$ inches is closer to $2\frac{1}{2}$ inches than to 3 inches. Therefore, to the nearest one-half inch, the length of the line segment is $2\frac{1}{2}$ inches.

As shown in the following example, you should round a number down when it is better for a measure to be too small than too large.

Example

APPLICATION

4 **Cooking** Suppose you are camping and you only have one cup of water left to use for cooking. You think the recipe for a pancake mix needs $\frac{2}{3}$ cup of water. Should you round down to one-half cup or round up to one cup?

To avoid making the mix too thin, you should round down to one-half cup of water. You can always add more.

Sometimes it is necessary to round a number up despite what the rule says.

Example

INTEGRATION

5 **Measurement** Ahn is being fitted for a tuxedo to wear in his sister's wedding. His neck measured $14\frac{3}{4}$ inches around. Shirt collars are measured in half inch increments. What size shirt should Ahn order?

If Ahn rounds down, his shirt collar will be too tight. Ahn should round up and order a shirt with a 15-inch collar.

Communicating Mathematics

Read and study the lesson to answer each question.

1. *Draw* a number line that shows how to round $1\frac{2}{3}$ to the nearest half.

2. *Describe* a situation where it would make sense to round a fraction up to the nearest unit.

3. *Write*, in your own words, how you know whether to round a fraction to $0, \frac{1}{2}$, or 1 when rounding to the nearest half.

Guided Practice

Round each number to the nearest half.

4. $\frac{5}{8}$ 5. $3\frac{15}{16}$ 6. $\frac{4}{10}$ 7. $\frac{7}{10}$ 8. $5\frac{2}{5}$

Tell whether each number should be rounded up or down.

9. the weight limit of a bridge

10. the capacity of a container needed to hold $4\frac{3}{4}$ liters of gasoline

11. *Measurement* Find the amount of water in the measuring cup to the nearest one-half cup.

EXERCISES

Practice

Round each number to the nearest half.

12. $1\frac{1}{10}$ 13. $\frac{5}{6}$ 14. $\frac{3}{8}$ 15. $6\frac{2}{3}$ 16. $\frac{1}{5}$

17. $\frac{9}{16}$ 18. $2\frac{4}{5}$ 19. $12\frac{1}{6}$ 20. $7\frac{3}{10}$ 21. $4\frac{2}{9}$

22. $\frac{1}{8}$ 23. $10\frac{7}{10}$ 24. $\frac{1}{9}$ 25. $5\frac{3}{7}$ 26. $\frac{27}{32}$

Tell whether each number should be rounded up or down.

27. the amount of chili pepper needed for a pot of chili

28. a patch for a $2\frac{1}{4}$-inch tear in a pair of jeans

29. the depth of the stream where you are fishing while wearing hip boots

30. the width of blinds to fit in a window $63\frac{3}{4}$ inches wide

31. the weight of cargo on an airplane

32. *Cooking* A recipe to make tacos calls for $1\frac{1}{4}$ pounds of ground beef. Should you buy a $1\frac{1}{2}$-pound package or a 1-pound package? Explain.

33. *Food* The graph shows the results of a survey of the food served at 28 major league baseball stadiums.

a. Write a fraction to represent the number of stadiums that sell each type of food.

b. Which foods are sold in about half of the stadiums?

c. Which foods are sold in almost all of the stadiums?

Take Me Out to the Ballgame

Food	Number of Ballparks
Frozen Yogurt	21
Italian sausage	20
Turkey	17
Chicken	22
Salads	14
Vegetables	5
Fruit	8
Popcorn	23

Number of Ballparks

34. *Working on the* **CHAPTER Project** Refer to the table on page 227. Begin drawing your map. Choose the first two landmarks so that the distance between them is $1\frac{1}{2}$ inches when rounded to the nearest half inch. However, on your map, use the actual distance.

35. *Critical Thinking* Name three mixed numbers that round to $5\frac{1}{2}$.

Mixed Review

36. Write $1\frac{5}{18}$ as a decimal using bar notation. *(Lesson 5-10)*

37. *Measurement* One acre is about 0.0016 square mile. Write this decimal in words. *(Lesson 3-1)*

38. *Statistics* Make a horizontal bar graph and one other type of graph for the set of data. *(Lesson 2-3)*

Raul's Earnings				
Monday	Tuesday	Wednesday	Thursday	Friday
$21.50	$13.75	$19.15	$20.00	$25.50

39. **Standardized Test Practice** A telephone operator answered 500 calls during the first three weeks of a month. She answered 137 calls the fourth week. A reasonable conclusion would be that the telephone operator answered — *(Lesson 1-3)*

A less than 100 calls per week.

B between 100 and 125 calls per week.

C between 126 and 150 calls per week.

D between 151 and 175 calls per week.

E more than 175 calls per week.

For **Extra Practice,** see page 572.

Estimating Sums and Differences

What you'll learn

You'll learn to estimate sums and differences of fractions and mixed numbers.

When am I ever going to use this?

You'll estimate sums and differences when planning what songs to record on a cassette tape.

Maxine is making a cassette tape of some of her favorite songs from her CD collection so that she can listen to them while she jogs. She has already recorded about $49\frac{3}{4}$ minutes of music on her 60-minute cassette tape. The songs she wants to record are about $3\frac{3}{4}$ minutes, $4\frac{1}{6}$ minutes, and $3\frac{1}{3}$ minutes long. Is there enough tape left to record all three songs? *This problem will be solved in Example 3.*

A good way to estimate the sum or difference of fractions is to round each fraction to the nearest half and then add or subtract.

Examples

Estimate.

1 $\frac{9}{16} - \frac{1}{8}$

$\frac{9}{16}$ rounds to $\frac{1}{2}$.

$\frac{1}{8}$ rounds to 0.

Subtract: $\frac{1}{2} - 0 = \frac{1}{2}$

$\frac{9}{16} - \frac{1}{8}$ is about $\frac{1}{2}$.

2 $\frac{5}{6} + \frac{7}{12}$

$\frac{5}{6}$ rounds to 1.

$\frac{7}{12}$ rounds to $\frac{1}{2}$.

Add: $1 + \frac{1}{2} = 1\frac{1}{2}$

$\frac{5}{6} + \frac{7}{12}$ is about $1\frac{1}{2}$.

To estimate sums and differences of mixed numbers, round each number to the nearest whole number.

Example

Real World APPLICATION

3 **Music** Refer to the beginning of the lesson. Is there enough tape left to record all three songs?

Explore You know how long the tape is, how much has been used, and the length of each of the three songs Maxine wants to record.

Plan Subtract to estimate the amount of time left on the tape. Then compare it to an estimate of the total length of the three songs.

Solve $49\frac{3}{4}$ rounds to 50.

Subtract: $60 - 50 = 10$

There are about 10 minutes left on the tape.

$3\frac{3}{4}$ rounds to 4.

$4\frac{1}{6}$ rounds to 4.

$3\frac{1}{3}$ rounds to 3.

Add: $4 + 4 + 3 = 11$

The three songs need about 11 minutes of tape.

The time needed is greater than the time left on the tape. Maxine cannot record all three songs.

Examine   Check by adding the total time for the three songs to the time already used.

Sometimes when estimating sums and differences of fractions and mixed numbers, you need to round all fractions up.

Example ④

Real World **APPLICATION**

Crafts Leon is making a wallet in a craft class. He plans to trim the wallet with a different color of leather. The wallet is $6\frac{3}{4}$ inches long and $3\frac{1}{2}$ inches wide. About how much leather trim does he need?

Leon wants to make sure he buys enough leather. He rounds up.

Since he wants to trim the entire wallet, he needs to estimate the perimeter of the wallet.

$6\frac{3}{4}$ rounds to 7, and $3\frac{1}{2}$ rounds to 4.

Estimate: $7 + 4 + 7 + 4 = 22$

Leon needs about 22 inches of leather trim.

$6\frac{3}{4}$ in.

$3\frac{1}{2}$ in.

CHECK FOR UNDERSTANDING

Communicating Mathematics

Read and study the lesson to answer each question.

1. *Draw* a number line that shows about where $2\frac{3}{4}$ is located.

2. *Explain* how you would estimate $3\frac{3}{4}$ minus $1\frac{1}{3}$.

3. *You Decide* Nina wants to make a square picture frame that is $4\frac{3}{8}$ inches on each side. Should she buy 16 inches of framing material or 20 inches of framing material? Explain.

Guided Practice   **Estimate.**

4. $\dfrac{7}{8} - \dfrac{5}{16}$ 5. $6\dfrac{7}{10} + 3\dfrac{5}{8}$ 6. $\dfrac{1}{2} + \dfrac{4}{5}$

7. $2\dfrac{5}{12} - \dfrac{1}{3}$ 8. $7\dfrac{1}{4} - 2\dfrac{3}{16}$ 9. $5\dfrac{7}{12} + 9\dfrac{5}{6}$

10. *Write a Problem* in which you would need to estimate the difference between $5\frac{2}{3}$ and $1\frac{1}{4}$.

Practice

Estimate.

11. $\frac{7}{8} + \frac{5}{16}$

12. $5\frac{1}{3} - 4\frac{3}{4}$

13. $8\frac{1}{4} - \frac{9}{16}$

14. $7\frac{4}{5} + 3\frac{1}{3}$

15. $\frac{9}{10} + \frac{1}{2}$

16. $3\frac{3}{4} - 2\frac{7}{8}$

17. $\frac{2}{3} + 6\frac{3}{8}$

18. $2\frac{3}{10} - 1\frac{7}{8}$

19. $8\frac{5}{6} + \frac{1}{12}$

20. $11\frac{7}{16} - \frac{5}{9}$

21. $21\frac{5}{8} + 4\frac{3}{4}$

22. $12\frac{4}{5} + \frac{2}{3}$

23. About how much longer than $\frac{5}{6}$ minute is $4\frac{1}{2}$ minutes?

24. Estimate the sum $3\frac{3}{10} + 2\frac{4}{5} + 3\frac{1}{3}$.

25. About how much more is $19\frac{3}{4}$ cups than $10\frac{7}{8}$ cups?

26. Estimate the difference between $1\frac{3}{5}$ and $\frac{1}{6}$.

27. Estimate the sum of $7\frac{1}{3}$, $6\frac{4}{5}$, $6\frac{3}{4}$, $7\frac{1}{10}$, and $6\frac{15}{16}$.

Applications and Problem Solving

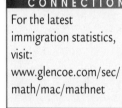

For the latest immigration statistics, visit:
www.glencoe.com/sec/math/mac/mathnet

28. *Geometry* Estimate the perimeter of the rectangle.

$12\frac{1}{8}$ in. | $2\frac{13}{16}$ in.

29. *Carpentry* A board that is $63\frac{5}{8}$ inches long is about how much longer than a board that is $62\frac{1}{4}$ inches long?

30. *Geography* In 1994, about $\frac{3}{8}$ of the U.S. immigrants came from Latin America, about $\frac{1}{5}$ came from Canada and Europe, and about $\frac{1}{25}$ came from Africa and Australia. Asians made up the rest of the immigrant population.

   a. About what fraction of immigrants were from Latin America, Canada, and Europe?

   b. Estimate to find the fraction of immigrants from Asia.

31. *Critical Thinking* The estimate for the sum of two fractions is 1. If 2 is added to each fraction, the estimate for the sum of the two mixed numbers is 4. What are examples of the fractions?

Mixed Review

32. **Standardized Test Practice** What is $6\frac{4}{7}$ rounded to the nearest half? *(Lesson 6-1)*

   A 6   **B** $6\frac{1}{2}$ **C** 7   **D** $7\frac{1}{2}$

33. *Physical Science* Helium has a mass of 0.17 kilogram per cubic meter. Express this mass as a fraction in simplest form. *(Lesson 5-9)*

34. Find the least common multiple of 7 and 21. *(Lesson 5-7)*

For Extra Practice, see page 572.

35. *Measurement* Change 4 kilometers to centimeters. *(Lesson 4-9)*

36. Estimate $39.7 - 28.561$ using rounding. *(Lesson 3-4)*

CARTOGRAPHY

Richard Jimenez
CARTOGRAPHER

Richard Jimenez is a cartographer for the U.S. Geological Survey. He creates maps of different areas of the United States. Mr. Jimenez is very skilled in accurately measuring distances and determining the location of any given place on a map by using coordinates.

A person who is interested in becoming a cartographer should take classes in geography, mathematics, mechanical drawing, and computer science. The mathematics courses should include algebra and trigonometry. A college degree in engineering or physical science is often required. College courses should include technical mathematics, drafting, and mapping.

For more information:
American Congress on Surveying
and Mapping
5410 Grosvenor Lane, Suite 210
Bethesda, MD 20814

interNET CONNECTION
www.glencoe.com/sec/
math/mac/mathnet

I use maps when I go hiking. Someday I'd like to create maps that people can use to travel around the world!

U.S. DEPARTMENT OF THE INTERIOR
GEOLOGICAL SURVEY

Your Turn

Plan a one-week vacation. Write a report about where you will travel, the sites you will visit along the way, and the distance and driving time for your trip. Include in your report a map to refer to while reporting your plan to the class.

6-2B Eliminate Possibilities

A Follow-Up of Lesson 6-2

Cesar and Franco are planning to take the bus to the City Center Mall. They have a map and timetables for the bus routes through their neighborhood. Let's listen in.

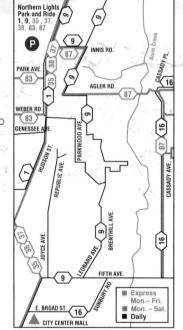

There are seven different bus routes that leave the Northern Lights Park and Ride. Which one do we want?

Since it's Saturday, we can eliminate the three express routes. They only run on weekdays.

And the number 1 and number 83 are east-west routes. We want to go south.

Cesar

Number 87 doesn't go downtown, but we could pick up number 16 at Cassady Avenue.

But then we would have to pay twice. That leaves us with the number 9 bus. Let's check the timetables.

Franco

THINK ABOUT IT

Work with a partner.

1. **Tell** how Cesar and Franco eliminated possibilities to solve the problem.

2. **Explain** how estimating could help you to **eliminate possibilities.**

3. **Apply** the eliminate possibilities strategy to solve the following problem.

   Mrs. Danko bought a ham that weighed $5\frac{1}{4}$ pounds and a turkey that weighed $13\frac{1}{2}$ pounds. About how many pounds of meat did she buy?

   A 8 lb   B 10 lb   C 19 lb   D 20 lb

For **Extra Practice,** see page 573.

ON YOUR OWN

4. The fourth step of the 4-step plan for problem solving asks you to *examine.* **Explain** how you can use the strategy of eliminating possibilities to examine a solution.

5. *Write a Problem* that could be solved by eliminating possibilities.

6. *Reflect Back* Tell how the eliminating possibilities strategy would help you on a multiple-choice test.

MIXED PROBLEM SOLVING

STRATEGIES

Look for a pattern.
Solve a simpler problem.
Act it out.
Guess and check.
Draw a diagram.
Make a chart.
Work backward.

Solve. Use any strategy.

7. Viho's father is 4 times as old as Viho. His grandfather is twice as old as Viho's father. The sum of their three ages is 104. How old is Viho, his father, and his grandfather?

8. **Standardized Test Practice** Last year, Fred's car odometer read 45,500.4. A year later the odometer reads 57,200.9. About how many miles did Fred drive over the past year?

A 1,000,000   **B** 200,000

C 10,000   **D** 1,000

E 100

9. *Money Matters* Candy bars are priced at 2 for 99¢. What will one candy bar cost?

10. *Geometry* Mrs. Coe wants to buy fence to put around her rectangular flower bed. About how much fence will she need?

$13\frac{1}{4}$ ft
$8\frac{1}{2}$ ft

11. *Food* Ani is preparing a meal for her friends. She wants to make three desserts — a cake, some cookies, and an apple pie. To make the cake, she needs $1\frac{1}{2}$ cups of flour. She needs $1\frac{1}{4}$ cups for the cookies, and $2\frac{3}{4}$ cups for the pie. About how much flour does she need to make the three desserts?

12. *Sports* Jill, Kai, Lyndsi, and Mykia are friends. Each of them is on one of the following school teams: basketball, golf, soccer, or tennis. Use the following information to determine who plays on each team.

Jill is shorter than the girl who plays basketball. Kai only likes to play games played on a court. Lyndsi has a problem with her knee and cannot run. Mykia practices kicking a ball as part of her training.

13. *Fashion* A manufacturer offers four different styles of tennis shoes in white, black, and blue. How many combinations of style and color are possible?

14. *Statistics* Is the mean of 123.9, 43.6, 120.89, 502.9, 12.7, and 72.34 about 150 or 15? Explain.

15. **Standardized Test Practice** Anita is making curtains for her room. She needs $12\frac{3}{4}$ yards of material for a larger window and $7\frac{1}{2}$ yards for a smaller window. She bought 30 yards of material. About how much will she have left after making the curtains?

A 21 yd

B 20 yd

C 19 yd

D 10 yd

E 5 yd

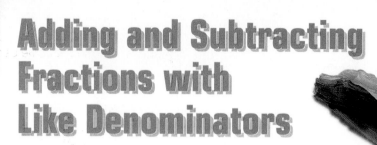

Adding and Subtracting Fractions with Like Denominators

What you'll learn

You'll learn to add and subtract fractions with like denominators.

When am I ever going to use this?

You'll add and subtract fractions with like denominators when you work with equal parts of a whole, such as sections of a garden.

Word Wise

like fractions

John prepared an area for a vegetable garden. He used rope to make a 3-by-5 grid of squares. He planted green beans in the back three squares and corn in two squares. How much of the garden has he planted so far?

You can use grid paper to model this problem.

HANDS-ON MINI-LAB

Work with a partner. 🃏 grid paper 📝 colored pencils

Try This

- On your grid paper, draw a rectangle like the one shown. Since a 3-by-5 grid has 15 squares, each one represents $\frac{1}{15}$.
- With a colored pencil, color three squares to represent the green beans.
- With a different colored pencil, color two more squares to represent the corn.

Talk About It

1. How many squares are colored?
2. What fraction represents the number of colored squares inside the rectangle?
3. If you color four more squares, what fraction would that represent?

Fractions with the same denominator are called **like fractions**. You add and subtract the numerators of like fractions the same way you add and subtract whole numbers. The denominator of the fraction names the units being added or subtracted.

From the Mini-Lab, we know that

3 fifteenths *plus* *2 fifteenths* *equals* *5 fifteenths.*

$$\frac{3}{15} \quad + \quad \frac{2}{15} \quad = \quad \frac{5}{15} \quad \text{or} \quad \frac{1}{3}$$

John has planted $\frac{1}{3}$ of the garden so far.

Adding Like Fractions	To add fractions with like denominators, add the numerators. Use the same denominator in the sum.

Example ❶ Find the sum of $\frac{3}{5}$ and $\frac{4}{5}$.

Estimate: $\frac{1}{2} + 1 = 1\frac{1}{2}$

$$\frac{3}{5} + \frac{4}{5} = \frac{3 + 4}{5}$$

$$= \frac{7}{5}$$

$$= 1\frac{2}{5} \quad \textit{Compared to the estimate, the answer is reasonable.}$$

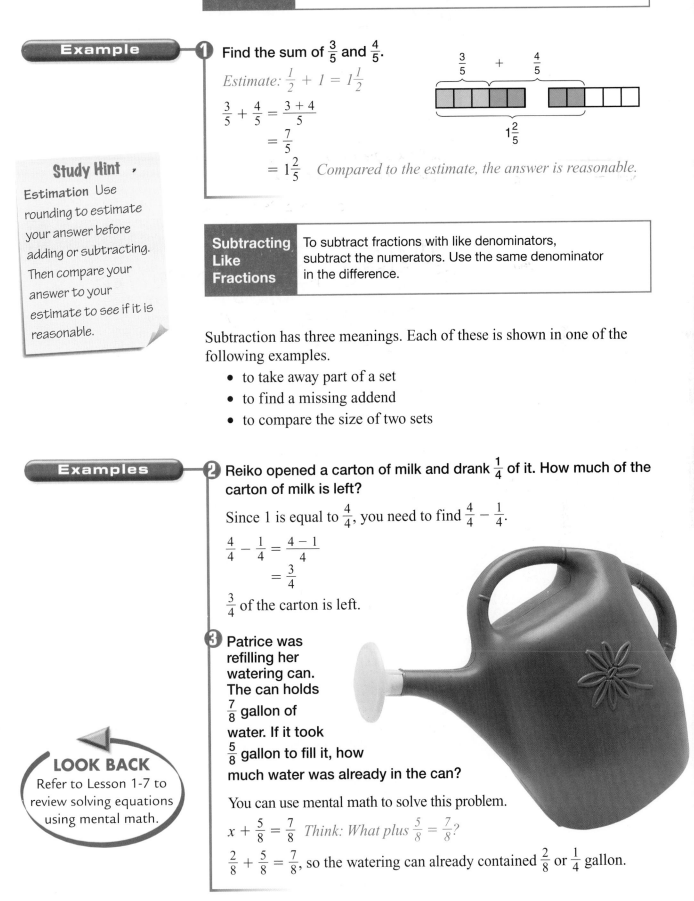

Study Hint

Estimation Use rounding to estimate your answer before adding or subtracting. Then compare your answer to your estimate to see if it is reasonable.

Subtracting Like Fractions	To subtract fractions with like denominators, subtract the numerators. Use the same denominator in the difference.

Subtraction has three meanings. Each of these is shown in one of the following examples.

- to take away part of a set
- to find a missing addend
- to compare the size of two sets

Examples ❷ Reiko opened a carton of milk and drank $\frac{1}{4}$ of it. How much of the carton of milk is left?

Since 1 is equal to $\frac{4}{4}$, you need to find $\frac{4}{4} - \frac{1}{4}$.

$$\frac{4}{4} - \frac{1}{4} = \frac{4 - 1}{4}$$

$$= \frac{3}{4}$$

$\frac{3}{4}$ of the carton is left.

❸ Patrice was refilling her watering can. The can holds $\frac{7}{8}$ gallon of water. If it took $\frac{5}{8}$ gallon to fill it, how much water was already in the can?

LOOK BACK
Refer to Lesson 1-7 to review solving equations using mental math.

You can use mental math to solve this problem.

$x + \frac{5}{8} = \frac{7}{8}$ *Think: What plus $\frac{5}{8} = \frac{7}{8}$?*

$\frac{2}{8} + \frac{5}{8} = \frac{7}{8}$, so the watering can already contained $\frac{2}{8}$ or $\frac{1}{4}$ gallon.

Lesson 6-3 Adding and Subtracting Fractions with Like Denominators **239**

Example

CONNECTION

④ **Geography** According to the 1990 census, about $\frac{12}{100}$ of the population of the United States lives in California. Another $\frac{7}{100}$ of the population lives in Texas. How much more of the population lives in California than in Texas?

$$\frac{12}{100} - \frac{7}{100} = \frac{12-7}{100} \quad \textit{Subtract the numerators.}$$

$$= \frac{5}{100} \text{ or } \frac{1}{20} \quad \textit{Simplify.}$$

About $\frac{1}{20}$ more of the population of the United States lives in California than in Texas.

CHECK FOR UNDERSTANDING

Communicating Mathematics

Read and study the lesson to answer each question.

1. *Tell* a simple rule for adding and subtracting like fractions.

2. *Write* the addition sentence shown by the model.

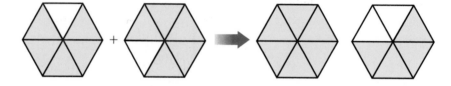

HANDS-ON MATH

3. *Make a model* to show the sum of $\frac{5}{8}$ and $\frac{5}{8}$. Write the sum as a mixed number.

Guided Practice

Add or subtract. Write the answer in simplest form.

4. $\frac{3}{8} + \frac{1}{8}$ 5. $\frac{3}{5} - \frac{2}{5}$ 6. $\frac{7}{10} - \frac{3}{10}$ 7. $\frac{5}{6} + \frac{5}{6}$

8. Find the sum of $\frac{2}{3}$ and $\frac{2}{3}$.   9. What is $\frac{11}{12}$ minus $\frac{5}{12}$?

10. *Food* Angie is making a punch mixture that calls for $\frac{3}{4}$ quart of grapefruit juice, $\frac{3}{4}$ quart of orange juice, and $\frac{3}{4}$ quart of pineapple juice. How much punch does the recipe make?

EXERCISES

Practice

Add or subtract. Write the answer in simplest form.

11. $\frac{7}{8} + \frac{5}{8}$ 12. $\frac{9}{16} - \frac{5}{16}$ 13. $\frac{3}{4} - \frac{3}{4}$ 14. $\frac{3}{5} + \frac{4}{5}$

15. $\frac{1}{3} + \frac{2}{3}$ 16. $\frac{3}{10} + \frac{5}{10}$ 17. $\frac{5}{6} - \frac{1}{6}$ 18. $\frac{11}{12} - \frac{7}{12}$

19. $\frac{5}{9} - \frac{2}{9}$ 20. $\frac{5}{8} - \frac{1}{8}$ 21. $\frac{11}{16} + \frac{13}{16}$ 22. $\frac{7}{15} + \frac{8}{15}$

Solve each equation mentally. Write the solution in simplest form.

23. $a + \frac{2}{5} = \frac{4}{5}$ **24.** $b = \frac{7}{8} + \frac{3}{8}$ **25.** $\frac{8}{9} - c = \frac{4}{9}$

26. How much longer than $\frac{7}{16}$ inch is $\frac{15}{16}$ inch?

27. Find the sum of $\frac{3}{10}$, $\frac{9}{10}$, and $\frac{7}{10}$.

28. How much more is $\frac{3}{4}$ cup than $\frac{1}{4}$ cup?

Applications and Problem Solving

29. *Carpentry* In industrial technology class, Namid made a plaque by gluing a piece of $\frac{3}{8}$-inch oak to a piece of $\frac{5}{8}$-inch poplar. What was the total thickness of the plaque?

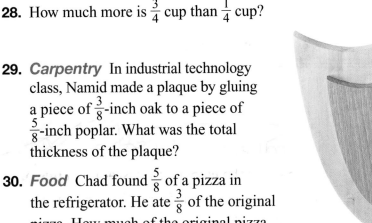

30. *Food* Chad found $\frac{5}{8}$ of a pizza in the refrigerator. He ate $\frac{3}{8}$ of the original pizza. How much of the original pizza is left?

31. *Life Science* The inner organs of an electric eel are located in the first $\frac{1}{5}$ of its body. The rest of the eel contains the organs that produce an electric current. How much of an eel produces an electric current?

32. *Critical Thinking* Find the sum $\frac{1}{20} + \frac{19}{20} + \frac{2}{20} + \frac{18}{20} + \frac{3}{20} + \frac{17}{20} + \ldots + \frac{10}{20} + \frac{10}{20}$. Look for a pattern to help you.

Mixed Review

33. Estimate $4\frac{1}{5} + 1\frac{7}{8}$. *(Lesson 6-2)*

34. **Standardized Test Practice** Order the fractions $\frac{7}{9}, \frac{2}{3}, \frac{2}{6}, \frac{5}{2}$, and $\frac{4}{9}$ from greatest to least. *(Lesson 5-8)*

 A $\frac{7}{9}, \frac{5}{2}, \frac{4}{9}, \frac{2}{6}, \frac{2}{3}$

 B $\frac{7}{9}, \frac{4}{9}, \frac{2}{6}, \frac{2}{3}, \frac{4}{9}$

 C $\frac{5}{2}, \frac{7}{9}, \frac{2}{3}, \frac{4}{9}, \frac{2}{6}$

 D $\frac{2}{6}, \frac{4}{9}, \frac{2}{3}, \frac{7}{9}, \frac{5}{2}$

35. Name a fraction to describe what part of the figure is shaded. *(Lesson 5-4)*

36. *Statistics* The number of days that it rained each month in Cincinnati, Ohio, was recorded for a complete year with the following results: 12, 17, 9, 21, 15, 7, 14, 7, 15, 22, 14, 19. Construct a stem-and-leaf plot for this data. *(Lesson 2-6)*

For **Extra Practice**, see page 573.

COOPERATIVE LEARNING

6-4A Renaming Sums

A Preview of Lesson 6-4

 pennies and nickels

pencils

pens

If you had 7 cassette tapes and 5 CDs, how would you tell someone how many you have all together? Since you can't add tapes and CDs, you need to find a common unit name for them. You could name them 12 things, 12 objects, 12 recordings, or 12 albums. Probably the best unit name would be albums.

In this lab, you will find a common unit name for other common objects.

TRY THIS

Work with a partner.

- Choose one person to be the recorder.
- Put 4 pennies and 3 nickels together. Write as many unit names as you can think of to describe the sum of the pennies and nickels.
- Look at your list. Choose the best unit name for the pennies and nickels.
- Repeat the steps using pencils and pens.

ON YOUR OWN

1. Explain why you need a common unit name to find the sum.

2. Did you find that some unit names fit better than others? Explain why or why not.

3. Make a graph showing the different units' names used by your class for the pennies and nickels.

4. Make a list of different objects that could have a common unit name.

5. *Look Ahead* What do you think you need to do to find the sum of $\frac{1}{2}$ and $\frac{3}{4}$?

242 **Chapter 6** Adding and Subtracting Fractions

Adding and Subtracting Fractions with Unlike Denominators

What you'll learn

You'll learn to add and subtract fractions with unlike denominators.

When am I ever going to use this?

You'll add and subtract fractions with unlike denominators when you analyze graphs.

LOOK BACK

You can refer to Lesson 5-8 to review LCD.

What would you do if you won $100,000? The graph shows how adults ages 35–50 said they would spend a $100,000 prize. What fraction of the money did the people surveyed say they would spend on a new car and a new home? *This problem will be solved in Example 1.*

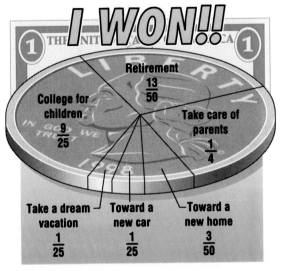

I WON!!

Retirement $\frac{13}{50}$

College for children $\frac{9}{25}$

Take care of parents $\frac{1}{4}$

Take a dream vacation $\frac{1}{25}$

Toward a new car $\frac{1}{25}$

Toward a new home $\frac{3}{50}$

Source: Market Research Institute Survey

To find the sum, you need a common unit name. In Lesson 6-4A, you came up with common unit names for a group of different objects. When you work with fractions with different, or unlike, denominators, you do the same thing.

To find the sum or difference of two fractions with unlike denominators, rename the fractions using the least common denominator (LCD). Then add or subtract and simplify.

Examples

APPLICATION
Real World

1 **Money Matters** **Refer to the beginning of the lesson. What part of the $100,000 prize did people say they would spend on a new home and a new car?**

Add $\frac{3}{50}$ and $\frac{1}{25}$. *The LCD of $\frac{3}{50}$ and $\frac{1}{25}$ is 50.*

$\frac{3}{50} + \frac{1}{25} = \frac{3}{50} + \frac{2}{50}$ *Rename $\frac{1}{25}$ as $\frac{2}{50}$.*

$= \frac{3+2}{50}$

$= \frac{5}{50}$ or $\frac{1}{10}$

The people surveyed said they would spend $\frac{1}{10}$ of the prize on a new home and a new car.

2 Find $\frac{2}{3} - \frac{1}{2}$.

$\frac{2}{3} - \frac{1}{2} = \frac{4}{6} - \frac{3}{6}$ *The LCD of $\frac{2}{3}$ and $\frac{1}{2}$ is 6. Rename $\frac{2}{3}$ as $\frac{4}{6}$ and $\frac{1}{2}$ as $\frac{3}{6}$.*

$= \frac{4-3}{6}$ *Subtract the numerators.*

$= \frac{1}{6}$

In Chapter 1, you evaluated expressions when the variables were whole numbers. Now you can also evaluate expressions when the variables are fractions.

Example 3

INTEGRATION

Algebra Evaluate $p - q$ if $p = \frac{5}{6}$ and $q = \frac{3}{4}$.

$$p - q = \frac{5}{6} - \frac{3}{4} \quad \text{Replace } p \text{ with } \frac{5}{6} \text{ and } q \text{ with } \frac{3}{4}.$$

$$= \frac{10}{12} - \frac{9}{12} \quad \text{Rename } \frac{5}{6} \text{ as } \frac{10}{12} \text{ and } \frac{3}{4} \text{ as } \frac{9}{12}.$$

$$= \frac{10 - 9}{12} \quad \text{Subtract the numerators.}$$

$$= \frac{1}{12}$$

CHECK FOR UNDERSTANDING

Communicating Mathematics

Read and study the lesson to answer each question.

1. **Explain** why you must rename fractions with unlike denominators when you add or subtract them.

2. **Tell** how to find the LCD for $\frac{2}{3}$ and $\frac{5}{8}$.

Guided Practice

Add or subtract. Write the answer in simplest form.

3. $\frac{3}{8} + \frac{1}{16}$ 4. $\frac{3}{5} - \frac{1}{10}$ 5. $\frac{1}{3} - \frac{1}{4}$ 6. $\frac{5}{6} + \frac{3}{4}$

7. What is $\frac{2}{3}$ minus $\frac{5}{12}$?   8. Find the sum of $\frac{1}{2}$ and $\frac{4}{5}$.

9. **Algebra** Evaluate $c + d$ if $c = \frac{1}{4}$ and $d = \frac{5}{8}$.

EXERCISES

Practice

Add or subtract. Write the answer in simplest form.

10. $\frac{5}{8} + \frac{1}{4}$ 11. $\frac{9}{16} - \frac{1}{2}$ 12. $\frac{7}{8} - \frac{3}{4}$ 13. $\frac{3}{5} + \frac{1}{2}$

14. $\frac{2}{3} - \frac{1}{5}$ 15. $\frac{7}{10} - \frac{1}{6}$ 16. $\frac{1}{6} + \frac{3}{4}$ 17. $\frac{11}{12} - \frac{5}{8}$

18. $\frac{1}{4} + \frac{2}{3}$ 19. $\frac{3}{4} - \frac{2}{5}$ 20. $\frac{5}{8} + \frac{5}{6}$ 21. $\frac{5}{9} - \frac{1}{12}$

22. Find the sum of $\frac{3}{8}$ and $\frac{5}{6}$.

23. How much is $\frac{7}{8}$ minus $\frac{1}{2}$?

24. How much more is $\frac{3}{4}$ cup than $\frac{2}{3}$ cup?

25. How much longer than $\frac{9}{16}$ inch is $\frac{7}{8}$ inch?

26. What is the sum of $\frac{2}{3}$, $\frac{5}{8}$, and $\frac{7}{12}$?

Solve each equation. Write the solution in simplest form.

27. $x = \dfrac{3}{10} + \dfrac{9}{20}$ **28.** $\dfrac{4}{5} + \dfrac{1}{2} = y$ **29.** $\dfrac{5}{6} - \dfrac{3}{4} = d$

30. *Algebra* Evaluate $j - k$ if $j = \dfrac{3}{4}$ and $k = \dfrac{1}{3}$.

31. *Algebra* Evaluate $r + s$ if $r = \dfrac{3}{5}$ and $s = \dfrac{7}{10}$.

Applications and Problem Solving

Family Activity

Find the recipes for two of your family's favorite desserts. Make a list of the total amount of each ingredient you would need to make both recipes.

32. *Money Matters* Refer to the graph at the beginning of the lesson. How much bigger is the part that would be spent on taking care of parents than on a dream vacation?

33. *Life Science* Proper diet is essential to raising champion horses. Their diet usually consists of grain and hay. The table shows the prescribed amounts that three horses are fed each time, two times a day. How many cans of grain are needed to feed all three horses per feeding?

Horse's Name	Amount of Grain	Amount of Hay
Penny	$\dfrac{3}{4}$ can	2 slices
Fancy Free	$\dfrac{1}{4}$ can	3 slices
Max	$\dfrac{1}{2}$ can	2 slices

34. *Carpentry* The Cabinet Shop made a desktop by gluing a sheet of oak veneer to a sheet of $\dfrac{3}{4}$-inch plywood. The total thickness of the desktop is $\dfrac{13}{16}$ inch. What was the thickness of the oak veneer?

35. *Critical Thinking* A piece of wire was cut into thirds. One-third was used. One-fifth of the remaining wire was used. The piece that remains is 16 feet long. What was the original length of the wire?

Mixed Review

36. Find $\dfrac{3}{5} - \dfrac{1}{5}$. *(Lesson 6-3)*

37. *Measurement* Find the width of your pencil to the nearest eighth inch. *(Lesson 5-6)*

38. *Geometry* Find the perimeter of the figure. *(Lesson 4-4)*

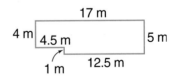

39. **Standardized Test Practice** Alphonse Bielevich caught an Atlantic codfish in New Hampshire that weighed 98.75 pounds. Donald Vaughn caught a Pacific codfish in Alaska that weighed 30 pounds. What is the difference in the weights of the two fish? *(Lesson 3-6)*

   A 98.45 lb

   B 95.75 lb

   C 70 lb

   D 68.75 lb

   E Not Here

For **Extra Practice**, see page 573.

40. *Algebra* Solve the equation $17 - m = 4$. *(Lesson 1-7)*

Lesson 6-4 Adding and Subtracting Fractions with Unlike Denominators   **245**

Adding and Subtracting Mixed Numbers

What you'll learn

You'll learn to add and subtract mixed numbers.

When am I ever going to use this?

You'll add and subtract mixed numbers when you work with lumber dimensions.

Sandy is taking a flight from Atlanta to San Antonio. Travel time is about $2\frac{3}{4}$ hours. When she reports to the gate, they tell her the flight has a $1\frac{1}{2}$-hour delay. In how many hours will she arrive in San Antonio?

You need to find $2\frac{3}{4} + 1\frac{1}{2}$. You can use models to solve the problem.

HANDS-ON MINI-LAB

Work with a partner. ⬭ paper plates ✂ scissors

Try This

- Place two paper plates in front of you to show the whole number 2 in the mixed number $2\frac{3}{4}$. Cut another paper plate into fourths. Place three fourths of the plate in front of you to show the fraction $\frac{3}{4}$ in the mixed number $2\frac{3}{4}$.

- Use more paper plates to show $1\frac{1}{2}$.

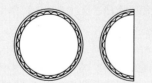

- Combine the pieces to make as many whole paper plates as you can.

Talk About It

1. How many whole paper plates do you have?
2. What fraction is represented by the pieces of paper plate that you have left over?
3. How long will it be before Sandy arrives in San Antonio?
4. Use paper plates to find $2\frac{3}{4} - 1\frac{1}{2}$.

The Mini-Lab suggests the following rule.

Adding and Subtracting Mixed Numbers	1. Add or subtract the fractions. 2. Then add or subtract the whole numbers. 3. Rename and simplify if necessary.

Examples

1 Find $4\frac{2}{3} - 2\frac{1}{3}$. *Estimate: $5 - 2 = 3$*

Subtract the fractions.   Subtract the whole numbers.

$$\begin{array}{r} 4\frac{2}{3} \\ -2\frac{1}{3} \\ \hline \frac{1}{3} \end{array} \quad \rightarrow \quad \begin{array}{r} 4\frac{2}{3} \\ -2\frac{1}{3} \\ \hline 2\frac{1}{3} \end{array}$$

2 Find $12\frac{1}{4} + 3\frac{5}{6}$. *Estimate: $12 + 4 = 16$*

Add the fractions.   Add the whole numbers.

$$\begin{array}{r} 12\frac{1}{4} \\ +3\frac{5}{6} \end{array} \rightarrow \begin{array}{r} 12\frac{3}{12} \\ +3\frac{10}{12} \\ \hline \frac{13}{12} \end{array} \rightarrow \begin{array}{r} 12\frac{3}{12} \\ +3\frac{10}{12} \\ \hline 15\frac{13}{12} \end{array}$$

Rename $\frac{13}{12}$ as $1\frac{1}{12}$.

$$15 + 1\frac{1}{12} = 16\frac{1}{12}$$

INTEGRATION

3 **Algebra** Evaluate $x - y$ if $x = 5\frac{9}{10}$ and $y = 2\frac{1}{2}$.

$$\begin{aligned} x - y &= 5\frac{9}{10} - 2\frac{1}{2} && \textit{The LCM of 2 and 10 is 10.} \\ &= 5\frac{9}{10} - 2\frac{5}{10} && \textit{Rename } 2\frac{1}{2} \textit{ as } 2\frac{5}{10}. \\ &= 3\frac{4}{10} \textit{ or } 3\frac{2}{5} \end{aligned}$$

INTEGRATION

4 **Geometry** Find the perimeter of the triangle.

$6\frac{3}{4}$ in. $2\frac{1}{2}$ in. $7\frac{1}{8}$ in.

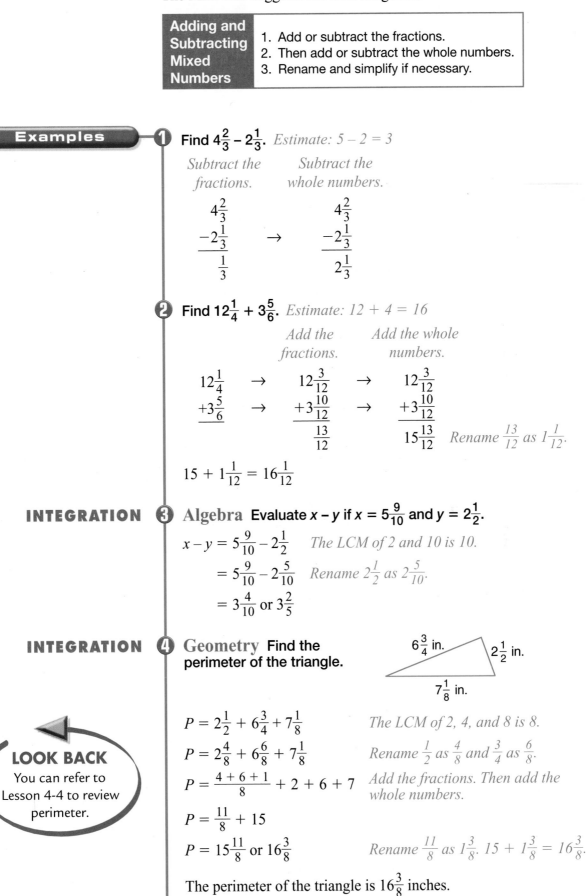

$$P = 2\frac{1}{2} + 6\frac{3}{4} + 7\frac{1}{8} \qquad \textit{The LCM of 2, 4, and 8 is 8.}$$

$$P = 2\frac{4}{8} + 6\frac{6}{8} + 7\frac{1}{8} \qquad \textit{Rename } \frac{1}{2} \textit{ as } \frac{4}{8} \textit{ and } \frac{3}{4} \textit{ as } \frac{6}{8}.$$

$$P = \frac{4 + 6 + 1}{8} + 2 + 6 + 7 \qquad \textit{Add the fractions. Then add the whole numbers.}$$

$$P = \frac{11}{8} + 15$$

$$P = 15\frac{11}{8} \textit{ or } 16\frac{3}{8} \qquad \textit{Rename } \frac{11}{8} \textit{ as } 1\frac{3}{8}. \; 15 + 1\frac{3}{8} = 16\frac{3}{8}.$$

The perimeter of the triangle is $16\frac{3}{8}$ inches.

LOOK BACK
You can refer to Lesson 4-4 to review perimeter.

Communicating Mathematics

Read and study the lesson to answer each question.

1. *Tell* the first step you should take when adding or subtracting mixed numbers.

2. *Write a Problem* where you need to subtract $1\frac{1}{2}$ from $3\frac{5}{8}$.

HANDS-ON MATH

3. *Make a model* to show $2\frac{1}{4} + 2\frac{3}{4}$.

Guided Practice

Add or subtract. Write the answer in simplest form.

4. $6\frac{3}{4} - 2\frac{1}{4}$

5. $7\frac{5}{8} + 3\frac{1}{8}$

6. $8\frac{9}{10} + 6\frac{1}{4}$

7. $21\frac{4}{5} + 6\frac{3}{10}$

8. $11\frac{1}{3} - 8\frac{1}{6}$

9. $7\frac{2}{3} - \frac{3}{5}$

10. *Algebra* Evaluate $j + k$ if $j = 9\frac{3}{5}$ and $k = 11\frac{3}{4}$.

11. *Life Science* A rare African giant frog was captured in 1989 on the Sanaga River, in Cameroon. The body of the frog measured $14\frac{1}{2}$ inches. With its legs fully extended, the frog was $34\frac{1}{2}$ inches long. How much longer was the frog when its legs were extended?

EXERCISES

Practice

Add or subtract. Write the answer in simplest form.

12. $4\frac{3}{5} + 2\frac{2}{5}$

13. $3\frac{5}{6} - 1\frac{1}{6}$

14. $2\frac{1}{2} + 3\frac{2}{3}$

15. $13\frac{1}{3} + 5\frac{3}{4}$

16. $8\frac{9}{10} - 6\frac{1}{4}$

17. $5\frac{3}{4} - 3\frac{1}{2}$

18. $17\frac{3}{10} + 9\frac{1}{4}$

19. $7\frac{4}{5} + \frac{3}{5}$

20. $8\frac{1}{3} - 2\frac{1}{6}$

21. $18\frac{5}{6} - 18\frac{5}{8}$

22. $11\frac{3}{4} + 9\frac{3}{5}$

23. $14\frac{3}{5} - 5\frac{7}{16}$

24. How much longer is $28\frac{1}{2}$ seconds than $23\frac{3}{10}$ seconds?

25. Find the sum of $4\frac{1}{5}$, $8\frac{7}{8}$, and $1\frac{7}{10}$.

Evaluate each expression if $a = 2\frac{3}{4}$, $b = 5\frac{1}{6}$, and $c = 7\frac{2}{3}$.

26. $a + b$

27. $c - b$

28. $a + c$

Applications and Problem Solving

29. *Health* Dakota's baby sister weighed $7\frac{1}{2}$ pounds at birth. She weighed $8\frac{3}{4}$ pounds at her one-month checkup. How much weight did the baby gain?

30. Agriculture The table shows the acreage of crops planted in the United States in two different years.

Acreage (millions)		
Crop	Year 1	Year 2 (est.)
corn	$79\frac{1}{2}$	$81\frac{1}{2}$
wheat	$75\frac{3}{5}$	$69\frac{1}{5}$
soybeans	$64\frac{1}{5}$	$68\frac{4}{5}$
cotton	$14\frac{7}{10}$	$14\frac{1}{2}$
sorghum	$13\frac{1}{5}$	$10\frac{9}{10}$
barley	$7\frac{1}{5}$	7
oats	$4\frac{7}{10}$	$5\frac{3}{10}$
rice	$2\frac{4}{5}$	$2\frac{9}{10}$

Source: Agriculture Department Economic Research Service

  a. In year 1, how many more acres of wheat were planted than barley?

  b. Find the total acreage used to raise corn, wheat, cotton, and rice in year 1.

  c. How many more acres of soybeans were projected to be planted in year 2?

  d. Were more or fewer acres of rice projected to be planted in year 2? Explain.

31. Write a Problem that involves the two amounts of acreage of cotton given in the table in Exercise 30.

32. Critical Thinking Use the numbers 1, 2, 3, 4, 1, and 2 to create an addition of mixed numbers problem so that the sum is $4\frac{1}{4}$.

Mixed Review

33. Standardized Test Practice What is the perimeter of the figure? *(Lesson 6-4)*

A $\frac{16}{24}$ in.   **B** $\frac{16}{8}$ in.

C 2 in.   **D** $2\frac{1}{2}$ in.

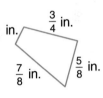

34. Express $\frac{100}{6}$ as a mixed number. *(Lesson 5-5)*

35. Measurement If a rabbit has a mass of 800 grams, what is its mass in kilograms? *(Lesson 4-9)*

For **Extra Practice**, see page 574.

36. Estimate $21.1 + 19 + 20 + 20.3 + 20.1 + 18.8$. *(Lesson 3-5)*

CHAPTER 6

Mid-Chapter Self Test

1. Food You are not sure how much salt to put in a pot of chili. Should you put one-half teaspoon or one teaspoon of salt in the pot? Explain. *(Lesson 6-1)*

Estimate. *(Lesson 6-2)*

2. $4\frac{3}{4} - 2\frac{2}{5}$ **3.** $22\frac{3}{8} - 14\frac{9}{10}$ **4.** $9\frac{1}{3} + 5\frac{5}{6}$

Add or subtract. Write the answer in simplest form. *(Lessons 6-3, 6-4, and 6-5)*

5. $\frac{13}{20} - \frac{7}{20}$ **6.** $\frac{4}{5} + \frac{3}{5}$ **7.** $\frac{2}{5} - \frac{1}{6}$

8. $\frac{3}{4} + \frac{7}{10}$ **9.** $13\frac{2}{3} + 6\frac{1}{2}$ **10.** $11\frac{5}{8} - 9\frac{2}{5}$

Subtracting Mixed Numbers with Renaming

What you'll learn

You'll learn to subtract mixed numbers involving renaming.

When am I ever going to use this?

You'll subtract mixed numbers involving renaming when you work with stock market prices.

LOOK BACK
You can refer to Lesson 5-5 to review renaming mixed numbers.

Brenda belongs to a cycling club. The club meets and rides every weekend. Brenda rides during the week to stay in shape. On Tuesday she rode her bike $4\frac{1}{2}$ miles, and on Thursday she rode $6\frac{1}{4}$ miles. How much farther did she ride on Thursday? You need to subtract to compare the mileage.

$$\begin{array}{r}6\frac{1}{4}\\-4\frac{1}{2}\\\hline\end{array} \quad \rightarrow \quad \begin{array}{r}6\frac{1}{4}\\-4\frac{2}{4}\\\hline\end{array}$$

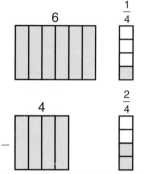

Notice that you cannot subtract $\frac{2}{4}$ from $\frac{1}{4}$. Rename $6\frac{1}{4}$ as $5\frac{5}{4}$.

$$\begin{array}{r}6\frac{1}{4}\\-4\frac{1}{2}\\\hline\end{array} \rightarrow \begin{array}{r}6\frac{1}{4}\\-4\frac{2}{4}\\\hline\end{array} \rightarrow \begin{array}{r}5\frac{5}{4}\\-4\frac{2}{4}\\\hline 1\frac{3}{4}\end{array}$$

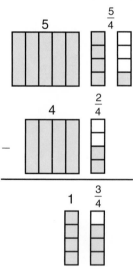

Brenda rode $1\frac{3}{4}$ miles farther on Thursday.

As shown above, sometimes it is necessary to rename the fraction of a mixed number as an improper fraction in order to subtract.

Example

1 Find $4 - 1\frac{2}{3}$.   *Estimate: 4 − 2 = 2*

$$\begin{array}{r}4\\-1\frac{2}{3}\\\hline\end{array} \quad \rightarrow \quad \begin{array}{r}3\frac{3}{3}\\-1\frac{2}{3}\\\hline 2\frac{1}{3}\end{array} \quad \begin{array}{l}\textit{Rename 4 as } 3\frac{3}{3}.\\ \\ 4 - 1\frac{2}{3} = 2\frac{1}{3}\end{array}$$

2 Find $17\frac{1}{4} - 3\frac{5}{8}$. *Estimate: 17 − 4 = 13*

Step 1 $17\frac{1}{4}$ → $17\frac{2}{8}$ *The LCM of 4 and 8 is 8.*

$\underline{-3\frac{5}{8}}$ → $\underline{-3\frac{5}{8}}$

Step 2 $17\frac{2}{8}$ → $16\frac{10}{8}$ *Since $\frac{5}{8}$ is greater than $\frac{2}{8}$, you must*

$\underline{-3\frac{5}{8}}$ → $\underline{-3\frac{5}{8}}$ *rename $17\frac{2}{8}$ as $16\frac{10}{8}$.*

$13\frac{5}{8}$

$17\frac{1}{4} - 3\frac{5}{8} = 13\frac{5}{8}$

CONNECTION

3 **Life Science** A male California sea lion grows to be between $6\frac{1}{2}$ and 8 feet long. Find the difference between the greatest and least lengths.

Subtract $8 - 6\frac{1}{2}$.   *Estimate: 8 − 7 = 1*

8 → $7\frac{2}{2}$ *Rename 8 as $7\frac{2}{2}$.*

$\underline{-6\frac{1}{2}}$ → $\underline{-6\frac{1}{2}}$

$1\frac{1}{2}$

The difference between the greatest and least lengths is $1\frac{1}{2}$ feet.

CHECK FOR UNDERSTANDING

Communicating Mathematics

Read and study the lesson to answer each question.

1. *Draw* a model or use paper plates to show how to subtract $4\frac{1}{2} - 2\frac{3}{8}$.

2. *Tell* why you might rename $3\frac{5}{8}$ as $2\frac{13}{8}$ in a subtraction problem.

3. *You Decide* Gail subtracted $10\frac{1}{5} - 3\frac{4}{5}$. Her answer was $6\frac{7}{5}$. Does Gail's answer make sense? How do you think Gail got her answer?

Guided Practice

Complete.

4. $4\frac{1}{8} = 3\frac{\blacksquare}{8}$

5. $\blacksquare = 29\frac{4}{4}$

Subtract. Write the answer in simplest form.

6. $6\frac{3}{10}$
      $\underline{-1\frac{4}{5}}$

7. $11\frac{1}{6} - 8\frac{1}{3}$

8. $7\frac{1}{4} - \frac{3}{5}$

9. $3\frac{5}{8} - 2\frac{3}{4}$

10. *Transportation* The U.S. Department of Transportation regulations prohibit a truck driver from driving more than 70 hours in any 8-day period. Mr. Galvez has driven $53\frac{3}{4}$ hours in the last 6 days. How many more hours is he allowed to drive during the next 2 days?

EXERCISES

Practice

Complete.

11. $3\frac{5}{8} = \blacksquare\frac{13}{8}$

12. $\blacksquare\frac{3}{10} = 5\frac{13}{10}$

13. $9\frac{2}{5} = 8\frac{\blacksquare}{5}$

14. $7\frac{\blacksquare}{3} = 6\frac{4}{3}$

15. $7\frac{13}{16} = 6\frac{\blacksquare}{16}$

16. $\blacksquare\frac{11}{12} = 7\frac{23}{12}$

Subtract. Write the answer in simplest form.

17. $\begin{array}{r} 30\frac{1}{4} \\ -24\frac{3}{4} \\ \hline \end{array}$

18. $\begin{array}{r} 20\frac{1}{5} \\ -4\frac{3}{5} \\ \hline \end{array}$

19. $\begin{array}{r} 5\frac{2}{5} \\ -1\frac{7}{10} \\ \hline \end{array}$

20. $\begin{array}{r} 18\frac{3}{8} \\ -5\frac{3}{4} \\ \hline \end{array}$

21. $7\frac{1}{3} - 2\frac{3}{4}$

22. $15\frac{3}{8} - 9\frac{5}{6}$

23. $16\frac{1}{6} - 8\frac{3}{4}$

24. $13 - 8\frac{7}{8}$

25. $30 - 20\frac{1}{4}$

26. $6\frac{4}{5} - 1\frac{5}{6}$

27. Find the difference between $11\frac{4}{7}$ and $3\frac{3}{5}$.

28. What is $2\frac{1}{2}$ less than $8\frac{1}{5}$?

Solve each equation. Write the solution in simplest form.

29. $f = 9\frac{1}{3} - 8\frac{2}{3}$

30. $p = 18\frac{5}{8} - 8\frac{3}{4}$

31. $12\frac{3}{5} - 10\frac{5}{6} = t$

Evaluate each expression if $a = 4\frac{1}{2}$, $b = 3\frac{2}{3}$, and $c = 2\frac{5}{6}$.

32. $a - b$

33. $a - c$

34. $b - c$

Applications and Problem Solving

35. *Sports* Some of the differences between Olympic softball and Olympic baseball are listed in the table.

  a. Which sport's ball has the larger circumference? How much larger is it compared to the other sport's ball?

  b. Which ball is heavier? How much heavier is it?

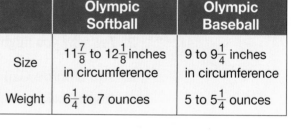

	Olympic Softball	Olympic Baseball
Size	$11\frac{7}{8}$ to $12\frac{1}{8}$ inches in circumference	9 to $9\frac{1}{4}$ inches in circumference
Weight	$6\frac{1}{4}$ to 7 ounces	5 to $5\frac{1}{4}$ ounces

36. *Geometry* Suppose you fold the short side of an $8\frac{1}{2}$ inch-by-11 inch piece of paper to align with the long side. What are the dimensions of the rectangle at the side of the paper?

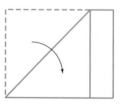

37. *Working on the* **CHAPTER Project** Refer to the table on page 227.

  a. Complete your map using the remaining landmarks.

  b. How much longer is the path from the start, through the landmarks, to the treasure than the shortest distance between the start and the treasure?

38. *Critical Thinking* Write a problem where you must subtract with renaming and the difference is between $\frac{1}{3}$ and $\frac{1}{2}$.

Mixed Review

39. Find the sum of $\frac{7}{8}$ and $1\frac{5}{6}$. Write the answer in simplest form. *(Lesson 6-5)*

40. Standardized Test Practice Train A runs every 8 minutes and Train B runs every 6 minutes. If they both leave the station at 9:00 A.M., at what time will they next leave the station together? *(Lesson 5-7)*

A 9:24 A.M.   **B** 9:48 A.M.

C 10:48 A.M.   **D** 12:00 P.M.

41. Find 586.1 ÷ 0.58 to the nearest tenth. *(Lesson 4-7)*

42. *Statistics* Find the mean and the median for the following set of low temperatures for a city: 45, 41, 28, 42, 44, 40, 40. *(Lesson 2-7)*

For **Extra Practice,** see page 574.

43. *Algebra* Evaluate $3x - y$ if $x = 4$ and $y = 1$. *(Lesson 1-5)*

Let the Games Begin

What Difference Does It Make?

Math Skill

Adding and Subtracting Mixed Numbers

Get Ready This game is for two players.

2 number cubes   spinner

Get Set Make the spinner shown.

Go
- Each player creates a mixed number by rolling the number cubes and spinning the spinner. For example, suppose a player rolls a 1 and a 6 and spins $\frac{2}{3}$. The player would record the mixed number $7\frac{2}{3}$.

- Each player creates a second mixed number using the same method.

- If the second mixed number is less than the first, the player should subtract it from the first.

- If the second mixed number is greater than the first, the player should add it to the first.

- Players continue to create mixed numbers and either add to or subtract from the previous result.

- The winner is the player with the smallest number after six rounds.

interNET CONNECTION Visit www.glencoe.com/sec/math/mac/mathnet for more games.

Integration: Measurement
Adding and Subtracting Measures of Time

Demetria's school uses block scheduling. She has an A schedule and a B schedule. For example, she goes to math class this week on Monday, Wednesday, and Friday (A Schedule). Next week she will go to math class on Tuesday and Thursday (B Schedule). How much time will she spend in math class during the week when she only goes two days if each class is 1 hour and 30 minutes?

To add or subtract measures of time:

1. Add or subtract the seconds.

2. Add or subtract the minutes.

3. Add or subtract the hours.

Rename if necessary in each step.

1 hour (h) = 60 minutes (min)
1 minute = 60 seconds (s)

Tuesday: 1 h 30 min
Thursday: + 1 h 30 min
                   2 h 60 min

Since 2 hours 60 minutes equals 3 hours, Demetria spends 3 hours in math class during weeks when she only goes two days.

Example
APPLICATION
Real World

1 **Aerospace** The countdown for the launching of a spacecraft was stopped at 12 hours, 7 minutes, 36 seconds before launch. Another stop occurred 1 hour, 28 minutes, 10 seconds before launch. How far did the countdown progress between stops?

Estimate: 12 hours − 1 hour = 11 hours

  12 h   7 min 36 s   *Since you cannot subtract 28 min from 7 min,*
− 1 h 28 min 10 s   *you must rename 12 h 7 min as 11 h 67 min.*
                26 s

  12 h   7 min 36 s   →   11 h 67 min 36 s
− 1 h 28 min 10 s   − 1 h 28 min 10 s
                26 s   10 h 39 min 26 s

The countdown progressed 10 hours, 39 minutes, 26 seconds between the two stops.

APPLICATION

2 **Travel** Josefina's flight is scheduled to leave Chicago at 11:14 A.M. and arrive in Houston at 1:36 P.M. How long will the flight last?

To answer this question, you need to find how much time has elapsed. Think of a clock.

11:14 A.M. to 12:00 noon is 46 minutes.

12:00 noon to 1:36 P.M. is 1 hour 36 minutes.

The elapsed time is 46 min + 1 h 36 min or 1 h 82 min. Now rename 82 min as 1 h and 22 min.

1 h + 1 h 22 min = 2 h 22 min

Josefina's flight will last 2 hours and 22 minutes.

APPLICATION

3 **Sports** Joan Benoit of the United States won the first women's Summer Olympic Games marathon in 1984. Her time was 2 hours, 24 minutes, 52 seconds. In the 1996 Summer Olympic Games, Fatuma Roba of Ethiopia won the marathon with a time of 2 hours, 26 minutes, 5 seconds. How much faster was Benoit's time?

Roba's time: 2 h 26 min 5 s *Since you cannot subtract 52 s*
Benoit's time: − 2 h 24 min 52 s *from 5 s, you must rename*
                                   26 min 5 s as 25 min 65 s.

$$
\begin{array}{r}
2\text{ h }26\text{ min }\ 5\text{ s} \\
-\ 2\text{ h }24\text{ min }52\text{ s} \\
\end{array}
\rightarrow
\begin{array}{r}
2\text{ h }25\text{ min }65\text{ s} \\
-\ 2\text{ h }24\text{ min }52\text{ s} \\
\hline
1\text{ min }13\text{ s} \\
\end{array}
$$

Benoit ran the marathon 1 minute, 13 seconds faster than Roba.

CHECK FOR UNDERSTANDING

Communicating Mathematics

Read and study the lesson to answer each question.

1. **Write** the number of minutes in 660 seconds.

2. **Draw** clocks to show how much time elapses between when you leave home to go to school and when you return home from school. Then determine how much time has elapsed.

3. **Write**, in your own words, how adding and subtracting measures of time is similar to adding and subtracting mixed numbers.

Complete.

4. 8 h 18 min = 7 h _?_ min   **5.** 9 h 93 min 8 s = _?_ h 33 min 8 s

Add or subtract. Rename if necessary.

6.   5 h 34 min   **7.**   2 min 35 s   **8.**   9 h 20 min
      − 4 h 9 min   + 8 min 10 s   − 3 h 45 min

9.   6 h   20 s   **10.**   12 h 40 min 30 s
      − 2 h 9 min 40 s   + 4 h 30 min 50 s

Find the elapsed time.

11. 7:50 A.M. to 2:35 P.M.   **12.** 10:30 P.M. to 6:15 A.M.

13. *Money Matters* Kenny worked for the Moore family by baby-sitting this weekend. On Friday, he worked 3 hours and 15 minutes. On Saturday, he worked 4 hours and 45 minutes. If he gets paid $3.50 an hour, how much money did he make this weekend?

EXERCISES

Practice   **Complete.**

14. 17 min 15 s = 16 min _?_ s

15. 4 h 8 min = 3 h _?_ min

16. 2 min 85 s = _?_ min 25 s

17. 42 h 5 min 87 s = 42 h _?_ min _?_ s

18. 2 h 23 min 28 s = 2 h 22 min _?_ s

19. 26 h 83 min 8 s = _?_ h _?_ min 8 s

Add or subtract. Rename if necessary.

20.   15 min 54 s   **21.**   7 h 20 min   **22.**   19 h 30 min
      − 9 min 50 s   + 3 h 18 min   − 12 h 40 min

23.   5 h 25 min   **24.**   12 h 38 min   **25.**   15 h 8 min
      + 10 h 36 min   − 3 h 46 min   − 14 h 48 min

26.   4 h 54 min   **27.**   4 min   **28.**   4 h 20 s
      + 12 h 6 min   + 12 min 55 s   − 3 h 45 s

29.   18 h 25 min   **30.**   7 h 20 min   **31.**   3 h 35 min 40 s
      − 2 h 9 min 40 s   + 2 h 48 min 10 s   + 6 h 50 min 40 s

32.   8 h   **33.**   6 h   20 s **34.**   2 h 9 min 23 s
      − 3 h 20 min 15 s   − 2 h 9 min 40 s   − 1 h 10 min 32 s

Find the elapsed time.

35. 10:20 A.M. to 11:19 A.M. **36.** 6:30 A.M. to 12:35 P.M.

37. 8:15 P.M. to 6:45 A.M. **38.** 8:05 A.M. to 4:15 P.M.

39. Find the sum of 7 hours, 25 minutes, 10 seconds and 13 hours 55 minutes.

40. How much time is there between 6:15 A.M. and 7:15 P.M.?

Applications and Problem Solving

41. *Music* Orlando's piano lesson started at 4:45 P.M. and ended at 5:30 P.M. How long was his lesson?

42. *Cooking* Rosalinda is making biscuits. She put the biscuits in the oven at 8:45 A.M. They need to bake for about 20 minutes. When should she check them?

43. *Travel* Peter flew from Tampa, Florida, to New York City. His plane left Tampa at 6:34 A.M., and the flight took 3 hours 55 minutes. What time did he arrive in New York City?

44. **Critical Thinking** Bridgette is flying from Philadelphia to San Francisco to visit her grandmother. Her flight leaves at 8:15 A.M. The non-stop flight takes about 6 hours. About what time will it be in San Francisco when Bridgette arrives? (*Hint:* Remember that there is a time difference between Philadelphia and San Francisco.)

Mixed Review

45. *Health* Joaquin is $65\frac{1}{2}$ inches tall. Jesús is $61\frac{3}{4}$ inches tall. How much taller is Joaquin than Jesús? *(Lesson 6-6)*

46. *Cooking* Paul needs $2\frac{1}{4}$ cups of flour for making cookies, $1\frac{2}{3}$ cups for almond bars, and $3\frac{1}{2}$ cups for cinnamon rolls. How much flour does he need in all? *(Lesson 6-5)*

47. **Standardized Test Practice** What is the greatest 3-digit number that is not divisible by 3 or 7? *(Lesson 5-1)*

  A 999

  B 998

  C 997

  D 996

48. *Measurement* Which is the better estimate for the capacity of a glass of milk: 360 liters or 360 milliliters? *(Lesson 4-8)*

For **Extra Practice,** see page 574.

49. *Astronomy* The distance from Earth to the Sun is close to 10^8 miles. How many miles is this? *(Lesson 1-6)*

Study Guide and Assessment

interNET **CONNECTION** Chapter Review **For additional lesson-by-lesson review, visit:** www.glencoe.com/sec/math/mac/mathnet

Vocabulary

After completing this chapter, you should be able to define each term, concept, or phrase and give an example or two of each.

Number and Operations
like fractions (p. 238)

Problem Solving
eliminate possibilities (pp. 236–237)

Understanding and Using the Vocabulary

Choose the correct term or number to complete each sentence.

1. Rounded to the nearest half, $5\frac{1}{5}$ rounds to $\left(5, 5\frac{1}{2}\right)$.

2. An estimate for $\frac{9}{20} + \frac{8}{9}$ is $\left(\frac{1}{2}, 1\frac{1}{2}\right)$.

3. The sum of $\frac{2}{5}$ and $\left(\frac{3}{5}, \frac{4}{5}\right)$ is $1\frac{1}{5}$.

4. $\left(\text{Denominators, Numerators}\right)$ are subtracted when you subtract fractions with like denominators.

5. The $\left(\text{LCD, GCF}\right)$ is the least common multiple of the denominators of unlike fractions.

6. The LCD of $\frac{1}{8}$ and $\frac{3}{10}$ is $\left(80, 40\right)$.

7. The solution of $x = 6\frac{7}{10} - 1\frac{1}{5}$ is $\left(5\frac{1}{2}, 5\frac{3}{5}\right)$.

8. The mixed number $9\frac{1}{4}$ can be renamed as $\left(8\frac{3}{4}, 8\frac{5}{4}\right)$.

9. The number of minutes in one hour and twenty minutes is $\left(120, 80\right)$.

In Your Own Words

10. *Explain* how to subtract 5 hours 50 minutes from 8 hours 15 minutes.

Objectives & Examples

Upon completing this chapter, you should be able to:

● round fractions and mixed numbers
(Lesson 6-1)

Round to the nearest half.

$\frac{13}{16}$ → 1   *13 is close to 16.*

$\frac{5}{9}$ → $\frac{1}{2}$ *5 is about half of 9.*

$\frac{2}{11}$ → 0   *2 is much smaller than 11.*

Review Exercises

Use these exercises to review and prepare for the chapter test.

Round each number to the nearest half.

11. $\frac{9}{16}$ **12.** $9\frac{2}{9}$

13. $11\frac{4}{7}$ **14.** $\frac{9}{11}$

Tell whether each number should be rounded up or down.

15. the diameter of a plant to put in an $11\frac{7}{8}$-inch pot

16. the weight limit of an elevator

17. the capacity of a pitcher needed to hold $\frac{7}{8}$ gallon of lemonade

18. the height of a truck that can go under a bridge

● estimate sums and differences of fractions and mixed numbers *(Lesson 6-2)*

$\frac{13}{16} - \frac{5}{12}$ → $1 - \frac{1}{2} = \frac{1}{2}$

$\frac{7}{8} + 8\frac{1}{5}$ → $1 + 8 = 9$

Estimate.

19. $8\frac{3}{16} - 4\frac{1}{4}$

20. $\frac{5}{8} + \frac{15}{16}$

21. $\frac{1}{8} + 4\frac{2}{3}$

22. $\frac{7}{8} - \frac{5}{12}$

● add and subtract fractions with like denominators *(Lesson 6-3)*

$\frac{7}{8} + \frac{5}{8} = \frac{7+5}{8}$

$= \frac{12}{8}$

$= 1\frac{4}{8}$ or $1\frac{1}{2}$

Add or subtract. Write the answer in simplest form.

23. $\frac{8}{15} + \frac{11}{15}$

24. $\frac{4}{7} - \frac{1}{7}$

25. $\frac{7}{8} - \frac{1}{8}$

26. $\frac{5}{6} + \frac{5}{6}$

Objectives & Examples

Review Exercises

• add and subtract fractions with unlike denominators *(Lesson 6-4)*

$$\frac{7}{12} - \frac{11}{24} = \frac{14}{24} - \frac{11}{24}$$
$$= \frac{14 - 11}{24}$$
$$= \frac{3}{24} \text{ or } \frac{1}{8}$$

Add or subtract. Write the answer in simplest form.

27. $\frac{7}{10} - \frac{1}{5}$ **28.** $\frac{2}{3} + \frac{1}{4}$

29. $\frac{4}{9} + \frac{5}{6}$ **30.** $\frac{2}{3} - \frac{1}{5}$

• add and subtract mixed numbers *(Lesson 6-5)*

$$11\frac{2}{3} \quad \rightarrow \quad 11\frac{16}{24}$$
$$+ 2\frac{3}{8} \quad \rightarrow \quad +2\frac{9}{24}$$
$$13\frac{25}{24} = 13 + 1\frac{1}{24} \text{ or } 14\frac{1}{24}$$

Add or subtract. Write the answer in simplest form.

31. $7\frac{3}{7} - 2\frac{1}{7}$ **32.** $8\frac{2}{3} + 1\frac{2}{3}$

33. $9\frac{4}{5} - 8\frac{1}{2}$ **34.** $11\frac{5}{8} - 3\frac{1}{6}$

• subtract mixed numbers involving renaming *(Lesson 6-6)*

Find $4\frac{1}{4} - \frac{2}{3}$.

$$4\frac{1}{4} \rightarrow 4\frac{3}{12} \rightarrow 3\frac{15}{12}$$
$$-\frac{2}{3} \rightarrow -\frac{8}{12} \rightarrow -\frac{8}{12}$$
$$3\frac{7}{12}$$

$$4\frac{1}{4} - \frac{2}{3} = 3\frac{7}{12}$$

Subtract. Write the answer in simplest form.

35. $8\frac{1}{8}$ **36.** $7\frac{1}{6}$
$\quad\;\; -3\frac{1}{4}$ $\quad\;\; -4\frac{1}{3}$

37. $8 - 2\frac{2}{3}$ **38.** $18\frac{7}{16} - 8\frac{3}{4}$

Solve each equation. Write the solution in simplest form.

39. $f = 6\frac{1}{8} - \frac{1}{4}$ **40.** $5\frac{2}{5} - 1\frac{7}{10} = h$

• add and subtract measures of time *(Lesson 6-7)*

$$\begin{array}{r} 3 \text{ h } 50 \text{ min} \\ + 2 \text{ h } 15 \text{ min} \\ \hline 5 \text{ h } 65 \text{ min} \end{array}$$

Rename 65 min as 1 h and 5 min.

5 h + 1 h 5 min = 6 h 5 min

Complete.

41. 6 h 8 min = 5 h __?__ min

42. 27 min 75 s = __?__ min __?__ s

Add or subtract. Rename if necessary.

43. 5 h 20 min   **44.** 7 h 45 min
$\quad\;\;$ + 2 h 16 min $\quad\;\;$ − 4 h 32 min

45. 9 h 7 min   **46.** 2 h 35 min
$\quad\;\;$ − 8 h 7 min 8 s $\quad\;\;$ + 6 h 41 min

Applications & Problem Solving

47. *Money Matters* How much money should Danielle get from her purse to pay for a tape that costs $8.10: $8 or $8.50? *(Lesson 6-1)*

48. *Food* Lana wants to buy her lunch at school. She may choose one entree, one side dish, and one drink. How many lunch combinations are possible from today's menu? *(Lesson 6-2B)*

Entree	Side Dish	Drink
cheeseburger	chips	milk
pizza	salad	juice
taco		water

49. *Measurement* Melinda walks $\frac{1}{2}$ mile to school in the morning and then $\frac{2}{3}$ mile to work after school. She gets a ride home after work from her dad. How many miles does she walk each day? *(Lesson 6-4)*

50. *Jobs* Gina began working at 9:15 A.M. and worked until 3:30 P.M. How long did she work? *(Lesson 6-7)*

Alternative Assessment

Open Ended

Suppose you are in charge of making costumes for a class play. You have determined you need $12\frac{2}{3}$ yards of blue fabric, $10\frac{5}{6}$ yards of pinstripe fabric, and $18\frac{1}{2}$ yards of white fabric. Tell how to estimate the sum or difference of mixed numbers. Estimate the total amount of fabric you need.

Suppose you have 6 yards of floral fabric and you know you will use $4\frac{1}{8}$ yards. How much fabric will you have left?

The play starts at 7:30 P.M. and ends at 9:18 P.M. How can you find the elapsed time without drawing clocks? How is finding elapsed time similar to subtracting mixed numbers involving renaming?

A practice test for Chapter 6 is provided on page 600.

Completing the CHAPTER Project

Use the following checklist to make sure your project is complete.

☑ You have included a list of instructions on how to use your map.

☑ You have included your map showing the location of the starting point, the four landmarks, and the treasure.

Add any finishing touches that you would like to make your project attractive.

 PORTFOLIO   Select one of the assignments from this chapter that you found particularly challenging. Place it in your portfolio.

Section One: Multiple Choice

There are twelve multiple-choice questions in this section. Choose the best answer. If a correct answer is *not here*, mark the letter for Not Here.

1. Which number sentence belongs to the same family of facts as $5 + 9 = 14$?

   A $19 - 5 = 14$

   B $14 - 9 = 5$

   C $19 - 14 = 5$

   D $9 - 5 = 4$

2. Julio measures 250 grams of salt for a science experiment. What is the mass in kilograms?

   F 0.0025 kg

   G 0.25 kg

   H 25 kg

   J 2,500 kg

3. What is the median of the following test scores?

   80   100   90   60   80   90   100
   90   90   60   70   90   100   80

   A 75

   B 80

   C 85

   D 90

4. Which ordered pair is represented by point *M*?

   F (3, 4)

   G (4, 2)

   H (1, 4)

   J (2, 4)

5. What is the LCD of $\frac{3}{8}$ and $\frac{5}{6}$?

   A 14

   B 15

   C 24

   D 48

6. Which is the list of prime factors of 32?

   F 2

   G 2, 16

   H 1, 32

   J 1, 2, 4, 8, 16, 32

Please note that Questions 7–12 have five answer choices.

7. Andy has asked one of his friends to feed his three dogs while he is away on vacation. Each dog eats $1\frac{1}{2}$ bags of dog food per week. What else do you need to know to find how many bags of food Andy should leave for his dogs to eat while he is away on vacation?

   A the size of each of Andy's dogs

   B the price of each bag of dog food

   C the number of weeks Andy will be away on vacation

   D the amount of exercise the dogs need each day

   E the amount of dog food in 1 bag

8. A football game started at 2:40 P.M. and ended at 5:15 P.M. How long was the game?

   F 2 h 15 min

   G 2 h 35 min

   H 3 h 15 min

   J 3 h 25 min

   K Not Here

9. Which is a reasonable remainder when a number is divided by 5?

  A 8

  B 7

  C 6

  D 5

  E 4

10. Pedro bought 6 batteries for his radio. Each battery cost $1.79, including tax. What was the total cost of the batteries?

  F $6.24

  G $10.24

  H $10.74

  J $11.74

  K Not Here

11. Each day Matias drives his delivery truck 3.25 miles to his first stop, 4.23 miles to his second stop, 8.8 miles to his third stop, and 6 miles back to headquarters. How many miles does Matias drive the truck each day?

  A 22.28 miles

  B 16.28 miles

  C 8.42 miles

  D 7.48 miles

  E Not Here

12. Ron was making curtains for his room. He had 7 yards of material. It took $5\frac{1}{2}$ yards to make the curtains. How much material was not used for the curtains?

  F $12\frac{1}{2}$ yd

  G $2\frac{1}{2}$ yd

  H 2 yd

  J $1\frac{1}{2}$ yd

  K Not Here

Test-Taking Tip   THE PRINCETON REVIEW

Most standardized tests have a time limit, so you must use your time wisely. Some questions will be easier than others. If you cannot answer a question quickly, go on to another. If there is time remaining when you are finished, go back to the questions that you skipped.

Section Two: Free Response

This section contains seven questions for which you will provide short answers. Write your answers on your paper.

13. When is the product of two decimals less than both factors?

14. **Geometry** Find the perimeter of the rectangle.

$6\frac{1}{2}$ in.

$12\frac{1}{2}$ in.

15. Write $\frac{35}{63}$ in simplest form.

16. Express $\frac{18}{11}$ as a mixed number.

17. **Algebra** Evaluate $a + b$ if $a = \frac{7}{8}$ and $b = \frac{3}{16}$.

18. What is $12\frac{3}{15}$ minus $\frac{2}{5}$?

19. Find the elapsed time from 8:30 P.M. to 10:40 A.M.

Interdisciplinary Investigation

JUST "STATE" THE FACTS!

Do you know the population of the largest city in the United States? According to the 1990 census, New York City had a population of 7,322,564. That is more than twice as many people as Los Angeles, the second largest city. In 1990, eight U. S. cities had populations over 1,000,000.

What You'll Do

In this investigation, you will choose a state and investigate what part of that state's population live in its largest cities.

Materials 📖 almanac 🖩 calculator

Procedure

1. Work in groups of four. Each member of your group should choose a different state. Find a resource book, map, or website that lists the populations of the largest cities.

2. Make a table, similar to the one below, for your chosen state. List the ten largest cities in order from greatest to least population. Determine how many people live in the rest of the state and call that category "other."

Montana Cities – 1990 Population			
City	Population	Population (rounded)	Fraction
Billings	81,125	81,000	$\frac{81}{799}$
Great Falls	55,125	55,000	$\frac{55}{799}$
⋮	⋮	⋮	⋮
Other	498,357	498,000	$\frac{498}{799}$
State Total	799,065	799,000	

In column 2, write the population of each city. In column 3, round the population to the nearest thousand. In column 4, write a fraction to show the population of each city compared to the total population of the state. (Round the total population to the nearest thousand.)

Technology Tips

- Use a **spreadsheet** to record populations.

- Use a **graphing calculator** or **graphing software** to make graphs.

- Use a **word processor** to prepare your display.

- Surf the **Internet** for more information.

3. For your chosen state, use a map and the scale for the map to find the perimeter in miles. Place a string along each side and then add the lengths of string.

4. Prepare a display of the facts about your group's four states. Include some similarities and differences between the states.

Making the Connection

Use the data you collected about state population as needed to help in these investigations.

Language Arts

Imagine that the people of your chosen state stand on the perimeter and hold hands. Write directions for finding out whether there are enough people in the state to form a human border.

Science

Research the climate of your chosen state. Make a poster describing the climate. Include the high and low temperatures and draw a graph showing the amount of precipitation for each month.

Social Studies

Research the geography of your chosen state. Write a short report about the major geographic features of the state.

Go Further

- Make two or more different types of graphs comparing the populations of the cities in your chosen state and another student's state.

- Suppose that the state you chose has a random drawing for a prize. The names of all people in the state are included. What is the probability that the winner lives in the largest city?

 **Research** **For current information from the U.S. Bureau of the Census, visit:**

www.glencoe.com/sec/math/mac/mathnet

 You may want to place your work on this investigation in your portfolio.

What you'll learn in Chapter 7

- to multiply and divide fractions and mixed numbers,

- to find the circumference of circles,

- to change units within the customary system,

- to recognize and extend sequences, and

- to solve real-world problems by finding and extending a pattern.

CHAPTER Project

BUILD A PAPER AIRPLANE

In this project, you will use multiplication and division of fractions to help you find flight distances and flight times of a paper airplane. You can create any type of airplane you wish, such as a bomber or a fighter plane.

Getting Started

- Look at the flight table. In which flight did Mitiku's paper airplane travel the farthest?

- Use the table to determine how far Mitiku's plane could have traveled in flight 3 if it had remained in the air for 2 seconds.

Mitiku's Flight and Distance Table		
Flight Number	Distance (ft)	Time (s)
1	$11\frac{1}{4}$	2
2	$13\frac{1}{2}$	3
3	$5\frac{3}{4}$	1

- Use a sheet of notebook paper and tape to construct a paper airplane.

  - Throw your airplane three times. Use a tape measure to measure the distance of each flight, in feet, your airplane travels. Use a stopwatch to time each flight. Round to the nearest second. Record your flight results in a table similar to the one above.

Technology Tips

- Use a **spreadsheet** to record your flight results and to find the speed of your airplane.

 **Research** For up-to-date information on paper airplanes, visit the following website.

Data Collection and Comparison To share and compare your data with other students in the U.S., visit:

www.glencoe.com/sec/math/mac/mathnet

Working on the Project

You can use what you'll learn in Chapter 7 to help you find flight distances

Page	Exercise
279	42
291	42
305	Alternative Assessment

and flight times of your paper airplane.

Estimating Products

You'll learn to estimate fraction products using compatible numbers and rounding.

When am I ever going to use this?

You'll use estimation to approximate the discount on a sale item.

Word Wise

compatible numbers

Where do you like to study? The circle graph shows where students prefer to study. If there are 16 students in Taylor's class, about how many students prefer to study in their bedrooms?

To solve the problem, you can estimate the product of $\frac{1}{3}$ and 16.

One way to estimate the product is to use **compatible numbers**. Compatible numbers are easy to divide mentally.

Where Students Prefer to Study

Source: Federal National Mortgage Association

$\frac{1}{3} \times 16 = ?$ *$\frac{1}{3} \times 16$ means $\frac{1}{3}$ of 16.*

$\frac{1}{3} \times 15 = ?$ *Think: For 16, the nearest multiple of 3 is 15. 3 and 15 are compatible numbers since $15 \div 3 = 5$.*

$\frac{1}{3} \times 15 = 5$ *Multiplying by $\frac{1}{3}$ is the same as dividing by 3.*

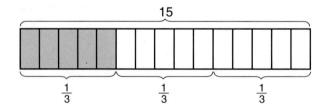

So, the product of $\frac{1}{3}$ and 16 is about 5.

About 5 of the 16 students prefer to study in their bedrooms.

Example 1   Estimate $\frac{5}{8} \times 25$.

$\frac{1}{8} \times 24 = 3$ *For 25, the nearest multiple of 8 is 24. $\frac{1}{8}$ of 24 is 3.*

$\frac{5}{8} \times 24 = 15$ *Since $\frac{1}{8}$ of 24 is 3, it follows that $\frac{5}{8}$ of 24 is 5×3 or 15.*

So, $\frac{5}{8} \times 25$ is about 15.

You can also estimate products by rounding fractions to 0, $\frac{1}{2}$, or 1.

LOOK BACK

You can refer to Lesson 6-1 for information on rounding fractions.

 **Example 2** Estimate $\frac{2}{9} \times \frac{5}{6}$.

Think: $\frac{5}{6}$ is close to 1.

$$\begin{array}{ccccccc} \frac{0}{6} & \frac{1}{6} & \frac{2}{6} & \frac{3}{6} & \frac{4}{6} & \frac{5}{6} & \frac{6}{6} \end{array}$$

$\frac{2}{9} \times 1 = \frac{2}{9}$ *Round $\frac{5}{6}$ to 1. Any number multiplied by 1 is the number.*

So, $\frac{2}{9} \times \frac{5}{6}$ is about $\frac{2}{9}$.

To estimate the product of mixed numbers, you can round each mixed number to the nearest whole number and then multiply.

Example 3

APPLICATION

Cooking Kim needs $3\frac{1}{2}$ batches of cookies. If one recipe calls for $2\frac{1}{4}$ cups of flour, about how many cups of flour are needed?

You need to estimate $3\frac{1}{2} \times 2\frac{1}{4}$.

$4 \times 2 = 8$ *Round $3\frac{1}{2}$ to 4. Round $2\frac{1}{4}$ to 2.*

So, $3\frac{1}{2} \times 2\frac{1}{4}$ is about 8.

Kim will need about 8 cups of flour.

CHECK FOR UNDERSTANDING

Communicating Mathematics

Read and study the lesson to answer each question.

1. *Explain* how you would use rounding to estimate $3\frac{2}{3} \times 9\frac{1}{8}$.

2. *Tell* how the model shows the use of compatible numbers to estimate $\frac{3}{4} \times 9$.

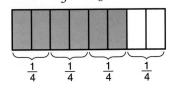

3. *You Decide* Juan says that $8\frac{1}{2} \times 6\frac{1}{4}$ is about 54. Odina says that the product is about 48. Whose estimate is better? Explain your reasoning.

Guided Practice

Round each fraction to 0, $\frac{1}{2}$, or 1 and each mixed number to the nearest whole number.

4. $\frac{1}{8}$

5. $\frac{5}{9}$

6. $7\frac{2}{3}$

Estimate each product.

7. $\frac{1}{8} \times 15$

8. $\frac{2}{5} \times \frac{6}{7}$

9. $3\frac{1}{4} \times 8\frac{6}{7}$

10. *Geometry* Estimate the area of the rectangle.

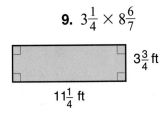

$3\frac{3}{4}$ ft

$11\frac{1}{4}$ ft

Practice

Round each fraction to 0, $\frac{1}{2}$, or 1 and each mixed number to the nearest whole number.

11. $\frac{6}{11}$ **12.** $\frac{4}{5}$ **13.** $\frac{4}{9}$ **14.** $\frac{2}{15}$

15. $6\frac{1}{8}$ **16.** $12\frac{3}{4}$ **17.** $3\frac{1}{2}$ **18.** $15\frac{2}{9}$

Family Activity

Ask a family member to help you make a list of situations in your daily life that involve a fraction or a mixed number. Then estimate each number by rounding.

Estimate each product.

19. $\frac{1}{4} \times 31$ **20.** $\frac{3}{5} \times \frac{1}{2}$ **21.** $\frac{1}{7} \times 22$

22. $2\frac{1}{6} \times 5\frac{1}{2}$ **23.** $\frac{1}{6} \times 40$ **24.** $\frac{7}{8} \times 3\frac{1}{4}$

25. $\frac{4}{9} \times 46$ **26.** $\frac{1}{6} \times \frac{7}{8}$ **27.** $7\frac{3}{4} \times 3\frac{1}{8}$

28. Estimate $4\frac{5}{6}$ multiplied by $16\frac{7}{8}$.

29. Estimate $14\frac{8}{9} \times 6\frac{3}{5}$.

Applications and Problem Solving

30. *Travel* The circle graph shows when people pack for a vacation. Suppose 96 people were surveyed. About how many people pack the day they leave?

31. *Recreation* There are about 7 million pleasure boats in the United States. About $\frac{2}{3}$ of these boats are motorboats. About how many motorboats are in the United States?

32. *Critical Thinking* Which point on the number line could be the graph of the product of the numbers graphed at C and D?

When Vacationers Pack

1 day before leaving $\frac{1}{2}$

2 or more days before leaving $\frac{9}{20}$

The day they leave $\frac{1}{20}$

Source: Carlson Wagonlit Travel Survey

Mixed Review

33. *Measurement* Add 26 min 37 s and 37 min 14 s. *(Lesson 6-7)*

34. Find the prime factorization of 300. *(Lesson 5-2)*

35. *Measurement* How many kilograms are in 498 grams? *(Lesson 4-9)*

36. **Standardized Test Practice** A bus travels at a speed of 47 miles per hour. About how long will it take the bus to travel 316 miles? *(Lesson 1-3)*

    A 4 hours

    B 5 hours

    C 6 hours

    D 8 hours

    E 10 hours

For **Extra Practice**, see page 575.

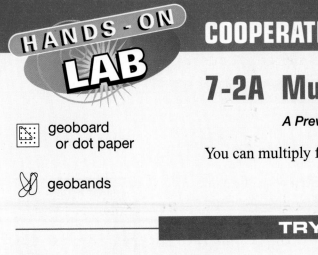
COOPERATIVE LEARNING

7-2A Multiplying Fractions

A Preview of Lesson 7-2

geoboard
or dot paper

geobands

You can multiply fractions by using geoboards.

TRY THIS

Work with a partner.

1 Model $\frac{1}{3} \times \frac{1}{2}$ using a geoboard.

Step 1 Place one geoband in a straight line along the bottom of the geoboard to show thirds.

Step 2 Place one geoband perpendicular to the first geoband to show halves.

Step 3 Use geobands to make a rectangle.

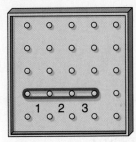

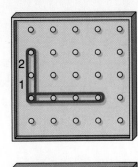

Step 4 Place one geoband on the peg as shown. This represents $\frac{1}{3}$.

Step 5 Place one geoband on the peg as shown. This represents $\frac{1}{2}$.

Step 6 Connect the geobands to show a small rectangle.

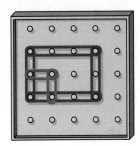

ON YOUR OWN

1. The area of the small rectangle has an area of 1 square unit. What is the area of the large rectangle?

2. What fraction of the area of the large rectangle is the area of the small rectangle?

3. What is $\frac{1}{3} \times \frac{1}{2}$?

Work with a partner.

2 Model $\frac{1}{4} \times \frac{1}{3}$ using a geoboard.

Step 1 Place one geoband in a straight line along the bottom of the geoboard to show fourths.

Step 4 Place one geoband on the peg as shown. This represents $\frac{1}{4}$.

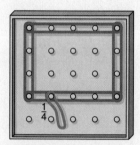

Step 2 Place one geoband perpendicular to the first geoband to show thirds.

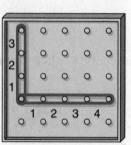

Step 5 Place one geoband on the peg as shown. This represents $\frac{1}{3}$.

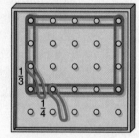

Step 3 Use geobands to make a rectangle.

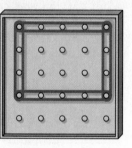

Step 6 Connect the geobands to show a small rectangle.

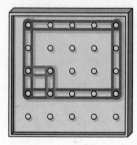

4. The small rectangle has an area of 1 square unit. What is the area of the large rectangle?

5. What fraction of the area of the large rectangle is the area of the small rectangle?

6. What is $\frac{1}{4} \times \frac{1}{3}$?

7. Write a rule you can use to multiply fractions.

8. *Look Ahead* Use your rule to find $\frac{3}{4} \times \frac{1}{2}$. Check your answer using a geoboard.

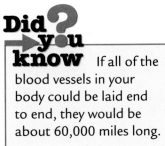

What you'll learn

You'll learn to multiply fractions.

When am I ever going to use this?

Knowing how to multiply fractions can help you make half a batch of chocolate chip cookies.

Did you know? If all of the blood vessels in your body could be laid end to end, they would be about 60,000 miles long.

Do you know your blood type? The human body can contain any one of four different blood types. Suppose the employees of a very large company have the fractions of blood types shown in the table. If $\frac{1}{2}$ of the company's employees donate blood, what fraction will donate type A blood?

Blood Type	Fraction of People With Each Blood Type
A	$\frac{2}{5}$
B	$\frac{1}{10}$
AB	$\frac{1}{20}$
O	$\frac{9}{20}$

You need to find $\frac{2}{5}$ of $\frac{1}{2}$, which means $\frac{2}{5} \times \frac{1}{2}$.

Method 1 Use a model.

Separate a rectangle into halves. Then color $\frac{1}{2}$ of the rectangle blue.

$\frac{1}{2}\Big\{$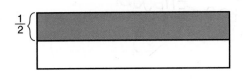

Next, separate the rectangle into fifths, and color $\frac{2}{5}$ of the rectangle yellow.

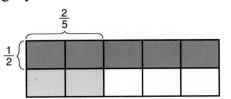

2 parts are green.
10 parts in all.

Since 2 out of 10 parts are green, the product is $\frac{2}{10}$ or $\frac{1}{5}$.

Method 2 Use a rule.

Find the product by multiplying the numerators and multiplying the denominators.

$\frac{2}{5} \times \frac{1}{2} = \frac{2 \cdot 1}{5 \cdot 2}$ *Multiply the numerators.*
                  Multiply the denominators.

       $= \frac{2}{10}$ or $\frac{1}{5}$ *Simplify.*

So, $\frac{1}{5}$ of the company's employees will donate type A blood.

You can use the following rule to multiply fractions.

Multiplying Fractions	To multiply fractions, multiply the numerators. Then multiply the denominators. Simplify if necessary.

Examples

① Find $\frac{1}{4} \times \frac{2}{3}$.

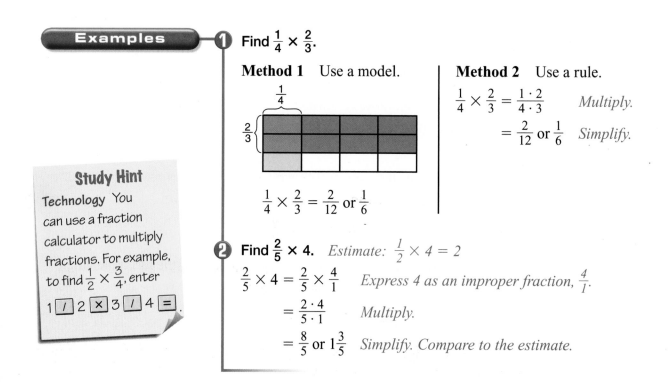

Method 1   Use a model.

$$\frac{1}{4} \times \frac{2}{3} = \frac{2}{12} \text{ or } \frac{1}{6}$$

Method 2   Use a rule.

$$\frac{1}{4} \times \frac{2}{3} = \frac{1 \cdot 2}{4 \cdot 3} \quad \textit{Multiply.}$$

$$= \frac{2}{12} \text{ or } \frac{1}{6} \quad \textit{Simplify.}$$

Study Hint

Technology You can use a fraction calculator to multiply fractions. For example, to find $\frac{1}{2} \times \frac{3}{4}$, enter 1 / 2 × 3 / 4 =.

② Find $\frac{2}{5} \times 4$.   *Estimate: $\frac{1}{2} \times 4 = 2$*

$$\frac{2}{5} \times 4 = \frac{2}{5} \times \frac{4}{1} \quad \textit{Express 4 as an improper fraction, } \frac{4}{1}.$$

$$= \frac{2 \cdot 4}{5 \cdot 1} \quad \textit{Multiply.}$$

$$= \frac{8}{5} \text{ or } 1\frac{3}{5} \quad \textit{Simplify. Compare to the estimate.}$$

If the numerator of one fraction and the denominator of another fraction have a common factor, you can simplify *before* you multiply.

Examples

③ Find $\frac{5}{9} \times \frac{3}{4}$.   *Estimate: $\frac{1}{2} \times 1 = \frac{1}{2}$*

Simplify before multiplying.

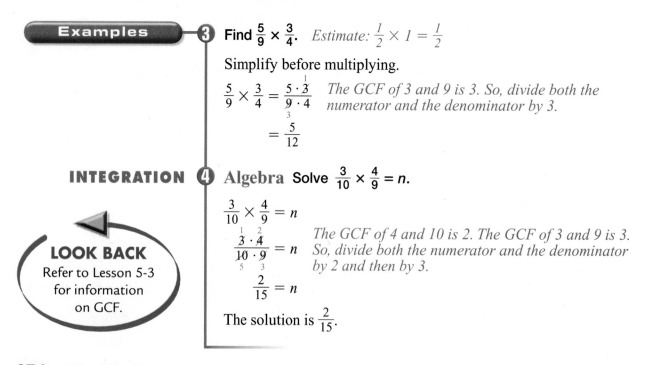

$$\frac{5}{9} \times \frac{3}{4} = \frac{5 \cdot \overset{1}{3}}{\underset{3}{9} \cdot 4} \quad \textit{The GCF of 3 and 9 is 3. So, divide both the numerator and the denominator by 3.}$$

$$= \frac{5}{12}$$

INTEGRATION ④ **Algebra** Solve $\frac{3}{10} \times \frac{4}{9} = n$.

$$\frac{3}{10} \times \frac{4}{9} = n$$

$$\frac{\overset{1}{3} \cdot \overset{2}{4}}{\underset{5}{10} \cdot \underset{3}{9}} = n \quad \textit{The GCF of 4 and 10 is 2. The GCF of 3 and 9 is 3. So, divide both the numerator and the denominator by 2 and then by 3.}$$

$$\frac{2}{15} = n$$

The solution is $\frac{2}{15}$.

LOOK BACK

Refer to Lesson 5-3 for information on GCF.

**Communicating
Mathematics**

Read and study the lesson to answer each question.

1. **Explain** how the model shows the product of $\frac{1}{3}$ and $\frac{1}{2}$.

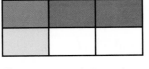

2. **Draw** a model to show that $\frac{2}{3} \times \frac{1}{5} = \frac{2}{15}$.

Math Journal

3. **Write** how to find the product of any two fractions.

Guided Practice

Find each product. Write in simplest form.

4. $\frac{1}{6} \times \frac{1}{2}$　　5. $\frac{3}{5} \times \frac{1}{4}$　　6. $\frac{1}{5} \times \frac{5}{7}$　　7. $\frac{2}{3} \times \frac{1}{5}$

Solve each equation. Write the solution in simplest form.

8. $14 \times \frac{2}{7} = a$　　9. $w = \frac{3}{4} \times \frac{4}{9}$　　10. $\frac{6}{7} \times \frac{5}{12} = x$

11. **Life Science** About $\frac{7}{10}$ of the human body is water. If a person weighs 150 pounds, how many pounds are water?

Practice

Find each product. Write in simplest form.

12. $\frac{1}{3} \times \frac{3}{5}$　　13. $\frac{8}{9} \times \frac{1}{3}$　　14. $\frac{7}{8} \times \frac{3}{4}$　　15. $\frac{2}{3} \times \frac{1}{6}$

16. $\frac{2}{3} \times 5$　　17. $\frac{3}{5} \times \frac{2}{9}$　　18. $6 \times \frac{5}{8}$　　19. $\frac{7}{9} \times \frac{6}{7}$

20. $\frac{1}{2} \times \frac{8}{9}$　　21. $\frac{2}{3} \times \frac{5}{9}$　　22. $\frac{3}{4} \times \frac{8}{9}$　　23. $\frac{1}{9} \times \frac{2}{3}$

Solve each equation. Write the solution in simplest form.

24. $x = \frac{5}{12} \times \frac{3}{7}$　　25. $10 \times \frac{2}{3} = h$　　26. $y = \frac{5}{8} \times \frac{2}{3}$

27. $m = \frac{5}{16} \times \frac{4}{15}$　　28. $\frac{1}{2} \times \frac{4}{9} = b$　　29. $d = \frac{5}{6} \times \frac{8}{15}$

30. $c = \frac{7}{10} \times \frac{5}{14}$　　31. $18 \times \frac{7}{12} = a$　　32. $p = \frac{9}{13} \times \frac{8}{9}$

33. **Algebra** Evaluate mn if $m = \frac{1}{3}$ and $n = \frac{12}{15}$.

**Applications and
Problem Solving**

34. **Broadcasting** A television network sells 16 minutes of air time to advertisers. A company purchases $\frac{3}{8}$ of the total ad time. How many minutes of ad time did the company buy?

35. *Algebra* The expression 3^2 means 3×3, or 9.

  a. If $m = \frac{3}{8}$, what is the value of m^2?

  b. Evaluate h^2 if $h = \frac{4}{5}$.

36. *Critical Thinking* Find $\frac{1}{2} \times \frac{2}{3} \times \frac{3}{4} \times \frac{4}{5} \times \ldots \times \frac{99}{100}$.

Mixed Review

37. Estimate the product of $\frac{3}{4}$ and 37. *(Lesson 7-1)*

38. *Algebra* Solve the equation $x - \frac{11}{12} = \frac{5}{12}$. *(Lesson 6-3)*

39. Determine whether 786 is divisible by 2, 3, 5, 6, 9, or 10. *(Lesson 5-1)*

40. **Standardized Test Practice** Hayley works 8 hours a week baby-sitting and earns $22.40. How much does she earn per hour? *(Lesson 4-5)*

  A $3.20   **B** $2.50   **C** $3.40   **D** $2.80   **E** Not Here

For **Extra Practice,** see page 575.

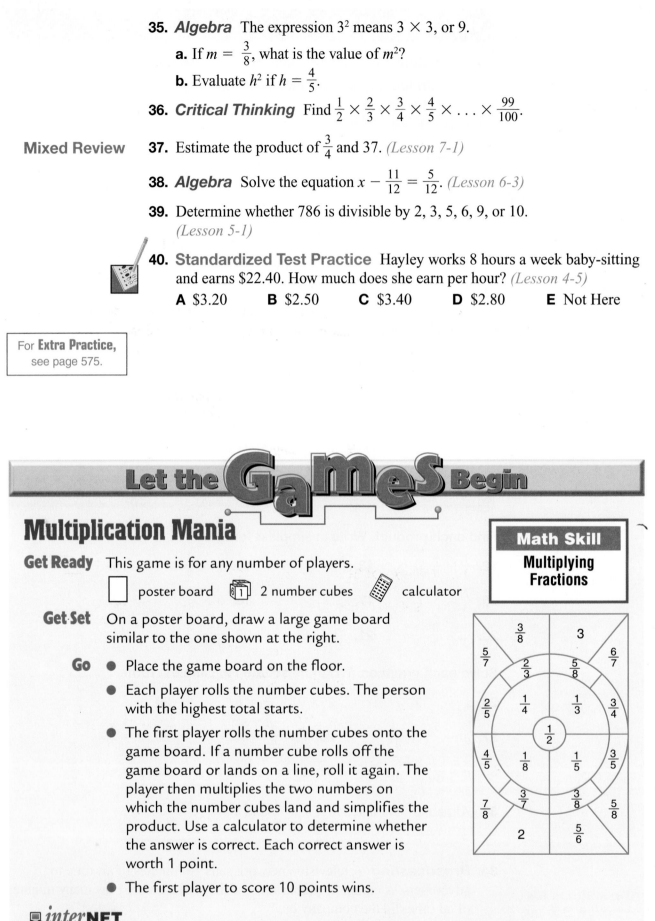

Let the GameS Begin

Multiplication Mania

Math Skill

Multiplying Fractions

Get Ready This game is for any number of players.

  poster board   2 number cubes   calculator

Get·Set On a poster board, draw a large game board similar to the one shown at the right.

Go ● Place the game board on the floor.

  ● Each player rolls the number cubes. The person with the highest total starts.

  ● The first player rolls the number cubes onto the game board. If a number cube rolls off the game board or lands on a line, roll it again. The player then multiplies the two numbers on which the number cubes land and simplifies the product. Use a calculator to determine whether the answer is correct. Each correct answer is worth 1 point.

  ● The first player to score 10 points wins.

interNET CONNECTION Visit www.glencoe.com/sec/math/mac/mathnet for more games.

7-3 Multiplying Mixed Numbers

What you'll learn

You'll learn to multiply mixed numbers.

When am I ever going to use this?

Knowing how to multiply mixed numbers can help you determine the ingredients needed to make a cake.

> **LOOK BACK**
> You can refer to Lesson 5-5 for information on improper fractions.

A dot following a musical note (○·) means that the note gets $1\frac{1}{2}$ times as many beats as the same note without a dot (○). If a whole note gets four beats, how many beats would a dotted whole note get?

You need to find $1\frac{1}{2}$ of 4, or $1\frac{1}{2} \times 4$. To multiply mixed numbers, express the mixed numbers as improper fractions and then multiply as with fractions.

$$1\frac{1}{2} \times 4 = \frac{3}{2} \times \frac{4}{1} \quad \textit{Express the mixed numbers as improper fractions.}$$

$$= \frac{3 \cdot 4}{2 \cdot 1} \quad \textit{Multiply the numerators. Multiply the denominators.}$$

$$= \frac{12}{2} \text{ or } 6 \quad \textit{Simplify.}$$

A dotted whole note gets 6 beats.

Examples

1 Find $\frac{1}{3} \times 3\frac{3}{8}$. *Estimate: $\frac{1}{3} \times 3 = 1$*

$$\frac{1}{3} \times 3\frac{3}{8} = \frac{1}{3} \times \frac{27}{8} \quad \textit{Express } 3\frac{3}{8} \textit{ as an improper fraction.}$$

$$= \frac{1 \cdot \overset{9}{27}}{\underset{1}{3} \cdot 8} \quad \textit{Divide 27 and 3 by the GCF, 3. Then multiply.}$$

$$= \frac{9}{8} \text{ or } 1\frac{1}{8} \quad \textit{Compare with your estimate.}$$

INTEGRATION **2** **Algebra** if $m = 2\frac{2}{5}$ and $n = 1\frac{7}{8}$, what is the value of *mn*?

$$mn = 2\frac{2}{5} \cdot 1\frac{7}{8} \quad \textit{Replace m with } 2\frac{2}{5} \textit{ and n with } 1\frac{7}{8}.$$

$$\textit{Estimate: } 2 \times 2 = 4$$

$$= \frac{12}{5} \cdot \frac{15}{8} \quad \textit{Express the mixed numbers as improper fractions.}$$

$$= \frac{\overset{3}{12} \cdot \overset{3}{15}}{\underset{1}{5} \cdot \underset{2}{8}} \quad \textit{Divide 12 and 8 by the GCF, 4. Divide 15 and 5 by the GCF, 5.}$$

$$\textit{Then multiply.}$$

$$= \frac{9}{2} \text{ or } 4\frac{1}{2} \quad \textit{Compare with your estimate.}$$

Communicating Mathematics

Read and study the lesson to answer each question.

1. *Tell* how to multiply mixed numbers.

2. *Explain* to a classmate how to find the product of $3\frac{1}{2}$ and $1\frac{1}{2}$. What is the product?

Guided Practice

Express each mixed number as an improper fraction.

3. $6\frac{1}{3}$
4. $2\frac{5}{8}$
5. $5\frac{3}{4}$

Find each product. Write in simplest form.

6. $6 \times 2\frac{1}{4}$
7. $\frac{2}{3} \times 5\frac{2}{5}$
8. $2\frac{4}{5} \times 3\frac{3}{4}$

Solve each equation. Write the solution in simplest form.

9. $12 \times 3\frac{1}{6} = n$
10. $a = 2\frac{4}{5} \times 3\frac{1}{8}$
11. $1\frac{1}{4} \times \frac{8}{9} = x$

12. **Algebra** Evaluate the expression st if $s = 2\frac{2}{3}$ and $t = 4\frac{2}{3}$.

13. **Music** Refer to the beginning of the lesson.
    a. If an eighth note gets half of a beat, how many beats would a dotted eighth note get?
    b. A sixteenth note gets one-fourth of a beat. How many beats would a dotted sixteenth note get?

EXERCISES

Practice

Express each mixed number as an improper fraction.

14. $4\frac{1}{2}$
15. $7\frac{2}{3}$
16. $3\frac{1}{4}$
17. $4\frac{2}{3}$

18. $5\frac{5}{7}$
19. $6\frac{3}{5}$
20. $1\frac{5}{8}$
21. $8\frac{3}{4}$

Find each product. Write in simplest form.

22. $1\frac{4}{5} \times 3$
23. $1\frac{7}{9} \times 3\frac{3}{4}$
24. $12\frac{1}{2} \times \frac{1}{5}$

25. $4\frac{3}{8} \times 2\frac{2}{5}$
26. $4 \times \frac{2}{5}$
27. $3\frac{3}{8} \times 1\frac{4}{9}$

28. $\frac{5}{11} \times 6\frac{2}{7}$
29. $7\frac{7}{8} \times 5\frac{1}{3}$
30. $3\frac{5}{8} \times 4$

Solve each equation. Write the solution in simplest form.

31. $6 \times 3\frac{1}{2} = y$
32. $h = \frac{1}{3} \times 2\frac{1}{4}$
33. $k = 3\frac{1}{6} \times 2\frac{4}{7}$

34. $j = \frac{5}{12} \times 1\frac{4}{5}$
35. $2\frac{3}{4} \times 2\frac{2}{3} = d$
36. $8 \times 1\frac{5}{6} = c$

37. **Algebra** What is the value of $\frac{2}{3}a$ if $a = 45$?

38. **Algebra** Evaluate ab if $a = \frac{2}{3}$ and $b = 3\frac{1}{2}$.

39. Algebra If $x = \frac{9}{10}$ and $y = 8\frac{1}{3}$, find the value of xy.

Applications and Problem Solving

40. Cooking A recipe for a two-layer, 8-inch cake calls for a box of cake mix, 2 eggs, and $1\frac{1}{3}$ cups of water. How much of each ingredient is needed to make a three-layer, 8-inch cake?

41. Geometry To find the area of a parallelogram, you can use the formula, $A = b \times h$, where b is the length of the base and h is the height. Find the area of the parallelogram.

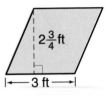

$2\frac{3}{4}$ ft

$\leftarrow$ 3 ft $\rightarrow$

42. Working on the CHAPTER Project Refer to the table on page 267.

a. In flight 2, Mitiku's airplane flew $13\frac{1}{2}$ feet in 3 seconds. How far would his airplane travel if it remained in the air for 1 minute? Find the distance the other two airplanes would travel in 1 minute.

b. For each of your flights, find the distance the plane would travel in 1 minute.

43. Critical Thinking Is $2\frac{2}{3} \times 4\frac{1}{2}$ more or less than 10? Explain how you know without actually multiplying.

Mixed Review

44. Standardized Test Practice Evaluate ab if $a = \frac{7}{8}$ and $b = \frac{2}{3}$. *(Lesson 7-2)*

A $\frac{72}{83}$ **B** $\frac{21}{16}$ **C** $\frac{9}{11}$ **D** $\frac{7}{12}$ **E** Not Here

45. Find $7\frac{1}{6} - 3\frac{3}{4}$. Write the answer in simplest form. *(Lesson 6-6)*

46. Round $\frac{8}{11}$ to the nearest half. *(Lesson 6-1)*

47. Money Matters Mr. Maldonado kept track of the phone calls each of his four children made during the month of September. Make a bar graph for the set of data. *(Lesson 2-3)*

Name	Number of Calls
Luis	45
Lorena	40
Diana	25
Mirna	25

For **Extra Practice**, see page 575.

48. Evaluate the expression $12 - 8 \div 2 + 1$. *(Lesson 1-4)*

MATH IN THE MEDIA

1. Explain why the comic is funny.

2. Suppose Sally multiplies $4\frac{1}{2}$ by $6\frac{5}{8}$. What is the product?

Integration: Geometry
Circles and Circumference

What you'll learn

You'll learn to find the circumference of circles.

When am I ever going to use this?

Knowing how to find the circumference of a circle can help you determine the amount of fencing needed to surround a circular swimming pool.

Word Wise

circle
center
radius
diameter
circumference

The first African-American woman to win an Olympic gold medal was Alice Coachman. She received a gold medal for the high jump in the 1948 Olympics. An Olympic medal is a model of a circle.

A **circle** is a set of points in a plane, all of which are the same distance from a fixed point in the plane called the **center**.

The distance from the center to any point on the circle is called the **radius (r)**. The distance across the circle through its center is called the **diameter (d)**. The diameter of a circle is twice the length of its radius. The **circumference (C)** is the distance around the circle.

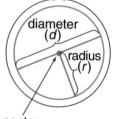

Circumference and diameter are related in a special way.

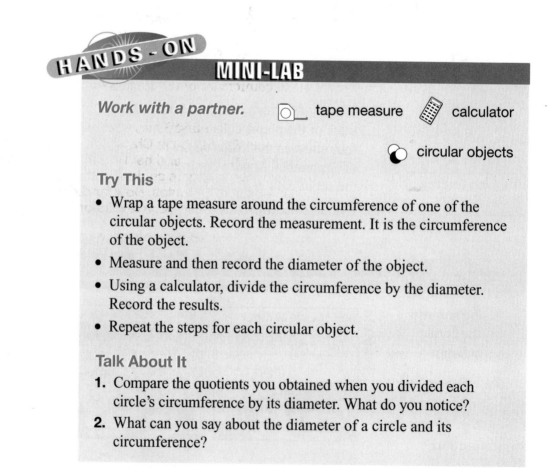

HANDS-ON
MINI-LAB

Work with a partner.   tape measure   calculator
   circular objects

Try This

- Wrap a tape measure around the circumference of one of the circular objects. Record the measurement. It is the circumference of the object.
- Measure and then record the diameter of the object.
- Using a calculator, divide the circumference by the diameter. Record the results.
- Repeat the steps for each circular object.

Talk About It

1. Compare the quotients you obtained when you divided each circle's circumference by its diameter. What do you notice?
2. What can you say about the diameter of a circle and its circumference?

In the Mini-Lab, you discovered that the circumference of a circle is always a little more than three times its diameter. The exact number of times is represented by the Greek letter π (pi).

Circumference	**Words:** The circumference of a circle is equal to π times its diameter or π times twice its radius.	
	Symbols: $C = \pi d$ or $C = 2\pi r$	**Model:** 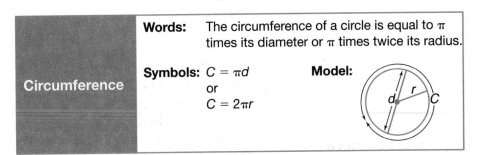

The decimal 3.14 and the fraction $\frac{22}{7}$ are used as approximations for π.

Examples

1 Find the circumference of a circle with a diameter of $1\frac{1}{2}$ feet.

$C = \pi d$

$\approx \frac{22}{7} \cdot 1\frac{1}{2}$ *Replace π with $\frac{22}{7}$ and d with $1\frac{1}{2}$.*

$\approx \frac{22}{7} \cdot \frac{3}{2}$ *Write $1\frac{1}{2}$ as an improper fraction.*

$\approx \frac{\overset{11}{22}}{7} \cdot \frac{3}{\underset{1}{2}}$ *Divide 2 and 22 by the GCF, 2.*

$\approx \frac{33}{7}$ or $4\frac{5}{7}$ *Simplify.*

The circumference of the circle is $4\frac{5}{7}$ feet.

APPLICATION

2 Inventions The first Ferris wheel was built for the 1893 World's Columbian Exposition in Chicago. It had a radius of 125 feet, weighed 1,200 tons, and held about 2,100 people. For 50 cents, fairgoers could ride the Ferris wheel for twenty minutes. How far did the passengers travel on each revolution?

You need to find the circumference of the Ferris wheel. The radius of the wheel is 125 feet.

$C = 2\pi r$

$\approx 2 \cdot 3.14 \cdot 125$ *Replace π with 3.14 and r with 125.*

2 ⊠ 3.14 ⊠ 125 ⊟ *785*

The passengers traveled about 785 feet on each revolution.

Communicating Mathematics

Read and study the lesson to answer each question.

1. *Write*, in your own words, how to find the circumference of a circle with a radius of 3 centimeters.

2. *Tell* how you know when to use either $\frac{22}{7}$ or 3.14 for π.

HANDS-ON MATH

3. *Locate* two circular objects in your home. Measure the circumference and diameter of each object. How does the quotient of each circumference and its diameter compare to π?

Guided Practice

Find the circumference of each circle shown or described. Use $\frac{22}{7}$ or 3.14 for π. Round decimal answers to the nearest tenth.

4.
14 ft

5.
$2\frac{1}{2}$ yd

6. $d = \frac{7}{8}$ yd

7. Find the circumference of a circle with a radius of 0.95 meter.

8. *Earth Science* Scientists in Goldstone, California, use a radio telescope to search for signs of life in space. This radio telescope has a circular dish with a diameter of 112 feet. What is the circumference of the radio dish?

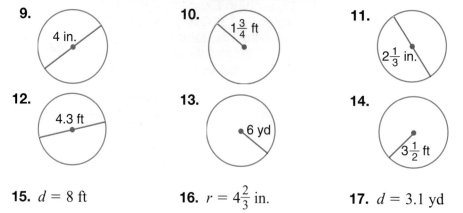

EXERCISES

Practice

Find the circumference of each circle shown or described. Use $\frac{22}{7}$ or 3.14 for π. Round decimal answers to the nearest tenth.

9.
4 in.

10.
$1\frac{3}{4}$ ft

11.
$2\frac{1}{3}$ in.

12.
4.3 ft

13.
6 yd

14.
$3\frac{1}{2}$ ft

15. $d = 8$ ft

16. $r = 4\frac{2}{3}$ in.

17. $d = 3.1$ yd

18. Find the circumference of a circle with a radius of 9.6 inches.

19. Find the measure of the circumference of a circle that has a diameter of 16.8 centimeters.

20. The radius of a circle measures $5\frac{1}{4}$ feet. What is the measure of its circumference?

Applications and Problem Solving

21. Transportation A certain model of a "penny-farthing" bicycle had a large front wheel with a radius of $2\frac{1}{2}$ feet. How far would the bicycle travel on each rotation of the wheel?

22. Entertainment The tallest Ferris wheel in the United States today is the Texas Star Ferris wheel at Fair Park in Dallas, Texas. Its diameter is about 212 feet. What is the difference between the circumference of this Ferris wheel and the Ferris wheel described in Example 2 on page 281?

23. Critical Thinking Tell how the circumference of two circles compare if the diameter of one is twice as long as the diameter of the other.

Mixed Review

24. Find $1\frac{5}{7} \times 2\frac{5}{8}$. Write in simplest form. *(Lesson 7-3)*

25. Algebra Solve $x - \frac{1}{12} = \frac{5}{12}$. Write the solution in simplest form. *(Lesson 6-3)*

26. Standardized Test Practice In July, an average of 209 books were checked out each day at the Strongville Public Library. Which is the best estimate for the number of books checked out for the whole month of July? *(Lesson 1-3)*

   A 2,000   **B** 4,000   **C** 5,000   **D** 6,000   **E** 8,000

For **Extra Practice**, see page 576.

Mid-Chapter Self Test

Estimate each product. *(Lesson 7-1)*

1. $17 \times \frac{3}{4}$

2. $3\frac{4}{5} \times 6\frac{1}{7}$

3. $2\frac{3}{4} \times \frac{2}{5}$

Find each product. *(Lessons 7-2 and 7-3)*

4. $\frac{1}{6} \times \frac{4}{5}$

5. $\frac{3}{10} \times \frac{5}{6}$

6. $\frac{1}{3} \times 4\frac{2}{7}$

7. $2\frac{5}{8} \times 1\frac{2}{7}$

Find the circumference of each circle shown or described. Use $\frac{22}{7}$ or 3.14 for π. Round decimal answers to the nearest tenth. *(Lesson 7-4)*

8.

$1\frac{2}{5}$ yd

9. $d = 13.6$ cm

10. Pets If a hamster wheel has a radius of 4 inches, how far does the hamster run in one rotation of the wheel? *(Lesson 7-4)*

Lesson 7-4 Integration: Geometry Circles and Circumference **283**

COOPERATIVE LEARNING

7-5A Dividing Fractions

A Preview of Lesson 7-5

geoboard or
dot paper

geobands

In Lesson 7-2A, you learned how to use a geoboard to multiply fractions. You can also use a geoboard to divide fractions.

TRY THIS

Work with a partner.

Step 1 Use geobands to make a rectangle as shown at the right. This rectangle represents 1.

Step 2 Use geobands to show fourths. There are four $\frac{1}{4}$'s in 1. Therefore, $1 \div \frac{1}{4} = 4$.

Step 3 Use geobands to show eighths. There are eight $\frac{1}{8}$'s in 1. Therefore, $1 \div \frac{1}{8} = 8$.

Examine the geoboard. What is $\frac{1}{4} \div \frac{1}{8}$? *Think: $\frac{1}{4} \div \frac{1}{8}$ means:*

How many $\frac{1}{8}$'s are in $\frac{1}{4}$?

ON YOUR OWN

Use a geoboard and geobands to find each quotient.

1. $1 \div \frac{1}{2}$ **2.** $\frac{1}{2} \div \frac{1}{8}$ **3.** $\frac{2}{3} \div \frac{1}{6}$ **4.** $\frac{3}{4} \div \frac{1}{8}$

5. Look Ahead Find the quotient of $\frac{1}{3}$ and $\frac{1}{6}$ without using a geoboard.

Dividing Fractions

What you'll learn

You'll learn to divide fractions.

When am I ever going to use this?

Knowing how to divide fractions can help you decide how much food to order for a party.

Word Wise

reciprocal

Dan ordered 5 large pepperoni pizzas for his birthday party. There will be 16 guests at the party. He estimates that each guest will eat about $\frac{1}{4}$ of a pizza. Did Dan order enough pizza?

You need to find the number of $\frac{1}{4}$ servings that are in 5 pizzas.

The model shows $5 \div \frac{1}{4}$.

The dashed lines show that each pizza contains four $\frac{1}{4}$-pizza servings. Five pizzas contain 5 times 4, or 20 servings of pizza. That is, $5 \div \frac{1}{4} = 20$. So, Dan will have enough pizza.

$5 \div \frac{1}{4} = 20 \Leftrightarrow 5 \times 4 = 20$ *Dividing by $\frac{1}{4}$ gives the same result as multiplying by 4.*

The numbers $\frac{1}{4}$ and 4 have a special relationship. Their product is 1.

$$\frac{1}{4} \times 4 = 1$$

Any two numbers whose product is 1 are called **reciprocals**.

Examples

Find the reciprocal of each number.

1 6

Since $6 \times \frac{1}{6} = 1$,

the reciprocal of 6 is $\frac{1}{6}$.

2 $\frac{3}{8}$

Since $\frac{3}{8} \times \frac{8}{3} = 1$,

the reciprocal of $\frac{3}{8}$ is $\frac{8}{3}$.

Study Hint

Mental Math The reciprocal of a number is found by "inverting" the fraction. That is, the numerator and the denominator are interchanged.

You can use reciprocals to divide fractions.

Dividing Fractions	To divide by a fraction, multiply by its reciprocal.

Example **3** Find $\frac{5}{8} \div \frac{3}{4}$.

$$\frac{5}{8} \div \frac{3}{4} = \frac{5}{8} \times \frac{4}{3} \qquad \textit{Multiply by the reciprocal of } \frac{3}{4}.$$

$$= \frac{5}{8} \times \frac{\overset{1}{\cancel{4}}}{3} \qquad \textit{Divide 4 and 8 by the GCF, 4.}$$

$$= \frac{5}{2} \times \frac{1}{3} \qquad \begin{array}{l}\textit{Multiply the numerators.}\\ \textit{Multiply the denominators.}\end{array}$$

$$= \frac{5}{6}$$

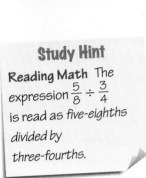

Study Hint

Reading Math The expression $\frac{5}{8} \div \frac{3}{4}$ is read as *five-eighths divided by three-fourths.*

Recall that division is related to multiplication. Sometimes it is helpful to write a multiplication sentence first and then write the related division sentence.

Example **4**

Real World APPLICATION

Food Mr. Tadashi had $\frac{3}{4}$ of a pan of lasagna left for dinner. He decided to divide the remaining lasagna into 6 equal pieces for his family. What part of the pan of lasagna will each person get?

Explore You know the number of pieces, 6, and the total amount of lasagna remaining, $\frac{3}{4}$ of a pan.

Plan

$$6 \quad \times \quad ? \quad = \quad \frac{3}{4} \quad \Leftrightarrow \quad \frac{3}{4} \quad \div \quad 6 \quad = \quad ?$$

| number of pieces | size of each piece | product or total | product or total | number of pieces | size of each piece |

Solve

$$\frac{3}{4} \div 6 = \frac{3}{4} \times \frac{1}{6} \qquad \textit{Multiply by the reciprocal of 6.}$$

$$= \frac{\overset{1}{\cancel{3}}}{4} \times \frac{1}{\underset{2}{\cancel{6}}} \qquad \textit{Divide 3 and 6 by the GCF, 3.}$$

$$= \frac{1}{4} \times \frac{1}{2} \qquad \begin{array}{l}\textit{Multiply the numerators.}\\ \textit{Multiply the denominators.}\end{array}$$

$$= \frac{1}{8}$$

Each person will get $\frac{1}{8}$ of a whole pan of lasagna.

Examine The number of pieces of lasagna times the size of each piece should equal $\frac{3}{4}$.

$$6 \times \frac{1}{8} = \frac{3}{4}$$

$$\frac{6}{8} = \frac{3}{4}$$

$$\frac{3}{4} = \frac{3}{4} \quad \checkmark$$

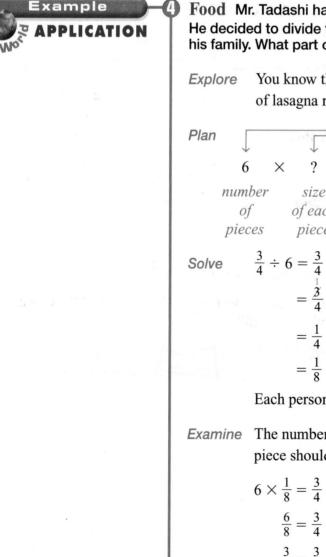

There are 8 parts and 6 are shaded. So, $\frac{6}{8}$ or $\frac{3}{4}$ of the lasagna was left.

5 Solve the equation $\frac{3}{4} \div \frac{2}{5} = m$.

$$\frac{3}{4} \div \frac{2}{5} = m$$

$$\frac{3}{4} \times \frac{5}{2} = m \quad \textit{Multiply by the reciprocal of } \frac{2}{5}.$$

$$\frac{15}{8} = m \quad \textit{Multiply the numerators.}$$
$$\phantom{\frac{15}{8} = m} \quad \textit{Multiply the denominators.}$$

$$1\frac{7}{8} = m$$

The solution is $1\frac{7}{8}$.

CHECK FOR UNDERSTANDING

Communicating Mathematics

Read and study the lesson to answer each question.

1. **Explain** how you would use the reciprocal to find $\frac{1}{2} \div \frac{2}{3}$.

2. **Write** a sentence that tells how the model shows $2 \div \frac{1}{3} = 6$.

Guided Practice

Find the reciprocal of each number.

3. $\frac{1}{2}$

4. $\frac{3}{5}$

5. 8

Find each quotient. Write in simplest form.

6. $\frac{1}{5} \div \frac{1}{4}$

7. $\frac{2}{3} \div 4$

8. $\frac{5}{8} \div \frac{1}{2}$

9. $\frac{3}{4} \div \frac{2}{3}$

Solve each equation. Write the solution in simplest form.

10. $\frac{1}{6} \div \frac{1}{2} = y$

11. $h = 8 \div \frac{2}{5}$

12. $\frac{8}{9} \div \frac{1}{3} = a$

13. **Algebra** What is the value of $m \div n$ if $m = \frac{2}{3}$ and $n = \frac{2}{5}$?

14. **Food** Each serving of a peach pie is $\frac{1}{12}$ of the pie. If $\frac{1}{2}$ of the peach pie is left, how many servings are left?

EXERCISES

Practice

Find the reciprocal of each number.

15. 5

16. $\frac{2}{5}$

17. $\frac{1}{3}$

18. $\frac{5}{6}$

19. $\frac{1}{7}$

20. 4

21. 1

22. $\frac{3}{8}$

Find each quotient. Write in simplest form.

23. $\frac{1}{3} \div \frac{3}{5}$

24. $\frac{5}{6} \div \frac{5}{8}$

25. $\frac{3}{5} \div \frac{3}{4}$

26. $2 \div \frac{1}{6}$

27. $\frac{2}{5} \div 4$

28. $\frac{1}{5} \div \frac{1}{6}$

29. $\frac{2}{9} \div \frac{2}{3}$

30. $\frac{1}{2} \div \frac{1}{3}$

31. $\frac{5}{8} \div 2$

32. If you divide $\frac{5}{6}$ by $\frac{1}{4}$, what is the quotient?

Solve each equation. Write the solution in simplest form.

33. $\frac{2}{3} \div \frac{3}{4} = m$

34. $k = \frac{3}{4} \div 12$

35. $\frac{1}{5} \div \frac{3}{10} = z$

36. $d = \frac{1}{4} \div \frac{1}{8}$

37. $\frac{3}{4} \div \frac{5}{6} = v$

38. $a = \frac{1}{3} \div \frac{1}{2}$

39. $\frac{8}{9} \div 4 = h$

40. $x = \frac{5}{9} \div \frac{2}{3}$

41. $3 \div \frac{6}{7} = w$

42. *Algebra* Find the value of $a \div b$ if $a = \frac{1}{2}$ and $b = 9$.

Applications and Problem Solving

43. *Money Matters* Mr. Barojas parked his car at a 4-hour parking meter. If each half-hour of parking costs 25 cents, how many quarters does he need for 4 hours of parking?

44. *Geography* The table lists the continents' approximate sizes relative to Earth's total landmass.

  a. About how many times larger is North America than South America?

  b. About how many times larger is Asia than North America?

  c. About how many times larger is Asia than Africa?

Continent	Fraction of Earth's Landmass
Asia	$\frac{3}{10}$
Africa	$\frac{1}{5}$
North America	$\frac{1}{6}$
South America	$\frac{1}{8}$
Antarctica	$\frac{1}{10}$
Europe	$\frac{1}{16}$
Australia	$\frac{1}{20}$

45. *Critical Thinking* Solve mentally.

  a. $\frac{6{,}978}{3{,}012} \times \frac{11}{10} \div \frac{6{,}978}{3{,}012}$

  b. $\frac{6{,}978}{10} \times \frac{11}{3{,}012} \div \frac{6{,}978}{3{,}012}$

Mixed Review

46. *Geometry* Find the circumference of a circle with a radius of 4 meters. *(Lesson 7-4)*

For **Extra Practice**, see page 576.

47. Express $\frac{7}{9}$ as a decimal. Use bar notation to show a repeating decimal. *(Lesson 5-10)*

48. **Standardized Test Practice** Warren's dog weighs 25.9 kilograms. How many grams does the dog weigh? *(Lesson 4-9)*

  A 25,900 g **B** 0.0259 g
  C 1025.9 g **D** 259,000 g

For the latest weather statistics, visit: www.glencoe.com/sec/math/mac/mathnet

49. *Statistics* The average annual precipitation in Georgia is 48.61 inches. About how many inches of precipitation does Georgia average each month? *(Lesson 4-5)*

50. Add 15.783 and 390.81. *(Lesson 3-6)*

Dividing Mixed Numbers

What you'll learn

You'll learn to divide mixed numbers.

When am I ever going to use this?

Knowing how to divide mixed numbers can help you find the average speed you travel on a trip.

A tsunami, (soo-NAH-mee), or tidal wave, can travel from one side of the Pacific Ocean to the other in less than a day. If a tsunami traveled 1,400 miles from a point in the Pacific Ocean to the Alaskan coastline in $2\frac{1}{2}$ hours, how many miles per hour did it travel?

You need to find $1,400 \div 2\frac{1}{2}$.

To divide mixed numbers, express each mixed number as an improper fraction. Then divide as with fractions.

$$1,400 \div 2\frac{1}{2} = \frac{1,400}{1} \div \frac{5}{2} \quad \textit{Express each mixed number as an improper fraction.}$$

$$= \frac{1,400}{1} \times \frac{2}{5} \quad \textit{Multiply by the reciprocal.}$$

$$= \frac{\overset{280}{1,400}}{1} \times \frac{2}{\underset{1}{5}} \quad \textit{Divide 5 and 1,400 by the GCF, 5.}$$

$$= 560$$

The tsunami traveled 560 miles per hour.

Examples

1 Find $4\frac{1}{2} \div 3\frac{3}{4}$. *Estimate: $4 \div 4 = 1$*

$$4\frac{1}{2} \div 3\frac{3}{4} = \frac{9}{2} \div \frac{15}{4} \quad \textit{Express each mixed number as an improper fraction.}$$

$$= \frac{9}{2} \times \frac{4}{15} \quad \textit{Multiply by the reciprocal.}$$

$$= \frac{\overset{3}{9}}{\underset{1}{2}} \times \frac{\overset{2}{4}}{\underset{5}{15}} \quad \begin{array}{l}\textit{Divide 9 and 15 by the GCF, 3.}\\\textit{Divide 2 and 4 by the GCF, 2.}\end{array}$$

$$= \frac{6}{5} \text{ or } 1\frac{1}{5} \quad \textit{Simplify. Compare with your estimate.}$$

INTEGRATION **2** **Algebra** Solve $w = 2\frac{4}{5} \div \frac{7}{8}$. *Estimate: $3 \div 1 = 3$*

$$w = \frac{14}{5} \div \frac{7}{8} \quad \textit{Express each mixed number as an improper fraction.}$$

$$w = \frac{\overset{2}{14}}{5} \times \frac{8}{\underset{1}{7}} \quad \begin{array}{l}\textit{Multiply by the reciprocal.}\\\textit{Divide 7 and 14 by the GCF, 7.}\end{array}$$

$$w = \frac{16}{5} \text{ or } 3\frac{1}{5} \quad \textit{Simplify. Compare with your estimate.}$$

Communicating Mathematics

Read and study the lesson to answer each question.

1. *Explain* to a classmate how to find the reciprocal of $4\frac{5}{8}$.

2. *Tell* whether $4 \div \frac{1}{2}$ is more or less than 4. Explain your reasoning.

3. *Write*, in your own words, the steps to follow when finding the quotient of 12 and $2\frac{2}{3}$.

Guided Practice

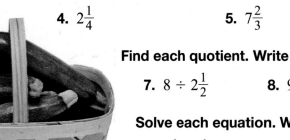

Write each mixed number as an improper fraction. Then write its reciprocal.

4. $2\frac{1}{4}$　　　　　5. $7\frac{2}{3}$　　　　　6. $6\frac{1}{3}$

Find each quotient. Write in simplest form.

7. $8 \div 2\frac{1}{2}$　　　8. $9\frac{1}{2} \div 4\frac{3}{4}$　　　9. $3\frac{3}{4} \div \frac{5}{6}$

Solve each equation. Write the solution in simplest form.

10. $6\frac{1}{4} \div \frac{1}{2} = g$　　　　　11. $5\frac{1}{3} \div 4\frac{2}{3} = n$

12. If $9\frac{1}{2}$ is divided by $\frac{1}{2}$, what is the quotient?

13. *Food* Mrs. Golubec needs to cut a zucchini into slices that measure $\frac{3}{8}$ inch thick. If the zucchini measures $13\frac{1}{2}$ inches long, how many slices can she cut?

EXERCISES

Practice

Write each mixed number as an improper fraction. Then write its reciprocal.

14. $4\frac{1}{2}$　　　15. $6\frac{2}{5}$　　　16. $1\frac{3}{8}$　　　17. $2\frac{4}{5}$

18. $5\frac{5}{6}$　　　19. $3\frac{3}{4}$　　　20. $7\frac{1}{2}$　　　21. $9\frac{1}{4}$

Find each quotient. Write in simplest form.

22. $7\frac{1}{3} \div 6$　　　　23. $4\frac{3}{4} \div \frac{5}{8}$　　　　24. $5\frac{1}{4} \div 3\frac{1}{2}$

25. $10\frac{1}{2} \div \frac{7}{8}$　　　26. $1\frac{5}{9} \div 2\frac{1}{3}$　　　27. $5 \div 6\frac{1}{4}$

28. $2\frac{4}{5} \div 5\frac{3}{5}$　　　29. $10 \div 2\frac{2}{7}$　　　30. $13 \div 2\frac{3}{5}$

Solve each equation. Write the solution in simplest form.

31. $3 \div 4\frac{1}{2} = t$　　　32. $a = 1\frac{3}{8} \div 2\frac{3}{4}$　　　33. $6\frac{1}{2} \div \frac{1}{4} = m$

34. $y = 4\frac{4}{7} \div 2\frac{2}{3}$　　　35. $\frac{4}{5} \div 1\frac{1}{5} = q$　　　36. $c = 5\frac{1}{3} \div 4$

37. Algebra What is the value of $m \div n$ if $m = 6\frac{2}{3}$ and $n = \frac{4}{5}$?

38. Algebra Evaluate $p \div q$ if $p = 9$ and $q = 3\frac{3}{5}$.

39. Find the following for each pair of fractions.

$\frac{2}{5}$ and $\frac{3}{10}$, $3\frac{1}{4}$ and $2\frac{2}{3}$, 8 and $1\frac{5}{8}$

   a. sum   **b.** difference

   c. product   **d.** quotient in simplest form

Applications and Problem Solving

40. Statistics The table shows how many hours Tiarri watched cartoons each day after school. What is the mean number of hours Tiarri watched cartoons in the 5 days?

Day	Hours
Monday	$1\frac{1}{4}$
Tuesday	$\frac{1}{2}$
Wednesday	$1\frac{1}{2}$
Thursday	$\frac{1}{2}$
Friday	$1\frac{3}{4}$

41. Measurement Marnee is working on her history project. She wants to place photographs of people from various countries in one vertical row on a poster board that is $17\frac{1}{2}$ inches long. If each photograph is $2\frac{3}{4}$ long, how many photographs can Marnee place on the poster board?

42. Working on the CHAPTER Project   Refer to the table on page 267.

   a. Use the information from Mitiku's first flight to determine how long it would take his airplane to fly 100 miles.

   b. Use the information from your first flight to determine how long it would take your airplane to fly 100 miles.

43. Critical Thinking Tell whether $\frac{8}{10} \div \frac{2}{3}$ is greater than or less than $\frac{8}{10} \div \frac{3}{4}$ without solving. Explain your reasoning.

Mixed Review

44. Find $\frac{3}{7} \div \frac{3}{5}$. Write in simplest form. *(Lesson 7-5)*

45. Find $\frac{5}{7} \times \frac{2}{3}$. Write in simplest form. *(Lesson 7-2)*

46. Algebra Solve $y - \frac{7}{12} = \frac{5}{12}$. *(Lesson 6-3)*

47. Standardized Test Practice What is the GCF of 96 and 172? *(Lesson 5-3)*

   A 2   **B** 4   **C** 24   **D** 43

48. Geometry What is the area of a rectangle with a length of 12.8 feet and a width of 7.4 feet? *(Lesson 4-4)*

49. Life Science A 150-pound human body contains about 0.2 ounce of iron, 0.07 ounce of zinc, and 0.004 ounce of copper. Write each of these decimals in words. *(Lesson 3-1)*

For **Extra Practice**, see page 576.

50. Technology Bob's answering machine can record fifty 30-second messages. How many minutes of tape does his machine have? *(Lesson 1-1)*

Integration: Measurement
Changing Customary Units

What you'll learn

You'll learn to change units within the customary system.

When am I ever going to use this?

You'll often need to change customary units when you are comparing labels in the grocery store.

Word Wise

fluid ounce   gallon
cup   ounce
pint   pound
quart   ton

Kijana (meaning "little boy" in Swahili) is the first African elephant since 1984 to survive birth in captivity. His mother rejected him so the keepers at the Oakland Zoo in California decided to raise him. Kijana drinks about 25 quarts of formula a day. How many gallons of formula does he drink each day?

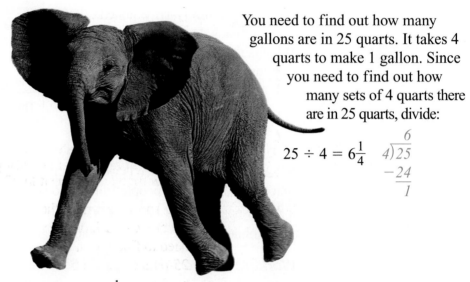

You need to find out how many gallons are in 25 quarts. It takes 4 quarts to make 1 gallon. Since you need to find out how many sets of 4 quarts there are in 25 quarts, divide:

$$25 \div 4 = 6\frac{1}{4} \qquad \begin{array}{r} 6 \\ 4\overline{)25} \\ -24 \\ \hline 1 \end{array}$$

Kijana drinks $6\frac{1}{4}$ gallons of formula a day.

The most commonly used customary units of capacity are the **fluid ounce, cup, pint, quart,** and **gallon.**

> 1 cup (c) = 8 fluid ounces (fl oz)
> 1 pint (pt) = 2 cups
> 1 quart (qt) = 2 pints
> 1 gallon (gal) = 4 quarts

The most commonly used customary units of weight are **ounce, pound,** and **ton.**

> 1 pound (lb) = 16 ounces (oz)
> 1 ton (T) = 2,000 pounds

To change customary units of capacity and weight,

1. Determine whether you are changing from smaller to larger units or from larger to smaller units.

2. To change from smaller to larger units, divide. To change from larger to smaller units, multiply.

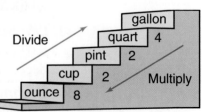

To solve Example 1, you can use the ratio of 2 cups to 1 pint.

1 5 pt = _?_ c   *Think: Each pint equals 2 cups.*

$5 \times 2 = 10$ *Multiply to change from a larger unit (pt) to smaller unit (c)*

5 pt = 10 c

2 5,000 lb = _?_ T   *Think: It takes 2,000 pounds to make 1 ton.*

$5,000 \div 2,000 = 2\frac{1}{2}$ *Divide to change from smaller units (lb) to larger units (T).*

$5,000 \text{ lb} = 2\frac{1}{2} \text{ T}$

3 9 gal = _?_ pt   *Think: Each gallon equals 4 quarts. Each quart equals 2 pints. You need to multiply twice.*

$9 \times 4 = 36$ *Multiply to change from gallons to quarts.*

$36 \times 2 = 72$ *Multiply to change from quarts to pints.*

9 gal = 72 pt

CONNECTION

4 **Life Science** Miquel estimates that the finches eat 8 ounces of bird seed a day at his feeder. If he buys a 25-pound bag of bird seed, about how many days will it last?

Explore   You know how much the finches eat each day. You need to find how many days a 25-pound bag of bird seed will last.

Plan   First, find the total number of ounces in 25 pounds. Then find how many sets of 8 ounces there are in the 25-pound bag of bird seed.

Solve   25 lb = _?_ oz   *Think: Each pound equals 16 ounces.*

$25 \times 16 = 400$ *Multiply to change from pounds to ounces.*

There are 400 ounces in a 25-pound bag of bird seed.

$400 \div 8 = 50$

So, the bag of bird seed will last about 50 days.

Examine   To see if your answer makes sense, think $8 \times 50 = 400$. So, 50 days is a reasonable answer.

CHECK FOR UNDERSTANDING

Communicating Mathematics

Read and study the lesson to answer each question.

1. **List** the common customary units of capacity and weight.

2. **Tell** how to change 3 tons to pounds.

Complete.

3. 2 gal = _?_ qt **4.** 14 c = _?_ pt **5.** 64 oz = _?_ lb

6. $3\frac{1}{2}$ T = _?_ lb **7.** 32 fl oz = _?_ pt **8.** 4 T = _?_ oz

9. ***Life Science*** Kijana, the baby elephant, will weigh about $6\frac{1}{2}$ tons when fully grown. About how many pounds will he weigh?

EXERCISES

Practice **Complete.**

10. 8 qt = _?_ pt **11.** 24 qt = _?_ gal **12.** 7 c = _?_ pt

13. 9 lb = _?_ oz **14.** 6 gal = _?_ qt **15.** 16 fl oz = _?_ c

16. 10 pt = _?_ qt **17.** 5 c = _?_ fl oz **18.** 500 lb = _?_ T

19. $2\frac{1}{4}$ T = _?_ lb **20.** $4\frac{1}{2}$ pt = _?_ c **21.** $2\frac{1}{2}$ qt = _?_ c

22. 32 c = _?_ gal **23.** 12 pt = _?_ gal **24.** 12 c = _?_ qt

25. How many tons are in 80,000 pounds?

26. Find how many quarts there are in 40 fluid ounces.

Applications and Problem Solving

27. ***Geography*** Giant clams live along the Great Barrier Reef off the coast of Australia. They measure up to 6 feet long and weigh as much as 0.25 ton. How many pounds is this?

28. ***Money Matters*** Tashauna rides her moped from her home in Connecticut to her ballet lessons in New York. She uses about 3 quarts of gasoline a week. How much will she save a year in taxes if she buys her gasoline in New York rather than Connecticut?

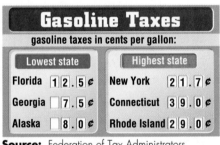

Gasoline Taxes	
gasoline taxes in cents per gallon:	
Lowest state	**Highest state**
Florida 12.5¢	New York 21.7¢
Georgia 7.5¢	Connecticut 39.0¢
Alaska 8.0¢	Rhode Island 29.0¢

Source: Federation of Tax Administrators

For **Extra Practice,** see page 577.

29. ***Critical Thinking*** What can you divide by to change 256 cups directly to gallons?

Mixed Review **30.** Find $2\frac{3}{5} \div 1\frac{2}{3}$. Write the quotient in simplest form. *(Lesson 7-6)*

31. State whether 405 is divisible by 2, 3, 5, 6, 9, or 10. *(Lesson 5-1)*

32. ***Measurement*** Change 12,237 milliliters to liters. *(Lesson 4-9)*

33. ***Standardized Test Practice*** Frank purchased $13.72 worth of gasoline. He gives the gas station attendant a $20 bill. How much change should he receive? *(Lesson 3-6)*

  A $7.72 **B** $7.58 **C** $7.28 **D** $6.72 **E** Not Here

measuring cups

containers

water

COOPERATIVE LEARNING

7-7B Measurement

A Follow-Up of Lesson 7-7

Have you ever dreamed of visiting another planet or the moon? You would discover many things to be different, even your weight!

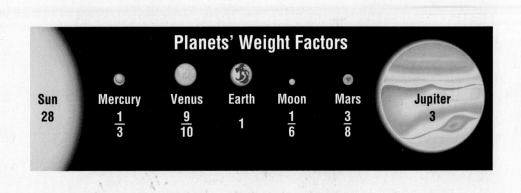

Planets' Weight Factors

| Sun 28 | Mercury $\frac{1}{3}$ | Venus $\frac{9}{10}$ | Earth 1 | Moon $\frac{1}{6}$ | Mars $\frac{3}{8}$ | Jupiter 3 |

TRY THIS

Work with a partner.

Step 1 Choose a work station. Record the name of the planet and its weight factor relative to Earth.

Step 2 Fill the 1-cup measuring cup with water. This represents the weight of one cup of water on Earth. Pour this amount into the container labeled *Earth*.

Step 3 Use the planet's weight factor to fill as many measuring cups with water as needed to represent the weight of one cup of water on this planet. For example, Jupiter's weight factor is 3 times that of the Earth's. So, empty 3 cups of water into the container labeled *Jupiter*. Record your results.

Step 4 Repeat Steps 1–3 at each work station.

ON YOUR OWN

1. Analyze your results. Is the weight of 1 cup of water on each planet more or less than the weight of one cup of water on Earth? more or less than one cup of water on Jupiter? Record your results.

2. Which planet's container weighs the most? Explain your reasoning.

3. How much would a 22-pound dog weigh on Jupiter?

4. How much would you weigh on the moon?

5. *Reflect Back* On Earth, a certain object weighs 1 pound. How many ounces would the same object weigh on Mars?

Lesson 7-7B HANDS-ON **295**

7-8A Look for a Pattern

A Preview of Lesson 7-8

Kerri and her brother Eric are at the museum. They are learning about the history and culture of the Hopi Indians. Let's listen in!

Kerri

It says here that the Hopi Indians began making pottery about 1,500 years ago. Wow, that's a long time ago!

Here is one of the designs they used to decorate the pottery. It appears that they liked to use geometric shapes in the designs.

I wonder what the design would look like if the pattern was continued to the right.

Well, since the Hopi Indians liked to use geometric shapes, maybe the design would include circles and triangles.

I disagree. The beauty of the design is in the pattern. I think that the same pattern is probably continued to the right.

Eric

THINK ABOUT IT

Work with a partner.

1. **Compare and contrast** Kerri's and Eric's thinking. Whose thinking do you think is accurate? Explain your reasoning.

2. **Describe** the pattern the Hopi Indians used in the design.

3. **Draw** the next three segments to the right in the pattern.

4. **Apply** the **look for a pattern** strategy to find the next two numbers in each pattern below.

   a. 3, 6, 9, 12, _?_ , _?_

   b. 5, 10, 20, 40, _?_ , _?_

For **Extra Practice**, see page 577.

ON YOUR OWN

5. The third step of the 4-step plan for problem solving asks you to *solve* the problem. *Tell* how you can use the look for a pattern strategy to help you solve a problem.

6. *Write a Problem* in which you would use the look for a pattern strategy to solve. Then ask a classmate to describe the pattern and solve the problem.

7. *Look Ahead* Explain how you could use the look for a pattern strategy to find the next number in the sequence in Exercise 17 on page 300.

MIXED PROBLEM SOLVING

STRATEGIES

Look for a pattern.
Solve a simpler problem.
Act it out.
Guess and check.
Draw a diagram.
Make a chart.
Work backward.

Solve. Use any strategy.

8. *Fashion* Matthew is choosing a shirt and a pair of jeans to wear. He has black, blue, and stonewashed jeans, and blue, gray, and white shirts. How many different combinations of jeans and shirts are possible?

9. *Food* Richard eats half of a ham sandwich at 1:00 P.M. At 3:00 P.M., he eats half of what was left of his sandwich from lunch. At 5:00 P.M., he eats half of what was left from his ham sandwich.

a. If Richard continues eating at this rate for four more hours, how much of the sandwich has been eaten?

b. At this rate, will Richard ever eat the entire sandwich? Explain.

10. *Measurement* What is the missing measurement in the pattern?

$\ldots, \underline{\quad?\quad}, \frac{1}{4}$ in., $\frac{1}{8}$ in., $\frac{1}{16}$ in., $\ldots$

11. *Geometry* Find the area of the figure.

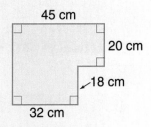

12. *Money Matters* Roger and Celina began working for the same company in 1997. Celina earned $18,000 per year, and Roger earned $14,500. Each year Roger received a $1,500 raise, and Celina received a $1,000 raise.

a. In what year will they earn the same amount of money?

b. What will be their annual salary in that year?

13. *Standardized Test Practice* At Hershal Middle School, the bell rings at 8:55, 9:40, 10:25, and 11:10 each morning. If this pattern continues, when would the next bell ring?

A 11:50 A.M.

B 11:45 A.M.

C 11:55 A.M.

D 10:40 A.M.

7-8

Integration: Patterns and Functions

Sequences

On Earth, gravity causes all falling objects to accelerate at 32 ft/s². This means that if gravity is the only force acting on a falling object, its speed will increase about 32 feet per second each second.

The table shows the effect of gravity. What will the speed of the object be after 5 seconds?

Seconds After Object is Dropped	0	1	2	3	4	5
Speed of Object (ft/s)	0	32	64	96	128	?

The numbers 0, 32, 64, 96, and 128 form a **sequence**. A sequence is a list of numbers in a specific order.

In the sequence, notice that 32 is added to each number.

$$0, \quad 32, \quad 64, \quad 96, \quad 128$$
$$+32 \quad +32 \quad +32 \quad +32$$

The next number in the sequence is 128 + 32, or 160. So, after 5 seconds, the speed of the object will be 160 ft/s.

Examples

Find the next number in each sequence.

1 18, 24, 30, 36, . . .
$$+6 \quad +6 \quad +6$$

In this sequence, 6 is added to each number. The next number is 36 + 6, or 42.

2 21, $18\frac{1}{2}$, 16, $13\frac{1}{2}$, . . .
$$-2\frac{1}{2} \quad -2\frac{1}{2} \quad -2\frac{1}{2}$$

In this sequence, $2\frac{1}{2}$ is subtracted from each number. The next number is $13\frac{1}{2} - 2\frac{1}{2}$ or 11.

Another type of sequence is one where the numbers are found by multiplying by the same number.

Examples

Study Hint

Problem Solving

You may want to use the look for a pattern strategy to help you with these sequences.

Find the next number in each sequence.

③ 2,   8,   32,   128,...

     × 4   × 4   × 4

Each number in the sequence is multiplied by 4. The next number is 128 × 4, or 512.

④ 243,   81,   27,   9,...

     $\times \frac{1}{3}$ $\times \frac{1}{3}$ $\times \frac{1}{3}$

Each number in the sequence is multiplied by $\frac{1}{3}$. The next number is $9 \times \frac{1}{3}$, or 3.

APPLICATION

⑤ **Carpentry** A roof rafter of a building is to be braced as shown. The length of braces *A*, *B*, and *C* are $10\frac{1}{2}$ inches, 21 inches, and $31\frac{1}{2}$ inches, respectively. Find the lengths of braces *D* and *E*.

$10\frac{1}{2}$,   21, $31\frac{1}{2}$, ...

    $+ 10\frac{1}{2}$ $+ 10\frac{1}{2}$

In the sequence, $10\frac{1}{2}$ is added to each number.

The length of brace *D* is $31\frac{1}{2} + 10\frac{1}{2}$, or 42 inches.

The length of brace *E* is $42 + 10\frac{1}{2}$, or $52\frac{1}{2}$ inches.

CHECK FOR UNDERSTANDING

Communicating Mathematics

Read and study the lesson to answer each question.

1. **Write**, in your own words, a definition for sequence.

2. **Tell** how the numbers are related in the sequence 16, 4, 1, $\frac{1}{4}$.

3. **You Decide** Janet says that the next number in the sequence 5, $7\frac{1}{2}$, 10, $12\frac{1}{2}$, is 14. Joshua says the next number is 15. Who is correct? Explain your reasoning.

Guided Practice

Find the next two numbers in each sequence.

4. 6, 18, 54, 162, . . .   5. 45, 38, 31, 24, . . .

Find the missing number in each sequence.

6. 2, _?_, 9, $12\frac{1}{2}$, . . .   7. _?_, 25, 5, 1, . . .

8. **Geometry** Draw the next two figures in the sequence.

Practice   **Find the next two numbers in each sequence.**

9. 15, 30, 45, 60, . . .   10. 4, 12, 36, 108, . . .

11. 12, 6, 3, $1\frac{1}{2}$, . . .   12. 19, $18\frac{1}{2}$, 18, $17\frac{1}{2}$, . . .

13. $\frac{1}{3}$, 2, 12, 72, . . .   14. $28\frac{1}{2}$, 30, $31\frac{1}{2}$, 33, . . .

Find the missing number in each sequence.

15. 49, 41, $\underline{\ ?\ }$, 25, . . .   16. 64, 16, $\underline{\ ?\ }$, 1, . . .

17. $\frac{2}{3}$, $\underline{\ ?\ }$, $1\frac{1}{3}$, $1\frac{2}{3}$, . . .   18. $\underline{\ ?\ }$, $56\frac{1}{2}$, 53, $49\frac{1}{2}$, . . .

19. . . . , 4, 40, $\underline{\ ?\ }$, 4,000, . . .   20. 1, $\frac{3}{4}$, $\underline{\ ?\ }$, $\frac{27}{64}$, . . .

21. What is the next term in the sequence x, x^2, x^3, x^4, . . .?

22. Find the missing term in the sequence $x + 5$, $x + 4$, $\underline{\ ?\ }$, $x + 2$, $x + 1$.

Applications and Problem Solving

23. *Music* The diagram shows the most common notes used in music. The names of the first four notes are whole note, half note, quarter note, and eighth note. What are the names of the next three notes?

Notes

whole (1) $\frac{1}{2}$ $\frac{1}{4}$ $\frac{1}{8}$

24. *Money Matters* Roberta Salgado rents an apartment for $565 a month. Each year, the monthly rent is expected to increase $12. What will be the monthly rent at the end of four years?

25. *Critical Thinking* The large square shown represents 1.

   a. Find the first ten numbers of the sequence represented by the model. The first number is $\frac{1}{2}$.

   b. Estimate the sum of the first ten numbers without actually adding.

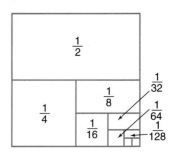

Mixed Review

26. *Measurement* How many pints are in 64 fluid ounces? *(Lesson 7-7)*

27. **Standardized Test Practice** Mrs. Matthews bought $3\frac{1}{4}$ pounds of caramels and $2\frac{1}{2}$ pounds of chocolate. How many pounds of candy did she buy altogether? *(Lesson 6-5)*

   A $5\frac{3}{4}$ **B** $5\frac{2}{6}$ **C** $1\frac{1}{2}$ **D** $\frac{3}{4}$ **E** Not Here

For **Extra Practice**, see page 577.

28. *Geometry* Find the perimeter of a triangle whose sides are each 23 centimeters long. *(Lesson 4-4)*

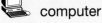
7-8B Sequences

A Follow-Up of Lesson 7-8

💻 computer

💾 spreadsheet software

As you learned in Lesson 4-3B, you can use a spreadsheet to prepare tables easily. You can use a spreadsheet to project results, make calculations, and print almost anything that can be arranged in a table. Spreadsheets can also be used to simulate experiments when the variables can be described algebraically.

TRY THIS

Work with a partner.

Use a spreadsheet to solve the following problem.

On each bounce, a ball rebounds and goes back up 0.8 of the way to its starting point. If the ball is dropped from a height of 20 meters, how high will it rebound on the 5th bounce?

Input the initial height of the ball in cell B1 and the rebound ratio in cell B2. The values of the remaining cells are determined by the formulas. The computer does the calculations.

The screen shows the results of running the spreadsheet. Cell B9 shows that after the 5th bounce, the height is approximately 6.554 meters.

B4*B2 means multiply the value in cell B4 by the value in cell B2.

	BALL BOUNCE	
	A	B
1	Initial Ht (m) =	20
2	Rebound Ratio =	0.8
3	Number of bounces	RETURN HT
4	zero ⟶ 0	= B1
5	A4 + 1	= B4*B2
6	A5 + 1	= B5*B2
7	A6 + 1	= B6*B2

A4 + 1 means add 1 to the value in cell A4.

	BALL BOUNCE	
	A	B
1	Initial Ht (m) =	20
2	Rebound Ratio =	0.8
3	Number of bounces	RETURN HT
4	0	20.000
5	1	16.000
6	2	12.800
7	3	10.240
8	4	8.192
9	5	6.554

ON YOUR OWN

1. Explain the meaning of the formula in cell B6.
2. Explain the meaning of the formula in cell A7.
3. If the ball rebounds 0.4 of the way to its starting point, what are the heights of its first six bounces?
4. Suppose a certain ball rebounds $\frac{1}{2}$ of the way to its starting point. How would you change the spreadsheet to show the heights of the bounces?

 **interNET** **CONNECTION** Chapter Review   For additional lesson-by-lesson review, visit:
www.glencoe.com/sec/math/mac/mathnet

Vocabulary

After completing this chapter, you should be able to define each
term, concept, or phrase and give an example or two of each.

Number and Operations
compatible numbers (p. 268)
reciprocals (p. 285)

Geometry
center (p. 280)
circle (p. 280)
circumference (p. 280)
diameter (p. 280)
radius (p. 280)

Problem Solving
look for a pattern (p. 296)

Measurement
cup (p. 292)
fluid ounce (p. 292)
gallon (p. 292)
ounce (p. 292)
pint (p. 292)
pound (p. 292)
quart (p. 292)
ton (p. 292)

Patterns and Functions
sequence (p. 298)

Understanding and Using the Vocabulary

Choose the letter of the term that best matches each phrase.

1. 2,000 pounds
2. the distance across a circle through the center
3. 2 cups
4. the distance from the center of a circle to any point on the circle
5. 8 fluid ounces
6. a list of numbers in a specific order
7. the distance around a circle
8. any two numbers whose product is 1
9. the set of all points in a plane that are the same distance from a given point

a. reciprocals
b. center
c. radius
d. circle
e. diameter
f. circumference
g. sequence
h. cup
i. pint
j. gallon
k. ton

In Your Own Words

10. *Explain* how you would change customary units of capacity such as cups to gallons or tons to pounds.

Objectives & Examples

Upon completing this chapter, you should be able to:

● estimate fraction products using compatible numbers and rounding *(Lesson 7-1)*

Estimate $\frac{4}{5} \times 11$.

For 11, the nearest multiple of 5 is 10.
$\frac{4}{5}$ *of 10 is 8.*

So, $\frac{4}{5} \times 11$ is about 8.

● multiply fractions *(Lesson 7-2)*

$\frac{5}{9} \times \frac{3}{10} = \frac{\overset{1}{5} \cdot \overset{1}{3}}{\underset{3}{9} \cdot \underset{2}{10}}$ *Divide by the GCF.*

$= \frac{1}{6}$

● multiply mixed numbers *(Lesson 7-3)*

$3\frac{1}{2} \times 4\frac{2}{3} = \frac{7}{2} \times \frac{\overset{7}{14}}{\underset{1}{3}}$

$= \frac{49}{3}$ or $16\frac{1}{3}$

● find the circumference of circles *(Lesson 7-4)*

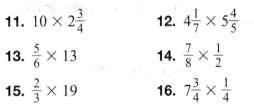

$C = 2\pi r$
$C \approx 2 \cdot 3.14 \cdot 2.4$
$C \approx 15.072$

The circumference of the circle is about 15.1 meters.

Review Exercises

Use these exercises to review and prepare for the chapter test.

Estimate each product.

11. $10 \times 2\frac{3}{4}$ **12.** $4\frac{1}{7} \times 5\frac{4}{5}$

13. $\frac{5}{6} \times 13$ **14.** $\frac{7}{8} \times \frac{1}{2}$

15. $\frac{2}{3} \times 19$ **16.** $7\frac{3}{4} \times \frac{1}{4}$

Find each product. Write in simplest form.

17. $\frac{7}{8} \times \frac{4}{5}$ **18.** $\frac{14}{15} \times \frac{10}{21}$

Solve each equation. Write the solution in simplest form.

19. $y = 12 \times \frac{5}{8}$ **20.** $\frac{4}{5} \times \frac{3}{8} = t$

Find each product. Write in simplest form.

21. $3\frac{3}{4} \times 1\frac{1}{5}$ **22.** $3\frac{2}{3} \times 6$

Solve each equation. Write the solution in simplest form.

23. $7\frac{1}{5} \times 1\frac{7}{8} = x$ **24.** $w = 3\frac{1}{8} \times 2\frac{2}{5}$

Find the circumference of each circle shown or described. Use $\frac{22}{7}$ or 3.14 for π. Round decimal answers to the nearest tenth.

25.

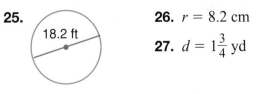

26. $r = 8.2$ cm

27. $d = 1\frac{3}{4}$ yd

Objectives & Examples

Review Exercises

divide fractions *(Lesson 7-5)*

$$\frac{3}{8} \div \frac{2}{3} = \frac{3}{8} \times \frac{3}{2} \quad \text{Multiply by the reciprocal of } \frac{2}{3}.$$

$$= \frac{9}{16}$$

Find each quotient. Write in simplest form.

28. $\frac{4}{9} \div 8$

29. $\frac{5}{6} \div \frac{3}{4}$

Solve each equation. Write the solution in simplest form.

30. $14 \div \frac{7}{8} = m$

31. $\frac{4}{9} \div \frac{2}{3} = d$

divide mixed numbers *(Lesson 7-6)*

$$5\frac{1}{2} \div 1\frac{5}{6} = \frac{11}{2} \div \frac{11}{6}$$

$$= \frac{\overset{1}{\cancel{11}}}{2} \times \frac{\overset{3}{\cancel{6}}}{\underset{1}{\cancel{11}}}$$

$$= \frac{3}{1} \text{ or } 3$$

Find each quotient. Write in simplest form.

32. $1\frac{7}{8} \div 7\frac{1}{2}$

33. $3\frac{1}{3} \div 10$

Solve each equation. Write the solution in simplest form.

34. $x = 6 \div 2\frac{2}{3}$

35. $2\frac{1}{4} \div 4\frac{2}{7} = d$

change units within the customary system
(Lesson 7-7)

a. 5 qt = _?_ pt   *2 pt = 1 qt*
    larger to smaller → multiply

   $5 \times 2 = 10$
   5 qt = 10 pt

b. 64 oz = _?_ lb   *16 oz = 1 lb*
    smaller to larger → divide

   $64 \div 16 = 4$
   64 oz = 4 lb

Complete.

36. 45 c = _?_ qt

37. $5\frac{1}{2}$ T = _?_ lb

38. 2.75 gal = _?_ pt

39. 3 pt = _?_ fl oz

40. 12 oz = _?_ lb

recognize and extend sequences *(Lesson 7-8)*

Find the next number in the sequence
24, 28, 32, 36, ...

In this sequence, 4 is added to each number.
The next number is 36 + 4, or 40.

Find the next two numbers in each sequence.

41. 256, 236, 216, 196, . . .

42. $\frac{1}{4}, \frac{1}{2}, 1, 2, 4, \ldots$

Find the missing number in each sequence.

43. $\frac{2}{5}, \underline{\ ?\ }, 10, 50, \ldots$

44. $6, 7\frac{1}{3}, \underline{\ ?\ }, 10, \ldots$

Applications & Problem Solving

45. Statistics Yana conducted a survey to find out how many students would participate in a school play. Out of 76 students, $\frac{3}{4}$ said they would participate in a play. How many students would participate in the play? *(Lesson 7-2)*

46. Geometry To find the perimeter of a square, you can use the formula $P = 4s$, where s is the length of one side of the square. Find the perimeter of the square. *(Lesson 7-3)*

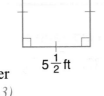

$5\frac{1}{2}$ ft

47. Food Adwoa bought 9 gallons of apple cider for the school party. How many 1-cup servings will he be able to serve? *(Lesson 7-7)*

48. Look for a Pattern Matsuko is using a computer to type her term paper. The table shows how many pages she has typed each day. If the pattern continues, how many pages long is the term paper? (Assume that she completes the term paper on the fourth day). *(Lesson 7-8A)*

Day	1	2	3	4
Number of Pages Typed	3	5	7	?

Alternative Assessment

Open Ended

Suppose you use the following ingredients to make one batch of modeling dough for your science project.

1 T oil	1 c salt
$2\frac{1}{2}$ c water	$2\frac{1}{2}$ c flour
2 pkg. soft drink mix	$1\frac{1}{2}$ t cream of tartar

How will you decide the amount of each ingredient needed to make 3 batches of the dough?

Suppose you need to make $2\frac{1}{2}$ batches of the dough. How much of each ingredient will you need?

A practice test for Chapter 7 is provided on page 601.

Completing the CHAPTER Project

Use the following checklist to complete your project.

☑ Your paper airplane is included.

☑ Directions for making the paper airplane are included.

☑ The flight table containing the flight distances and flight times of your paper airplane is included.

Add any finishing touches that you would like to make your project attractive.

PORTFOLIO   Select one of the assignments from this chapter and place it in your portfolio. Attach a note to it explaining why you selected it.

Section One: Multiple Choice

There are twelve multiple-choice questions in this section. Choose the best answer. If a correct answer is *not here*, mark the letter for Not Here.

1. What are the next two numbers in the sequence 1,024, 256, 64, 16, . . . ?

   A 8, 4

   B 8, 2

   C 4, 1

   D 4, 0

2. What is the area of the rectangle?

   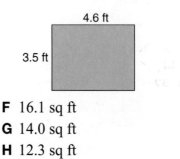

   4.6 ft

   3.5 ft

   F 16.1 sq ft

   G 14.0 sq ft

   H 12.3 sq ft

   J 12.0 sq ft

3. Jamonte's room is 8 feet wide. How many inches is this?

   A 92 inches

   B 96 inches

   C 20 inches

   D 44 inches

4. Evaluate $p \div m - 4$ if $p = 10$ and $m = 2$.

   F 1

   G 0

   H 5

   J 2

5. Express $3 \times 4 \times 4 \times 4 \times 5$ in exponential notation.

   A $3^2 \times 4^2 \times 5$

   B $3 \times 4^3 \times 5$

   C $3 \times 4^2 \times 5$

   D $3 \times 4^4 \times 5$

Please note that Questions 6–12 have five answer choices.

6. How many 1-foot square tiles are needed to cover the floor of a kitchen that is 13 feet by 10 feet?

   F 13

   G 130

   H 1,300

   J 13,000

   K 130,000

7. A sign at The Gift Gallery reads:

   Clearance Sale
   $2.99
   to
   $9.99

   Suppose a customer purchases four items that are on sale. What is a reasonable total cost of the items, without tax?

   A $5

   B $30

   C $50

   D $10

   E $75

8. A car travels at an average speed of 64 miles per hour. At this rate, how many hours will it take to travel 384 miles?

   F 2 hours

   G 4 hours

   H 3 hours

   J 6 hours

   K 5 hours

9. Suppose a car is traveling at a speed of 51 miles per hour. Which is a good estimate for the number of hours it will take to travel 248 miles?

    A 2 h

    B 3 h

    C 4 h

    D 5 h

    E 6 h

10. Mrs. Alvarez bought groceries for $19.58. How much change should she receive from a $20 bill?

    F $1.42

    G $0.42

    H $1.52

    J $0.52

    K Not Here

11. Mr. Meuser is planning to travel from Dallas to New York. The flight will take about 3 hours. It is time to leave, but the flight has been delayed for $2\frac{1}{2}$ hours. About how long will it be before he arrives in New York?

    A $5\frac{1}{2}$ hours

    B $6\frac{1}{2}$ hours

    C $4\frac{1}{2}$ hours

    D 5 hours

    E Not Here

12. Mrs. Miles purchased a ham that weighed $6\frac{1}{4}$ pounds and a turkey that weighed $11\frac{1}{2}$ pounds. How many pounds of meat did she purchase?

    F $17\frac{1}{2}$ pounds   **G** $18\frac{1}{4}$ pounds

    H $17\frac{3}{4}$ pounds   **J** $16\frac{1}{4}$ pounds

    K Not Here

Test-Taking Tip THE PRINCETON REVIEW

If you're working on a group of questions and find that the questions are getting too difficult, quickly read through the rest of the questions in the section and answer the ones that you know. Then come back to the ones you skipped.

Section Two: Free Response

This section contains six questions for which you will provide short answers. Write your answers on your paper.

13. Jamaal works 22 hours a week and earns $111.10. How much money does he earn per hour?

14. Order the fractions $\frac{2}{7}, \frac{1}{3}, \frac{2}{9}, \frac{2}{5}, \frac{4}{1}$ from greatest to least.

15. Andrea had purchased $6\frac{1}{2}$ yards of material to make curtains for her dining room. If she used 6 yards of material to make the curtains, how much of the material was not used?

16. How many fluid ounces are in 3 quarts?

17. Draw the next figure in the sequence.

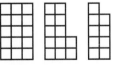

18. Four dogs have weights of 17, 24, 18, and 25 pounds. What is the average weight of the dogs?

interNET **CONNECTION** Test Practice For additional test practice questions, visit:

www.glencoe.com/sec/math/mac/mathnets

CHAPTER 8

Exploring Ratio, Proportion, and Percent

What you'll
learn in Chapter 8

- to express ratios and rates as fractions,
- to solve proportions by using cross products,
- to solve problems by drawing a diagram,
- to find actual length from a scale drawing,
- to express percents as fractions and as decimals, and
- to find the percent of a number.

CHAPTER Project

THE WONDERFUL WORLD OF TOYS

Many toys are small replicas of actual objects. For example, Matchbox cars are small replicas of cars. In this project, you will use ratios and proportions to find and compare the size of toys and the actual size of the objects that they represent. You will then make a poster to illustrate your findings.

Getting Started

- Look at the drawings of the two race cars. What is the length of stock car 88? What is the height of stock car 16?

- How could you use these dimensions to calculate the actual length and height of the stock cars that they represent?

Stock Car 88 Scale:
1 cm = 144 cm

3.6 cm

1.1 cm

88

|←1.5 cm→|

Stock Car 16 Scale:
1 cm = 64 cm

2.3 cm

16

3.0 cm

16

←——— 7.7 cm ———→

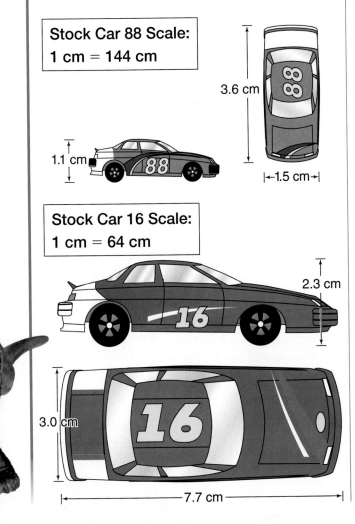

Technology Tips

- Use a **spreadsheet** to record the measurements of toys and actual objects and to calculate measurements.

- Use **publishing software** to make parts of your poster such as graphs, tables, and drawings.

- Use a **calculator** to help you find the actual dimensions of the stock cars.

■ *inter***NET** **CONNECTION** For up-to-date information on toy cars, visit: **www.glencoe.com/sec/math/mac/mathnet**

Working on the Project

You can use what you'll learn in Chapter 8 to help you make your poster.

Page	Exercise
315	35
327	11
333	37
347	Alternative Assessment

COOPERATIVE LEARNING

8-1A Ratios

2 sheets of
patty paper

scissors

A Preview of Lesson 8-1

Have you ever put together a tangram? The tangram was first developed in China and was very popular in the 1800s. In this lab, you will construct a tangram and use ratios to compare the areas of the tangram pieces. Recall that a ratio is a comparison of two numbers by division.

TRY THIS

Work with a partner.

Step 1 Using one sheet of patty paper, fold the top left corner to the bottom right corner. Crease the paper and unfold the square. Cut along the fold.

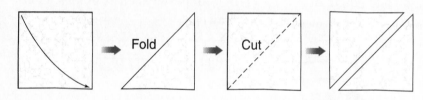

The ratio 1 to 2 compares the area of one of the triangles to the area of the uncut square.

Step 2 Using one of the triangles, fold the bottom left corner to the bottom right corner. Make a crease and unfold the triangle. Cut along the fold. Label the triangles A and B.

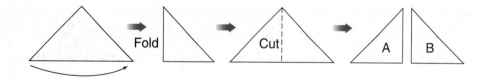

Step 3 Using the other triangle, fold the bottom left corner to the bottom right corner as shown. Make a crease and unfold the triangle. Then fold the top corner to the bottom edge along the crease. Crease and cut on this second crease line. Label the small triangle C.

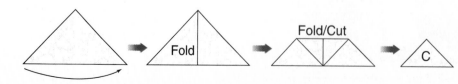

The ratio 1 to 3 compares the area of triangle C to the area of the remaining piece.

Step 4 Cut on the crease of the remaining piece as shown. Using the piece on the left, fold the bottom left corner to the bottom right corner. Make a crease and unfold. Cut on this fold. Label the triangle D and the square E.

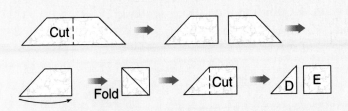

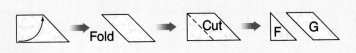

The ratio 1 to 6 compares the area of triangle D to the area of the original figure in this step. The ratio 1 to 3 compares the area of square E to the area of the original figure.

Step 5 Using the remaining piece, fold the bottom left corner to the top right corner. Make a crease and unfold. Cut along the fold. Label the triangle F and the other figure G.

The ratio 1 to 3 compares the area of triangle F to the area of the original figure in this step.

ON YOUR OWN

1a. If the area of triangle B is 1 square unit, what is the area of triangle C?

  b. How does the area of triangle C compare to the area of triangle B?

2a. If the area of triangle B is 1 square unit, what is the area of triangle F?

  b. How does the area of triangle F compare to the area of triangle B?

3. Write a ratio that compares the areas of the tangram pieces. Express each ratio as a fraction.

  a. C to A   **b.** F to A   **c.** A to B

  d. B to C   **e.** D to E   **f.** F to G

4. *Look Ahead* Write a ratio that compares the number of yellow squares to the total number of squares. Express the ratio as a fraction in simplest form.

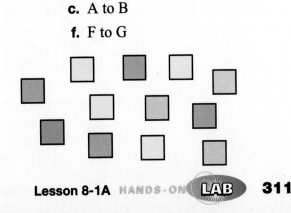

Ratios and Rates

What you'll learn

You'll learn to express ratios and rates as fractions.

When am I ever going to use this?

Knowing how to express ratios as fractions can help you determine better buys in a grocery store.

Word Wise

rate

The table shows the cards in a card game. Notice that 24 out of the 108 cards are purple.

You can compare these two numbers by using a ratio. Recall that the ratio that compares 24 to 108 can be written in several ways.

Type of Card	Amount
Heart	24
Yellow	24
Purple	24
Green	24
Wild Draw	10
Extra Cards	2

24 to 108   24:108

24 out of 108 $\frac{24}{108}$

A common way to express a ratio is as a fraction in simplest form.

$$\frac{24}{108} = \frac{2}{9}$$

The GCF of 24 and 108 is 12. Divide the numerator and the denominator by the GCF, 12.

So, $\frac{2}{9}$ of the cards are purple.

The ratio can also be expressed as 2 to 9, 2:9, or 2 out of 9.

Example

1 Express the ratio that compares the number of strawberries to the total number of berries as a fraction in simplest form.

strawberries → $\frac{4}{10} = \frac{2}{5}$ *The GCF of 4 and 10 is 2.*
berries →   *Divide the numerator and the denominator by the GCF, 2.*

The ratio in simplest form is $\frac{2}{5}$.

If the two quantities that you are comparing have different units of measure, the ratio is called a **rate**. For example, $\frac{125 \text{ miles}}{2 \text{ hours}}$ compares the number of miles traveled to the number of hours the trip took.

Rate	A rate is a ratio of two measurements that have different units.

Rates are usually expressed in a *per unit* form, where the number in the denominator is 1.

2 Express the ratio *7 inches of rain in 28 days* as a rate.

$$\frac{7 \text{ inches}}{28 \text{ days}} = \frac{0.25 \text{ inch}}{1 \text{ day}}$$

← ÷ 28 →
← ÷ 28 →

Divide the numerator and denominator by 28 to get a denominator of 1.

The total rainfall is equivalent to 0.25 inch each day.

APPLICATION

Real World

3 **Money Matters** Dhara purchased a 14.5-ounce bag of chips for $3.19. What was the cost per ounce?

Explore You know that the 14.5-ounce bag of chips costs $3.19. You need to find the cost per ounce.

Plan Use a rate to compare the cost to the weight.

Solve

$$\frac{\$3.19}{14.5 \text{ ounces}} = \frac{\$0.22}{1 \text{ ounce}}$$

← ÷ 14.5 →
← ÷ 14.5 →

Divide the numerator and denominator by 14.5 to get a denominator of 1.
$3.19 \div 14.5 = 0.22$

The cost is 22 cents per ounce.

Examine $\$0.22 \times 14.5 = \$3.19$

So, the answer is correct.

CHECK FOR UNDERSTANDING

Communicating Mathematics

Read and study the lesson to answer each question.

1. *Explain* the difference between a ratio and a rate.

2. *Draw* a picture in which the ratio of blue circles to the total number of circles is $\frac{3}{5}$.

Math Journal

3. *Write* about a situation in which rates would be helpful.

Guided Practice

Write each ratio in three different ways.

4. 7 carnations out of 24 flowers

5. 15 mint chip cookies in a bag of 34 cookies

Express each ratio as a fraction in simplest form.

6. 16 beagles out of 24 dogs

7. 10 emeralds out of 25 gems

Express each ratio as a rate.

8. 120 words in 3 minutes

9. 5 soft drinks for $3.25

10. *Life Science* The fastest running bird is the ostrich. An ostrich can run 240 miles in 6 hours. What is the average speed of an ostrich?

Practice

Write each ratio in three different ways.

11. 9 pairs of boots out of 16 are yellow.

12. 13 bikes out of 16 bikes have combination locks.

13. 18 out of 29 centimeter cubes are pink.

14. 11 brownies out of 15 brownies contain walnuts.

15. 23 out of 25 ants are red.

16. 7 out of 19 days were cloudy.

Express each ratio as a fraction in simplest form.

17. 19 games won out of 57 games played

18. 25 brick houses out of 45 houses

19. 14 tabby cats out of 18 cats

20. 16 states visited out of 50 states

21. 32 cows out of 72 animals

22. 10 drums out of 75 instruments

Express each ratio as a rate.

23. 395 kilometers in 5 hours

24. 3 CDs for $41.91

25. 4 tickets for $35

26. 63 million albums in 7 years

27. $1.32 for a dozen eggs

28. 79.8 miles on 3 gallons of gas

29. Write the ratio *13 quarters out of 24 coins* in three different ways.

30. Express the ratio *ten oranges for two dollars* as a rate.

Applications and Problem Solving

31. **Geography** Refer to the map.

   a. Write the ratio that compares the number of states in the Southwestern States region to the total number of states.

   b. Write the ratio that compares the number of states in the Midwestern States to the total number of states.

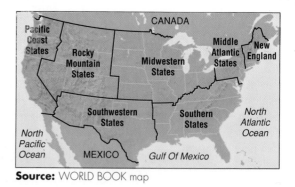

Source: WORLD BOOK map

32. **Geometry** Draw a 1-by-1 unit square and a 2-by-2 unit square on grid paper. Then write the ratio that compares each of the following.

   a. the length of the side of the small square to the length of the side of the large square

   b. the perimeter of the small square to the perimeter of the large square

   c. the area of the small square to the area of the large square

   d. Draw a 3-by-3 unit square. Compare the same measures between the 1-by-1 unit square and the 3-by-3 unit square. Describe the relationship that exists between the ratios of the sides, perimeters, and areas.

33. **Life Science** Scientists say that a pterodactyl could fly 75 miles in three hours. How far could a pterodactyl travel in one hour?

34. **Money Matters** At the supermarket, the cost of a 16-ounce bag of gum balls is $2.56. A 32-ounce bag costs $3.52. Which is the better buy? Explain.

35. **Working on the** CHAPTER Project   Refer to page 309.
   a. Write the ratio that compares the size of the toy stock car 88 to the actual size of the race car as a fraction.
   b. Write the ratio that compares the size of the toy stock car 16 to the actual size of the race car as a fraction.
   c. Find three toys that represent real-life objects. Measure the height and width of each toy to the nearest centimeter. Then find out the actual height and width of the objects that the toys represent. Write the ratios that compare the height, width, and length of each toy to the actual height, width, and length of each real object as a fraction in simplest form.

36. **Critical Thinking** Mr. Sarmiento stated that 18 out of 24 students in his class scored 85% or higher on the last test. He instructed the class to write a ratio comparing the number of students scoring below 85% to the total number of students. Explain why each answer is incorrect.
   a. 18:24   b. 24:6   c. 24:18

Mixed Review

37. **Patterns** Find the missing number in the sequence $3\frac{7}{8}$, $4\frac{1}{4}$, _?_ , 5, $5\frac{3}{8}$.
   (Lesson 7-8)

38. **Geometry** What is the circumference of a circle that has a diameter of 13.4 centimeters? *(Lesson 7-4)*

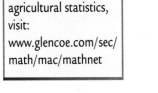

For the latest agricultural statistics, visit: www.glencoe.com/sec/math/mac/mathnet

39. **Agriculture** Refer to the graph. *(Lesson 3-6)*
   a. How many more pounds of potatoes were harvested in Washington than in Colorado and North Dakota?
   b. Find the total amount of potatoes harvested in these eight states.

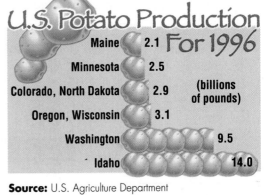

U.S. Potato Production For 1996

	(billions of pounds)
Maine	2.1
Minnesota	2.5
Colorado, North Dakota	2.9
Oregon, Wisconsin	3.1
Washington	9.5
Idaho	14.0

Source: U.S. Agriculture Department

40. **Standardized Test Practice**   Find the next figure in the pattern. *(Lesson 1-2)*

A   B   C   D

For **Extra Practice**, see page 578.

8-1B Ratios and Probability

A Follow-Up of Lesson 8-1

⊛ spinner

Suppose the probability that a basketball player makes a basket from the free throw line is 0.75. Since the decimal 0.75 means seventy-five hundredths, the probability can also be expressed as the ratio $\frac{75}{100}$. This means that the player makes a free throw 75 out of 100 times. In this lab, you will explore ratios and probability. *You can refer to Lesson 5-4B to review experimental probability.*

TRY THIS

Work with a partner.

1 • Make a spinner with four equal sections.

• Mark one section of the spinner "make" and the other sections "miss". Each section represents the player making or missing a basket.

• Spin the spinner 100 times. Record the number of baskets made as the ratio $\frac{baskets\ made}{100}$. Write the ratio as a decimal. What is the probability that the player makes a basket from the free throw line?

2 • Mark three sections of the spinner "make" and one section "miss". Again, each section represents the player making or missing a basket.

• Spin the spinner 100 times. Record the number of baskets made as the ratio $\frac{baskets\ made}{100}$. Write the ratio as a decimal. What is the probability that the player makes a basket from the free throw line?

ON YOUR OWN

1. Write a ratio that compares the area of each region marked "make" to the total area of the circle for each spinner.

2. What is the probability of making a basket on each spinner? Express the probability as a decimal.

3. Compare your results from the activity to each probability in Exercise 2.

4. Describe how ratio and probability are related.

5. **Look Back** A player makes a basket every 50 out of 80 times. What is the probability that he will make a basket on his next attempt? Express the probability as a decimal.

Solving Proportions

Have you ever played dominoes? In the standard set of dominoes, 7 out of 28 tiles, or one-fourth of the tiles, are doubles.

The ratios $\frac{7}{28}$ and $\frac{1}{4}$ are equivalent. That is, $\frac{7}{28} = \frac{1}{4}$. The equation $\frac{7}{28} = \frac{1}{4}$ is an example of a **proportion**.

Proportion	**Words:** A proportion is an equation that shows that two ratios are equivalent.
	Symbols: $\frac{a}{b} = \frac{c}{d}$, $b \neq 0$, $d \neq 0$

In the following Mini-Lab, you will use pattern blocks to explore ratios that are equivalent and ratios that are not equivalent.

HANDS-ON MINI-LAB

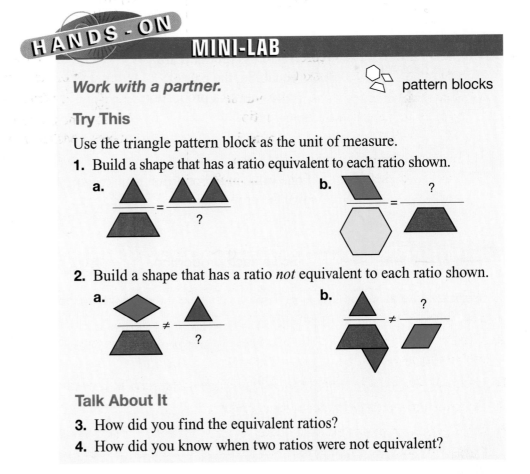

Work with a partner.

pattern blocks

Try This

Use the triangle pattern block as the unit of measure.

1. Build a shape that has a ratio equivalent to each ratio shown.

2. Build a shape that has a ratio *not* equivalent to each ratio shown.

Talk About It

3. How did you find the equivalent ratios?
4. How did you know when two ratios were not equivalent?

One way to determine whether two ratios form a proportion is to find their **cross products**. If the cross products of two ratios are equal, then the ratios form a proportion. In the proportion shown, notice that the cross products, 6×3 and 9×2, are equal.

$$\frac{6}{9} \diagdown \frac{2}{3} \qquad \begin{array}{l} 6 \times 3 = 18 \\ 9 \times 2 = 18 \end{array}$$

Property of Proportions	**Words:** The cross products of a proportion are equal.
	Symbols: If $\frac{a}{b} = \frac{c}{d}$, then $ad = bc$.

Examples

Use cross products to determine whether each pair of ratios forms a proportion.

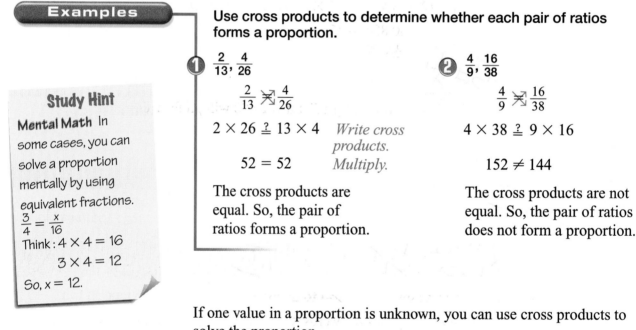

① $\frac{2}{13}, \frac{4}{26}$

$$\frac{2}{13} \diagdown \frac{4}{26}$$

$2 \times 26 \overset{?}{=} 13 \times 4$ *Write cross products.*

$52 = 52$ *Multiply.*

The cross products are equal. So, the pair of ratios forms a proportion.

② $\frac{4}{9}, \frac{16}{38}$

$$\frac{4}{9} \diagdown \frac{16}{38}$$

$4 \times 38 \overset{?}{=} 9 \times 16$

$152 \neq 144$

The cross products are not equal. So, the pair of ratios does not form a proportion.

If one value in a proportion is unknown, you can use cross products to solve the proportion.

Examples

Solve each proportion.

③ $\frac{5}{9} = \frac{z}{54}$

$5 \times 54 = 9 \times z$ *Write cross products.*

$270 = 9z$ *Multiply.*

$\frac{270}{9} = \frac{9z}{9}$ *Divide.*

$30 = z$

The solution is 30.

④ $\frac{1.4}{1.8} = \frac{3.5}{w}$

$1.4 \times w = 1.8 \times 3.5$

$1.4w = 6.3$

$\frac{1.4w}{1.4} = \frac{6.3}{1.4}$

$w = 4.5$

The solution is 4.5.

You can use proportions to make predictions.

Technology According to the results of a test conducted by a popular magazine, 12 out of 17 kids prefer video game A to video game B. Suppose there are 3,400 kids in your community. Predict how many will prefer video game A.

Set up a proportion that compares the number of kids who preferred video game A in the test to the number of kids in your community. Let *a* represent the number of kids out of 3,400 who will prefer video game A.

Test		Your Community

prefer video game A $\rightarrow$ $\dfrac{12}{17} = \dfrac{a}{3,400}$ $\leftarrow$ *prefer video game A*
total $\rightarrow$ $\leftarrow$ *total*

The ratios are equivalent so the cross products are equal.

$$\frac{12}{17} \bowtie \frac{a}{3,400}$$

$12 \times 3,400 = 17 \times a$ *Write the cross products.*

$40,800 = 17a$ *Multiply.*

$\dfrac{40,800}{17} = \dfrac{17a}{17}$ *Divide each side by 17.*

$2,400 = a$

You can predict that 2,400 will prefer video game A.

CHECK FOR UNDERSTANDING

Communicating Mathematics

Read and study the lesson to answer each question.

1. *Explain*, in your own words, the meaning of proportion.

2. *Tell* how you can determine whether two ratios form a proportion.

HANDS-ON MATH

3. *Use* pattern blocks to build a proportion. Draw a picture of your model and explain why the two ratios are proportional.

Guided Practice

Use cross products to determine whether each pair of ratios forms a proportion.

4. $\dfrac{1}{8}, \dfrac{8}{64}$

5. $\dfrac{7}{12}, \dfrac{8}{15}$

Solve each proportion.

6. $\dfrac{6}{7} = \dfrac{y}{42}$

7. $\dfrac{7}{12} = \dfrac{42}{h}$

8. $\dfrac{2.4}{a} = \dfrac{1.2}{1.5}$

9. *Advertising* According to an advertisement for Brand X toothpaste, 8 out of 10 dentists prefer Brand X. There are 150 dentists in a certain city. Predict how many of them prefer Brand X.

Practice

Use cross products to determine whether each pair of ratios forms a proportion.

10. $\frac{6}{8}, \frac{9}{15}$

11. $\frac{4}{5}, \frac{16}{20}$

12. $\frac{5}{6}, \frac{30}{36}$

13. $\frac{21}{28}, \frac{3}{7}$

14. $\frac{24}{75}, \frac{8}{25}$

15. $\frac{1.3}{3.5}, \frac{2.8}{1.6}$

Solve each proportion.

16. $\frac{d}{5} = \frac{24}{40}$

17. $\frac{w}{10} = \frac{4}{5}$

18. $\frac{6}{9} = \frac{r}{72}$

19. $\frac{5}{q} = \frac{25}{55}$

20. $\frac{1.7}{3} = \frac{85}{c}$

21. $\frac{23}{20} = \frac{115}{m}$

22. $\frac{1.6}{2.4} = \frac{2.8}{k}$

23. $\frac{25}{n} = \frac{12}{48}$

24. Suppose a car travels 174 miles in 3 hours. How long will it take to travel 290 miles?

25. Suppose you can buy 3 pairs of sunglasses for $38.91. How many pairs can you buy for $77.82?

Applications and Problem Solving

26. *School* At Market Street Middle School, the teacher to student ratio is 3 to 85. If there are 510 students enrolled at the school, how many teachers are there at the school?

27. *Health* According to the results of a survey, 27 out of 50 people exercise regularly. Suppose there are 2,600 people in a community. How many people can be expected to exercise regularly?

28. *Geometry* A series of rectangles are cut so that the ratio of the length of the short side to the long side is 3:5.

   a. Suppose the short side of one of the rectangles measures 6 units. What is the measure of the long side?

   b. Suppose the long side of one of the rectangles measures 25 units. What is the measure of the short side?

29. *Critical Thinking* Suppose 24 out of 180 people said they liked hiking, and 5 out of every 12 hikers buy Acme hiking shoes. In a group of 270 people, how many would you expect to have Acme hiking shoes?

Mixed Review

30. *Money Matters* The Shutter Bug Camera Shop charges $5.04 to develop 24 pictures. What is the cost of developing each picture? *(Lesson 8-1)*

For **Extra Practice,** see page 578.

31. *Algebra* Evaluate $w \div v$ if $w = 5\frac{3}{8}$ and $v = \frac{3}{4}$. *(Lesson 7-6)*

32. Find the value of m if $m = 15.64 \div 0.34$. *(Lesson 4-6)*

33. **Standardized Test Practice** Which shows the decimal twenty-two and four hundred five ten-thousandths? *(Lesson 3-1)*

   A 22.00405   **B** 2.2405   **C** 22.405   **D** 22.0405

SPREADSHEETS

8-2B Proportions

A Follow-Up of Lesson 8-2

computer

spreadsheet
software

Cara and Jeff are going to cater the Nielson family reunion. Their recipe for potato salad serves 10 people. How much of each ingredient will they need to make enough potato salad for 75 people?

You can find the amount of each ingredient needed by using a spreadsheet.

Mustard Potato Salad

$\frac{1}{2}$ c light mayonnaise	$\frac{1}{4}$ t salt
2 T Dijon mustard	$\frac{1}{8}$ t pepper
2 T sweet pickle relish	5 c cooked potatoes
1 T white vinegar	$\frac{1}{4}$ c minced parsley

Combine the first six ingredients in a large bowl. Add potatoes and toss. Cover and chill overnight. Garnish with parsley. Serves 10.

TRY THIS

Work with a partner.

B1 is the number of people, and the number of servings per batch is 10. So, the number of batches needed is $\frac{B1}{10}$, which is given in cell B2. The formula in cell B4 takes the number of batches needed and multiplies it by $\frac{1}{2}$ or 0.5, the amount of mayonnaise needed for one batch. The result is the total amount of mayonnaise needed to make enough potato salad for 75 people.

The printout shows the results of entering the number of servings needed, 75, in cell B1.

POTATO SALAD RECIPE

	A	B	C
1	People To Serve	= B1	
2	Batches needed	= B1/10	
3	INGREDIENT	NUMBER	
4	mayonnaise	= B2 * 0.5	cups
5	mustard	= B2 * 2	T
6	relish	= B2 * 2	T
7	vinegar	= B2 * 1	T
8	salt	= B2 * 0.25	t
9	pepper	= B2 * 0.125	t
10	potatoes	= B2 * 5	cups
11	parsley	= B2 * 0.25	cups

POTATO SALAD RECIPE

	A	B	C
1	People To Serve	75	
2	Batches needed	7.5	
3	INGREDIENT	NUMBER	
4	mayonnaise	3.75	cups
5	mustard	15.0	T
6	relish	15.0	T
7	vinegar	7.5	T
8	salt	1.875	t
9	pepper	0.9375	t
10	potatoes	37.5	cups
11	parsley	1.875	cups

ON YOUR OWN

1. Explain the formula in cell B6.

2. Using the results of the spreadsheet, write each ingredient amount needed as a fraction in simplest form.

3. Use the spreadsheet to find the amount of each ingredient needed to make enough potato salad for 120 people. Write each ingredient amount as a fraction in simplest form.

4. How could you change the spreadsheet if one batch of potato salad served 12 people?

8-3A Draw a Diagram

A Preview of Lesson 8-3

Alison and Ben are the first guests to arrive at Cindy's party. They are discussing how many handshakes there will be if every guest at the party shakes hands with everyone else once. Let's listen in!

If we shake hands, then there would be one handshake.

Alison

If Cindy joined us, then she would shake hands with you and me. That's two, a total of three.

Let's draw a diagram to picture the situation.

Ben

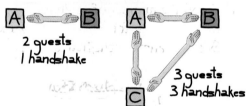

2 guests
1 handshake

3 guests
3 handshakes

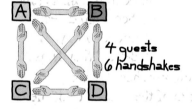

4 guests
6 handshakes

For five people, we can add the handshakes for the fifth person. Add four more handshakes. A total of 10!

Wow! We can use this method to find the number of handshakes for any number of people.

THINK ABOUT IT

Work with a partner.

1. **Draw a diagram** to illustrate the number of handshakes with five guests.

2. **Describe** the pattern that exists between the number of guests and the number of handshakes.

3. **Extend** the pattern to predict how many handshakes there will be with six guests.

4. **Apply** the **draw a diagram** strategy to solve the following problem.

   Kayla lives in Littleton and works in Parker. There is no direct route from Littleton to Parker, so Kayla goes through either Stoney Creek or Castle Rock. There is one road between Stoney Creek and Castle Rock. How many different ways can Kayla drive to work?

For **Extra Practice,** see page 578.

ON YOUR OWN

5. The third step of the 4-step plan for problem solving asks you to *solve* the problem. **Explain** how the draw a diagram strategy can help you solve a problem.

6. *Write a Problem* that can be solved by using the draw a diagram strategy. Then ask a classmate to determine the answer by drawing a diagram.

7. *Look Ahead* Draw a diagram that you can use to answer Exercise 8 on page 326.

MIXED PROBLEM SOLVING

Solve. Use any strategy.

STRATEGIES

Look for a pattern.
Solve a simpler problem.
Act it out.
Guess and check.
Draw a diagram.
Make a chart.
Work backward.

8. *Number Sense* A number multiplied by itself is 676. What is the number?

9. *School* At the Science Festival's bridge-building competition, Juliet came in second, and Daniel finished behind Pedro. Keela finished ahead of Pedro, and Reynelda won first place. In what order did they finish?

10. *Recreation* Robin purchased a tent for camping. Each side of the four sides of the tent needs 3 stakes to secure properly to the ground. How many stakes should there be in the box?

11. *Life Science* The chart shows the average weight in kilograms of certain bears. What is the mean of these weights?

Weight of Bears

Polar Bear 410 kg
Grizzly Bear 360 kg
Asiatic Black Bear 120 kg
Kodiak Bear 780 kg

Source: Encyclopedia Americana

12. *Patterns* What are the next two figures in the pattern?

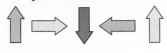

13. *Construction* India has a piece of lumber that is 140 inches long. She wants to cut it into 20-inch pieces. How many cuts does she need to make?

14. *Standardized Test Practice* The graph shows the number of students who signed up to volunteer at the Rush Creek Middle School's annual spaghetti dinner.

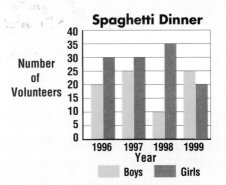

Spaghetti Dinner

Which is a conclusion that can be drawn from the data on the graph?

A There were more volunteers in 1998 than in 1996.

B The number of volunteers increased each year.

C More boys signed up to volunteer in 1997 than in any other year.

D There were fewer volunteers in 1999 than in 1996.

E More girls signed up to volunteer in 1997 than in 1998.

What you'll learn

You'll learn to find actual length from a scale drawing.

When am I ever going to use this?

Knowing how to find actual length from a scale drawing can help you determine mileage between cities on a map.

Word Wise

scale drawing

One of the largest meat-eating dinosaurs to have ever lived was the Tyrannosaurus Rex. Since it would be difficult to draw the dinosaur actual size, you can make a **scale drawing**. A scale drawing shows an object exactly as it looks, but it is generally larger or smaller. The scale gives the ratio that compares the lengths on the drawing to the actual lengths of the object.

Suppose a drawing of a dinosaur had a scale of 1 inch = 20 feet. If the length of the dinosaur on the drawing is 2 inches, what is the actual length?

You can write a proportion to find the actual length, ℓ.

	Scale		**Dinosaur**

$$\begin{array}{rl} \textit{length in drawing} \rightarrow & \dfrac{1 \text{ inch}}{20 \text{ feet}} = \dfrac{2 \text{ inches}}{\ell \text{ feet}} \leftarrow \textit{length in drawing} \\ \textit{actual length} \rightarrow & \qquad\qquad\qquad\quad \leftarrow \textit{actual length} \end{array}$$

$$1 \times \ell = 20 \times 2 \quad \textit{Find the cross products.}$$
$$\ell = 40 \quad \textit{Multiply.}$$

The actual length of the dinosaur was 40 feet or 480 inches.

Example ① **CONNECTION**

Life Science Suppose a drawing of a dinosaur had a scale of 1 inch = 18 feet. If the height of the dinosaur at the hips is $\frac{1}{2}$ inch, what is the actual height at the hips?

Let h represent the height of the dinosaur at the hips.
Write and solve a proportion.

	Scale		**Dinosaur**

$$\begin{array}{rl} \textit{length in drawing} \rightarrow & \dfrac{1 \text{ inch}}{18 \text{ feet}} = \dfrac{\frac{1}{2} \text{ inch}}{h \text{ feet}} \leftarrow \textit{length in drawing} \\ \textit{actual length} \rightarrow & \qquad\qquad\qquad\quad \leftarrow \textit{actual length} \end{array}$$

$$1 \times h = 18 \times \frac{1}{2} \quad \textit{Find the cross products.}$$
$$h = 9 \quad \textit{Multiply.}$$

The height of the dinosaur at the hips is 9 feet or 108 inches.

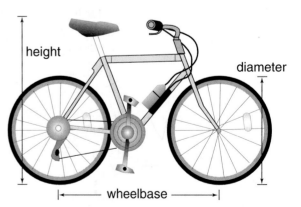

Example

CONNECTION

History In 1936, a meat company first produced a vehicle shaped like a hot dog. The 1995 version of the vehicle is 8 feet wide and 27 feet long. If a scale model of the vehicle is built with a width of 4 inches, what will be its length?

Let ℓ represent the length of the model.
Write and solve a proportion.

Model		**Actual Vehicle**

$$\begin{array}{ll} \textit{width of model} \rightarrow & \dfrac{4 \text{ inches}}{8 \text{ feet}} = \dfrac{\ell \text{ inches}}{27 \text{ feet}} \leftarrow \textit{length of model} \\ \textit{actual width} \rightarrow & \qquad\qquad\qquad \leftarrow \textit{actual length} \end{array}$$

$4 \times 27 = 8 \times \ell$ *Find the cross products.*

$108 = 8\ell$ *Multiply.*

$\dfrac{108}{8} = \dfrac{8\ell}{8}$ *Divide each side by 8.*

$13.5 = \ell$

The length of the model is 13.5 inches or $1\frac{1}{8}$ feet.

CHECK FOR UNDERSTANDING

Communicating Mathematics

Read and study the lesson to answer each question.

1. ***Explain*** why an architect would make a scale drawing or a scale model of a building.

2. ***Tell*** whether a scale drawing of a ladybug would be smaller or larger than the actual ladybug. Explain your reasoning.

Guided Practice

3. The drawing of the bicycle has a scale of 1 inch = 2 feet. Use a ruler to measure each dimension to the nearest eighth inch. Then find the actual measure of each dimension.
   a. the diameter of the wheels
   b. the height of the bicycle
   c. the distance between the centers of the wheels

4. ***Architecture*** An architect's drawing of a building has a height of $15\frac{3}{4}$ inches. If the scale on the drawing is $\frac{1}{2}$ inch = 1 foot, how tall is the building?

Practice

5. The drawing of the pup tent has a scale of $\frac{3}{4}$ inch = 1 yard. Use a ruler to measure each dimension to the nearest fourth inch. Then find the actual dimensions.

a. the height

b. the width

c. the length

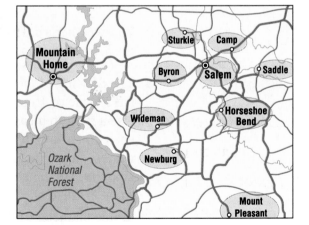

6. The floor plan shown has a scale of $\frac{1}{2}$ inch = 5 feet. Use a ruler to measure each room to the nearest eighth of an inch. Then find the actual measurements.

Rooms	Drawing Length	Width	Actual Length	Width
a. porch				
b. kitchen				
c. bath				
d. dining				
e. living				

Family *Activity*

Ask a family member to help you make a scale drawing of a room in your home. Include a scale that shows how the size of the drawing compares to the actual size.

7. A map of the northern portion of Arkansas is shown. It has a scale of 1 inch = 19 miles. Use a ruler to measure each map distance. Then find the actual distance.

a. Wideman to Horseshoe Bend

b. Camp to Newburg

c. Sturkie to Saddle

d. Mountain Home to Mount Pleasant

e. Salem to Byron

Applications and Problem Solving

8. *Transportation* On a drawing of a truck, the truck has a height of $\frac{7}{8}$ inch. What is the actual height of the truck if the drawing has a scale of $\frac{1}{4}$ inch = 28 inches?

9. *History* During World War II, America's main strategic weapon was the B-17 Flying Fortress. This aircraft has a length of about 80 feet and a wingspan of about 104 feet. If a scale model of the airplane is built with a wingspan of 26 feet, what will be its length?

10. *School* An original 5-inch by 8-inch photograph must be reduced to $1\frac{1}{4}$ by 2 inches to fit in the school yearbook. What is the scale of the reduced photograph to the original in simplest form?

11. *Working on the* **CHAPTER Project** Refer to the drawings on page 309.
   a. The drawing of stock car 88 has a scale of 1 cm = 144 cm. Find the length, width, and height of the actual stock car 88 race car.
   b. The drawing of stock car 16 has a scale of 1 cm = 64 cm. Find the length, width, and height of the actual stock car 16 race car.
   c. For each of the three toys that you selected, write a scale that can be used to compare the size of the toy to the actual size of the object.

For **Extra Practice**, see page 579.

12. *Critical Thinking* If you were asked to make a scale drawing of the Statue of Liberty, which scale would you use: 1 cm = 1 m or 1 cm = 1 mm? Explain.

Mixed Review   13. Solve $\frac{7.3}{h} = \frac{14.6}{10.8}$. *(Lesson 8-2)*

14. **Standardized Test Practice** Valerie purchased $40.06 worth of groceries. She gives the cashier a $50 bill. How much change should she receive? *(Lesson 3-6)*

   A $9.84   **B** $9.34   **C** $9.94   **D** $10.84   **E** Not Here

---

CHAPTER 8   Mid-Chapter Self Test

Write each ratio in three different ways. *(Lesson 8-1)*

   1. 11 rulers out of 19 are blue.   2. 7 students out of 39 are boys.

Express each ratio as a fraction in simplest form. *(Lesson 8-1)*

   3. 12 parrots out of 28 birds   4. 15 green apples out of 48 apples

Express each ratio as a rate. *(Lesson 8-1)*

   5. 270 miles in 4.5 hours   6. 8 kiwis for $1.00

Solve each proportion. *(Lesson 8-2)*

   7. $\frac{8}{9} = \frac{w}{108}$ 8. $\frac{m}{17} = \frac{35}{42.5}$ 9. $\frac{1.3}{h} = \frac{5.2}{11.2}$

10. *Geography* On a map, the distance between the towns of Gorden and Florin measures $\frac{3}{4}$ inch. Suppose the map has a scale of $\frac{1}{4}$ inch = 10 miles. What is the actual distance between the two towns? *(Lesson 8-3)*

Arts and Crafts

Carol Larsen
DOLL MAKER

Carol Larsen, mother of nine children, has turned her hobby of doll collecting into a career. In addition to designing and making her own dolls, she also sews the clothing for the dolls, and teaches classes on doll making.

There are various types of crafts in the arts and crafts industry. The skills that you will need depend on the type of craft that you would like to make. To design and make dolls or any other toys, you need to have a good understanding of mathematics. Having a good knowledge of proportions and scale drawings will enable you to create finished products that have the correct dimensions.

For more information:
Toy Manufacturers of
America, Inc.
200 5th Avenue, Suite 740
New York, NY 10010

*inter***NET**
CONNECTION
www.glencoe.com/sec/math/mac/mathnet

Someday, I'd like to design and create toys.

Your Turn
Interview a person in your community that is involved in the arts and crafts industry. Find out how they began their career and how they use mathematics when making their craft.

8-4A Modeling Percents

A Preview of Lesson 8-4

grid paper

colored pencils

You know that a 10 × 10 grid can be used to represent hundredths. Since the word *percent* means *out of one hundred,* you can also use a 10 × 10 grid to model percents.

TRY THIS

Work with a partner.

1 Model 25%.

Shade 25 of the 100 squares.

25% means "25 out of 100". By shading 25 out of 100 squares, you can show 25%. What decimal does the model represent?

2 Shade two fifths of the 10 × 10 grid. What percent have you modeled?

Separate the model into fifths. Shade two of the fifths.

There are 40 squares shaded, so 40% is modeled. Two fifths is equivalent to 40%.

ON YOUR OWN

1. Draw five 10 × 10 squares on a sheet of grid paper. Shade a 10 × 10 grid to represent each percent.
   a. 1%   **b.** 75%   **c.** 30%   **d.** 100%   **e.** 60%

2. Draw five more 10 × 10 squares on another sheet of grid paper. Shade a 10 × 10 grid to represent each fraction or number.

   a. 0.3   **b.** $\frac{1}{100}$ **c.** 0.75   **d.** 1   **e.** $\frac{3}{5}$

3. How do the two sets of shaded 10 × 10 grids compare?

4. *Look Ahead* Express 36% as a fraction in simplest form. Use a model if necessary.

Percents and Fractions

What you'll learn

You'll learn to express percents as fractions and vice versa.

When am I ever going to use this?

Knowing how to express a number in a different form can help you interpret a monthly budget.

Word Wise

percent

Study Hint

Reading Math The symbol % means percent.

Did you know that the word pasta is an Italian term meaning dough? Pasta comes in more than 100 shapes and sizes. What is your favorite type of pasta: spaghetti, ravioli, vermicelli, or possibly gnocchi? According to the results of a survey, 10% of people eat lasagna at least once every two weeks.

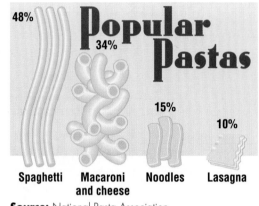

Popular Pastas

48% Spaghetti 34% Macaroni and cheese 15% Noodles 10% Lasagna

Source: National Pasta Association

A **percent** is a ratio that compares a number to 100. All of the percents in the graph can be expressed as fractions. To express a percent as a fraction, follow these steps.

• Express the percent as a fraction with a denominator of 100.
• Simplify.

Examples

Express each percent as a fraction in simplest form.

1 28%

28% means "28 out of 100."

$28\% = \dfrac{28}{100}$ *Express the percent as a fraction with a denominator of 100.*

$= \dfrac{\overset{7}{\cancel{28}}}{\underset{25}{\cancel{100}}}$ *Simplify. Divide the numerator and the denominator by 4, the GCF of 28 and 100.*

$= \dfrac{7}{25}$

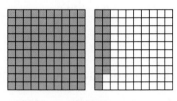

28%

2 118%

118% means "118 out of 100."

$118\% = \dfrac{118}{100}$ *Express the percent as a fraction with a denominator of 100.*

$= 1\dfrac{18}{100} \text{ or } 1\dfrac{9}{50}$ *Simplify.*

118%

3 **Health** Refer to the graph. What fraction of teen-age boys always or often use sunscreen before going out into the sun?

From the graph, you can see that 30% of teen-age boys use sunscreen before going into the sun.

$$30\% = \frac{30}{100}$$

$$= \frac{3}{10}$$

So, $\frac{3}{10}$ of teen-age boys always or often use sunscreen.

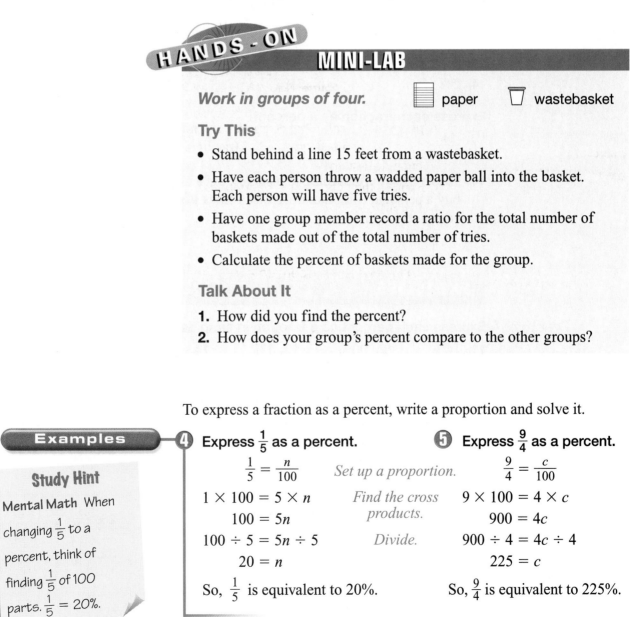

How regularly teenagers use sunscreen

					46%
36%		34%	33%		
	21%			30%	
Boys	Girls	Boys	Girls	Boys	Girls
Rarely/never		**Sometimes**		**Always/often**	

Source: Sovereign Research for Seventeen, Nivea, and the American Academy of Dermatology

You can also express fractions as percents.

HANDS-ON

MINI-LAB

Work in groups of four. 🗒 paper 🗑 wastebasket

Try This

- Stand behind a line 15 feet from a wastebasket.
- Have each person throw a wadded paper ball into the basket. Each person will have five tries.
- Have one group member record a ratio for the total number of baskets made out of the total number of tries.
- Calculate the percent of baskets made for the group.

Talk About It

1. How did you find the percent?
2. How does your group's percent compare to the other groups?

To express a fraction as a percent, write a proportion and solve it.

Examples

4 Express $\frac{1}{5}$ as a percent.

$\frac{1}{5} = \frac{n}{100}$	*Set up a proportion.*
$1 \times 100 = 5 \times n$	*Find the cross*
$100 = 5n$	*products.*
$100 \div 5 = 5n \div 5$	*Divide.*
$20 = n$	

So, $\frac{1}{5}$ is equivalent to 20%.

5 Express $\frac{9}{4}$ as a percent.

$$\frac{9}{4} = \frac{c}{100}$$

$$9 \times 100 = 4 \times c$$

$$900 = 4c$$

$$900 \div 4 = 4c \div 4$$

$$225 = c$$

So, $\frac{9}{4}$ is equivalent to 225%.

Study Hint

Mental Math When changing $\frac{1}{5}$ to a percent, think of finding $\frac{1}{5}$ of 100 parts. $\frac{1}{5} = 20\%$.

Communicating Mathematics

Read and study the lesson to answer each question.

1. *Identify* the percent that is represented by each model shown.

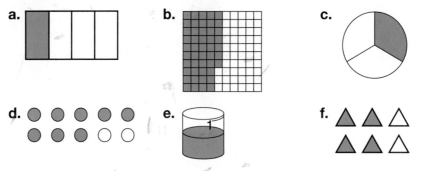

   a.   b.   c.

   d.   e.   f.

2. *Tell* what fraction is equivalent to 75%.

HANDS-ON MATH

3. Toss a coin 50 times. Record the number of times the coin shows heads. Then calculate the percent that represents the number of times the coin showed heads.

Guided Practice

Express each percent as a fraction in simplest form.

4. 2%   5. 55%   6. 120%

Express each fraction as a percent.

7. $\frac{34}{100}$ 8. $\frac{13}{20}$ 9. $\frac{5}{4}$

10. *Technology* According to a survey, 36% of consumers prefer to buy a personal computer at an electronics store. What fraction of consumers is this?

Practice

Express each percent as a fraction in simplest form.

11. 14%   12. 90%   13. 1%   14. 65%   15. 130%

16. 96%   17. 105%   18. 66%   19. 17%   20. 112%

Use a 10 × 10 grid to shade the amount stated in each fraction. Then express each fraction as a percent.

21. $\frac{4}{5}$ 22. $\frac{4}{10}$ 23. $\frac{9}{20}$ 24. $\frac{13}{50}$

Express each fraction as a percent.

25. $\frac{63}{100}$ 26. $\frac{1}{20}$ 27. $\frac{3}{2}$ 28. $\frac{12}{25}$

29. $\frac{37}{50}$ 30. $\frac{17}{20}$ 31. $\frac{19}{25}$ 32. $\frac{17}{10}$

33. How is forty-five hundredths written as a percent?

34. Express eighty-five percent as a fraction in simplest form.

35. *Travel* The graph shows why RV owners like to travel in an RV.

    a. What percent of those surveyed believe that traveling in a recreational vehicle is the best way to see the United States?

    b. What percent of those surveyed have other reasons why they like to travel in an RV?

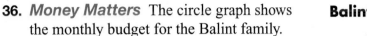

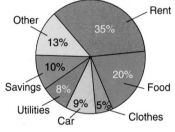

Source: Allstate Motor Club Survey of 400 RV owners

36. *Money Matters* The circle graph shows the monthly budget for the Balint family.

    a. What fraction, in simplest form, represents the portion of the family budget that is spent on rent?

    b. What fraction, in simplest form, represents the portion of the family budget that is spent on food and utilities?

Balint Family Budget

- Rent 35%
- Food 20%
- Clothes 5%
- Car 9%
- Utilities 8%
- Savings 10%
- Other 13%

37. *Working on the* **CHAPTER Project** Refer to Exercise 35 on page 315.

    a. Express the ratio in part a as a fraction, decimal, and as a percent.

    b. Express the ratio in part b as a fraction, decimal, and as a percent.

    c. Express each ratio in part c as a fraction, decimal, and as a percent.

38. *Critical Thinking* A woman made a will leaving $\frac{1}{2}$ of her fortune to her sister, $\frac{1}{3}$ to her brother, $\frac{1}{7}$ to her nephew, and the remainder to her cat. What percent of the woman's fortune was left to the cat? Round to the nearest whole percent.

Mixed Review

39. *Geography* A map has a scale of 1 inch = 20 miles. The distance from Hartville to Dixon is $2\frac{3}{4}$ inches. What is the actual distance between these two cities? *(Lesson 8-3)*

40. *Measurement* Painters used 170 gallons of white topcoat to paint the famous Hollywood sign. How many quarts are in 170 gallons? *(Lesson 7-7)*

41. Find $\frac{3}{8} + \frac{7}{8}$. *(Lesson 6-3)*

42. **Standardized Test Practice** What is the perimeter of the figure? *(Lesson 4-4)*

    A 74.3 m
    B 73.2 m
    C 72.2 m
    D 75.3 m

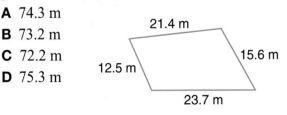

For **Extra Practice,** see page 579.

43. *Algebra* Evaluate $gh + j$ if $g = 4$, $h = 7$, and $j = 2$. *(Lesson 1-5)*

Percents and Decimals

What you'll learn

You'll learn to express percents as decimals and vice versa.

When am I ever going to use this?

Knowing how to express a percent as a decimal can help you determine the amount of sales tax paid on a dollar.

Five years after it opened, a theme park became France's top tourist attraction. The graph shows where the 11.7 million visitors came from. What part of the visitors came from Italy? *This problem will be solved in Example 3.*

Theme Park Visitors

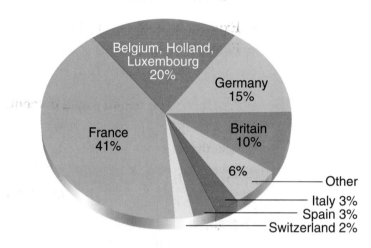

Notice that the sections of the graph are labeled as percents. In Lesson 8-4, you learned how to express percents as fractions. Percents can also be written as decimals. To express a percent as a decimal, follow these steps.

- Rewrite the percent as a fraction with a denominator of 100.
- Express the fraction as a decimal.

Examples

Express each percent as a decimal.

Study Hint

Mental Math To express a percent as a decimal, you can use this shortcut. Move the decimal point two places to the left, which is the same as dividing by 100.

1 65%

$65\% = \dfrac{65}{100}$ *Rewrite the percent as a fraction with a denominator of 100.*

$\quad = 0.65$ *Express the fraction as a decimal.*

2 0.5%

$0.5\% = \dfrac{0.5}{100}$ *Rewrite the percent as a fraction with a denominator of 100.*

$\quad = \dfrac{5}{1,000}$ $\dfrac{0.5}{100} \times \dfrac{10}{10} = \dfrac{5}{1,000}$

$\quad = 0.005$ *Express the fraction as a decimal.*

3 **Statistics** Refer to the beginning of the lesson. What part of the visitors came from Italy? Express the answer as a decimal.

From the graph, you can see that 3% of the visitors came from Italy.

$$3\% = \frac{3}{100}$$
$$= 0.03$$

The graph shows that 0.03 of the visitors came from Italy.

To express a decimal as a percent, follow these steps.

- Express the decimal as a fraction.
- Express the fraction as a percent.

Examples

Express each decimal as a percent.

4 **0.72**

$0.72 = \frac{72}{100}$ *Express the decimal as a fraction.*

$= 72\%$ *Express the fraction as a percent.*

5 **0.257**

$0.257 = \frac{257}{1,000}$ *Express the decimal as a fraction.*

$= \frac{25.7}{100}$ $\frac{257 \div 10}{1,000 \div 10} = \frac{25.7}{100}$

$= 25.7\%$ *Express the fraction as a percent.*

Study Hint

Mental Math To express a decimal as a percent you can use another shortcut. Move the decimal point two places to the right, which is the same as multiplying by 100.

CHECK FOR UNDERSTANDING

Communicating Mathematics

Read and study the lesson to answer each question.

1. *Draw* a model that shows the decimal equivalent of 36%.

2. *Explain* how to express 0.008 as a percent.

3. *You Decide* Kellile says that the decimal 3.78 is less than 100%. Betty says that the decimal is more than 100%. Who is correct? Explain your reasoning.

Guided Practice

Express each percent as a decimal.

4. 46%   **5.** 81%   **6.** 7%   **7.** 0.8%

Express each decimal as a percent.

8. 0.52   **9.** 0.9   **10.** 0.175   **11.** 0.02

12. *Life Science* About 95% of all species of fish have skeletons made of bone. Express 95% as a decimal.

EXERCISES

Practice

Express each percent as a decimal.

13. 32%   **14.** 84%   **15.** 1%   **16.** 0.9%

17. 6%   **18.** 17%   **19.** 0.3%   **20.** 63%

21. 39%   **22.** 3.5%   **23.** 4%   **24.** 26%

Express each decimal as a percent.

25. 0.96   **26.** 0.1   **27.** 0.364   **28.** 0.27

29. 0.716   **30.** 0.66   **31.** 0.07   **32.** 0.5

33. 0.08   **34.** 0.104   **35.** 0.2   **36.** 0.03

37. How is eighteen thousandths written as a percent?

38. Write two and four tenths percent as a decimal.

39. Express each number as a:

    a. decimal: $\frac{321}{10,000}$, $\frac{3}{25}$, 9%, $26\frac{1}{2}$

    b. fraction: 0.07, 1.129, 3%, 47

    c. percent: 0.04, 0.538, $\frac{12}{20}$, $\frac{3}{8}$

Applications and Problem Solving

40. *Taxes* In 1997, the residents of the state of Texas paid a 8.25% sales tax on purchases.

    a. Express 8.25% as a decimal.

    b. On each dollar, how much money did Texans pay in sales tax?

41. *Media* Refer to the graph.

    a. What percent of those surveyed spend more than an hour reading the Sunday paper?

    b. What percent of those surveyed do *not* spend more than an hour reading the Sunday paper?

42. *Critical Thinking* Arrange the following groups of numbers.

    a. from greatest to least: 23.4%, 2.34, 0.0234, 20.34%

    b. from least to greatest: $2\frac{1}{4}$, 0.6, 2.75, 40%, $\frac{7}{5}$

TIME SPENT READING THE

1998 **Sunday Paper**

0.46

0.18

0.12

0.24

| More than an hour | 26-60 minutes | 25 minutes or less | None |

Source: Impact Resources

Mixed Review

43. Express 156% as a fraction in simplest form. *(Lesson 8-4)*

44. *Money Matters* Ahmik purchased a package of four mechanical pencils for $6.48. How much does each pencil cost? *(Lesson 8-1)*

45. **Standardized Test Practice** What number is missing from the sequence 14, 56, __?__ , 896, 3,584? *(Lesson 7-8)*

    A 284   **B** 194   **C** 334   **D** 224

For **Extra Practice**, see page 579.

46. *Measurement* How many inches are in $2\frac{1}{2}$ yards? *(Lesson 5-6)*

47. Round 64.35 ÷ 12 to the nearest tenth. *(Lesson 4-5)*

8-6

Estimating with Percents

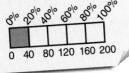

What you'll learn

You'll learn to estimate the percent of a number.

When am I ever going to use this?

Knowing how to estimate the percent of a number can help you determine the tip at a restaurant.

The Watch Store is having a sale on watches. Every watch is on sale for 60% of the regular price. About how much would you pay for a watch that normally sells for $34.99? *This problem will be solved in Example 3.*

The word *about* tells you that an exact answer is not needed. So, you can estimate the answer. When you estimate with percents, you should round the percent to a fraction that is easy to multiply. The chart shows some commonly-used percents and their fraction equivalents.

$20\% = \frac{1}{5}$	$25\% = \frac{1}{4}$	$12\frac{1}{2}\% = \frac{1}{8}$	$16\frac{2}{3}\% = \frac{1}{6}$
$40\% = \frac{2}{5}$	$50\% = \frac{1}{2}$	$37\frac{1}{2}\% = \frac{3}{8}$	$33\frac{1}{3}\% = \frac{1}{3}$
$60\% = \frac{3}{5}$	$75\% = \frac{3}{4}$	$62\frac{1}{2}\% = \frac{5}{8}$	$66\frac{2}{3}\% = \frac{2}{3}$
$80\% = \frac{4}{5}$	$100\% = 1$	$87\frac{1}{2}\% = \frac{7}{8}$	$83\frac{1}{3}\% = \frac{5}{6}$

Examples

Study Hint

Problem Solving You can also make a model to help you estimate 22% of 197.

Estimate each percent.

① 22% of 197

22% is close to 20% or $\frac{1}{5}$.

Round 197 to 200.

$\frac{1}{5} \times 200 = \frac{1}{5} \times \frac{\overset{40}{200}}{1}$

$\phantom{\frac{1}{5} \times 200} = 40$

So, 22% of 197 is about 40.

② 50% of 1,512

50% is $\frac{1}{2}$.

Round 1,512 to 1,500.

$\frac{1}{2} \times 1,500 = \frac{1}{2} \times \frac{\overset{750}{1,500}}{1}$

$\phantom{\frac{1}{2} \times 1,500} = 750$

So, 50% of 1,512 is about 750.

APPLICATION

Real World

③ Money Matters Refer to the beginning of the lesson. About how much would you pay for a watch that normally sells for $34.99?

Express 60% as a fraction. $60\% = \frac{60}{100} = \frac{3}{5}$ *Simplify.*

Then estimate the product of the fraction and the price.

$\frac{3}{5} \times 35 = \frac{3}{5} \times \frac{35}{1}$ *Think: $34.99 is close to $35.*

$\phantom{\frac{3}{5} \times 35} = \frac{3}{5} \times \frac{\overset{7}{35}}{1}$ *Divide 5 and 35 by their GCF, 5.*

$\phantom{\frac{3}{5} \times 35} = 21$ *Multiply.*

You will pay about $21 for the watch.

Sometimes you need to estimate a percent.

Example ─④ **Estimate the percent of the figure that is shaded.**

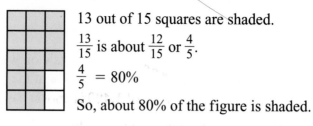

13 out of 15 squares are shaded.

$\frac{13}{15}$ is about $\frac{12}{15}$ or $\frac{4}{5}$.

$\frac{4}{5} = 80\%$

So, about 80% of the figure is shaded.

CHECK FOR UNDERSTANDING

Communicating Mathematics

Read and study the lesson to answer each question.

1. *Explain* how you would estimate 75% of 1,976.

2. *Draw* a figure in which the percent of the figure that is shaded is about 60%.

3. *Write* about a situation in which you would use estimating to find the percent of a number.

Guided Practice

Estimate each percent.

4. 17% of 34   5. 25% of 208   6. 8% of 15

Estimate the percent of each figure that is shaded.

7.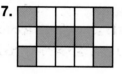

8.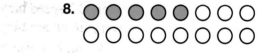

9. *Money Matters* When the Chou family went out to dinner, their bill was $35.98. Mrs. Chou wants to leave a 20% tip. About how much money should she leave?

EXERCISES

Practice

Estimate each percent.

10. 77% of 39   11. 34% of 60   12. 9% of 91

13. 47% of 52   14. 80% of 123   15. 38% of 104

16. 66% of 89   17. 97% of 302   18. 24% of 276

Estimate the percent of each figure that is shaded.

19.   20. 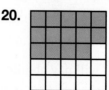   21.

22. 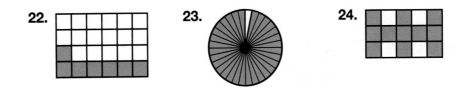 **23.** **24.**

25. Estimate 37.5% of 50.

26. *True* or *false*? 26% of 1,400 is about 350. Explain.

27. About how much is one and two tenths percent of ten?

Applications and Problem Solving

28. *Life Science* According to a survey, 13% of gardeners plant geraniums in their flowerbed. If 2,450 gardeners participated in the survey, about how many can be expected to plant geraniums in their flowerbed?

29. *Food* Refer to the graph. Suppose 5,012 people participated in the fast food survey. About how many more people can be expected to eat takeout food in their home than in their car?

Source: Beef Industry Council

30. *Money Matters* The Bike Shop is having their annual bike sale. Every 15-speed bicycle is on sale for 75% of the regular price, and every mountain bicycle is on sale for 80% of the regular price.

    a. About how much would you pay for a 15-speed bicycle that usually sells for $125?

    b. About how much would you save if you purchased a mountain bike that normally sells for $196?

31. *Critical Thinking* Which percent problem does not belong? Explain.

    a. 50% of 22   **b.** 22% of 50   **c.** $\frac{1}{2}$% of 22

Mixed Review

32. *Statistics* According to a survey, 56% of people brush their teeth after eating snacks. Express 56% as a decimal. *(Lesson 8-5)*

33. *Geometry* Find the area of the rectangle. *(Lesson 4-4)*

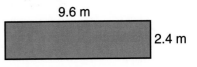
9.6 m
2.4 m

34. *Standardized Test Practice* Shalana is buying a jean shirt for $35.95, a pair of jeans for $25.98, and a T-shirt for $11.50. Find the total cost of the items not including tax. *(Lesson 3-6)*

    A $68.93   **B** $73.43   **C** $70.63   **D** $72.83   **E** Not Here

For **Extra Practice,** see page 580.

35. *Statistics* Find the mean of 14, 73, 25, 25, and 53. *(Lesson 2-7)*

Percent of a Number

***What* you'll learn**

You'll learn to find the percent of a number.

***When* am I ever going to use this?**

Knowing how to find the percent of a number can help you determine the sale price of a skateboard.

According to the graph, 20% of parents buy shoes for their children every four to five months. If 2,500 parents were surveyed, how many said that they purchase shoes for their children every four to five months?

1% ⊏ Don't know
6% ⊏ Once a month
36% ⊏ Every 2-3 months
20% ⊏ Every 4-5 months
27% ⊏ 2 times a year
10% ⊏ Once a year or less

How often parents buy shoes for their children

Source: Opinion Research for Payless Shoe Source

To find 20% of 2,500, you can change the percent to a fraction or to a decimal, and then multiply it by the number. You can also use a model or a calculator.

Method 1 Change the percent to a fraction.

$20\% = \dfrac{20}{100}$ or $\dfrac{1}{5}$

$\dfrac{1}{5}$ of $2{,}500 = \dfrac{1}{5} \times 2{,}500$

$= 500$

Method 2 Change the percent to a decimal.

$20\% = \dfrac{20}{100}$ or 0.2

0.2 of $2{,}500 = 0.2 \times 2{,}500$

$= 500$

Method 3 Use a model.

Since $20\% = \dfrac{1}{5}$, separate a rectangle into fifths. Label the top and bottom in equal intervals as shown.

0%	20%	40%	60%	80%	100%

0   500   1,000   1,500   2,000   2,500

20% of 2,500 is 500.

Method 4 Use a calculator.

20 [2nd] [%] [×] 2,500 [=] *500*

500 out of 2,500 parents surveyed buy shoes for their children every four to five months.

Example

1 Find 24% of 250 by changing the percent to a fraction.

$24\% = \dfrac{24}{100}$ or $\dfrac{6}{25}$ *Change the percent to a fraction.*

$\dfrac{6}{25} \times 250 = \dfrac{6}{\underset{1}{25}} \times \dfrac{\overset{10}{250}}{1}$ *Divide 25 and 250 by the GCF, 25.*

$= 60$

24% of 250 is 60.

2 Find 9% of 105 by changing the percent to a decimal.

$9\% = \dfrac{9}{100}$ or 0.09   *Change the percent to a decimal.*

$0.09 \times 105 = 9.45$ *Multiply 0.09 by 105.*

9% of 105 is 9.45.

3 Find 75% of 256 by using a model.

Since $75\% = \dfrac{3}{4}$, separate a rectangle into fourths.

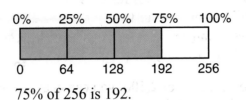

75% of 256 is 192.

4 Find 0.5% of 188 by using a calculator.

0.5 [2nd] [%] [×] 188 [=] *0.94*

0.5% of 188 is 0.94.

5 **Travel** According to a survey, 31% of parents named "road construction and traffic" when asked what they'd like to avoid on a family vacation. If 3,000 parents were surveyed, how many said they'd like to avoid road construction and traffic?

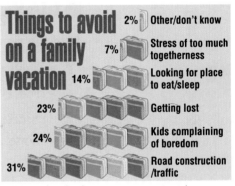

Source: Travel Industry Association, Amtrak

Explore   You need to know the number of parents, out of 3,000, who said to avoid road construction and traffic.

Plan   Multiply 31% by the number of parents surveyed, 3,000.

Estimate: 31% is close to 30% or $\dfrac{3}{10}$.

$$\dfrac{3}{10} \times 3{,}000 = 900$$

Solve   Change 31% to a decimal. Then multiply it by 3,000.

$31\% = \dfrac{31}{100}$ or 0.31

$$\begin{aligned} 0.31 \text{ of } 3{,}000 &= 0.31 \times 3{,}000 \\ &= 930 \end{aligned}$$

930 out of 3,000 parents said they'd like to avoid road construction and traffic.

Examine   Check the answer by comparing it to the estimate. 930 is close to the estimate. So, the answer is correct.

**Communicating
Mathematics**

Read and study the lesson to answer each question.

1. ***Describe*** the steps in three of the four methods for finding the percent of a number.

2. ***Tell*** what number the shaded portion of the model shows.

0%   100%

0   ?   950

3. ***You Decide*** Arturo says that 125% of 150 is 18.75. Kosey disagrees. He says that 125% of 150 is 187.5. Who is correct? Explain your reasoning.

Guided Practice

Find the percent of each number.

4. 40% of 65   **5.** 8% of 84   **6.** 0.3% of 500

7. What is 98% of 6?

8. ***Sports*** The Fitch High School softball team won 88% of their games. If they played 25 games, how many games did they win?

EXERCISES

Practice

Find the percent of each number.

9. 25% of 72	**10.** 7% of 7	**11.** 80% of 115
12. 3% of 156	**13.** 15% of 40	**14.** 33% of 390
15. 101% of 98	**16.** 0.5% of 85	**17.** 125% of 145
18. 0.4% of 20	**19.** 100% of 137	**20.** 0.1% of 250

21. What is 60% of 365?   **22.** Find 22% of 55.

23. What is eight-tenths percent of eight hundred?

24. A test had 64 problems. Anika got 48 correct.
    a. What percent were correct?
    b. What percent were incorrect?
    c. If Maria got 81.25% correct, how many problems did she get correct?

**Applications and
Problem Solving**

25. ***Money Matters*** Ali paid $31.50 for an item that was reduced by 30%.
    a. What was the original price?
    b. If the original price was reduced by 25%, what is the sale price?

26. ***Technology*** Rachel is using her computer to decode a secret message. The table shows the results of the scan. Suppose the secret message contains 1,500 vowels.
    a. How many of the vowels are a?
    b. How many of the vowels are o?
    c. How many of the vowels are e?

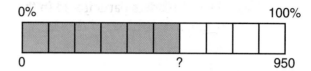

Vowel	Occurred (%)
A	25%
E	30%
I	20%
O	20%
U	5%

27. *Critical Thinking* Solve each problem.
    a. 14 is what percent of 70?
    b. What is 35% of 240?
    c. 45 is 15% of what?

Mixed Review

28. *Hobbies* The graph shows how much readers are willing to spend on a book. Suppose 932 readers participated in the survey. About how many of those surveyed are willing to spend between $5 and $14.99? *(Lesson 8-6)*

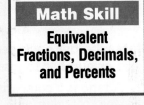

Buying a Book

Source: NPD Group

29. Estimate $\frac{3}{8} \times 65$. *(Lesson 7-1)*

30. **Standardized Test Practice** What is the greatest common factor of 35, 56, and 63? *(Lesson 5-3)*
    A 6   **B** 9   **C** 8   **D** 7

For **Extra Practice,**
see page 580.

Let the Games Begin

Fishin' for Matches

Math Skill
Equivalent Fractions, Decimals, and Percents

Get Ready This game is for two to four players.
    24 index cards   scissors

Get Set Cut two index cards in half. On one half, write a percent. On the other halves, write its decimal equivalent, its fraction equivalent, and its fraction equivalent in simplified form. If the fraction cannot be simplified, write the fraction on two cards. For example, $\frac{3}{4}$ cannot be simplified. So, $\frac{3}{4}$ will be written on two cards. Continue until you have 12 sets of cards.

Go
- Shuffle the cards and deal 7 cards to each player. Place the remaining cards face down on the table.
- The first player requests a matching card from any player. If a match is made, that player sets the match aside and it becomes the next player's turn. If a match is not made, the player chooses one card without looking from the remaining cards. If a match is made, the player sets the match aside, and it is the next player's turn.
- The game continues until a player runs out of cards to match. The player with the most matches wins.

25%

0.25

$\frac{25}{100}$

$\frac{1}{4}$

inter NET CONNECTION Visit www.glencoe.com/sec/math/mac/mathnet for more games.

🖳 *inter*NET
CONNECTION Chapter Review   **For additional lesson-by-lesson review, visit:**
www.glencoe.com/sec/math/mac/mathnet

Vocabulary

After completing this chapter, you should be able to define each term, concept, or phrase and give an example or two of each.

Number and Operations
cross products (p. 318)
percent (p. 330)
proportion (p. 317)
rate (p. 312)

Geometry
scale drawing (p. 324)

Problem Solving
draw a diagram (p. 322)

Understanding and Using the Vocabulary

State whether each sentence is *true* or *false*. If false, replace the underlined word or number to make a true sentence.

1. A ratio is a comparison of two numbers by <u>multiplication</u>.

2. The simplest form of the ratio $\frac{12}{28}$ is $\frac{3}{7}$.

3. A <u>rate</u> is a ratio of two measurements that have different units.

4. Three tickets for $7.50 expressed as a rate is <u>$1.50</u> per ticket.

5. A <u>percent</u> is an equation which shows that two ratios are equivalent.

6. The model shown represents <u>85%</u>.

7. The cross products of a proportion are <u>equal</u>.

8. 8% can be expressed as <u>0.008</u>.

9. $\frac{3}{25}$ can be expressed as <u>12%</u>.

10. 38% of 40 is <u>1,520</u>.

11. A <u>scale drawing</u> shows an object exactly as it looks, but it is generally larger or smaller.

12. A percent is a ratio that compares a number to <u>10</u>.

13. The decimal 0.346 can be expressed as <u>3.46%</u>.

14. $62\frac{1}{2}\%$ can be expressed as the fraction $\frac{5}{8}$.

In Your Own Words

15. *Explain* how to change a percent to a fraction.

Objectives & Examples

Upon completing this chapter, you should be able to:

● express ratios and rates as fractions *(Lesson 8-1)*

Express the ratio 3 winners out of 12 competitors as a fraction in simplest form.

$$\frac{3 \text{ winners}}{12 \text{ competitors}} \overset{\div\,3}{=} \frac{1 \text{ winner}}{4 \text{ competitors}}$$
$$\div\,3$$

● solve proportions by using cross products *(Lesson 8-2)*

Solve the proportion $\frac{9}{12} = \frac{g}{8}$.

$9 \times 8 = 12 \times g$ *Write cross products.*
$\quad 72 = 12g$ *Multiply.*
$\quad \frac{72}{12} = \frac{12g}{12}$ *Divide.*
$\quad\quad 6 = g$

● find actual length from a scale drawing *(Lesson 8-3)*

The drawing of the room has a scale of 1 unit = 2 feet. What is the actual length of the room?

6 units

11 units

drawing → $\dfrac{1 \text{ unit}}{2 \text{ feet}} = \dfrac{11 \text{ units}}{d \text{ feet}}$ ← *drawing*
actual → ← *actual*

$\quad\quad 1 \times d = 2 \times 11$
$\quad\quad\quad\quad d = 22$

The actual length is 22 feet.

Review Exercises

Use these exercises to review and prepare for the chapter test.

Express each ratio as a fraction in simplest form.

16. 18 out of 30 babies are girls.
17. 11 potatoes out of 20 are rotten.

Express each ratio as a rate.

18. 3 inches of rain in 6 months
19. 189 pounds of garbage in 12 weeks

Use cross products to determine whether each pair of ratios forms a proportion.

20. $\frac{4}{11}, \frac{7}{22}$ 21. $\frac{3}{9}, \frac{12}{36}$

Solve each proportion.

22. $\frac{7}{11} = \frac{m}{33}$ 23. $\frac{12}{20} = \frac{15}{k}$

24. $\frac{g}{20} = \frac{9}{12}$ 25. $\frac{10}{12} = \frac{25}{h}$

Refer to the scale drawing at the left.

26. What is the actual width of the room?

27. Suppose that the doorway to the room is actually 3 feet wide. How wide will the doorway be on the drawing?

28. *Geography* On a map, the measure from Baxter to Sidney is 2 inches. Suppose that the map has a scale of $\frac{3}{4}$ inch = 30 miles. What is the actual distance between the two cities?

Objectives & Examples

express percents as fractions and vice versa
(Lesson 8-4)

Express $\frac{3}{10}$ as a percent.

$$\frac{3}{10} \not\asymp \frac{n}{100}$$

$$3 \times 100 = 10 \times n$$
$$300 = 10n$$
$$300 \div 10 = 10n \div 10$$
$$30 = n$$

So, $\frac{3}{10}$ is equivalent to 30%.

express percents as decimals and vice versa
(Lesson 8-5)

Express 21% as a decimal.

$$21\% = \frac{21}{100}$$
$$= 0.21$$

estimate the percent of a number
(Lesson 8-6)

Estimate 33% of 60.

33% is close to $33\frac{1}{3}\%$ or $\frac{1}{3}$.

$$\frac{1}{3} \times 60 = \frac{1}{\cancel{3}} \times \frac{\cancel{60}^{20}}{1}$$
$$= 20$$

So, 33% of 60 is about 20.

find the percent of a number
(Lesson 8-7)

Find 20% of 50.

$$20\% = \frac{20}{100} \text{ or } \frac{1}{5}$$

$$\frac{1}{5} \times 50 = 10$$

20% of 50 is 10.

Review Exercises

Express each percent as a fraction in simplest form.

29. 3%

30. 17%

31. 150%

32. 48%

Express each fraction as a percent.

33. $\frac{3}{5}$

34. $\frac{9}{10}$

35. $\frac{11}{20}$

36. $\frac{5}{25}$

37. How is five hundredths written as a percent?

Express each percent as a decimal.

38. 2.2%

39. 38%

40. 140%

41. 66%

Express each decimal as a percent.

42. 0.003

43. 1.3

44. 0.65

45. 0.591

Estimate each percent.

46. 19% of 99

47. 27% of 82

48. 48% of 48

49. 41% of 243

50. Estimate the percent of the figure that is shaded.

Find the percent of each number.

51. 18% of 89

52. 40% of 150

53. 5% of 340

54. 0.8% of 132

55. What is one-tenth percent of 162?

56. Find two hundred thirty-six percent of 42.

Applications & Problem Solving

57. *Travel* Tatanka drove 300 miles in 6 hours. How many miles did he drive per hour? *(Lesson 8-1)*

58. *Statistics* According to a survey, 37% of Americans feel that public libraries will become the place to go to use computers. If 1,500 people participated in the survey, how many said that libraries would become the place to use computers? *(Lesson 8-2)*

59. *Draw a Diagram* A store manager displays 36 cans of cat food in a triangular shape. The display is one can deep. How many cans are on the bottom row? *(Lesson 8-3A)*

60. *School* Nick's class is planning a fund-raiser. The results of a class vote show that 75% of the students want to sell candy, 15% of the students want to sell wrapping paper, and 10% of the students want to sell calendars. *(Lessons 8-4 and 8-7)*

  a. What fraction of the class wants to sell calendars?

  b. What fraction of the class wants to sell candy?

  c. If there are 40 students in the class, how many voted to sell candy?

Alternative Assessment

Open Ended

Suppose your teacher assigns you to make a scale drawing of your school. What information would you need to collect? What decisions would you need to make?

Suppose your school is rectangular-shaped with a length of 220 feet and a width of 90 feet. You want to make the length of the school on the scale drawing 11 inches. Find the scale and the width of the school on the map.

Completing the CHAPTER Project

Use the following checklist to make sure your poster is complete.

☑ A sketch of each toy is included.

☑ The dimensions of each toy are labeled, and the dimensions of the actual objects are listed.

☑ For each sketch, include a scale that can be used to compare the size of the toy to the actual size of the object.

Add any finishing touches to make your poster unique.

 PORTFOLIO Select an item from the chapter that you found to be challenging. Place it in your portfolio and write a paragraph that explains why you found the item to be challenging.

A practice test for Chapter 8 is provided on page 602.

Section One: Multiple Choice

There are thirteen multiple-choice questions in this section. Choose the best answer. If a correct answer is *not here,* choose the letter for Not Here.

1. How could you estimate a 20% tip for a $23.25 restaurant bill?

A Find $\frac{1}{2}$ of $23.25.

B Find $\frac{1}{4}$ of $23.00.

C Find $\frac{1}{5}$ of $23.00.

D Find $\frac{1}{20}$ of $24.00.

2. Rogelio is 1.6 meters tall. How many centimeters is this?

F 0.016 cm   **G** 16 cm

H 160 cm   **J** 1,600 cm

3. The stem-and-leaf plot shows the speeds in miles per hour for the fastest birds in the world. What is the mean speed?

Stem	Leaf
6	5 5 8
7	0 2 7
8	0 8
9	5
10	6 $8|8 = 88\ mph$

A 78.6 mph   **B** 74.5 mph
C 75.6 mph   **D** 77.8 mph

4. Round $3\frac{3}{5}$ to the nearest half.

F $3\frac{1}{5}$ **G** 3

H $3\frac{1}{2}$ **J** 4

5. Find the value of $132 \div 6 + 4 \times 3$.

A 36   **B** 78
C 68   **D** 34

Please note that Questions 6–13 have five answer choices.

6. The table shows the cost for 4 brands of tennis shoes at two different stores.

Brand	Sporting Goods Store	Department Stores
R	$49.99	$45.65
S	$67.98	$65.00
T	$37.50	$37.95
U	$75.97	$72.98

At the department store, how much more is brand S than brand R?

F $20.45

G $19.45

H $20.35

J $19.35

K $19.25

7. Use the table in Question 6. Connie purchases brand T and brand R from the sporting goods store. What is the total cost before sales tax is added?

A $87.39

B $90.59

C $88.99

D $87.49

E $86.59

8. Tickets to Splash City cost $10 for adults and $8 for children. If 12 people paid a total of $102, how many were adults, and how many were children?

F 7 adults, 4 children

G 5 adults, 6 children

H 3 adults, 9 children

J 6 adults, 6 children

K 4 adults, 8 children

9. Solve $\frac{16}{25} = \frac{x}{100}$.

  A 4

  B 10

  C 16

  D 64

  E Not Here

10. Yoruba bought 6 packages of hot dogs for a picnic. Each package weighs 1.15 pounds. What is the total weight of the packages?

  F 7.15 lb

  G 6.90 lb

  H 8.40 lb

  J 6.55 lb

  K Not Here

11. Find the sum of 12.675 and 7.081.

  A 18.996

  B 19.566

  C 19.756

  D 18.796

  E Not Here

12. A bicycle costs $159, a helmet costs $59, and bike shorts cost $24. What is the cost of these items before tax is added?

  F $280   **G** $242

  H $220   **J** $275

  K Not Here

13. $3\frac{3}{4} \times 2\frac{3}{5} =$

  A $6\frac{9}{20}$ **B** $8\frac{1}{2}$

  C $7\frac{2}{5}$ **D** $9\frac{3}{4}$

  E Not Here

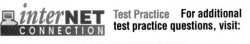 **Test Practice   For additional test practice questions, visit:**

www.glencoe.com/sec/math/mac/mathnet

Section Two: Free Response

This section contains seven questions for which you will provide short answers. Write your answers on your paper.

14. What is the value of $3c - 4d$ if $c = 8$ and $d = 3$?

15. Two-fifths of the registered voters voted in the town election. What percent of the voters did not vote?

16. Estimate 39% of 70.

17. To the nearest eighth of an inch, what is the length of the line segment shown?

  ├───────────────┤

18. Find the value of the expression $76 - 16 \div 4 \times 3 + 12$.

19. Dawit sells small bouquets of flowers at his flower shop for $2.75 each. On Saturday, he made $57.75 in sales from the bouquets. How many bouquets of flowers did he sell?

20. Add a number to the following set of data so that the mode, median, and mean of the new set of data are the same as those for the original set of data: 91, 93, 93, 95, 95, 98, 100.

Geometry: Investigating Patterns

 you'll
learn in Chapter 9

- to classify and measure angles,
- to solve problems by using logical reasoning,
- to bisect segments and angles,
- to name two-dimensional figures, and
- to determine congruence and similarity.

DESIGNING A DREAM HOUSE

In this project, you will use geometry to find the actual angle measures and the actual lengths of the walls of the dream house shown. You will need a ruler and a protractor to complete this project.

Getting Started

- Look at the floor plan shown. What shape is the house?

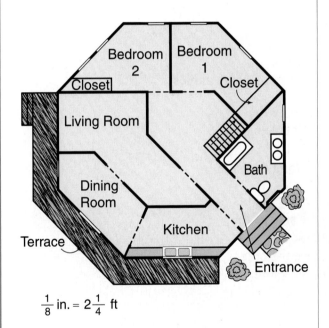

Bedroom 2

Bedroom 1

Closet

Closet

Living Room

Bath

Dining Room

Kitchen

Terrace

Entrance

$\frac{1}{8}$ in. $= 2\frac{1}{4}$ ft

Technology Tips

- Use geometry or drawing **software** to design a dream house.
- Surf the **Internet** for more information on designing houses.
- Use a **spreadsheet** to convert from inches to feet.

*inter***NET** CONNECTION Research **For up-to-date information on houses, visit:**

www.glencoe.com/sec/math/mac/mathnet

Working on the Project

You can use what you'll learn in Chapter 9 to help you design a dream house.

Page	Exercise
361	27
373	27
382	19
389	Alternative Assessment

Angles

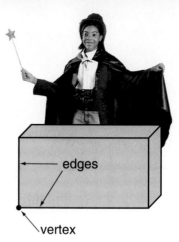

Have you ever watched a magician place a person in a box and then "saw the box in half?" The **edges** of the front of the box look like two lines that meet at a point called the **vertex**.

Vertices and edges of the box form **angles**.

The most common unit of measure for angles is the **degree**. A circle can be separated into 360 equal-sized parts. Each part would make up a one-degree (1°) angle as shown.

You can use a **protractor** to measure angles.

Step 1 Place the center of the protractor on the vertex of the angle with the straightedge along one side.

Step 2 Use the scale that begins with 0° on the side of the angle. Read the angle measure where the other side crosses the same scale. Extend the sides if needed.

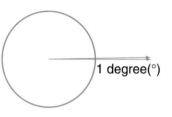

The angle measures 140°.

Notice the arrows on the sides of the angles. These tell you that you can extend the sides.

Angles can be classified according to their measure.

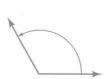

Acute angles measure between 0° and 90°.

Obtuse angles measure between 90° and 180°.

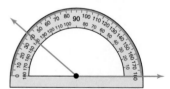

This mark indicates a right angle.

Right angles measure 90°.

Use a protractor to find the measure of each angle. Then classify the angle as acute, right, or obtuse.

Study Hint

Problem Solving Lay a corner of your notebook paper on top of the given angle to determine whether an angle is acute, right, or obtuse.

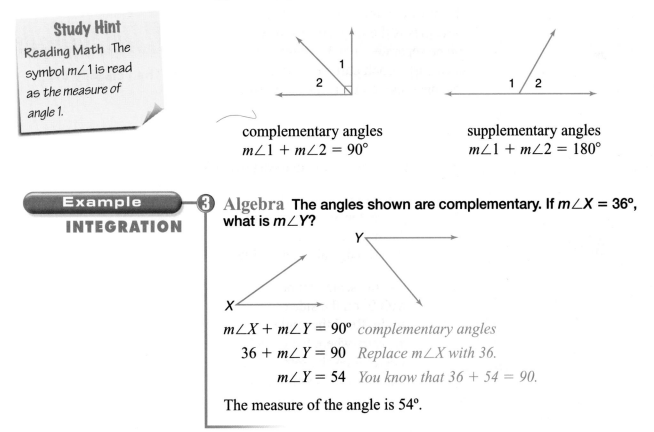

① 0° on one side

center of protractor

The angle measures 65°. It is an acute angle.

②

The angle measures 135°. It is an obtuse angle.

Some pairs of angles are **complementary** or **supplementary**. If the sum of the measures of two angles is 90°, the angles are complementary. If the sum of the measures of two angles is 180°, the angles are supplementary.

Study Hint

Reading Math The symbol $m\angle 1$ is read as the measure of angle 1.

complementary angles
$m\angle 1 + m\angle 2 = 90°$

supplementary angles
$m\angle 1 + m\angle 2 = 180°$

Example

INTEGRATION

③ Algebra The angles shown are complementary. If $m\angle X = 36°$, what is $m\angle Y$?

$m\angle X + m\angle Y = 90°$ *complementary angles*
$36 + m\angle Y = 90$ *Replace $m\angle X$ with 36.*
$m\angle Y = 54$ *You know that $36 + 54 = 90$.*

The measure of the angle is 54°.

CHECK FOR UNDERSTANDING

Communicating Mathematics

Read and study the lesson to answer each question.

1. *Draw* an obtuse angle. Then write the steps describing how to use a protractor to measure your obtuse angle.

2. *Tell* how you can determine whether two angles are complementary or supplementary.

Use a protractor to find the measure of each angle. Then classify the angle as *acute, right,* or *obtuse.*

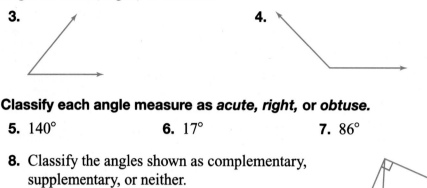

3.

4.

Classify each angle measure as *acute, right,* or *obtuse.*

5. 140°　　　　　　　6. 17°　　　　　　　7. 86°

8. Classify the angles shown as complementary, supplementary, or neither.

9. *Algebra* Angles A and B are supplementary. If $m\angle B = 106°$, find $m\angle A$.

EXERCISES

Use a protractor to find the measure of each angle. Then classify the angle as *acute, right,* or *obtuse.*

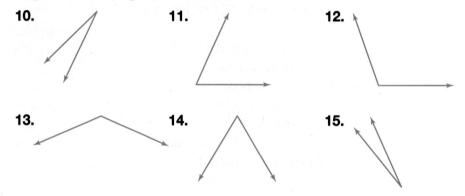

10.

11.

12.

13.

14.

15.

Classify each angle measure as *acute, right,* or *obtuse.*

16. 74°　　　　17. 90°　　　　18. 28°　　　　19. 174°

20. 168°　　　21. 54.6°　　　22. 137.5°　　　23. 4°

Classify each pair of angles as *complementary, supplementary,* or *neither.*

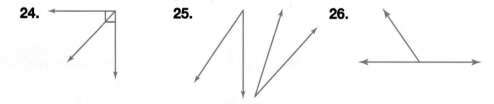

24.

25.

26.

27. An angle measures 89.9°. Is it an acute angle or a right angle?

28. If an angle measures 90.4°, is it an obtuse angle or a right angle?

29. *Algebra* Angles J and K are complementary angles. Find $m\angle K$ if $m\angle J = 72°$.

30. *Algebra* Angles M and N are supplementary angles. If angle M measures 119°, what is the measure of angle N?

Applications and Problem Solving

31. *Carpentry* A carpenter joins two pieces of molding as shown. What types of angles can you identify and where? Explain.

32. *Geometry* The geometric figure shown is a regular pentagon. Are the angles inside a regular pentagon acute, right, or obtuse?

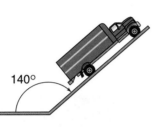

33. *Critical Thinking* Truck drivers find it difficult to accelerate when the grade of a hill is steep.

140°

a. How would you change the grade so that it is not so steep?

b. Would the obtuse angle with the ground become larger or smaller?

Mixed Review

34. Estimate 1.5% of $127,500. *(Lesson 8-6)*

35. *Statistics* Eight out of 10 mountain bicyclists have been injured at least once while riding their mountain bike. The graph shows where accidents occur. Express each decimal as a fraction in simplest form. *(Lesson 5-9)*

Mountain Bike Accidents

Downhill 0.76

Flat 0.15

0.09

Uphill

Source: *Acute Injuries from Mountain Biking,* by T. Chow, M. Bracker, K. Patrick, M.D.'s

36. Standardized Test Practice The table shows the results of a cost comparison for three compact discs at different stores. Which statement is true? *(Lesson 3-3)*

Compact Disc	Music Store	Department Store	Discount Store
X	$13.95	$18.95	$11.95
Y	$15.95	$14.50	$13.50
Z	$16.50	$14.98	$15.98

A Compact disc X costs the most at the department store.

B The discount store has the best price for any of the three compact discs.

C Compact disc Y costs the most at the department store.

D The least expensive place to get compact disc Z is at the discount store.

For **Extra Practice,** see page 580.

37. *Algebra* Evaluate $3 + xy - 5$ if $x = 6$ and $y = 9$. *(Lesson 1-5)*

38. Round 108 to the nearest ten. *(Lesson 1-3)*

9-1B Use Logical Reasoning

A Follow-Up of Lesson 9-1

Jesse and Estella are standing in front of the sign-up board for extra-curricular activities. There are 55 students who play sports, 60 who are involved in clubs, and 33 who are involved in both sports and clubs. They are trying to find out how many students participate only in sports. Let's listen in!

I made a Venn diagram using the information from the sign-up sheets.

Jesse

What's a Venn diagram?

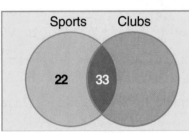

Sports   Clubs

22   33

Estella

You can use a Venn diagram to solve the problem. The circle on the left represents sports. The circle on the right represents clubs. The overlapping region represents both sports and clubs and the region outside the circles represents neither sports nor clubs.

I see! There are 55 students who play sports and 33 who are involved in both sports and clubs. 55 – 33 = 22. So, 22 students participate only in sports.

THINK ABOUT IT

Work with a partner.

1. *Analyze* the information in the *Venn diagram*. Do you agree or disagree with Estella's and Jesse's thinking? Explain your reasoning.

2. *Use logical reasoning* to find the number of students who are involved only in clubs.

3. *Find* the total number of students who participate in extracurricular activities by **using logical reasoning**.

4. *Apply* the logical reasoning strategy to solve the following problem.

   In a line of 24 students waiting to go on a field trip, 12 have lunch bags, 10 have backpacks, and 7 have both lunch bags and backpacks.

   a. *How many students have only lunch bags?*

   b. *How many have neither a lunch bag nor a backpack?*

For **Extra Practice,** see page 581.

ON YOUR OWN

5. The last step of the 4-step plan for problem solving asks you to *examine* the solution. *Explain* how you can use the solution in Exercise 4 to verify the number of students who are going on the field trip.

6. *Write a Problem* that can be solved by using logical reasoning. Then ask a classmate to solve the problem.

7. *Reflect back* Explain how you can use logical reasoning to answer Exercise 28 on page 354.

MIXED PROBLEM SOLVING

STRATEGIES

Look for a pattern.
Solve a simpler problem.
Act it out.
Guess and check.
Draw a diagram.
Make a chart.
Work backward.

Solve. Use any strategy.

8. *Communication* Lisa places a long distance phone call to Jason and talks for 33 minutes at a rate of 17 cents per minute. Did she spend about $4 or $6 for the call?

9. *Patterns* What is the missing number in the pattern? . . ., 234, 345, ?, 567, . . .

10. *School* Of the 150 students at Washington Middle School, 55 are in the orchestra, 75 are in marching band, and 25 are in both orchestra and marching band. How many students are in neither orchestra nor marching band?

11. *Patterns* The Sweepstakes Prize Company is having a contest. To enter the contest, you need to draw the next figure in the pattern.

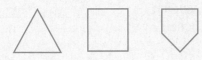

a. Draw a sample entry.

b. Explain why your entry would be the next figure in the pattern.

12. *Careers* Enrico, Angela, and Marcus are all engineers. Their specialties are electrical, mechanical, and civil engineering, but not in that particular order. The civil engineer and Enrico played golf together on Friday. Marcus, who works for the power company, is married to the civil engineer. Who is the mechanical engineer?

13. *Standardized Test Practice* The Venn diagram shows the relationship between the members in Moesha's scout troop.

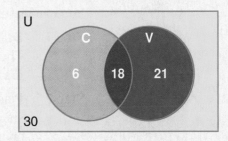

U = set of all members in Moesha's scout troop
C = set of members who have a camping badge
V = set of members who have a volunteer badge
Which of the following is *not* true?

A 30 scouts do not have a camping badge or a volunteer badge.

B 21 scouts have only a volunteer badge.

C 63 students have at least one badge.

D There are more scouts with a camping badge than those who have only a volunteer badge.

E 6 students have only a camping badge.

Using Angle Measures

Many people enjoy the relaxation of sailing. Others enjoy the excitement of sailboat racing. To determine their direction when sailing, sailors can use a compass.

If a sailboat traveling east increases its directional angle by 45°, in what direction is it now sailing? (*Hint*: The directional angle is measured clockwise from magnetic north.)

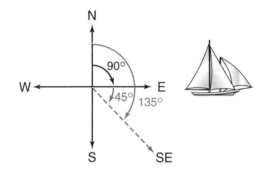

East (E) has a directional angle of 90° measured from the north (N). By increasing this angle by 45°, the new directional angle is 90° + 45°, or 135°. The sailboat is now traveling southeast (SE).

You can use a protractor and a straightedge to draw an angle.

Step 1 Draw one side of the angle. Then mark the vertex and draw an arrow.

Step 2 Place the protractor along the side as you would to measure an angle. On the protractor, find the number of degrees needed for the angle you are drawing and make a pencil mark as shown.

Step 3 A **straightedge** is a ruler or any object with a straight side, which can be used to draw a line. With a straightedge, draw the side that connects the vertex and the pencil mark. Draw an arrow on the end of the other side.

The angle drawn is a 150° angle.

Example ① **Draw a 68° angle.**

Draw one side. Mark the vertex and draw an arrow.

Find 68° on the appropriate scale. Make a pencil mark.

Draw the side that connects the vertex and the pencil mark.

You can estimate the measure of an angle by comparing it to an angle whose measure you know.

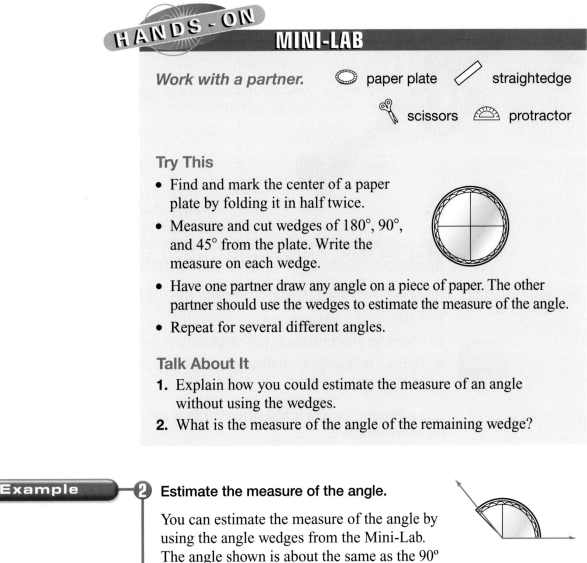

HANDS-ON

MINI-LAB

Work with a partner.   ◯ paper plate   ╱ straightedge

✂ scissors   ◠ protractor

Try This

- Find and mark the center of a paper plate by folding it in half twice.
- Measure and cut wedges of 180°, 90°, and 45° from the plate. Write the measure on each wedge.
- Have one partner draw any angle on a piece of paper. The other partner should use the wedges to estimate the measure of the angle.
- Repeat for several different angles.

Talk About It

1. Explain how you could estimate the measure of an angle without using the wedges.
2. What is the measure of the angle of the remaining wedge?

Example ② **Estimate the measure of the angle.**

You can estimate the measure of the angle by using the angle wedges from the Mini-Lab. The angle shown is about the same as the 90° angle wedge and the 45° angle wedge. So, the measure of the angle is about 90° + 45°, or 135°.

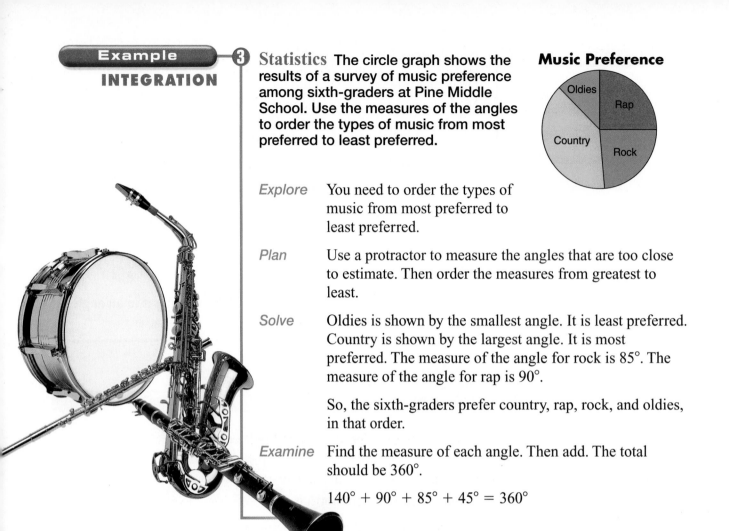

Example ─③ **Statistics** The circle graph shows the
INTEGRATION results of a survey of music preference
among sixth-graders at Pine Middle
School. Use the measures of the angles
to order the types of music from most
preferred to least preferred.

Music Preference

Explore You need to order the types of
music from most preferred to
least preferred.

Plan Use a protractor to measure the angles that are too close
to estimate. Then order the measures from greatest to
least.

Solve Oldies is shown by the smallest angle. It is least preferred.
Country is shown by the largest angle. It is most
preferred. The measure of the angle for rock is 85°. The
measure of the angle for rap is 90°.

So, the sixth-graders prefer country, rap, rock, and oldies,
in that order.

Examine Find the measure of each angle. Then add. The total
should be 360°.

$$140° + 90° + 85° + 45° = 360°$$

CHECK FOR UNDERSTANDING

Communicating **Read and study the lesson to answer each question.**
Mathematics
1. *Explain* to a classmate how to draw a 145° angle.

2. *Show* an angle of about 120° using two pencils as the sides.

3. *Write* a sentence explaining how to use paper folding to show an angle
of 45°.

Guided Practice **Use a protractor and a straightedge to draw angles having the following**
measurements.

4. 120°   **5.** 25°   **6.** 43°

Estimate the measure of each angle.

7.   **8.**

9. Is the measure of angle *ABC* greater than,
less than, or about equal to 155º?

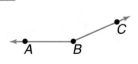

EXERCISES

Practice

Use a protractor and a straightedge to draw angles having the following measurements.

10. 165°   **11.** 90°   **12.** 85°   **13.** 8°

14. 95°   **15.** 145°   **16.** 32°   **17.** 66°

Estimate the measure of each angle.

18.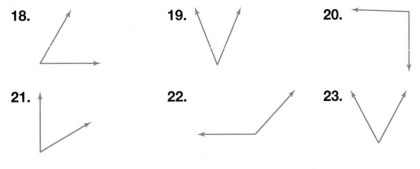

19.

20.

21.

22.

23.

24. Is the angle shown greater than, less than, or about equal to 79°?

Applications and Problem Solving

25. *Navigation* A pilot is flying northeast (45° east of north). He increases the directional angle of the plane by 135°.

    a. In which direction is the plane now traveling?

    b. How many degrees from north is the plane now traveling?

26. *Life Science* The branches on young trees should be spread to form angles of at least 60° with the tree trunk. This strengthens the branches and allows for more air circulation and light. Which branches on the tree shown need to be spread?

27. *Working on the* **CHAPTER Project** Refer to page 351. Trace the floor plan onto your paper. Use a protractor to measure each angle of the bathroom, the dining room, and the second bedroom. Then label the measures of the angles on your diagram. Save your diagram for use in other lessons.

28. *Critical Thinking* The sum of the measures of a triangle is 180°. Find the measure of all the angles of a right triangle if one angle measures 10°.

Mixed Review

29. *Geometry* Classify the angle shown as *acute, right,* or *obtuse.* *(Lesson 9-1)*

30. *Standardized Test Practice* If an airplane travels 438 miles per hour, how many miles will it travel in 5 hours? *(Lesson 8-2)*

    A 2,050 mi   **B** 2,090 mi   **C** 2,150 mi   **D** 2,190 mi   **E** Not Here

31. Find the LCM of 24 and 30. *(Lesson 5-7)*

32. *Measurement* How many milliliters are in 2.14 liters? *(Lesson 4-9)*

COOPERATIVE LEARNING

straightedge

compass

9-3A Constructing Congruent Segments and Angles

A Preview of Lesson 9-3

A *line segment* is a straight path between two endpoints. To indicate line segment *JK*, write $\overline{JK}$. Line segments that have the same length are called *congruent segments*.

You can construct congruent segments using a straightedge and a compass. A compass is used to draw circles or circular arcs and to measure distances.

TRY THIS

Work with a partner.

1 To construct a line segment congruent to $\overline{JK}$, follow these steps.

- Draw $\overline{JK}$. Then use a straightedge to draw a line segment longer than $\overline{JK}$. Label it $\overline{LM}$.

- Place the compass point at *J* and adjust the compass setting so that the pencil is on point *K*. The compass setting equals the length of $\overline{JK}$.

- Use this compass setting and place the compass point at *L*. Draw an arc that intersects $\overline{LM}$ at *P*. $\overline{LP}$ is congruent to $\overline{JK}$.

ON YOUR OWN

Trace each segment. Then construct a segment congruent to it.

1.   A   B   **2.**   S   T   **3.**   X   Y

4. *Look Ahead* The length of $\overline{MN}$ is 38 millimeters. If $\overline{MN}$ is separated into two congruent parts, what will be the length of each part?

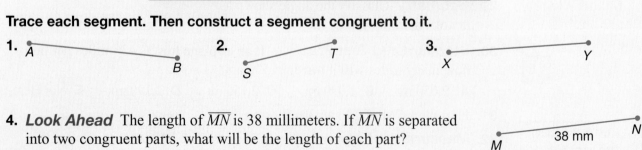

The symbol ∠ is used to indicate an angle. The angle shown can be named in two ways, ∠JKL or ∠LKJ. The middle letter is always the vertex.

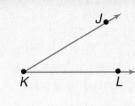

You can also use a straightedge and a compass to construct congruent angles. *Congruent angles* are angles that have the same measure.

TRY THIS

Work with a partner.

② To construct an angle congruent to ∠JKL, follow these steps.

- Draw ∠JKL. Then use a straightedge to draw side $\overrightarrow{BA}$. $\overrightarrow{BA}$ means ray BA. A ray is a path that extends infinitely from one point in a certain direction.

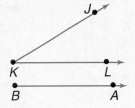

- With the compass point at K, draw an arc that intersects the sides of ∠JKL. Label the intersections M and N.

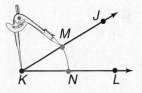

- With the same compass setting, move to side BA. Place the compass point at B and draw an arc that intersects side $\overrightarrow{BA}$. Label this intersection P.

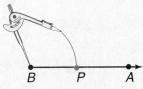

- On ∠JKL, set the compass at points M and N as shown. With that setting, go to side $\overrightarrow{BA}$. Place the compass point at P, and draw an arc that intersects the larger arc you drew before. Label the intersection Q.

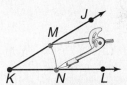

- Use a straightedge to draw side $\overrightarrow{BQ}$. ∠QBA is congruent to ∠JKL.

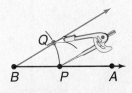

ON YOUR OWN

Trace each angle. Then construct an angle congruent to it.

5.   6.   7.

8. **Look Ahead** The measure of an angle is 113°. If the angle is divided into two congruent angles, what will be the measure of each angle?

Constructing Bisectors

EVCLIDE MEGAREN·

The Greek mathematician Euclid answered the following problem more than 2,000 years ago.

Given a line segment of any length, find a geometric method for dividing the line segment into two equal parts.

To **bisect** something means to separate it into two congruent parts.

HANDS-ON MINI-LAB

Work with a partner.

🖊 ruler

Try This

- Draw $\overline{AB}$ using a straightedge.

- Fold point A onto point B and make a crease as shown. The crease bisects $\overline{AB}$. Label the intersection point C.

Talk About It

Measure $\overline{AC}$ and $\overline{CB}$. What can you say about point C?

You can also use a straightedge and a compass to bisect a line segment.

Example ① Use a straightedge and a compass to bisect $\overline{YZ}$.

- Use a straightedge to draw $\overline{YZ}$.

- Place the compass point at Y. Set the compass to more than half the length of $\overline{YZ}$. Draw two arcs as shown.

- With the same compass setting, place the compass point at Z and draw two arcs as shown. These arcs should intersect the first arc at W and V.

- With a straightedge, draw $\overline{WV}$. $\overline{WV}$ bisects $\overline{YZ}$ at P. So, $\overline{PY}$ and $\overline{PZ}$ are congruent.

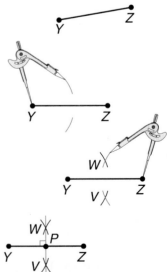

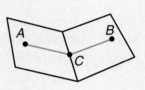

When segments meet to form right angles, they are perpendicular. You can say that $\overline{WV}$ is a perpendicular bisector of $\overline{YZ}$.

In the previous Mini-Lab and in Example 1, you learned how to bisect a line segment. You can also bisect an angle.

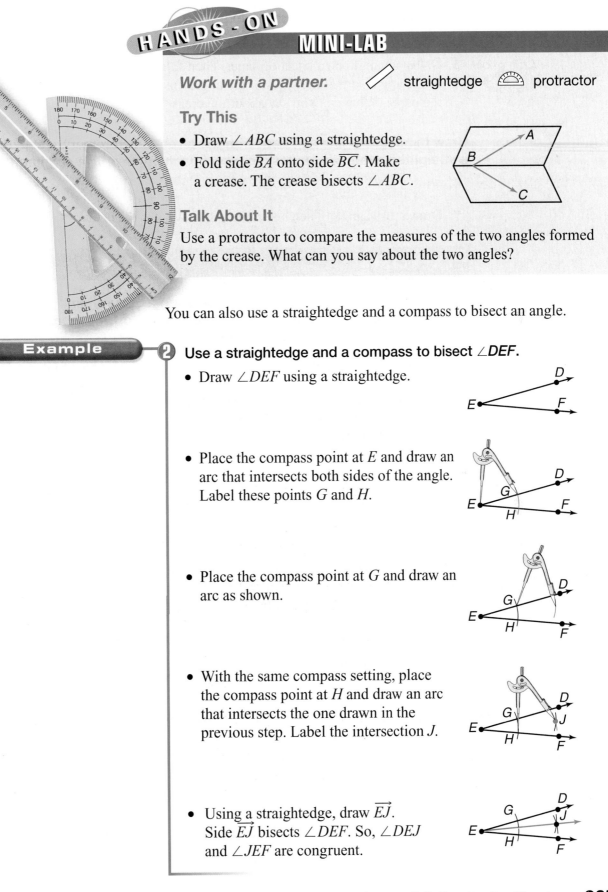

HANDS-ON MINI-LAB

Work with a partner.   ╱ straightedge   ⌓ protractor

Try This
- Draw ∠*ABC* using a straightedge.
- Fold side $\overline{BA}$ onto side $\overline{BC}$. Make a crease. The crease bisects ∠*ABC*.

Talk About It
Use a protractor to compare the measures of the two angles formed by the crease. What can you say about the two angles?

You can also use a straightedge and a compass to bisect an angle.

Example

2 Use a straightedge and a compass to bisect ∠*DEF*.
- Draw ∠*DEF* using a straightedge.

- Place the compass point at *E* and draw an arc that intersects both sides of the angle. Label these points *G* and *H*.

- Place the compass point at *G* and draw an arc as shown.

- With the same compass setting, place the compass point at *H* and draw an arc that intersects the one drawn in the previous step. Label the intersection *J*.

- Using a straightedge, draw $\overrightarrow{EJ}$. Side $\overrightarrow{EJ}$ bisects ∠*DEF*. So, ∠*DEJ* and ∠*JEF* are congruent.

Communicating Mathematics

Read and study the lesson to answer each question.

1. *Explain* the meaning of the word *bisect*.

2. *Tell* how to bisect a line segment using a straightedge and a compass.

HANDS-ON  3. *Draw* an angle on a sheet of paper. Then fold the paper through the vertex of the angle so that the sides of the angle match. Make a crease and unfold the paper. What can you say about the crease?

Guided Practice

Draw the angle or line segment with the given measurement. Then use a straightedge and a compass to bisect each angle or line segment.

4. 4 in.   5. 65°   6. 7 cm

7. Draw a 119° angle. Then bisect the angle. What is the measure of each angle?

Practice

Draw the angle or line segment with the given measurement. Then use a straightedge and a compass to bisect each angle or line segment.

8. 120°   9. 75°   10. 5 cm   11. 30°

12. 25 mm   13. 9 cm   14. 108°   15. 3 in.

16. In the figure shown, name the side that appears to bisect ∠*MNP*.

Applications and Problem Solving

17. *Geometry* Copy the figure. Use a protractor to measure each angle formed by one side of the square and a diagonal. What can you say about the diagonals of a square?

18. *Design* Draw a circle that is 6 inches in diameter. Draw a diameter. Construct a second diameter that bisects the first one. Construct two more diameters that bisect the angles of the intersection.

19. *Critical Thinking* Explain how to construct the following using a compass and straightedge.

For Extra Practice, see page 581.

     a. an equilateral triangle   b. an equilateral triangle

     c. a perpendicular bisector of a line segment

     d. a line passing through a given point and perpendicular to a given line

Mixed Review

20. *Geometry* Estimate the measure of the angle. *(Lesson 9-2)*

21. Express 0.74 as a fraction in simplest form. *(Lesson 5-9)*

22. **Standardized Test Practice** Three people bought pens for $11.55. How much did each person pay if they shared the price equally? *(Lesson 4-5)*

     A $3.25   **B** $3.45   **C** $3.65   **D** $3.85   **E** $3.55

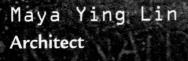

...UNTIL JUSTICE ROLLS DOWN LIKE WATERS
AND RIGHTEOUSNESS LIKE A MIGHTY STREAM
MARTIN LUTHER KING JR.

Architecture

Maya Ying Lin
Architect

When Maya Ying Lin designed the *Vietnam Veterans Memorial* in 1980, she became the first woman to design a major Washington, D.C., monument. Since then she has designed other memorials including the *Civil Rights Memorial* in Montgomery, Alabama.

To be an architect, you'll need at least a bachelor's degree in architecture and three years of practical experience. In addition to organizational skills and computer skills, math skills are needed to determine the dimensions of a design and to draw blue prints. Architects design shopping centers, schools, homes, museums, office buildings, and sports arenas, to name a few.

For more information:
American Institute of Architects
1735 New York Avenue, NW
Washington, DC 20006

interNET CONNECTION
www.glencoe.com/sec/
math/mac/mathnet

Someday, I'd like to design memorials and monuments like Maya Lin.

Your Turn
Research architecture as a career. Then write a report that describes the benefits of a career in architecture.

COOPERATIVE LEARNING

9-4A Triangles and Quadrilaterals

A Preview of Lesson 9-4

geoboard

geobands

dot paper

scissors

Polygons are simple closed figures formed by three or more line segments. The line segments, or sides, intersect at their endpoints. These points of intersection are called *vertices* (plural of vertex).

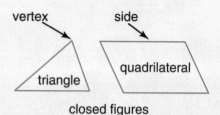

closed figures

In this lab, you will classify *triangles* and *quadrilaterals*. A triangle is a polygon with three sides. A quadrilateral is a polygon with four sides.

TRY THIS

Work with a partner.

1 Use geobands to make a triangle. A sample triangle is shown.

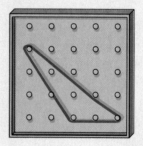

- Draw the triangle on dot paper and cut it out.
- Repeat these steps until you have ten different triangles.
- Every triangle has at least two acute angles. The triangle shown has two acute angles. Since the third angle is obtuse, the triangle is an obtuse triangle.
- Sort your triangles into three groups, based on the third angle. Name the groups acute, right, and obtuse triangles.

ON YOUR OWN

Classify each triangle as *acute, right,* or *obtuse*.

1.

2.

3.

4. *Look Ahead* Write a definition for each type of triangle.

Triangles can be classified by the type of angles they have while quadrilaterals are classified by the characteristics of their sides and angles.

TRY THIS

Work with a partner.

2 Use geobands to make a quadrilateral. Several samples are shown.

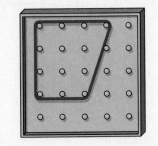

- Draw the quadrilateral on dot paper and cut it out.
- Repeat these steps until you have ten different quadrilaterals.
- You can classify quadrilaterals according to the characteristics of their sides and angles. The first quadrilateral shown can be classified as a quadrilateral with four congruent sides and four right angles.
- Sort your quadrilaterals into three groups, based on any characteristic you choose. Write a description of the quadrilaterals in each group.

ON YOUR OWN

Write a sentence that describes one characteristic of each quadrilateral. Then draw a different quadrilateral that has the same characteristic.

5.
6.
7.

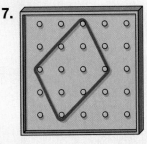

8. *Look Ahead* Draw a quadrilateral that has two obtuse angles and exactly one pair of opposite sides congruent.

Two-Dimensional Figures

Powwows are American Indian celebrations of heritage. Dancing, games, and parades are all part of the festivities. The colorful headdress shown is an example of what an American Indian may wear to a powwow. The figures on the headdress are examples of polygons.

A **polygon** is a simple closed figure formed by three or more sides. The number of sides determines the name of the polygon.

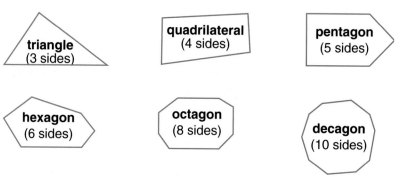

triangle (3 sides) **quadrilateral** (4 sides) **pentagon** (5 sides)

hexagon (6 sides) **octagon** (8 sides) **decagon** (10 sides)

Any polygon with all sides congruent and all angles congruent is called a **regular polygon**. Some examples of common regular polygons are shown below.

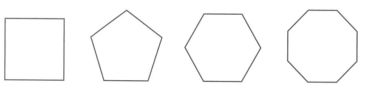

Examples

Name each polygon. Then tell if the polygon is a regular polygon.

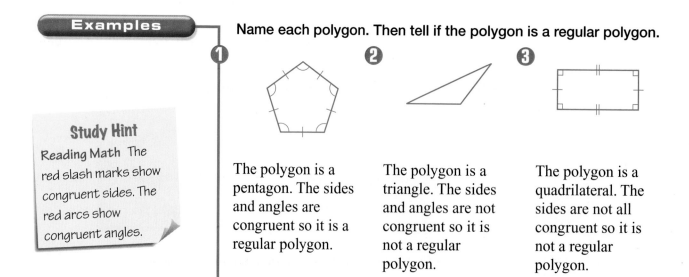

① The polygon is a pentagon. The sides and angles are congruent so it is a regular polygon.

② The polygon is a triangle. The sides and angles are not congruent so it is not a regular polygon.

③ The polygon is a quadrilateral. The sides are not all congruent so it is not a regular polygon.

Certain types of quadrilaterals have special characteristics.

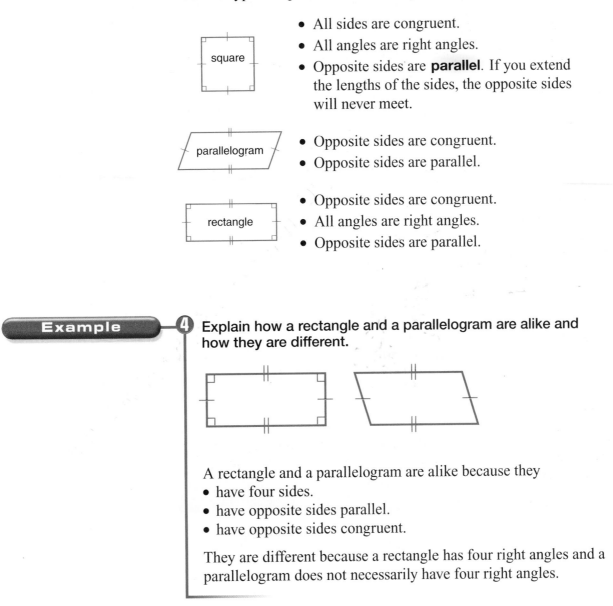

square
- All sides are congruent.
- All angles are right angles.
- Opposite sides are **parallel**. If you extend the lengths of the sides, the opposite sides will never meet.

parallelogram
- Opposite sides are congruent.
- Opposite sides are parallel.

rectangle
- Opposite sides are congruent.
- All angles are right angles.
- Opposite sides are parallel.

Example 4

Explain how a rectangle and a parallelogram are alike and how they are different.

A rectangle and a parallelogram are alike because they
- have four sides.
- have opposite sides parallel.
- have opposite sides congruent.

They are different because a rectangle has four right angles and a parallelogram does not necessarily have four right angles.

A triangle with three congruent sides is called an **equilateral triangle**.

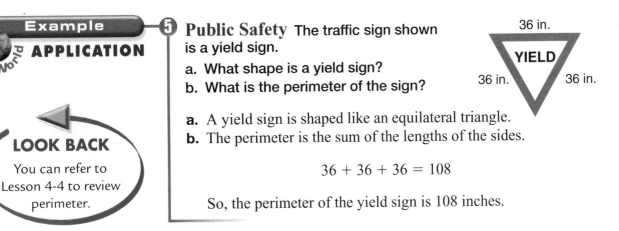

Example 5

APPLICATION

Public Safety The traffic sign shown is a yield sign.

a. What shape is a yield sign?
b. What is the perimeter of the sign?

36 in.

YIELD

36 in.   36 in.

a. A yield sign is shaped like an equilateral triangle.
b. The perimeter is the sum of the lengths of the sides.

$$36 + 36 + 36 = 108$$

So, the perimeter of the yield sign is 108 inches.

LOOK BACK

You can refer to Lesson 4-4 to review perimeter.

CHECK FOR UNDERSTANDING

Communicating Mathematics

Read and study the lesson to answer each question.

1. *Identify* two different polygons that you see in your classroom.

2. *Draw* a triangle that has exactly one pair of congruent sides.

3. *You Decide* Victoria says that all parallelograms are squares. Jonathan says that all squares are parallelograms. Who is correct? Explain your reasoning.

Guided Practice

Name each polygon. Then tell if the polygon is a regular polygon.

4.

5.

Explain how each pair of figures is alike and how each pair is different.

6.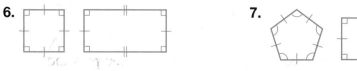

7.

Draw an example of each polygon. Mark any congruent sides, congruent angles, and right angles.

8. parallelogram

9. regular hexagon

10. *Design* Winona needs to design a table that will seat six people with equal comfort and work space. What shape should she make the table?

EXERCISES

Practice

Name each polygon. Then tell if the polygon is a regular polygon.

11.   12.   13.   14.

Explain how each pair of figures is alike and how each pair is different.

15.   16.   17.

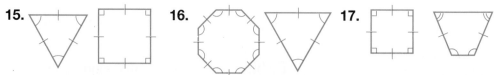

Draw an example of each polygon. Mark any congruent sides, congruent angles, and right angles.

18. quadrilateral   19. pentagon   20. equilateral triangle

21. regular octagon   22. triangle   23. rectangle

24. How many sides does a regular decagon have?

Applications and Problem Solving

25. *Sports* The most popular game in the world is soccer. A soccer ball is shown. Which two polygons make up the pattern on a soccer ball?

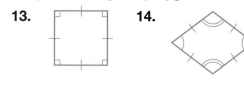

26. **Carpentry** A clubhouse is shaped like a regular pentagon. If the perimeter of the floor is 30 feet long, how long is each wall?

27. **Working on the** CHAPTER Project Refer to page 351.

   a. Name the different types of polygons in the floor plan.

   b. Find the sum of the measures of the angles of the bathroom, the dining room, and the second bedroom. What is the sum of the measures of the angles of an octagon? Explain.

28. **Critical Thinking** A point is in the *interior* of a polygon if it lies on the inside of the polygon and does not lie on the polygon itself. A point is in the *exterior* if it lies outside the polygon and does not lie on the polygon itself. Tell whether each point is in the interior of the polygon, the exterior of the polygon, or neither.

   a.

   b.

   c.

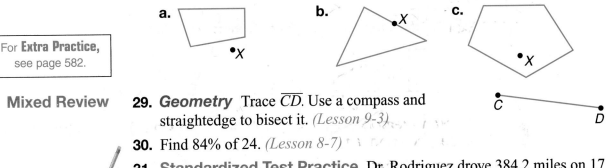

For **Extra Practice,**
see page 582.

Mixed Review

29. **Geometry** Trace $\overline{CD}$. Use a compass and straightedge to bisect it. *(Lesson 9-3)*

30. Find 84% of 24. *(Lesson 8-7)*

31. **Standardized Test Practice** Dr. Rodriguez drove 384.2 miles on 17 gallons of fuel. How many miles per gallon did his car get? *(Lesson 8-1)*

   A 22.5 mpg   **B** 22.6 mpg   **C** 126 mpg   **D** 226 mpg   **E** Not Here

32. Express 2.375 as a fraction. *(Lesson 5-9)*

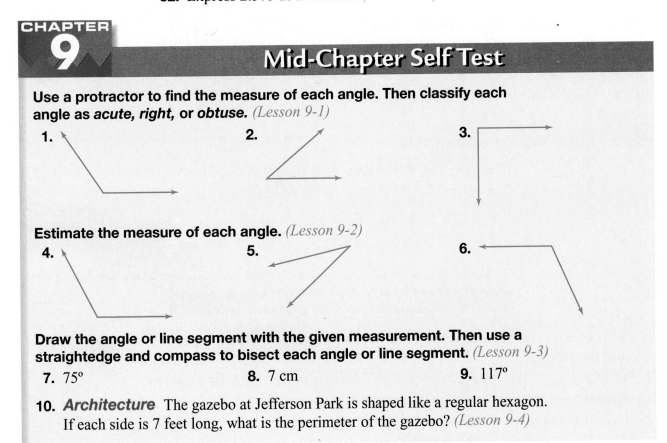

CHAPTER 9

Mid-Chapter Self Test

Use a protractor to find the measure of each angle. Then classify each angle as *acute*, *right*, or *obtuse*. *(Lesson 9-1)*

1.

2.

3.

Estimate the measure of each angle. *(Lesson 9-2)*

4.

5.

6.

Draw the angle or line segment with the given measurement. Then use a straightedge and compass to bisect each angle or line segment. *(Lesson 9-3)*

7. 75°

8. 7 cm

9. 117°

10. **Architecture** The gazebo at Jefferson Park is shaped like a regular hexagon. If each side is 7 feet long, what is the perimeter of the gazebo? *(Lesson 9-4)*

9-4B Using Nets to Wrap a Cube

A Follow-Up of Lesson 9-4

✂ scissors

▢ cube

In this lab, you will make a figure called a *net* and use it to wrap a cube. The faces of the cube are shaped like squares.

TRY THIS

Work with a partner.

Step 1 Count the number of faces of the cube.

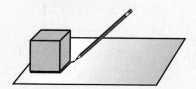

Step 2 Place the cube on the paper as shown and draw a square.

Step 3 Continue drawing squares to make a figure like the one shown. This figure is called a net.

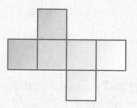

Step 4 Cut out the net and try to wrap the cube.

Step 5 Repeat Step 3 to make a net like the one shown. Cut out the net and try to wrap the cube.

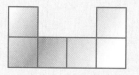

ON YOUR OWN

1. **Tell** if both nets covered the cube. If not, explain why the net or nets did not cover the cube.

2. **Find** three other nets of six squares that will cover the cube. Look for a pattern.

3. **Write** three sentences describing the net patterns you found.

4. **Reflect Back** Give an example of a net that can be used to cover a box whose faces are shaped like rectangles.

Lines of Symmetry

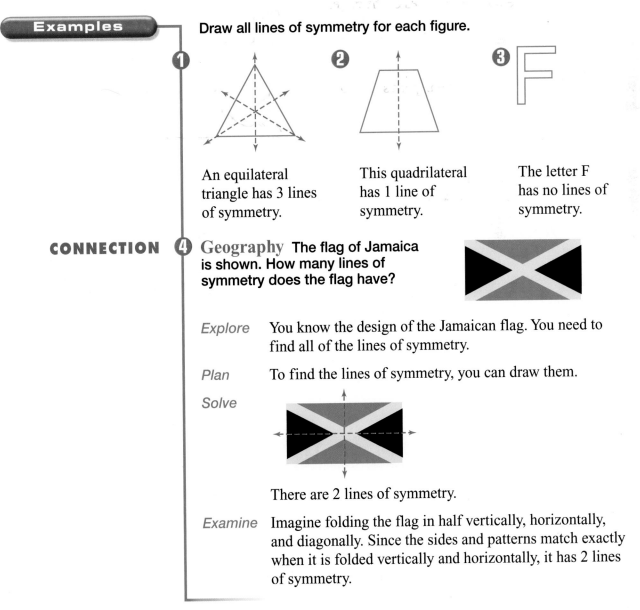

What you'll learn

You'll learn to describe and define lines of symmetry.

When am I ever going to use this?

Knowing how to describe lines of symmetry can help you design and build a kite.

Word Wise

line of symmetry
reflection

More than 2,000 years ago, the Chinese military used kites for military purposes. They attached bamboo pipes to kites and flew them over the enemy at night. As wind passed through the pipes, the whistling noise caused the enemy to flee.

If you draw a line down the center of a diamond-shaped kite, the two halves match. When this happens, the line is called a **line of symmetry**. Lines of symmetry can be found in certain figures.

Examples

Draw all lines of symmetry for each figure.

1

An equilateral triangle has 3 lines of symmetry.

2

This quadrilateral has 1 line of symmetry.

3 F

The letter F has no lines of symmetry.

CONNECTION

4 **Geography** The flag of Jamaica is shown. How many lines of symmetry does the flag have?

Explore   You know the design of the Jamaican flag. You need to find all of the lines of symmetry.

Plan   To find the lines of symmetry, you can draw them.

Solve

There are 2 lines of symmetry.

Examine   Imagine folding the flag in half vertically, horizontally, and diagonally. Since the sides and patterns match exactly when it is folded vertically and horizontally, it has 2 lines of symmetry.

The capital letter M has one line of symmetry. Notice that the right half is a **reflection** of the left half. A reflection is a mirror image of a figure across a line of symmetry. You can use a geomirror to draw reflections.

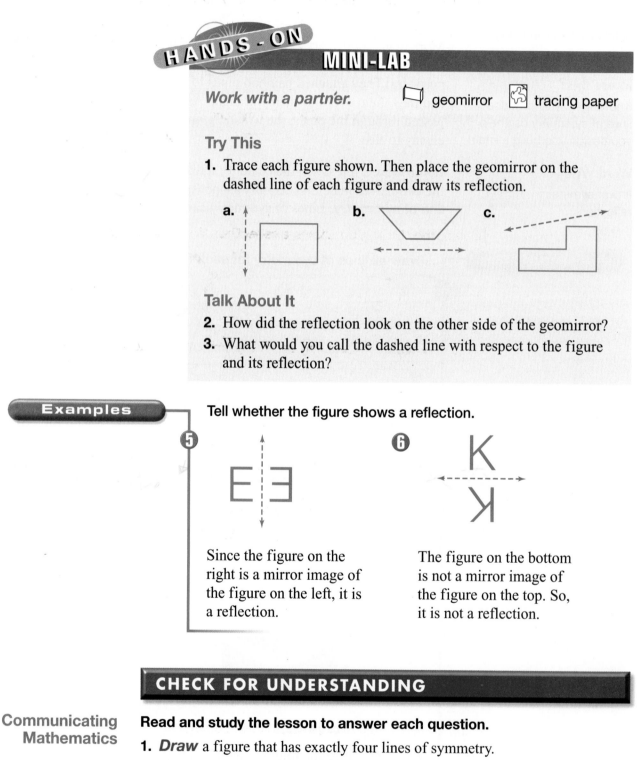

MINI-LAB

Work with a partner. ⬜ geomirror 🧩 tracing paper

Try This

1. Trace each figure shown. Then place the geomirror on the dashed line of each figure and draw its reflection.

   a.   **b.**   **c.**

Talk About It

2. How did the reflection look on the other side of the geomirror?
3. What would you call the dashed line with respect to the figure and its reflection?

Examples

Tell whether the figure shows a reflection.

5

Since the figure on the right is a mirror image of the figure on the left, it is a reflection.

6

The figure on the bottom is not a mirror image of the figure on the top. So, it is not a reflection.

CHECK FOR UNDERSTANDING

Communicating Mathematics

Read and study the lesson to answer each question.

1. *Draw* a figure that has exactly four lines of symmetry.

2. *Tell* which figure does not have at least one line of symmetry.

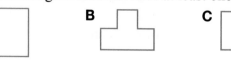

   A   B   C   D

3. *Draw* and cut out a rectangle and a square. How many ways can you fold each figure so that one half matches the other half? How many lines of symmetry does each figure have?

Guided Practice

Tell whether the dashed line is a line of symmetry. Write *yes* or *no*.

4. **5.** **6.**

Trace each figure. Draw all lines of symmetry.

7. **8.** **9.**

Tell whether the figure shows a reflection. Write *yes* or *no*.

10. **11.**

12. *Life Science* How many lines of symmetry does the starfish shown have?

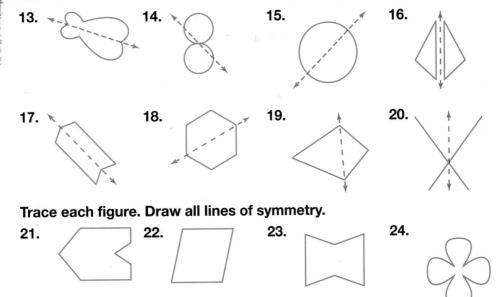

EXERCISES

Practice **Tell whether the dashed line is a line of symmetry. Write *yes* or *no*.**

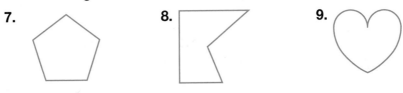

13. **14.** **15.** **16.**

17. **18.** **19.** **20.**

Trace each figure. Draw all lines of symmetry.

21. **22.** **23.** **24.**

Trace each figure. Draw all lines of symmetry.

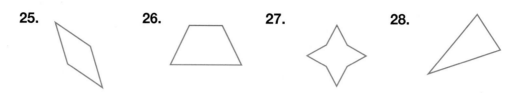

25.   26.   27.   28.

Tell whether the figure shows a reflection. Write *yes* or *no*.

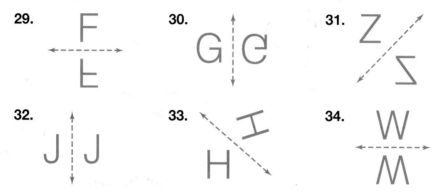

29.   30.   31.

32.   33.   34.

Applications and Problem Solving

35. How many lines of symmetry does a stop sign have?

36. *Geography* Switzerland is located in the Alps mountains in central Europe. How many lines of symmetry does the flag of Switzerland have?

37. *Life Science* There are more than 20,000 different types of bees. The mining bee is known for making its nest in loose ground. How many lines of symmetry does the mining bee shown at the left have?

38. *Critical Thinking* Copy the figure shown. Then shade enough squares so that the figure has a diagonal line of symmetry.

Mixed Review

39. *Geometry* Draw an example of an octagon. *(Lesson 9-4)*

40. **Standardized Test Practice** Three-fifths of the sixth-grade students went on the class trip. What percent of the sixth-grade students did *not* go on the class trip? *(Lesson 8-4)*

    A 40%   **B** 45%   **C** 50%   **D** 60%

41. *Algebra* If $x = 1\frac{2}{3}$, and $y = 3\frac{4}{5}$, what is the value of xy? *(Lesson 7-3)*

42. Express 0.84 as a fraction in simplest form. *(Lesson 5-9)*

43. *Entertainment* The graph shows the number of visitors who toured the Rock and Roll Hall of Fame in Cleveland on a Saturday. How many visitors altogether toured the Rock and Roll Hall of Fame before 1:00 P.M.? *(Lesson 2-3)*

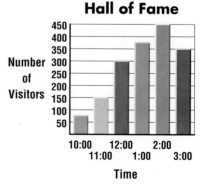

Rock and Roll Hall of Fame

For **Extra Practice,** see page 582.

Size and Shape

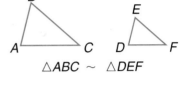

What you'll learn

You'll learn to determine congruence and similarity.

When am I ever going to use this?

Knowing how to determine congruence and similarity helps an architect design statues.

Word Wise

similar figures
congruent figures

The New York-New York Hotel in Las Vegas, Nevada, has a 150-foot replica of the Statue of Liberty. The height of the actual Statue of Liberty is 305 feet.

Figures that have the same shape and angles, but different size are called **similar figures**. The symbol $\sim$ means *is similar to*.

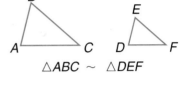

$\triangle ABC \sim \triangle DEF$

Suppose the replica was 305 feet tall. The statues would be the same size and shape.

Figures that are the same size and shape are called **congruent figures**. The symbol $\cong$ means *is congruent to*.

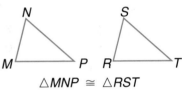

$\triangle MNP \cong \triangle RST$

In the following lab, you will explore some characteristics of congruent figures and similar figures.

> ### Study Hint
>
> **Reading Math** The symbol $\triangle ABC$ is read as triangle ABC.

HANDS-ON MINI-LAB

Work with a partner. ✂ scissors ▨ tracing paper

Try This

- Trace each pair of figures and cut them out.

  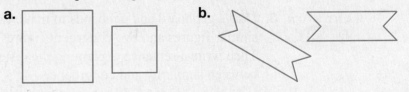

- Try to match each pair of figures.

Talk About It

1. Congruent figures are exact matches. Did either pair of figures match? If so, name the congruent figures.

2. Compare the angles and segments of each pair of figures. How are they alike? How are they different?

3. Did either pair of figures not match? A pair of figures may be similar without being congruent. Name the similar figures.

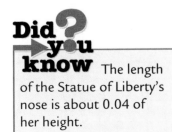

Did you know The length of the Statue of Liberty's nose is about 0.04 of her height.

Tell whether each quadrilateral is congruent or similar to the quadrilateral shown.

1 The quadrilaterals have the same shape and angles, but not the same size. They are similar.

2 The quadrilaterals are the same size and shape. They are congruent.

3 The quadrilaterals are not the same shape or size. They are neither congruent nor similar.

INTEGRATION **4** Geometry △*MNP* is congruent to △*RST*.

a. What side of △*RST* corresponds to side $\overline{MN}$?

b. What is the perimeter of △*RST*?

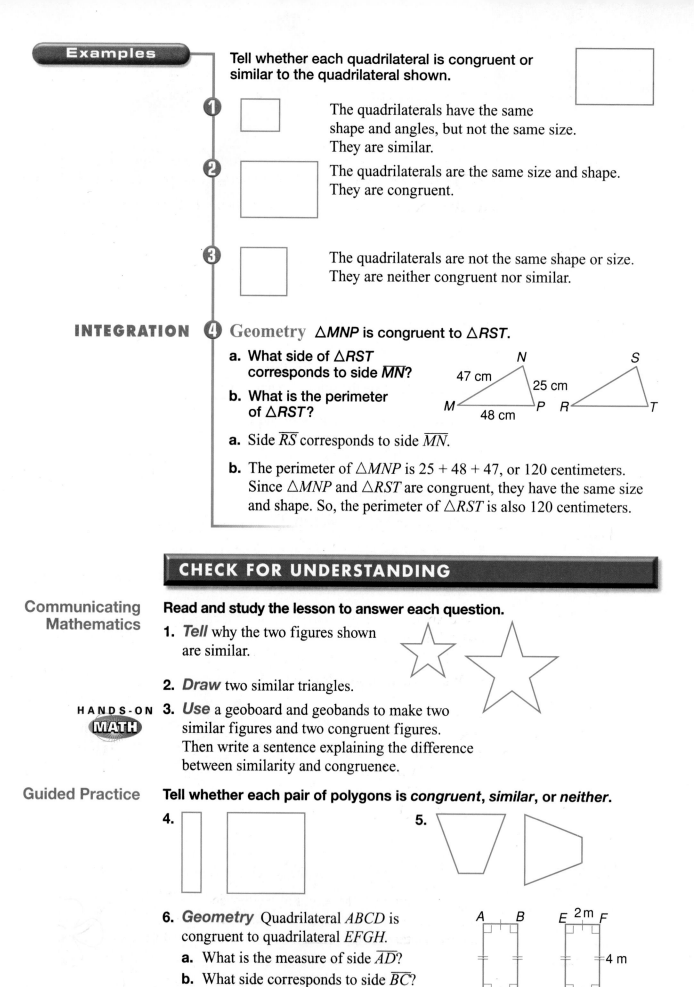

47 cm 25 cm 48 cm

a. Side $\overline{RS}$ corresponds to side $\overline{MN}$.

b. The perimeter of △*MNP* is 25 + 48 + 47, or 120 centimeters. Since △*MNP* and △*RST* are congruent, they have the same size and shape. So, the perimeter of △*RST* is also 120 centimeters.

CHECK FOR UNDERSTANDING

Communicating Mathematics

Read and study the lesson to answer each question.

1. *Tell* why the two figures shown are similar.

2. *Draw* two similar triangles.

HANDS-ON MATH 3. *Use* a geoboard and geobands to make two similar figures and two congruent figures. Then write a sentence explaining the difference between similarity and congruence.

Guided Practice

Tell whether each pair of polygons is *congruent*, *similar*, or *neither*.

4.

5.

6. *Geometry* Quadrilateral *ABCD* is congruent to quadrilateral *EFGH*.

a. What is the measure of side $\overline{AD}$?

b. What side corresponds to side $\overline{BC}$?

2 m 4 m

7. *Geography* The flag of the state of North Carolina is shown. How many pairs of congruent quadrilaterals are there?

EXERCISES

Practice

Tell whether each pair of polygons is *congruent*, *similar*, or *neither*.

8. **9.**

10. **11.**

12. **13.**

14. Which pair of polygons are neither similar nor congruent?

    a.   **b.**   **c.**

15. The triangles shown are acute triangles and $\triangle DEF \cong \triangle LMN$.

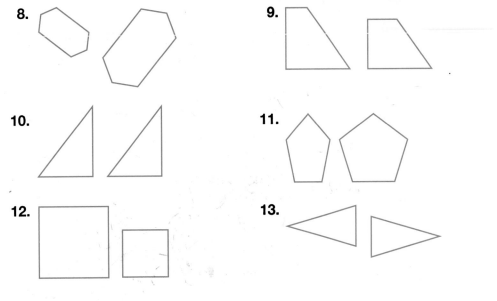

    a. What is the measure of side $\overline{DF}$?

    b. What side corresponds to side $\overline{FE}$?

16. Quadrilateral *MNOP* is congruent to quadrilateral *WXYZ*.

    a. Find the perimeter of quadrilateral *WXYZ*.

    b. What side corresponds to side $\overline{MP}$?

Applications and Problem Solving

17. *Puzzles* A tangram is a Chinese puzzle consisting of seven geometric shapes. The diagram shows how the pieces of the puzzle can form a square. Which of the geometric shapes in a tangram are similar?

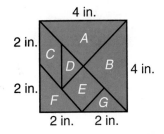

18. *Technology* A photocopier can make copies that are reduced or enlarged. At 50%, a photocopier makes a copy that is 50%, or one-half, of the original length and width. Nina needs to photocopy a picture that is 7 inches long and 5 inches wide.

  a. Draw a diagram that shows the dimensions of the original picture and the dimensions of the copy reduced by 50%.

   b. Are the two pictures similar or congruent?

19. *Working on the* 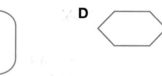 Refer to the floor plan on page 351. Use the given scale to find the actual length of each wall of the bathroom, dining room, and the second bedroom. Then find the actual perimeter of the house. Label each length on your diagram.

Mixed Review

20. *Critical Thinking* Are all squares similar? Explain your reasoning.

21. **Standardized Test Practice** Which figure does *not* have at least one line of symmetry? *(Lesson 9-5)*

  A   **B**   **C**   **D**

22. *Money Matters* One fourth of the 417 students in the sixth grade owe the library for fines. About how many sixth grade students owe the library for fines? *(Lesson 7-2)*

For **Extra Practice**, see page 582.

23. Find $7\frac{1}{2} - 1\frac{3}{4}$. *(Lesson 6-6)*

MATH IN THE MEDIA

Peanuts

1. What does it mean to be congruent to?

2. *Draw* two octagons that are congruent.

Pento

Get Ready   This game is for two players.

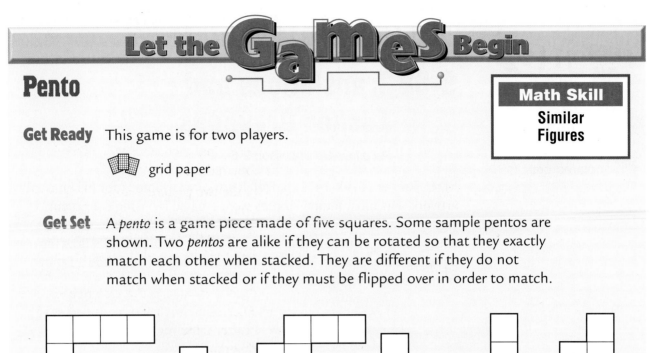

grid paper

Get Set   A *pento* is a game piece made of five squares. Some sample pentos are shown. Two *pentos* are alike if they can be rotated so that they exactly match each other when stacked. They are different if they do not match when stacked or if they must be flipped over in order to match.

alike   **different**   **different**

Go ● **Round 1:** Each player has 1 minute to draw as many different *pentos* as possible on grid paper. Each player scores 1 point for every *pento* drawn. Two points are scored for each *pento* found by only one of the players.

● **Round 2:** Cut out all of the *pentos* from round 1. Then choose one *pento* as a model. Using any four different *pentos*, each player builds a similar figure where the length of each side is twice the length of the side in the model. Sketch the results on grid paper.

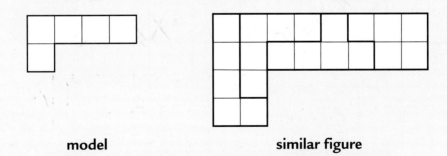

model   **similar figure**

Continue this round for 2 minutes. Each player scores 3 points for each similar figure drawn. The player with the most total points wins.

COOPERATIVE LEARNING

9-6B Translations and Escher Drawings

A Follow-Up of Lesson 9-6

M. C. Escher (1898-1972), a Dutch artist, was famous for his unusual artwork. His most famous pieces were created by using congruent figures shaped as birds, reptiles, or fish that fit together to cover the entire surface of the artwork. You can create an Escher-like drawing by using a *translation*. When you do a translation, you slide a figure from one location to another.

TRY THIS

Work with a partner.

1 To create an Escher-like drawing using a translation, follow these steps.

• Draw a square. Then draw a triangle on the top of the square as shown.

• Translate, or slide, the triangle to the opposite side of the square. The pattern unit for our Escher-like drawing is formed.

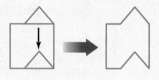

• Repeat the pattern unit you made in the steps above to create an Escher-like drawing on grid paper. Use colored pencils to decorate your drawing.

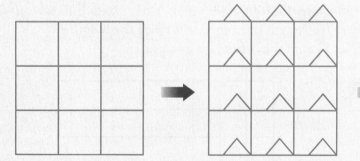

ON YOUR OWN

Create an Escher-like drawing for each pattern unit shown.

1.   **2.**   **3.**

4. *Reflect Back* Are the pattern units in your Escher-like drawing congruent?

You can also create an Escher-like drawing using two translations.

Work with a partner.

2 To create an Escher-like drawing using two translations, follow these steps.

- Draw a square. Then draw a rectangle on the top of the square as shown. Next, translate the rectangle to the opposite side of the square.

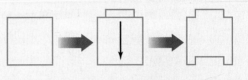

- Draw a triangle on the left side and translate the triangle to the right side.

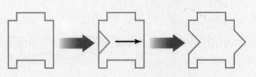

- Repeat the pattern unit you made above to create an Escher-like drawing on grid paper. Use colored pencils to decorate your drawing.

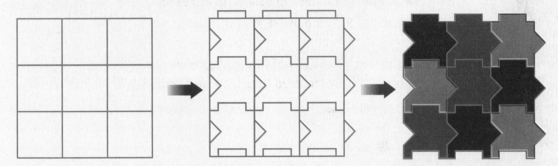

ON YOUR OWN

Create an Escher-like drawing for each pattern unit shown.

5.

6.

7.

8. ***Reflect Back*** Why do the pieces fit together exactly?

interNET
CONNECTION Chapter Review **For additional lesson-by-lesson review, visit:**
www.glencoe.com/sec/math/mac/mathnet

Vocabulary

After completing this chapter, you should be able to define each term, concept, or phrase and give an example or two of each.

Geometry
acute angle (p. 352)
angle (p. 352)
bisect (p. 364)
complementary (p. 353)
congruent angles (p. 363)
congruent figures
  (p. 379)
congruent segments
  (p. 362)
decagon (p. 370)
degree (p. 352)
edges (p. 352)
equilateral triangle
  (p. 371)

hexagon (p. 370)
line of symmetry (p. 375)
line segment (p. 362)
net (p. 374)
obtuse angle (p. 352)
octagon (p. 370)
parallel (p. 371)
parallelogram (p. 371)
pentagon (p. 370)
polygon (pp. 368, 370)
protractor (p. 352)
quadrilateral
  (pp. 368, 370)
ray (p. 363)

rectangle (p. 371)
reflection (p. 376)
regular polygon (p. 370)
right angle (p. 352)
similar figures (p. 379)
square (p. 371)
straightedge (p. 358)
supplementary (p. 353)
translation (p. 384)
triangle (pp. 368, 370)
vertex (p. 352)

Problem Solving
use logical reasoning
  (p. 356)

Understanding and Using the Vocabulary

Choose the letter of the term that best matches each phrase.

1. a line dividing a shape into two matching halves
2. a six-sided figure
3. a polygon with all sides and all angles congruent
4. a quadrilateral with opposite sides parallel
5. a simple closed figure formed by three or more sides
6. a ruler or any object with a straight side
7. the point where two edges of a polygon intersect
8. the most common unit of measure for an angle
9. an angle whose measure is between 0° and 90°
10. an angle whose measure is between 90° and 180°

a. regular polygon
b. vertex
c. straightedge
d. acute angle
e. polygon
f. line of symmetry
g. obtuse angle
h. degree
i. hexagon
j. parallelogram

In Your Own Words

11. *Explain* how to classify an angle as acute, right, or obtuse.

Objectives & Examples

Upon completing this chapter, you should be able to:

● classify and measure angles *(Lesson 9-1)*

The angle is an acute angle since it measures between 0° and 90°.

The angle is a right angle since it measures 90°.

The angle is an obtuse angle since it measures between 90° and 180°.

● draw angles and estimate measures of angles *(Lesson 9-2)*

You can use a protractor and a straightedge to draw an angle.

The angle drawn is a 47° angle.

● bisect line segments and angles *(Lesson 9-3)*

To *bisect* something means to divide it into two congruent parts. You can use paper folding or a straightedge and a compass to bisect a line segment or an angle.

Review Exercises

Use these exercises to review and prepare for the chapter test.

Use a protractor to find the measure of each angle. Then classify the angle as *acute, right,* or *obtuse*.

12.    13.

Classify each angle measure as *acute, right,* or *obtuse*.

14. 147°   15. 38.5°

16. 93°   17. 12.2°

Use a protractor and a straightedge to draw angles having the following measurements.

18. 36°   19. 127°

Estimate the measure of each angle.

20.    21.

Draw the angle or line segment with the given measurement. Then use a straightedge and a compass to bisect each angle or line segment.

22. 5 in.   23. 100°

24. 35°   25. 7 cm

Objectives & Examples

name two-dimensional figures *(Lesson 9-4)*

Name each polygon. Then tell if the polygon is a regular polygon.

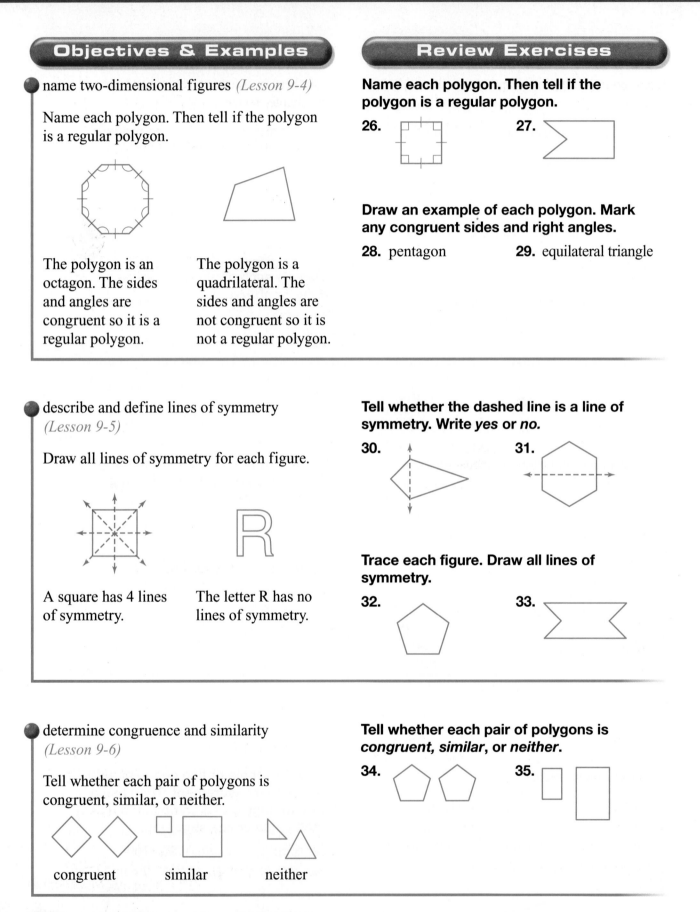

The polygon is an octagon. The sides and angles are congruent so it is a regular polygon.

The polygon is a quadrilateral. The sides and angles are not congruent so it is not a regular polygon.

describe and define lines of symmetry *(Lesson 9-5)*

Draw all lines of symmetry for each figure.

A square has 4 lines of symmetry.

The letter R has no lines of symmetry.

determine congruence and similarity *(Lesson 9-6)*

Tell whether each pair of polygons is congruent, similar, or neither.

congruent   similar   neither

Review Exercises

Name each polygon. Then tell if the polygon is a regular polygon.

26.

27.

Draw an example of each polygon. Mark any congruent sides and right angles.

28. pentagon

29. equilateral triangle

Tell whether the dashed line is a line of symmetry. Write *yes* or *no*.

30.

31.

Trace each figure. Draw all lines of symmetry.

32.

33.

Tell whether each pair of polygons is congruent, similar, or neither.

34.

35.

Applications & Problem Solving

36. Use Logical Reasoning There are 32 students in Mr. Miyar's literature class. Of these, 12 take Spanish, 15 take tennis, and 8 take both Spanish and tennis. How many take neither Spanish nor tennis? *(Lesson 9-1B)*

37. Geometry Kurano drew the figure shown as a design on a round tablecloth. She wants to draw three more stripes from the center that will bisect the three angles. Copy her design and bisect the angles to show what the new design will look like. *(Lesson 9-3)*

38. Remodeling A kitchen floor is to be covered with vinyl flooring. The pattern of the flooring is a square surrounded by four regular octagons. Draw a section of the flooring. Mark any congruent sides and angles. *(Lesson 9-4)*

39. School The student council voted to have a contest to design a school flag. The flag may be of any shape but must have at least one line of symmetry. Use any shapes, letters, objects, and colors to design a flag. *(Lesson 9-5)*

Alternative Assessment

● Open Ended

Suppose that you are a graphic designer. A client has asked you to make a sign for the grand opening of her store, The Sports Store. Design a logo that displays line symmetry. Use two-dimensional figures in the design of your sign.

Identify the types of angles in your design. Explain how your design meets all of the requirements.

● Completing the CHAPTER Project

Use the following checklist to make sure the design of the house is complete.

- ☑ The angle measures of the dining room, bathroom, and second bedroom are correct.
- ☑ The measurements of the walls of the dining room, bathroom, and second bedroom are correct.
- ☑ The floor plans of the house's actual angle measures and actual lengths are included.

PORTFOLIO Select one of the assignments from this chapter and place it in your portfolio. Attach a note to it explaining why you selected it.

A practice test for Chapter 9 is provided on page 603.

Section One: Multiple Choice

There are twelve multiple-choice questions in this section. Choose the best answer. If a correct answer is *not here,* mark the letter for Not Here.

1. What is the measure of the line segment shown?

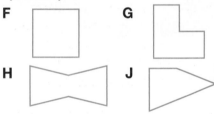

- **A** 60 km
- **B** 60 m
- **C** 60 mm
- **D** 60 cm

2. Which figure has only two lines of symmetry?

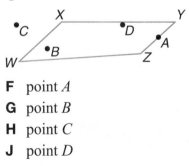

3. A pedometer measures the number of miles a person has walked. Which list shows the pedometer readings in order from least to greatest?

- **A** 4.75, 4.7, 4.04, 4.17
- **B** 4.17, 4.75, 4.7, 4.04
- **C** 4.04, 4.7, 4.17, 4.75
- **D** 4.04, 4.17, 4.7, 4.75

4. Which point is in the exterior of quadrilateral *WXYZ*?

- **F** point *A*
- **G** point *B*
- **H** point *C*
- **J** point *D*

Please note that Questions 5–12 have five answer choices.

5. Four students are 15, 11, 12, and 14 years old. What is an average age of the four students?

- **A** 13
- **B** 10
- **C** 14
- **D** 12
- **E** 15

6. $\triangle MNP \cong \triangle RST$. Which of the following is a true statement?

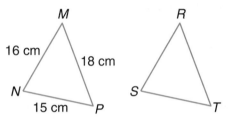

- **F** The perimeter of $\triangle RST$ is 59 centimeters.
- **G** The measure of side *RT* is 16 centimeters.
- **H** The measures of angles *M* and *T* are equal.
- **J** The measures of sides *RS* and *MN* are equal.
- **K** The measure of side *RS* is equal to 18 centimeters.

7. The lights on a street sign blink 5 times every 6 seconds. How many times do they blink in 3 minutes?

- **A** 900
- **B** 150
- **C** 75
- **D** 30
- **E** 450

8. Which is a reasonable remainder when a number is divided by 8?

- **F** 10
- **G** 8
- **H** 11
- **J** 7
- **K** 9

9. Find the product of $\frac{3}{4}$ and $\frac{4}{5}$.

A $\frac{3}{5}$

B $\frac{15}{16}$

C $1\frac{2}{5}$

D 12

E Not Here

10. Teisha works 17 hours a week and earns $99.45. How much money does she earn each hour?

F $5.65

G $4.75

H $5.85

J $4.95

K Not Here

11. The students at Glenwood Middle School sold 1,808 magazine subscriptions. The students at Center Middle School sold 2,046 subscriptions. How many more subscriptions did the students at Center Middle School sell than the students at Glenwood Middle School?

A 148

B 242

C 238

D 762

E Not Here

12. Solve $m = 26 \div 3.25$.

F 0.008

G 0.08

H 0.8

J 8

K Not Here

Section Two: Free Response

This section contains six questions for which you will provide short answers. Write your answers on your paper.

13. A dog is on a 20-foot leash attached to a stake in the ground. If the dog walks in a circle at the end of the leash, how far can he walk before returning to his starting point? Round to the nearest foot.

14. Determine whether 708 is divisible by 2, 3, 5, 6, 9, or 10.

15. Write the decimal 0.125 as a fraction in simplest form.

16. Evaluate $a + b$ if $a = 4\frac{1}{2}$ and $b = 7\frac{1}{4}$.

17. Brenda keeps her turtle in an aquarium that is 14 inches wide and 18 inches long. She buys a new aquarium that is 12 inches wider, but is the same length as the other cage. What are the dimensions of the new aquarium?

18. Angles C and D are supplementary angles. If angle C measures 101.3°, what is the measure of angle D?

 Test Practice **For additional test practice questions, visit:**

www.glencoe.com/sec/math/mac/mathnet

Interdisciplinary Investigation

THAT IS ONE HUMONGOUS PIE!

Did you know that the largest pie ever made weighed almost 38,000 pounds? The largest milkshake ever made was 2,000 gallons. Imagine the amount of each ingredient needed to make a record-setting food.

What You'll Do

In this investigation, you will write a recipe for a giant-sized food and determine the cost of making this food.

Materials   📖   a copy of a book of world records

                📖   cookbooks

                   supermarket advertisements

                   calculator

Procedure

1. Work in groups of 3 or 4. Find a record-setting food in a book of world records. Then find a recipe for that food in a cookbook. Make sure that the recipe has at least 6 ingredients and that at least 3 of the ingredients have measurements that are given as fractions or mixed numbers.

2. Tell how many times the recipe would need to be increased to make a record-setting food. Express this relationship as a ratio. Explain your reasoning. Then find the amount of each ingredient needed to make the record-setting food. Organize the information in a table.

3. Research the cost of each ingredient in your recipe. Which ingredient costs the most? Find the total cost of the ingredients.

4. Use your data to find the percent of the total cost that each ingredient represents. Make a graph of your data.

5. Make a booklet or brochure that contains your recipe, the costs, the graphs, and any other information on your project.

Technology Tips

- Use a **spreadsheet** to calculate the amount of each ingredient needed and the total cost of the ingredients.

- Use **graphing software** to display the data.

- Use **publishing software** to help make your booklet or brochure.

Making the Connection

Use the data collected about your record-setting food as needed to help in these investigations.

Language Arts

Suppose you want to make a record-setting food to put your school in a book of world records. Write a newspaper article convincing people to donate to your project. Use the data that you gathered in the investigation. Be creative and describe the cost involved.

Health

Find the number of Calories in one serving of your record-setting food, and also in the whole food.

Social Studies

Research the history of a book of world records. When and why was the first book published?

Go Further

- Estimate the area and perimeter of your record-setting food. Mark off an area in your school or on the school grounds to demonstrate the size of the food. How does the perimeter of the record-setting food compare to its area?

- How many people could have a serving of your record-setting food? Determine how much money you should charge for a serving so that enough is made to pay for the project.

*inter*NET CONNECTION Research **For current information on recipes, visit:**
www.glencoe.com/sec/math/mac/mathnet

 PORTFOLIO

You may want to place your work on this investigation in your portfolio.

CHAPTER 10

Geometry: Understanding Area and Volume

 you'll
learn in Chapter 10

- to estimate the areas of irregular figures,
- to find the areas of parallelograms, triangles, and circles,
- to construct circle graphs,
- to identify and draw three-dimensional figures,
- to solve problems by making a model, and
- to find the surface area and volume of rectangular prisms.

CHAPTER Project

MEASURING UP

You are familiar with inches, feet, yards, centimeters, and meters. In this project, you will design your own system of measure. You will take measurements using your system and compare your system to a standard system. You will present your findings in a poster.

Getting Started

- Early measurement systems in many cultures were based on readily available objects. For example, the cubit is one of the oldest units. A cubit is the length of your forearm from the point of your elbow to the tip of your middle finger. The table shows some other early units of measure.

Early Units of Measure	
Unit	**Measure**
girth	distance around the waist
span	length from tip of thumb to little finger of outstretched hand
fathom	distance between finger tips when arms are outstretched
grain	weight of one seed of grain
Sun or moon	time in one day

Select a unit of length to use for your project. You may choose a historical unit such as your span, or make your own unit from something you have close by. Some units could be the length of a pen, your little brother's height, or the length of your foot.

Technology Tips

- Use a **spreadsheet** to calculate areas and volumes using your own measurement system.
- Use an **electronic encyclopedia** to research historical or unusual measurements.
- Use a **word processor** to create all or part of your poster.

interNET CONNECTION Research For up-to-date information on the history of mathematics, visit:

www.glencoe.com/sec/math/mac/mathnet

Working on the Project

You can use what you'll learn in Chapter 10 to help you make a poster about your system of measurement.

Page	Exercise
420	17
424	22
429	Alternative Assessment

COOPERATIVE LEARNING

10-1A Area of Irregular Shapes

A Preview of Lesson 10-1

centimeter grid paper

ruler

In Chapter 4, you learned how to find the area of a rectangle. You can also find the areas of irregular shapes. There are two different methods you might use.

TRY THIS

Work with a partner.

1 Find the mean.

- Place your hand on a sheet of centimeter grid paper so that your fingers and thumb are close together. Then draw an outline.

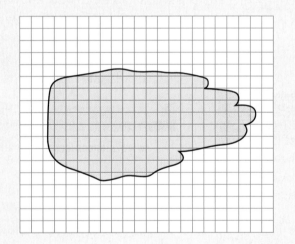

- Count the number of whole squares inside of your outline. Record the number.
- Count the number of squares that have the outline running through them. Add this number to the number of whole squares that you recorded. Then record this number.

ON YOUR OWN

1. Find the mean of the two numbers that you recorded to estimate the area of your hand. *Refer to Lesson 2-7 for information on finding the mean.*

2. Can you think of another way to find the approximate area of your hand? Explain your answer.

TRY THIS

Work with a partner.

❷ Use a rectangle.

- Draw an outline of an irregular shape like the one at the right on a piece of centimeter grid paper.

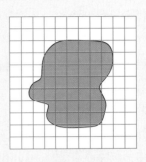

- Use the grid lines to draw a rectangle that encloses most of the figure.
- Count squares to find the length of the rectangle.
- Then count squares to find the width of the rectangle.

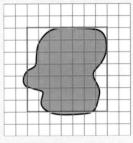

ON YOUR OWN

3. Find the area of the rectangle you drew to estimate the area of the irregular figure.

4. Use the method in the first activity to estimate the area of the irregular figure.

5. Which estimation method do you think is more accurate? Explain your reasoning.

Estimate the area of each figure. Use whichever method you prefer.

6.
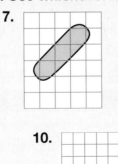

7.

8.

9.
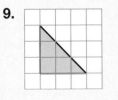

10.
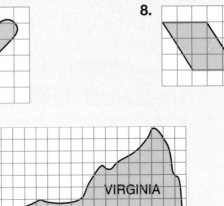

VIRGINIA

one square = 900 sq mi

11. **Look Ahead** Compare the area of the figure in Exercise 8 to the area of a square with sides 3 units long. What do you observe?

What you'll learn

You'll learn to find the area of parallelograms.

When am I ever going to use this?

Knowing how to find the area of a parallelogram can help you determine the amount of wood needed to build a deck.

Word Wise

base
height

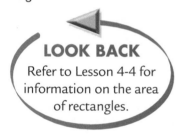

LOOK BACK
Refer to Lesson 4-4 for information on the area of rectangles.

If you watched the Olympic Games in Atlanta, you probably remember the way the United States women's basketball team rolled over the competition to win the gold medal.

A basketball court is rectangular, with a length of 94 feet and a width of 50 feet. Because the court is rectangular, you can find its area by multiplying the length and the width.

$$A = \ell \times w$$
$$A = 94 \times 50 \quad \text{Replace } \ell \text{ with 94 and } w \text{ with 50.}$$
$$A = 4{,}700$$

The area of a basketball court is 4,700 square feet.

A rectangle is a special type of parallelogram. A parallelogram is a quadrilateral with two pairs of parallel sides. The **base** of a parallelogram is any one of its sides. The shortest distance from the base to the opposite side is the **height** of the parallelogram.

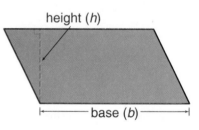

In the Mini-Lab, you will compare the area of a parallelogram to the area of a rectangle with the same height and base.

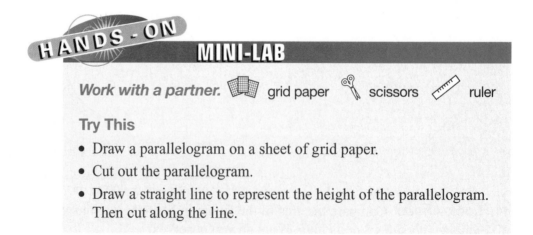

MINI-LAB

Work with a partner. grid paper ✂ scissors 📏 ruler

Try This

- Draw a parallelogram on a sheet of grid paper.
- Cut out the parallelogram.
- Draw a straight line to represent the height of the parallelogram. Then cut along the line.

- Reassemble the pieces to form a rectangle as shown.

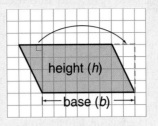

Talk About It

1. How do the areas of the parallelogram and the rectangle compare?

2. What part of the rectangle is the same as the height of the parallelogram? What part is the same as the parallelogram's base?

3. Use what you observed to write a formula for the area of a parallelogram.

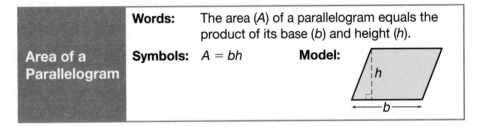

Area of a Parallelogram	**Words:** The area (A) of a parallelogram equals the product of its base (b) and height (h).	
	Symbols: $A = bh$ **Model:**	

Examples

Find the area of each parallelogram to the nearest tenth.

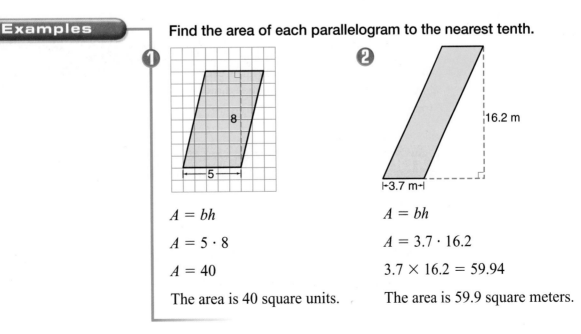

1

$A = bh$

$A = 5 \cdot 8$

$A = 40$

The area is 40 square units.

2

$A = bh$

$A = 3.7 \cdot 16.2$

$3.7 \times 16.2 = 59.94$

The area is 59.9 square meters.

Example 3

Real World APPLICATION

Manufacturing A company in Dallas, Texas, makes ceramic tile that is used to cover floors or walls. Find the area of a ceramic tile that needs to be glazed if it is shaped like a parallelogram with a height of $4\frac{1}{2}$ inches and a base of $6\frac{3}{4}$ inches.

Use the formula for the area of a parallelogram to find the area of the top of the tile.

$A = bh$

$A = \left(6\frac{3}{4}\right)\left(4\frac{1}{2}\right)$ *Replace b with $6\frac{3}{4}$ and h with $4\frac{1}{2}$.*

$A = \left(\frac{27}{4}\right)\left(\frac{9}{2}\right)$ $6\frac{3}{4} = \frac{27}{4}$, and $4\frac{1}{2} = \frac{9}{2}$

$A = \frac{243}{8}$ or $30\frac{3}{8}$ The area of the tile is $30\frac{3}{8}$ square inches.

CHECK FOR UNDERSTANDING

Communicating Mathematics

Read and study the lesson to answer each question.

1. *State* why the formula for finding the area of a rectangle is related to the formula for finding the area of a parallelogram.

2. *Explain* why a rectangle is a parallelogram, but a parallelogram may not be a rectangle.

HANDS-ON MATH

3. *Draw* a rectangle and a parallelogram that have the same area. Explain how you know they have the same area.

Guided Practice

Find the area of each parallelogram to the nearest tenth.

4.
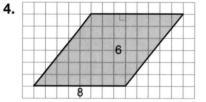
6

8

5.
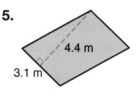
4.4 m

3.1 m

6. *Flags* The flag of Brunei Darussalam has white and black parallelograms on a field of yellow. If a flag is 40 inches wide and the white and black parallelograms each have an area of 240 square inches, how long is the base of each parallelogram?

EXERCISES

Practice

Find the area of each parallelogram. Round decimal answers to the nearest tenth.

7.
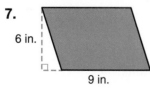
6 in.

9 in.

8.
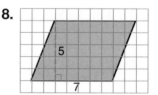
5

7

9.
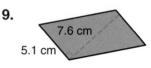
7.6 cm

5.1 cm

10.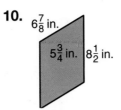
$6\frac{7}{8}$ in.
$5\frac{3}{4}$ in. $8\frac{1}{2}$ in.

11.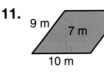
9 m
7 m
10 m

12.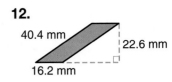
40.4 mm
22.6 mm
16.2 mm

13. Find the area of a parallelogram that is 5.4 centimeters wide and 7.9 centimeters high to the nearest tenth.

14. The area of a parallelogram is $20\frac{7}{12}$ square yards. What is the length of the parallelogram if the base is $3\frac{1}{4}$ yards long?

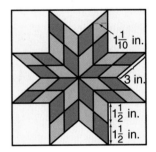

15. *Sports* Some alpine snowboards resemble parallelograms. Estimate the area of a snowboard if it is 29 centimeters wide and 159 centimeters long.

16. *Crafts* The quilt pattern at the right is called the Lone Star pattern.
   a. Find the area of one of the smallest parallelograms.
   b. If the fabric chosen is 36 inches wide, how long should the piece be to have enough to make all of the pieces for a quilt square?

$1\frac{1}{10}$ in.
3 in.
$1\frac{1}{2}$ in.
$1\frac{1}{2}$ in.

17. *Critical Thinking* Suppose you double the height of a parallelogram.
   a. How does the area change?
   b. What happens to the area if you double the base as well as the height?

18. **Standardized Test Practice** Which best describes the two triangles? *(Lesson 9-6)*
   A symmetrical
   B congruent
   C similar
   D regular
   E reflections

2 in. 4 in.
4 in.
6 in. 3 in.
6 in.

19. State whether 708 is divisible by 2, 3, 5, 6, 9, or 10 using divisibility rules. *(Lesson 5-1)*

Area of Triangles

What you'll learn

You'll learn to find the area of triangles.

When am I ever going to use this?

You can use the area of a triangle to solve problems in navigation and architecture.

Sailors have told stories of unusual occurrences in the area known as The Bermuda Triangle. This imaginary triangle has Melbourne, Florida; Bermuda; and Puerto Rico as its vertices as shown in the diagram. What is the area enclosed by the Bermuda Triangle? *You will solve this problem in Example 3.*

To answer this and other questions, you need to discover a formula for the area of a triangle. In the Mini-Lab, you will compare the area of a triangle to the area of a parallelogram.

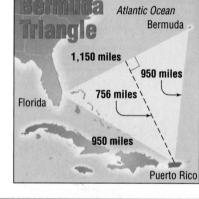

Bermuda Triangle

Atlantic Ocean
Bermuda
Florida
1,150 miles
950 miles
756 miles
950 miles
Puerto Rico

HANDS-ON

MINI-LAB

Work with a partner.   ☐ plain paper   ✂ scissors   📏 ruler

Try This

- Trace two triangles like the one at the right on a sheet of paper.
- Cut out both triangles.
- Place the triangles together to form a parallelogram.

Talk About It

1. How does the area of one of the triangles compare to the area of the parallelogram?
2. What part of the parallelogram is the same as the height of the triangle? What part is the same as the triangle's base?
3. Write a formula for finding the area of a triangle.
4. Rearrange the triangles into a different parallelogram. Does your formula still hold true?

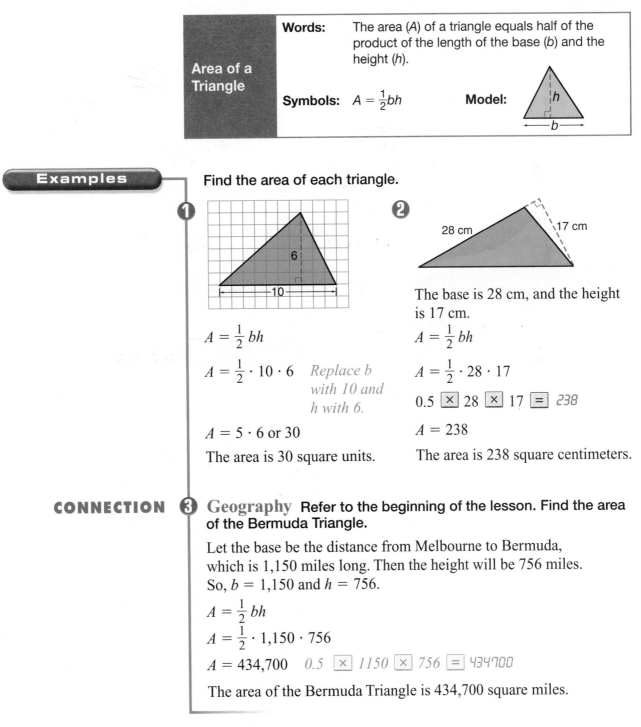

A parallelogram can be formed by two congruent triangles. You know that the area of a parallelogram is $A = bh$. The triangles have the same area. So, the area of a triangle is one-half the area of the parallelogram, or $A = \frac{1}{2} bh$.

The base of a triangle is any one of its sides. The height of the triangle is the distance from a base to the opposite vertex.

Area of a Triangle	**Words:** The area (A) of a triangle equals half of the product of the length of the base (b) and the height (h).
	Symbols: $A = \frac{1}{2}bh$ **Model:**

Examples

Find the area of each triangle.

1

$A = \frac{1}{2} bh$

$A = \frac{1}{2} \cdot 10 \cdot 6$ *Replace b with 10 and h with 6.*

$A = 5 \cdot 6$ or 30

The area is 30 square units.

2

The base is 28 cm, and the height is 17 cm.

$A = \frac{1}{2} bh$

$A = \frac{1}{2} \cdot 28 \cdot 17$

0.5 ⊠ 28 ⊠ 17 ⊟ *238*

$A = 238$

The area is 238 square centimeters.

CONNECTION **3** **Geography** Refer to the beginning of the lesson. Find the area of the Bermuda Triangle.

Let the base be the distance from Melbourne to Bermuda, which is 1,150 miles long. Then the height will be 756 miles. So, $b = 1{,}150$ and $h = 756$.

$A = \frac{1}{2} bh$

$A = \frac{1}{2} \cdot 1{,}150 \cdot 756$

$A = 434{,}700$ *0.5* ⊠ *1150* ⊠ *756* ⊟ *434700*

The area of the Bermuda Triangle is 434,700 square miles.

Communicating Mathematics

Read and study the lesson to answer these questions.

1. *Tell* why the area of a triangle can be defined as half the area of a parallelogram.

2. *Develop* a way to remember the formulas for finding the area of a parallelogram and a triangle.

3. *You Decide* Ebony says that for the triangle at the right, $A = \frac{1}{2} \cdot 7 \cdot 5.8$. Dália says that $A = \frac{1}{2} \cdot 8 \cdot 5.8$. Who is correct and why?

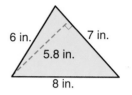

Guided Practice

Find the area of each triangle.

4.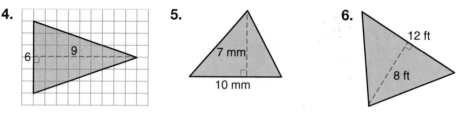

5.

6.

7. What is the area of a triangle with a height of 40 meters and a base 56.8 meters long?

8. *Architecture* The Rock and Roll Hall of Fame in Cleveland opened to fans in 1995. Part of the hall is a pyramid covered in glass. Each of the four sides of the pyramid is a triangle with a base of 241 feet and a height of 165 feet. How much glass was used to cover the entire pyramid?

Practice

Find the area of each triangle. Round decimal answers to the nearest tenth.

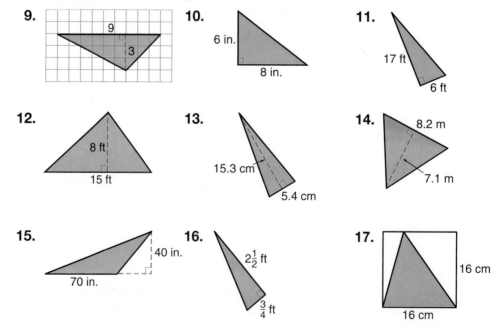

9.

10. 6 in. 8 in.

11. 17 ft 6 ft

12. 8 ft 15 ft

13. 15.3 cm 5.4 cm

14. 8.2 m 7.1 m

15. 40 in. 70 in.

16. $2\frac{1}{2}$ ft $\frac{3}{4}$ ft

17. 16 cm 16 cm

18. Find the area of a triangle with a base of 42 inches and a height of 35 inches.

19. Find the area and perimeter of the figure at the right.

2 ft
4 ft
5 ft
5 ft

Applications and Problem Solving

20. *Sailing* When comparing sailboats, sailors often consider the sail area. The sail area is the total of the areas of all the sails used on the boat. On a Precision 28, the jib sail is a triangle with a base of 10 feet 6 inches and a height of 31 feet. The base of the triangular mainsail is 11 feet 7 inches and the height is 33 feet 3 inches. What is the sail area of the boat?

21. *Life Science* African elephants have large ears that are roughly triangular in shape. They help the elephant to keep cool in the hot climate. Research the ways that the ears cool an elephant in a reference book or on the Internet.

22. *Sports* The triangular wing of a hang glider has an area of 27 square feet. If the wingspread is 9 feet, what is the height of the wing?

23. *Write a Problem* that can be solved by finding the area of a triangle.

24. *Critical Thinking* Draw two different triangles that have areas of 18 square feet.

Mixed Review

25. *Geometry* Find the area of the parallelogram at the right. *(Lesson 10-1)*

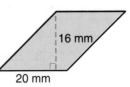
16 mm
20 mm

26. **Standardized Test Practice** At Langley High School, 19% of the 2,200 students walk to school. How many students walk to school? *(Lesson 8-7)*

    A 400   **B** 418   **C** 428   **D** 476

27. *Waste* Every person in the U.S. is responsible for about 2.9 pounds of waste that is disposed of per day. How many ounces is that? *(Lesson 7-7)*

For **Extra Practice**, see page 583.

What you'll learn

You'll learn to find the area of circles.

When am I ever going to use this?

You can use the area of a circle to find the area of your yard covered by a sprinkler.

The people of Kobe, Japan, were awakened at 5:46 the morning of January 17, 1995 by one of the most destructive earthquakes in history. The epicenter of the quake was near the island of Awaji-shima. Damage extended as far as Osaka, 27 miles away. What is the area of the region affected by the earthquake? *This problem will be solved in Example 3.*

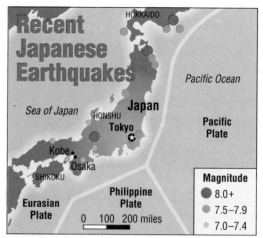

Source: U.S. Geological Survey, Agence France Presse

You need a formula for the area of a circle to answer this question. In the Mini-Lab, you will find the area of a circle by forming a parallelogram.

HANDS-ON MINI-LAB

Work with a partner.   paper plate   scissors

Try This

- Fold your paper plate into eighths.
- Unfold the plate and cut along the creases.
- Arrange the pieces to form a "parallelogram" as shown below.

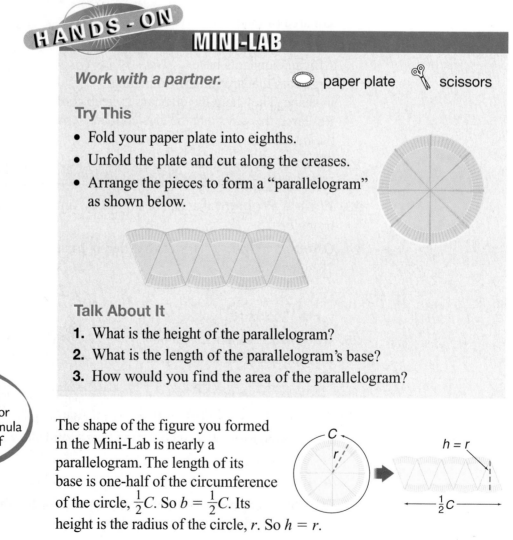

Talk About It

1. What is the height of the parallelogram?
2. What is the length of the parallelogram's base?
3. How would you find the area of the parallelogram?

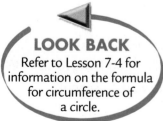

LOOK BACK
Refer to Lesson 7-4 for information on the formula for circumference of a circle.

The shape of the figure you formed in the Mini-Lab is nearly a parallelogram. The length of its base is one-half of the circumference of the circle, $\frac{1}{2}C$. So $b = \frac{1}{2}C$. Its height is the radius of the circle, r. So $h = r$.

$A = bh$ *Formula for the area of a parallelogram*

$A = \left(\frac{1}{2}C\right)r$ *Replace b with $\frac{1}{2}C$ and h with r.*

$A = \frac{1}{2}(2\pi r)r$ *Replace C with $2\pi r$.*

$A = \pi \cdot r \cdot r$ *Simplify. $\frac{1}{2} \cdot 2 = 1$*

$A = \pi r^2$ *Simplify. $r \cdot r = r^2$*

Area of a Circle	**Words:** The area (A) of a circle equals the product of π and the square of the radius (r).	
	Symbols: $A = \pi r^2$	**Model:** 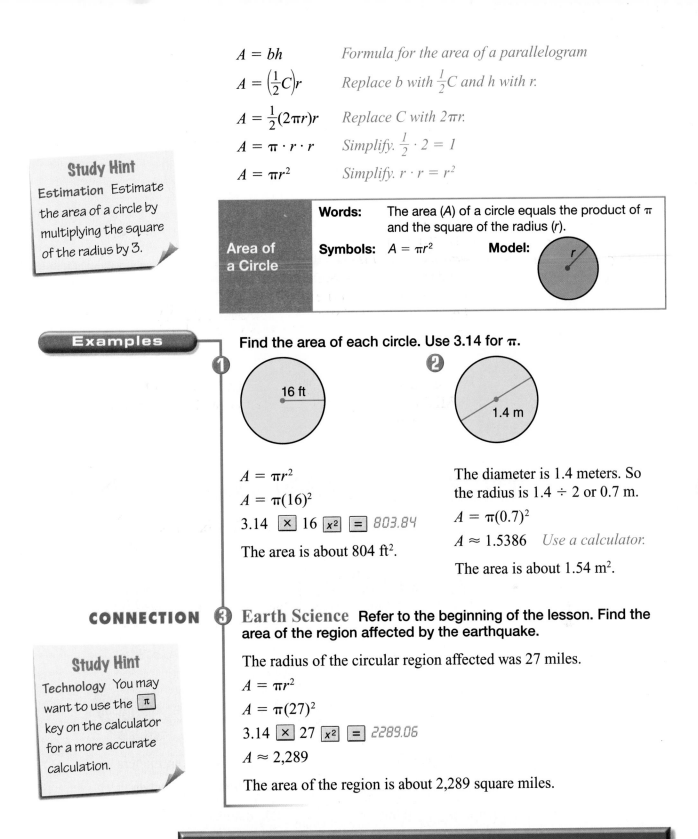

Examples

Find the area of each circle. Use 3.14 for π.

① 16 ft

$A = \pi r^2$

$A = \pi(16)^2$

3.14 [×] 16 [x²] [=] *803.84*

The area is about 804 ft².

② 1.4 m

The diameter is 1.4 meters. So the radius is 1.4 ÷ 2 or 0.7 m.

$A = \pi(0.7)^2$

$A \approx 1.5386$ *Use a calculator.*

The area is about 1.54 m².

CONNECTION **③** **Earth Science** **Refer to the beginning of the lesson. Find the area of the region affected by the earthquake.**

The radius of the circular region affected was 27 miles.

$A = \pi r^2$

$A = \pi(27)^2$

3.14 [×] 27 [x²] [=] *2289.06*

$A \approx 2,289$

The area of the region is about 2,289 square miles.

CHECK FOR UNDERSTANDING

Communicating Mathematics

Read and study the lesson to answer each question.

1. *Explain* how the area of a parallelogram is related to the area of a circle.

2. *Tell* why you can estimate the area of a circle by multiplying the square of the radius by 3.

HANDS-ON MATH

3. *Find* an object with a circular face such as a jar lid or the bottom of a mug. Then find the area of the circular region.

Guided Practice   **Find the area of each circle to the nearest tenth. Use 3.14 for π.**

4.
5 in.

5.
18 m

6. radius, 3.7 millimeters

7. diameter, $1\frac{1}{3}$ feet

8. *Gardening* The Roundabout water sprinkler can be adjusted to spray up to 25 feet. If the spray is in a circular pattern, what is the area watered by the sprinkler?

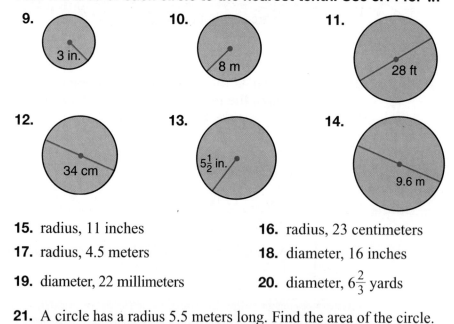

EXERCISES

Practice   **Find the area of each circle to the nearest tenth. Use 3.14 for π.**

9.
3 in.

10.
8 m

11.
28 ft

12.
34 cm

13.
$5\frac{1}{2}$ in.

14.
9.6 m

15. radius, 11 inches

16. radius, 23 centimeters

17. radius, 4.5 meters

18. diameter, 16 inches

19. diameter, 22 millimeters

20. diameter, $6\frac{2}{3}$ yards

21. A circle has a radius 5.5 meters long. Find the area of the circle.

22. What is the area of a circle with a diameter of $32\frac{1}{2}$ feet?

Applications and Problem Solving

23. *Physical Therapy* A therapy pool in the Central High School training room is in the shape of a circle. The diameter is 9 meters. The coach would like to have a cover made to conserve energy when the pool is not in use. How much material is needed to cover the pool?

24. *Sports* American Matt Ghaffari captured a silver medal in 286-pound Greco-Roman wrestling at the Atlanta Olympics. Wrestling uses a 12-by-12 meter mat with a circular ring inside. The ring has an inside radius of 4.5 meters and a width of 10 centimeters.

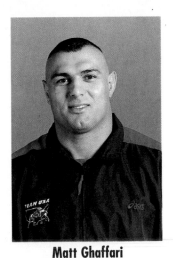

Matt Ghaffari

a. Draw and label a diagram of a wrestling mat and ring.

b. What is the area inside a wrestling ring?

c. What is the area of the ring itself? (*Hint:* Subtract the area of the inside circle from that of the outside circle.)

25. *Critical Thinking* If you double the radius of a circle, how is the area affected?

Mixed Review

For **Extra Practice**, see page 583.

26. What is the area of a triangle with a base 8 meters long and a height of 14 meters? (*Lesson 10-2*)

27. Standardized Test Practice A recipe calls for $1\frac{2}{3}$ cups water, $\frac{1}{3}$ cup oil, and $2\frac{1}{3}$ cups milk. How much liquid is used? (*Lesson 6–3*)

A $4\frac{1}{3}$ cups **B** $3\frac{1}{3}$ cups **C** 3 cups **D** $2\frac{3}{3}$ cups **E** Not Here

28. Change 3,400 centimeters to meters. (*Lesson 4-9*)

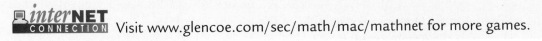

Tiddlywink Target

Math Skill

Area

Get Ready This game is for 2 or 3 players. ◐ counters

▢ felt squares ✎ markers ✎ compass

Get Set Draw three circles on a felt square to make a target like the one at the right. Give each player five counters of one color and a felt piece.

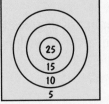

Go ● To shoot a counter, place it on a felt piece and press with a second counter. Slowly slide off the edge to shoot the counter onto the target. The score is the number in the area where most of the counter lands.

● Shoot counters toward the target in turns. A counter may be bumped off the target by another counter. Any counter that lands off of the target is out of play.

● After each player shoots four counters, the winner is the player with the highest total score.

interNET
CONNECTION Visit www.glencoe.com/sec/math/mac/mathnet for more games.

10-3B Making Circle Graphs

A Follow-Up of Lesson 10-3

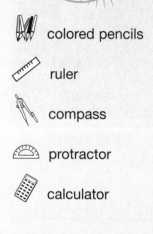

colored pencils

ruler

compass

protractor

calculator

Sun tops the list of things Americans look for in a vacation spot. The graph shows how people answered the question "How important is sunny weather in a vacation location?"

Circle graphs are used to compare parts of a whole. Usually the information is expressed in percents as in the circle graph at the right.

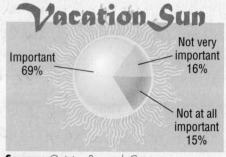

Vacation Sun

Important 69%

Not very important 16%

Not at all important 15%

Source: *Opinion Research Corp.*

TRY THIS

Work with a partner.

The chart shows the percent of toys that are sold in several price ranges. Use the information to make a circle graph.

Price	Percent of Toys
Under $4.00	47%
$4.00–$7.99	21%
$8.00 and above	32%

Source: The TMI Report

Step 1 Find the number of degrees for each price range. To do this, multiply the decimal equivalent of each percent by 360°, which is the total number of degrees in a circle. For example, for "Under $4.00," find the number of degrees as follows.

0.47 ⊠ 360 ⊟ *169.2*

The section of the circle graph for this price range should be about 169°.

Step 2 Use a compass to draw a circle. Then draw a radius of the circle with the ruler.

Step 3 Draw the angle for "Under $4.00" using your protractor. Repeat this step for each price range.

Step 4 Use the colored pencils to color each section of the graph. Label the sections. Then give the graph a title.

1. Compare the graph you made to the chart. Which do you think displays the data more clearly? Explain your reasoning.

2. Why were you able to display the data on toy prices in a circle graph?

3. What type of information *cannot* be displayed in a circle graph? How could you display this type of data?

4. *Hobbies* Say cheese! A camera film maker surveyed customers about how often they take pictures. Make a circle graph of the data.

How Long Does Film Spend in Your Camera Before Being Developed?					
2 weeks or less	2–4 weeks	1–3 months	3–12 months	more than a year	don't know
40%	19%	19%	12%	3%	7%

5. *Population* The table below shows the approximate percent of the population of Earth living in each region. Make a circle graph of the data.

Region	Percent of Population
Asia	59.39%
Africa	12.68%
Europe	8.78%
Latin America and the Caribbean	8.47%
Former USSR	5.07%
North America	5.11%
Oceania and Australia	0.50%

Source: Bureau of the Census

6. *Entertainment* Do you like to play video games? According to a sports magazine survey, 90% of kids do! The graph shows the amount of time kids who play video games spend playing each day.

   a. How much time do most of the kids surveyed spend playing video games each day?

   b. How much time do 15% of the video players spend each day?

   c. What percent of those surveyed spend one hour or less on video games each day?

Time for Video Games

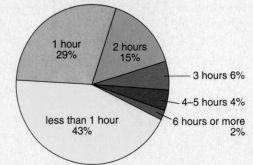

Source: Sports Illustrated

7. *Reflect Back* How is the area of a circle related to making a circle graph?

Three-Dimensional Figures

What you'll learn

You'll learn to identify three-dimensional figures.

When am I ever going to use this?

Knowing how to identify three-dimensional figures will help you to solve problems in social studies and earth science.

Word Wise

three-dimensional
  figure
face
edge
vertex (vertices)
prism
base
rectangular prism
pyramid
square pyramid
cone
cylinder
sphere
center

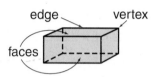

You can find Captain Eddie's Flying Circus Kite Team in competitions all year. A stunt kite like the team uses is a two-dimensional object. Other kites, like box kites, are three-dimensional objects.

A **three-dimensional figure** encloses a part of space. The flat surfaces of a three-dimensional figure are called **faces**. The **edges** are the segments formed by intersecting faces. The edges intersect at the **vertices**.

Boxes are examples of **prisms**. A prism has at least three lateral faces that are shaped like rectangles. The faces on the top and bottom are the **bases** and are parallel. The shape of the bases tells the name of the prism. Because its bases are rectangles, a cereal box is a **rectangular prism**.

A rectangular prism has 12 edges and 8 vertices.

This pyramid has 8 edges and 5 vertices.

A **pyramid**, like the Egyptian pyramids, has only one base. The base can be shaped like any polygon. The shape of the base gives the pyramid its name. For example, a pyramid with a square base is called a **square pyramid**. The other faces of the pyramid are triangles.

Example
CONNECTION

Geography How many faces, edges, and vertices are there in the Great Pyramid of Khufu?

The Great Pyramid of Khufu is a square pyramid. It has one square face and four triangular faces. So there are five faces in all.

There are four edges where the side faces meet the base. Four more edges are formed where the side faces meet each other. So, there are eight edges in the pyramid.

There are four vertices of the square base and the vertex where the other faces meet. So there are five vertices in the pyramid.

There are three-dimensional figures that have curved surfaces. One of these figures is a **cone**. Like a pyramid, a cone has one base. Its base is a circle.

A cone has one vertex, but no edges.

Another figure with a curved surface is a **cylinder**. A soup can is a model of a cylinder. A cylinder has two circular bases.

A cylinder has no vertices and no edges.

A basketball is a model of a sphere. A **sphere** is a three-dimensional figure with no faces, bases, edges, or vertices. All of the points on a sphere are the same distance from a given point called the **center**.

center

A sphere has no vertices or edges.

CHECK FOR UNDERSTANDING

Communicating Mathematics

Read and study the lesson to answer each question.

1. *Tell* what type of pyramid is shown at the right.

2. *Explain* the difference between a two-dimensional and a three-dimensional figure.

3. *List* each type of three-dimensional figure in this lesson. Then give an example of a real-world object with that shape.

Guided Practice

Name each figure.

4.

5.

State the number of faces, edges, and vertices in each figure.

6. rectangular prism

7. triangular prism

8. *Hobbies* What type of three-dimensional figure is the base for the box kite shown on page 412?

EXERCISES

Practice

Name each figure.

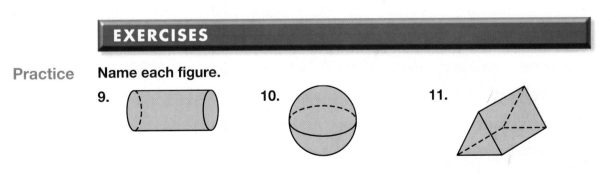

9.

10.

11.

Name each figure.

12.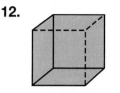

13.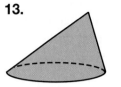

14.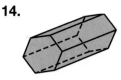

Copy and complete the chart for the numbers of faces, edges, and vertices in each figure.

	Figure	Faces	Edges	Vertices
15.	triangular pyramid			
16.	cylinder			
17.	sphere			
18.	hexagonal pyramid			
19.	hexagonal prism			

Applications and Problem Solving

20. *Number Theory* Swiss mathematician Leonard Euler (pronounced OY ler) found that in three-dimensional figures that have no curved surfaces $E = F + V - 2$ for the number of faces F, edges E, and vertices, V. Does this formula agree with what you found in Exercises 15, 18, and 19?

21. *Earth Science* A crystal of alum is in the form of two square pyramids sharing the same base.1
  a. How many faces, edges, and vertices are in the crystal?
  b. Does your answer to part a agree with Euler's formula above?

22. *Critical Thinking* A pyramid has four faces. What type of pyramid must it be?

Mixed Review

23. Find the area of a circle with a radius of 22 inches. *(Lesson 10-3)*

24. How many lines of symmetry are there in a square? *(Lesson 9-5)*

25. **Standardized Test Practice** If you work 22 hours a week and earn $139.70, how much money do you earn per hour? *(Lesson 8-2)*
  A $6.50   **B** $6.35   **C** $6.05   **D** $5.85   **E** Not Here

For **Extra Practice,** see page 584.

CHAPTER 10

Mid-Chapter Self Test

1. Find the area of a parallelogram with a base 50 feet long and a height of 77 feet. *(Lesson 10-1)*

2. What is the area of a triangle with a 21-inch base and a height of 15 inches? *(Lesson 10-2)*

3. What is the area of a circle with a 2-inch radius? *(Lesson 10-3)*

4. The diameter of a circle is 88 meters. Find the area of the circle. *(Lesson 10-3)*

5. A stack of compact discs is a model of what three-dimensional figure? *(Lesson 10-4)*

COOPERATIVE LEARNING

10-4B Three-Dimensional Figures

A Follow-Up of Lesson 10-4

isometric dot paper

ruler

Sometimes it helps to draw a sketch of a three-dimensional figure when you are trying to solve a problem. You can use isometric dot paper to help you draw a three-dimensional figure.

--- TRY THIS ---

Work with a partner.

A rectangular prism has a length of 3 units, a height of 4 units, and a width of 2 units. Use isometric dot paper to sketch the prism.

The rectangular surfaces of the prism are drawn as parallelograms to give a three-dimensional appearance.

Step 1 Draw a parallelogram with sides of 3 units and 2 units. This represents the top of the prism.

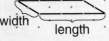

Step 2 Place your pencil at one of the vertices of the parallelogram. Then draw a line passing through four dots. Repeat for the other three vertices, drawing the hidden edges as dashed lines.

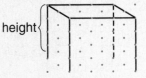

Step 3 Finally, connect the ends of the lines to complete the prism.

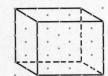

--- ON YOUR OWN ---

1. In the prism that you drew, which faces are the bases of the prism?
2. If you turned the drawing on its side, would the figure appear different?
3. Explain how to draw a prism with a hexagonal base.

Use isometric dot paper to draw a sketch of each figure.

4. a rectangular prism 4 units long, 2 units high, and 2 units wide
5. cube that is 3 units long, 3 units high, and 3 units wide
6. *Reflect Back* Do you think this method would work well for sketching a sphere? Explain.

10-5A Make a Model

A Preview of Lesson 10-5

Emilio and Terrence work at Brooks Sporting Goods. The store has just received a shipment of 72 basketballs packaged in 18-inch cubes. The store manager asked Emilio and Terrence to use as many of them as possible to build a pyramid with the boxes in the display window. Let's listen in as they plan.

Emilio

You know, Terrence, there are an awful lot of boxes here. Let's make sure we have a plan that will work before we start to build.

Terrence

You're right. Let's build a model with these smaller baseball boxes over here. The base of the pyramid should be a square. Then each layer above it can be a smaller square.

That sounds good. If the bottom layer is a 4-by-4 square, it will take 16 boxes.

The next layer would be 3 by 3, or 9 boxes. Then 2 by 2 or 4 boxes in the next layer. And 1 box on top.

That's 16 + 9 + 4 + 1 or 30 boxes. Since we have 72 boxes to use, we can probably add another layer.

OK. Then we'd start with a 5-by-5 layer. So we'd use 25 boxes in the first layer and 30 in the ones above it, for a total of 55 boxes.

Great! That leaves 72 − 55 or 17 boxes left over. 17 isn't enough for a 6-by-6 layer, so it will work.

Grab a box and let's go!

For **Extra Practice,** see page 584.

THINK ABOUT IT

Work with a partner.

1. *Tell* how making a model helped Terrence and Emilio plan their pyramid.

2. *Describe* another way that Terrence and Emilio could have displayed the basketballs.

ON YOUR OWN

3. The last step of the 4-step plan for problem solving is to *examine* the solution. How did Emilio and Terrence examine the solution?

4. *Write* a list of real-life situations where it would be helpful to **make a model** to solve a problem.

5. *Look Ahead* If you built a model of Emilio and Terrence's pyramid with cubes that are 1 cubic inch each, how many cubic inches would there be in the model?

MIXED PROBLEM SOLVING

STRATEGIES

Look for a pattern.
Solve a simpler problem.
Act it out.
Guess and check.
Draw a diagram.
Make a chart.
Work backward.

Solve. Use any strategy.

6. *Patterns* A number is doubled and then 9 is subtracted. If the result is 15, what was the original number?

7. *Sales* Karen is making a pyramid-shaped display of laundry detergent. Each box is a rectangular prism. The bottom layer of the pyramid has six boxes. If there is one less box in each layer and there are five layers in the pyramid, how many boxes will Karen need to make the display?

8. *Sports* Vicky, Benito, and Suzy play volleyball, soccer, and basketball. One of the girls is Benito's next door neighbor. No person's sport begins with the same letter as their first name. Benito's neighbor plays volleyball. Which sport does each person play?

9. *Geometry* A rectangular prism is made of exactly 8 cubes. Find the length, width, and height of the prism.

10. *Geometry* Find the area of the shaded region. Round to the nearest tenth.

11. *Number Theory* What is the least positive number that you can divide by 7 and get a remainder of 4, divide by 8 and get a remainder of 5, and divide by 9 and get a remainder of 6?

12. *Standardized Test Practice* An equilateral triangle has a base of 10 inches and a height of 8.7 inches. If you arranged six triangles like this into a hexagon as shown, what is the area of the hexagon?

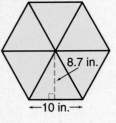

A 600 in^2 B 522 in^2
C 300 in^2 D 261 in^2

Volume of Rectangular Prisms

What you'll learn

You'll learn to find the volume of rectangular prisms.

When am I ever going to use this?

Knowing how to find the volume of a rectangular prism will help you find the amount of water needed to fill an aquarium.

Word Wise

volume

The amount of space inside a three-dimensional figure is called its **volume**. You can investigate the volume of a rectangular prism by building models.

MINI-LAB

Work with a partner.   ⬚ centimeter cubes

Try This

1. Build rectangular prism A so that it is 3 units long, 2 units wide, and 4 units high.
2. Count the number of cubes that you used to build the prism.
3. Find the product of the length, width, and height of the prism.
4. Copy the table below. Complete the first row.

Prism	Length	Width	Height	Number of Cubes	Length × Width × Height
A	3	2	4		
B	3	4	4		
C	4	1	3		

Talk About It

5. In prism A, how does the number of cubes compare to the product of its length, width, and height?
6. Build prisms B and C with the dimensions given. Record the results in your table.
7. Write a sentence about how the volume of a prism is related to the length, width, and height.

From the Mini-Lab, we can conclude that the volume of a rectangular prism is directly related to its dimensions.

Volume of a Rectangular Prism	**Words:**	The volume (V) of a rectangular prism equals the product of its length (ℓ), its width (w), and its height (h).
	Symbols: $V = \ell wh$	**Model:** 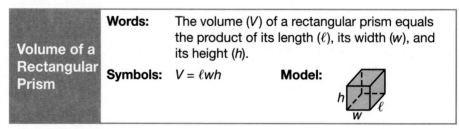

Volume is expressed in cubic units. For example, if the length, width, and height are measured in feet, the volume will be written in cubic feet, or ft^3.

1 Find the volume of the rectangular prism.

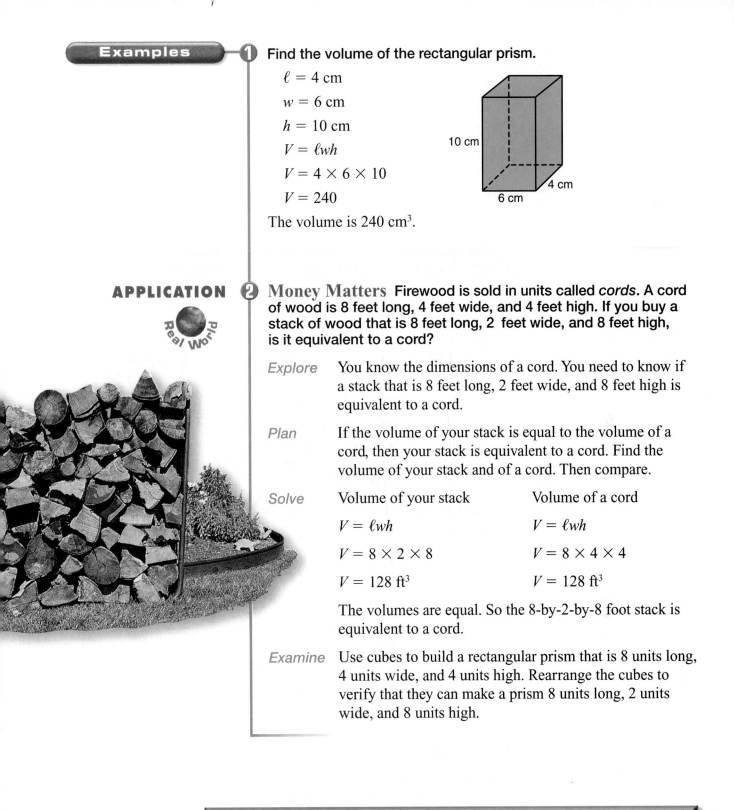

$\ell = 4$ cm

$w = 6$ cm

$h = 10$ cm

$V = \ell wh$

$V = 4 \times 6 \times 10$

$V = 240$

10 cm

4 cm

6 cm

The volume is 240 cm³.

APPLICATION

Real World

2 **Money Matters** Firewood is sold in units called *cords*. A cord of wood is 8 feet long, 4 feet wide, and 4 feet high. If you buy a stack of wood that is 8 feet long, 2 feet wide, and 8 feet high, is it equivalent to a cord?

Explore   You know the dimensions of a cord. You need to know if a stack that is 8 feet long, 2 feet wide, and 8 feet high is equivalent to a cord.

Plan   If the volume of your stack is equal to the volume of a cord, then your stack is equivalent to a cord. Find the volume of your stack and of a cord. Then compare.

Solve

Volume of your stack	Volume of a cord
$V = \ell wh$	$V = \ell wh$
$V = 8 \times 2 \times 8$	$V = 8 \times 4 \times 4$
$V = 128$ ft³	$V = 128$ ft³

The volumes are equal. So the 8-by-2-by-8 foot stack is equivalent to a cord.

Examine   Use cubes to build a rectangular prism that is 8 units long, 4 units wide, and 4 units high. Rearrange the cubes to verify that they can make a prism 8 units long, 2 units wide, and 8 units high.

CHECK FOR UNDERSTANDING

Communicating Mathematics

Read and study the lesson to answer each question.

1. *State* why volume is expressed in cubic units.

2. *You Decide* Maise says that when you find the volume of a prism, the order in which you multiply the length, width, and height does not matter. Craig disagrees. Who is correct and why?

HANDS-ON

3. *Build* a rectangular prism that has a volume of 18 cm³ using centimeter cubes.

Find the volume of each rectangular prism.

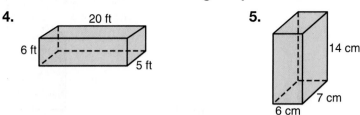

4. 20 ft, 6 ft, 5 ft

5. 14 cm, 7 cm, 6 cm

6. Find the volume of a rectangular prism 15 meters long, 22 meters wide, and 26 meters tall.

EXERCISES

Practice

Find the volume of each rectangular prism to the nearest tenth.

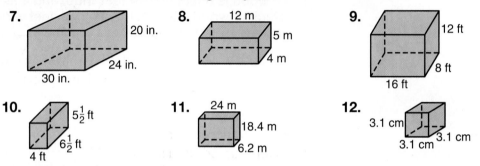

7. 20 in., 24 in., 30 in.

8. 12 m, 5 m, 4 m

9. 12 ft, 8 ft, 16 ft

10. $5\frac{1}{2}$ ft, $6\frac{1}{2}$ ft, 4 ft

11. 24 m, 18.4 m, 6.2 m

12. 3.1 cm, 3.1 cm, 3.1 cm

13. Find the volume of a rectangular prism that is 5 mm wide, 3 mm high, and 6 mm long.

14. What is the volume of a rectangular prism 15 by 12 by 3 feet?

Applications and Problem Solving

15. *Sports* An Olympic-sized pool is 25 meters wide, 50 meters long, and 3 meters deep.
  a. What is the pool's volume?
  b. A liter of water occupies 0.001 cubic meter. How many liters of water are needed to fill an Olympic-sized pool?

16. *World Records* Popcorn lovers in Albermarle, North Carolina, made the largest box of popcorn ever in the United States. The box was 52 feet $7\frac{1}{4}$ inches wide and 10 feet $1\frac{1}{2}$ inches long. The average depth of the popcorn was 10 feet $2\frac{1}{2}$ inches! What was the volume of the popcorn? Round to the nearest cubic foot.

For **Extra Practice**, see page 584.

17. *Working on the* CHAPTER Project Use your unit of length and a standard unit such as feet to measure the length, width, and height of a rectangular prism such as your classroom. Then find the volume of your prism using each cubic unit.

18. *Critical Thinking* A mailing box in the shape of a rectangular prism has a volume of 288 in³. What are the possible dimensions of the box?

Mixed Review

19. **Standardized Test Practice** The base of a cone is a — *(Lesson 10-4)*
  A triangle. **B** circle. **C** radius. **D** rectangle.

20. *Patterns* Find the next two numbers in the sequence 160, 80, 40, 20, ... *(Lesson 7-8)*

Surface Area of Rectangular Prisms

Lumpy Gravy *by John Long*

IT'S TIME TO MAKE BASKETBALL MORE OF A CHALLENGE

What you'll learn

You'll learn to find the surface area of rectangular prisms.

When am I ever going to use this?

Knowing how to find the surface area of a rectangular prism will help you find the amount of paint needed to complete a room.

Word Wise

surface area

Just imagine! The NBA changes the rules so that a basketball is a cube! Dribbling, passing, shooting . . . everything would be different. The current NBA ball uses about 284 square inches of material. Would they use more or less material to make a basketball that is shaped like a $9\frac{1}{4}$-inch cube? *This problem will be solved in Example 3.*

The **surface area** of a three-dimensional object is the total area of its faces and curved surfaces. In the Mini-Lab, you will construct a rectangular prism and find its surface area.

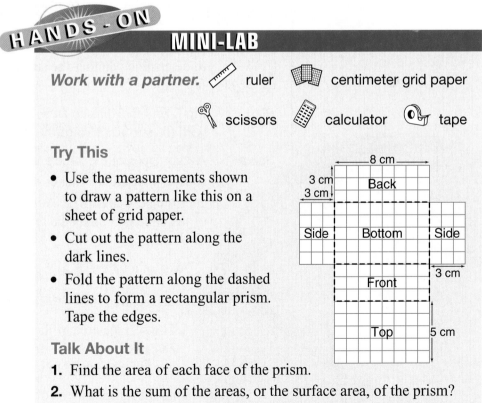

HANDS-ON

MINI-LAB

Work with a partner. ruler   centimeter grid paper

scissors   calculator   tape

Try This

- Use the measurements shown to draw a pattern like this on a sheet of grid paper.
- Cut out the pattern along the dark lines.
- Fold the pattern along the dashed lines to form a rectangular prism. Tape the edges.

Talk About It

1. Find the area of each face of the prism.
2. What is the sum of the areas, or the surface area, of the prism?
3. What do you observe about the opposite sides of the prism? How could this simplify finding the surface area?

Find the surface area of each rectangular prism to the nearest tenth.

1

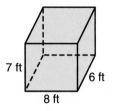

7 ft 6 ft
8 ft

2

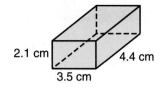

2.1 cm 4.4 cm
3.5 cm

In a rectangular prism, opposite sides have the same dimensions.

top and bottom
  $8 \times 6 = 48$ ft^2

front and back
  $8 \times 7 = 56$ ft^2

right and left sides
  $7 \times 6 = 42$ ft^2

Add the areas.
$2(48) + 2(56) + 2(42) = 292$

The surface area is 292 ft^2.

Opposite sides have the same dimensions.

top and bottom
  $3.5 \times 4.4 = 15.4$ cm^2

front and back
  $2.1 \times 3.5 = 7.35$ cm^2

right and left sides
  $2.1 \times 4.4 = 9.24$ cm^2

Find the total of the areas.

2 $\boxed{\times}$ 15.4 $\boxed{+}$ 2 $\boxed{\times}$ 7.35 $\boxed{+}$ 2
$\boxed{\times}$ 9.24 $\boxed{=}$ *63.98*

The surface area is about 64.0 cm^2.

APPLICATION

3 **Sports** Refer to the beginning of the lesson. Would the cube-shaped basketball require more or less material than the current professional basketball?

Explore You know the surface area of the spherical ball. You need to determine whether the surface area of the cube is larger or smaller.

Plan First, find the surface area of the cube. Then compare it with the surface area of the sphere.

Solve All six faces of the cube are squares that are $9\frac{1}{4}$ inches on each side. So the surface area of the cube is 6 times the area of one face.

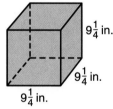

$9\frac{1}{4}$ in.
$9\frac{1}{4}$ in.
$9\frac{1}{4}$ in.

Area of a face: $9\frac{1}{4} \times 9\frac{1}{4} = \frac{37}{4} \times \frac{37}{4} = \frac{1,369}{16}$ in^2

Surface area: $6 \times \frac{1,369}{16} = \frac{8,214}{16}$ or $513\frac{3}{8}$ in^2

The surface area of the spherical ball is about 284 in^2. The cube's surface area is greater. It would use more material to make a cube-shaped basketball.

Examine Find the surface areas of several different cubes with edges between 9 and 10 inches long to verify that the solution is reasonable.

Communicating Mathematics

Read and study the lesson to answer each question.

1. *Identify* the relevant dimension as length, area, or volume.
   a. the perimeter of a rectangle
   b. the number of tiles needed to tile a floor
   c. the capacity of a swimming pool
   d. the height of a flagpole
   e. the capacity of a pitcher
   f. the garden space to be planted

2. *Tell* why you can find the surface area of a rectangular prism after finding the area of just three faces.

Math Journal

3. *Describe* a situation when you would have to find the surface area of a rectangular solid.

Guided Practice

Find the surface area of each rectangular prism.

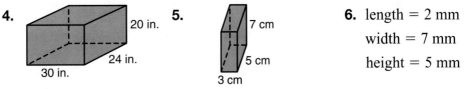

4. 20 in. 24 in. 30 in.

5. 7 cm 5 cm 3 cm

6. length = 2 mm
   width = 7 mm
   height = 5 mm

7. *Manufacturing* A cereal package is a rectangular prism 12 inches high, 8 inches wide, and 3 inches deep. How many square inches did the label designer have to cover on the package?

Practice

Find the surface area of each rectangular prism to the nearest tenth.

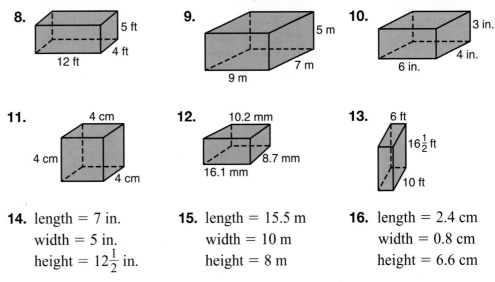

8. 5 ft 4 ft 12 ft

9. 5 m 7 m 9 m

10. 3 in. 4 in. 6 in.

11. 4 cm 4 cm 4 cm

12. 10.2 mm 8.7 mm 16.1 mm

13. 6 ft $16\frac{1}{2}$ ft 10 ft

14. length = 7 in.
    width = 5 in.
    height = $12\frac{1}{2}$ in.

15. length = 15.5 m
    width = 10 m
    height = 8 m

16. length = 2.4 cm
    width = 0.8 cm
    height = 6.6 cm

17. What is the surface area of a rectangular prism that is 10 inches long, 14 inches wide, and 20 inches high?

18. Find the volume of a rectangular prism with base 15 inches, height 25 inches, and width 30 inches. For the same rectangular prism, find its surface area.

19. *Write a Problem* involving a rectangular prism with a surface area of 720 in².

20. *Sports* Wallyball is a variation of volleyball invented in the 1970s. A wallyball court is 20 feet wide, 40 feet long, and 20 feet high. If a gallon of paint will cover 400 square feet, how many gallons will be needed to cover the walls, ceiling, and floor of a wallyball court?

21. *Life Science* The shark petting tank at Nauticus, in Norfolk, Virginia, is 20 feet long, 8 feet wide, and 3 feet deep. Sometimes the tanks must be resurfaced to prevent leakage. What is the area to be resurfaced if the top of the shark petting tank is open?

22. *Working on the* **CHAPTER Project**
Choose an object shaped like a rectangular prism such as a cabinet or your classroom. Find the surface area of the object using your unit of length and a standard unit such as square feet or square meters.

23. *Critical Thinking* In a certain cube, the measure of the surface area is the same as the measure of its volume. What are its dimensions?

24. **Standardized Test Practice** A rectangular fish tank measures 34 inches by 22 inches by 18 inches. If the tank is filled to a height of 15 inches, what is the volume of water in the tank? *(Lesson 10-5)*
A 13,464 in³
B 11,220 in³
C 9,180 in³
D 5,940 in³

25. On December 19, the sunrise was at 7:19 A.M., and the sunset was at 4:28 P.M. How many hours of daylight were there? *(Lesson 6-7)*

For **Extra Practice**, see page 585.

26. *Statistics* What scale would you use in making a frequency table for the following set of data? 21, 79, 11, 9, 55, 38, 111, 92 *(Lesson 2-2)*

10-6B Surface Area and Volume

A Follow-Up of Lesson 10-6

computer

spreadsheet software

The spreadsheet below can be used to find the volume and surface area of a rectangular prism. To use the spreadsheet, you enter the length, width, and height of the prism. Then the computer finds the volume and the surface area.

The formula in cell D2 tells the computer to multiply the values in cells A2, B2, and C2 together.

	A	B	C	D	E
1	LENGTH	WIDTH	HEIGHT	VOLUME	SURFACE AREA
2	A2	B2	C2	= A2*B2*C2	= 2*A2*B2 + 2*A2*C2 + 2*B2*C2
3	A3	B3	C3	= A3*B3*C3	= 2*A3*B3 + 2*A3*C3 + 2*B3*C3
4	A4	B4	C4	= A4*B4*C4	= 2*A4*B4 + 2*A4*C4 + 2*B4*C4
5	A5	B5	C5	= A5*B5*C5	= 2*A5*B5 + 2*A5*C5 + 2*B5*C5
6	A6	B6	C6	= A6*B6*C6	= 2*A6*B6 + 2*A6*C6 + 2*B6*C6
7	A7	B7	C7	= A7*B7*C7	= 2*A7*B7 + 2*A7*C7 + 2*B7*C7

TRY THIS

Work with a partner.

The result of using the spreadsheet to find the surface area and volume of the rectangular prism in row 2 is shown below.

	A	B	C	D	E
1	LENGTH	WIDTH	HEIGHT	VOLUME	SURFACE AREA
2	3	3	7	63	102
3	10.2	4.1	1.6		
4	4	4	4		
5	8	4	4		
6	8	8	4		
7	8	8	8		

ON YOUR OWN

1. Use the spreadsheet to determine the volume and surface area for the rectangular prisms in rows 3, 4, 5, 6, and 7. Print or record your results.

2. What does the formula in cell E2 tell the computer to do?

3. Show how to modify the spreadsheet to find the area and perimeter of a rectangle.

4. How do the dimensions of the prisms in rows 5, 6, and 7 compare to those of the prism in row 4?

5. Describe the pattern of volumes for prisms in rows 4, 5, 6, and 7.

6. Describe the pattern of surface areas of prisms in rows 4, 5, 6, and 7.

*inter*NET
CONNECTION Chapter Review **For additional lesson-by-lesson review, visit:**
www.glencoe.com/sec/math/mac/mathnet

Vocabulary

After completing this chapter, you should be able to define each
term, concept, or phrase and give an example or two of each.

Measurement
surface area (p. 421)
volume (p. 418)

Geometry
base (pp. 398, 412)
center (p. 413)
cone (p. 413)
cylinder (p. 413)
edge (p. 412)
face (p. 412)

height (p. 398)
prism (p. 412)
pyramid (p. 412)
rectangular prism (p. 412)
sphere (p. 413)
square pyramid (p. 412)
three-dimensional figure (p. 412)
vertex (p. 412)

Problem Solving
make a model (p. 416)

Understanding and Using the Vocabulary

Choose the correct term or number to complete each sentence.

1. The (height, edge) of a parallelogram is the distance from the base to the opposite side.

2. The flat surfaces of a three-dimensional figure are called (faces, vertices).

3. The faces of three-dimensional figures intersect in (bases, edges).

4. A (pyramid, cylinder) is a three-dimensional figure with one base where all other faces are triangles that meet at one point.

5. A (sphere, cone) is a three-dimensional figure with no faces, bases, edges, or vertices.

6. A three-dimensional figure with two circular bases is a (cone, cylinder).

7. The amount of space that a three-dimensional figure contains is called its (area, volume).

8. A rectangular prism with length 3 meters, width 4 meters, and height 2 meters has a volume of (14, 24) cubic meters.

9. The total area of a three-dimensional object's faces and curved surfaces is called its (surface area, volume).

In Your Own Words

10. **Explain** how to find the surface area of the rectangular prism.

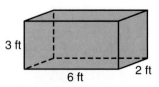

Objectives & Examples

Upon completing this chapter, you should be able to:

● find the area of parallelograms *(Lesson 10-1)*

Find the area.

$A = bh$

$A = 6 \times 5$

$A = 30$

5 in.

6 in.

The area of the parallelogram is 30 in².

Review Exercises

Use these exercises to review and prepare for the chapter test.

Find the area of each parallelogram.

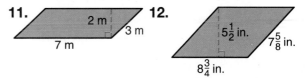

11. 2 m, 3 m, 7 m

12. $5\frac{1}{2}$ in., $7\frac{5}{8}$ in., $8\frac{3}{4}$ in.

13. Find the area of a parallelogram that is 12.5 centimeters wide and 7 centimeters high.

● find the area of triangles *(Lesson 10-2)*

Find the area.

$A = \frac{1}{2} bh$

$A = \frac{1}{2} (150 \times 50)$

$A = 3,750$

The area of the triangle is 3,750 m².

150 m

50 m

Find the area of each triangle. Round decimals to the nearest tenth.

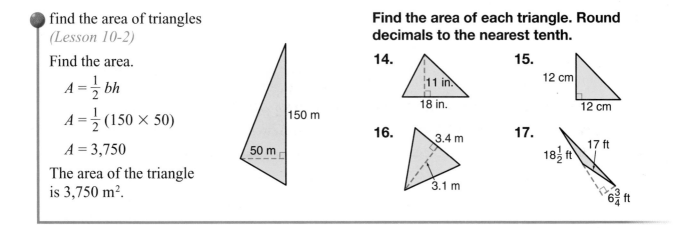

14. 11 in., 18 in.

15. 12 cm, 12 cm

16. 3.4 m, 3.1 m

17. 17 ft, $18\frac{1}{2}$ ft, $6\frac{3}{4}$ ft

● find the area of circles *(Lesson 10-3)*

What is the area of the circle?

Use 3.14 for π.

$A = \pi r^2$

$A = \pi \cdot 7^2$

$A = \pi \cdot 49$

$A \approx 153.9 \text{ cm}^2$

7 cm

The area is about 153.9 cm².

Find the area of each circle to the nearest tenth. Use 3.14 for π.

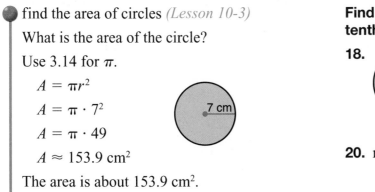

18. 11 m

19. 14 in.

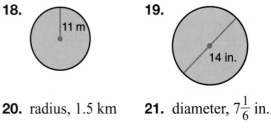

20. radius, 1.5 km

21. diameter, $7\frac{1}{6}$ in.

Chapter 10   Study Guide and Assessment

● identify three-dimensional figures
(Lesson 10-4)

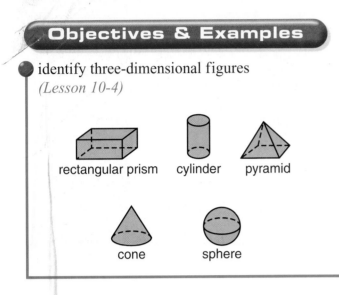

rectangular prism   cylinder   pyramid

cone   sphere

Name each figure.

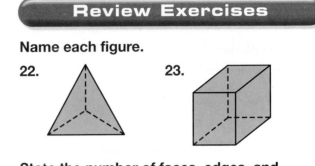

22.   **23.**

State the number of faces, edges, and vertices in each figure.

24. triangular pyramid

25. square prism

● find the volume of rectangular prisms
(Lesson 10-5)

Find the volume.

$V = \ell wh$

$V = 8 \times 4 \times 5$

$V = 160 \text{ in}^3$

5 in.

8 in.   4 in.

The volume is 160 cubic inches.

Find the volume of each rectangular prism.

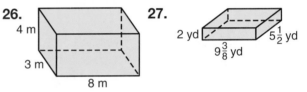

26.   **27.**

4 m

3 m

8 m

2 yd $5\frac{1}{2}$ yd

$9\frac{3}{8}$ yd

28. A rectangular prism has a length of 3.6 meters, a height of 4.1 meters, and a width of 8.2 meters. What is the volume of the prism to the nearest tenth?

● find the surface area of rectangular prisms
(Lesson 10-6)

What is the surface area of the prism above?

top and bottom: $8 \times 4 = 32 \text{ in}^2$
front and back: $8 \times 5 = 40 \text{ in}^2$
sides: $4 \times 5 = 20 \text{ in}^2$

Surface area $= 2(32) + 2(40) + 2(20)$
           $= 64 + 80 + 40 \text{ or } 184$

The surface area is 184 square inches.

Find the surface area of each rectangular prism.

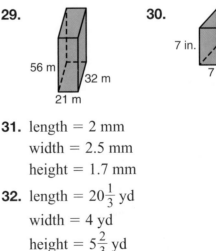

29.   **30.**

7 in.

56 m

32 m   6 in.

21 m   7 in.

31. length = 2 mm
    width = 2.5 mm
    height = 1.7 mm

32. length = $20\frac{1}{3}$ yd
    width = 4 yd
    height = $5\frac{2}{3}$ yd

Applications & Problem Solving

33. *Architecture* Each year, the Corn Palace in Mitchell, South Dakota, is covered in grain murals. Suppose one mural is as shown below. *(Lessons 10-1 and 10-2)*

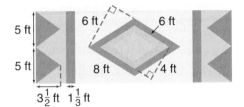

MITCHELL CORN PALACE

a. What is the area to be covered in dark grain?

b. If a bushel of grain covers 25 square feet, how many bushels of dark grain are needed?

34. *Gift Wrapping* Mauna Loa of Honolulu, Hawaii, sells gift boxes of macadamia nuts. The 15-pound box is 18 inches long, 8 inches wide, and 5 inches high. Not counting overlap, how much wrapping paper is needed for each 15-pound box? *(Lesson 10-6)*

35. *Plumbing* A plumber digs a rectangular hole in the ground to install water pipes. The hole is dug straight down and has parallel sides. If the hole is 3 meters deep, 1.5 meters wide, and 2 meters long, what volume of dirt must be removed? *(Lesson 10-5)*

36. *Make a Model* The outside of a cube made of 27 small cubes is painted. Find the number of small cubes that are unpainted. *(Lesson 10-5A)*

Alternative Assessment

Open Ended

Suppose you are planning to remodel your family room. The rectangular room is 12 feet wide, 10 feet long, and 8 feet high. How can you determine the number of square feet of carpet, wall paint, and ceiling paint needed? Find these amounts.

If carpet costs $5 per square foot and paint costs $15 per gallon (which is enough to cover 400 square feet), how much will these supplies cost?

A practice test for Chapter 10 is provided on page 604.

Completing the CHAPTER Project

Use the following checklist to make sure your poster is complete.

☑ Describe your unit of measure. Include it or an equivalent length of string.

☑ Show how you found the volume and surface area of your prisms.

☑ Explain why you think measurement systems are now standardized.

PORTFOLIO   Draw examples of the three-dimensional figures you studied in this chapter. Label the faces, edges, and vertices, or center. Place these diagrams in your portfolio.

Section One: Multiple Choice

There are ten multiple choice questions in this section. Choose the best answer. If a correct answer is *not here,* choose the letter for Not Here.

1. Name the percent shaded on the base-ten model.

   A 45%

   B 55%

   C 68%

   D 72%

2. Find the area of the parallelogram.

   F 22.826 cm²

   G 62.0 cm²

   H 91.53 cm²

   J 163.62 cm²

   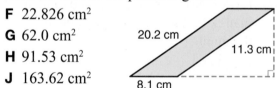

   20.2 cm
   11.3 cm
   8.1 cm

3. Two figures that are the same size and shape are —

   A similar.

   B congruent.

   C neither similar nor congruent.

   D bisectors.

4. Find the area of the triangle.

   F $\frac{15}{8}$ in²

   G $\frac{15}{16}$ in²

   H $\frac{30}{8}$ in²

   J 2 in²

   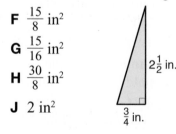

   $2\frac{1}{2}$ in.

   $\frac{3}{4}$ in.

Please note that Questions 5–10 have five answer choices.

5. Lenora bought pencils that cost 29¢ each and pens that cost 89¢ each. What do you need to know to find out how much she spent?

   A the cost of each pen and pencil

   B how much money she has

   C how much change she received

   D how many pens and pencils she bought

   E how much tax is in her area

6. Which is a reasonable remainder when a number is divided by 8?

   F 8

   G 9

   H 7

   J 12

   K 10

7. Which is the best estimate for the weight of a 1-year old child?

   A 8 g

   B 8 kg

   C 60 kg

   D 60 g

   E 60 mg

8. A map has a scale of 1 cm = 20 km. The map distance from Rawson to Delta is 7.2 cm. How far is it from Rawson to Delta?

   F 720 km

   G 27.2 km

   H 144 km

   J 20 km

   K Not Here

9. The graph shows the number of boys and girls who signed up for Moosehead Camp.

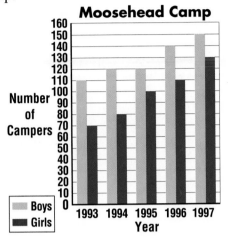

Moosehead Camp

Which is a reasonable conclusion that can be drawn from the information in the graph?

A The percent of campers who are girls increased every year after 1993.

B Girls are less interested in camp than boys.

C The number of boys who signed up for camp went down each year.

D More girls than boys signed up for camp in 1995.

E More people signed up for camp in 1996 than any other year.

10. The figure shows a triangle in the interior of a rectangle.

Which method would you use to find the area of the shaded region?

F perimeter of the rectangle minus perimeter of the triangle

G perimeter of the rectangle plus perimeter of the triangle

H area of the rectangle minus area of the triangle

J area of the rectangle plus area of the triangle

K area of the triangle minus area of the rectangle

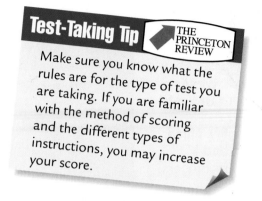

Test-Taking Tip THE PRINCETON REVIEW

Make sure you know what the rules are for the type of test you are taking. If you are familiar with the method of scoring and the different types of instructions, you may increase your score.

Section Two: Free Response

This section contains six questions for which you will provide short answers. Write your answers on your paper.

11. Dave, Judy, and Marco bought a large submarine sandwich for $19.05. How much did each pay if they shared the price of the submarine equally?

12. Which difference is greater, $28\frac{1}{7} - 6\frac{3}{4}$ or $30\frac{1}{8} - 8\frac{3}{4}$?

13. Name the number of faces, edges, and vertices of a triangular pyramid.

14. Find the volume of the rectangular prism.

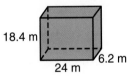

18.4 m   6.2 m   24 m

15. Find the surface area of a rectangular solid with a length of 8 ft, a width of 10 ft, and a height of $15\frac{1}{2}$ ft.

16. Each serving of pizza is $\frac{1}{16}$ of a pizza. If $\frac{3}{4}$ of the pizza is left, how many servings are left?

inter NET **Test Practice** For additional test
CONNECTION practice questions, visit:

www.glencoe.com/sec/math/mac/mathnet

Algebra: Investigating Integers

What you'll learn in Chapter 11

- to identify, name, graph, and compare integers,
- to add, subtract, multiply, and divide integers,
- to solve problems by working backward,
- to graph ordered pairs of numbers on a coordinate grid, and
- to graph transformations on a coordinate grid.

Congresswoman Marcy Kaptur, Ohio

CHAPTER Project

YOU WIN SOME, YOU LOSE SOME

In this project, you will draw a map of the United States and use the map to visually show how the number of members in the House of Representatives has changed for each state. You will also make a line plot and write a paragraph about these changes. You will display your map, line plot, and paragraph on a poster or in a brochure.

Getting Started

- The number of members in the House of Representatives for each state varies according to its population. As a result of the 1990 census, Florida gained 4 members to the House of Representatives, and Pennsylvania lost 2 members. Research how the number of representatives for each state changed after the last census.

- Draw or trace a map of the United States showing each state.

- Color all of the states that lost members in the House of Representatives red. Color all of the states that gained members in the House of Representatives blue. Leave the states that did not change the number of representatives white.

Technology Tips

- Use the **Internet** to find out which states gained and lost members in the House of Representatives.

- Use a **spreadsheet** to keep track of the data you collect.

- Use a **word processor** to write your paragraph about the changes.

 **interNET CONNECTION** For up-to-date information on the House of Representatives, visit:
www.glencoe.com/sec/math/mac/mathnet

Working on the Project

You can use what you'll learn in Chapter 11 to help you keep track of the changes in the House of Representatives.

Page	Exercise
436	41
444	34
461	34
471	Alternative Assessment

11·1 Integers

What you'll learn

You'll learn to identify, name, and graph integers.

When am I ever going to use this?

Knowing about integers can help you express temperatures.

Word Wise

integer
positive integer
negative integer
opposite

Each March, a dog-sled race, called the Iditarod, is held between Anchorage and Nome, Alaska. During this time of the year, the average daytime high temperature for Anchorage is 34°F. However, the racers and their dogs can face temperatures as low as 30°F below zero. You can write 30 below zero as −30.

The numbers 34 and −30 are integers. An **integer** is any number from the set {...−3, −2, −1, 0, 1, 2, 3,...} where ... means *continues without end*.

Integers that are greater than zero are called **positive integers**. Integers that are less than zero are called **negative integers**. Zero itself is neither positive nor negative. You can show positive and negative numbers on a number line.

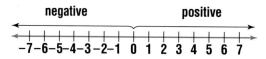

| negative | positive |

$$-7\ -6\ -5\ -4\ -3\ -2\ -1\ \ 0\ \ 1\ \ 2\ \ 3\ \ 4\ \ 5\ \ 6\ \ 7$$

Negative integers are written with a − sign.

Positive integers can be written with or without a + sign.

You can graph integers on a number line by drawing a dot.

Examples

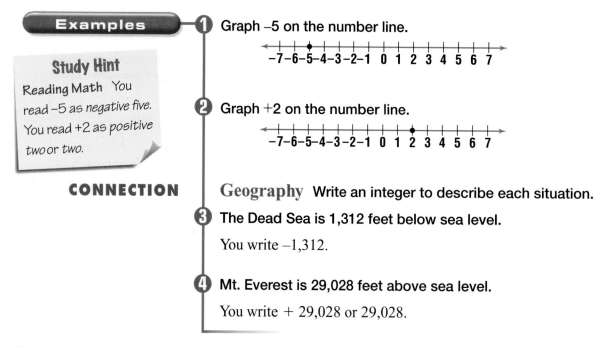

1 Graph −5 on the number line.

$$-7\ -6\ -5\ -4\ -3\ -2\ -1\ \ 0\ \ 1\ \ 2\ \ 3\ \ 4\ \ 5\ \ 6\ \ 7$$

Study Hint

Reading Math You read −5 as *negative five*. You read +2 as *positive two* or *two*.

2 Graph +2 on the number line.

$$-7\ -6\ -5\ -4\ -3\ -2\ -1\ \ 0\ \ 1\ \ 2\ \ 3\ \ 4\ \ 5\ \ 6\ \ 7$$

CONNECTION

Geography Write an integer to describe each situation.

3 The Dead Sea is 1,312 feet below sea level.

You write −1,312.

4 Mt. Everest is 29,028 feet above sea level.

You write + 29,028 or 29,028.

Each integer has an opposite. **Opposite** integers are the same distance from zero in opposite directions on the number line. Zero is considered to be the starting point.

Examples

Write the opposite of each integer.

5 +4

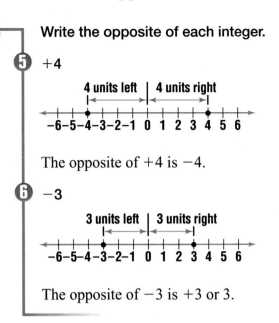

The opposite of +4 is −4.

6 −3

The opposite of −3 is +3 or 3.

CHECK FOR UNDERSTANDING

Communicating Mathematics

Read and study the lesson to answer each question.

1. *Identify* the integers graphed on the number line.

2. *Show* the number 6 and its opposite by graphing them on a number line. Explain why the two numbers are opposites.

Math Journal

3. *Write* about a situation that you could describe using positive and negative integers. Give examples using a positive integer and a negative integer. Explain what each integer means.

Guided Practice

Draw a number line from −10 to 10. Graph each integer on the number line.

4. +9   5. −7   6. 3

Write an integer to describe each situation.

7. The quarterback gained 6 yards on the play.

8. Cecilia lost 5 pounds.

Write the opposite of each integer.

9. 4   10. −9   11. +345

12. Graph the opposite of −10.

13. *Physical Science* Physicist Paul Ching-Wu Chu discovered materials that conduct electricity at 178°C below zero. Write this number as an integer.

EXERCISES

Practice

Draw a number line from −10 to 10. Graph each integer on the number line.

14. +4   **15.** −1   **16.** 0   **17.** −6

18. −8   **19.** 3   **20.** 7   **21.** −2

Write an integer to describe each situation.

22. A token is moved back 6 spaces on a game board.

23. A helicopter rises 75 feet.

24. A withdrawal of $45 is made from a bank account.

25. A submarine is 100 meters below the surface of the water.

26. The value of a stock decreases by $1.

27. An employee receives a $100 bonus.

Write the opposite of each integer.

28. −35   **29.** +23   **30.** −1   **31.** 45

32. −250   **33.** 77   **34.** −52   **35.** −110

36. Graph the opposite of 12.

37. Graph −6, −8, 0, 3, and 5 on the same number line.

38. Graph 7, −3, 4, and 0 on the same number line.

Applications and Problem Solving

39. *Earth Science* The temperature of the center of Earth is estimated to be 7,000°C. Express this number as an integer.

40. *Geography* Jacksonville, Florida, is at sea level. Express the elevation of Jacksonville as an integer.

41. *Working on the* **CHAPTER Project** Refer to the map of the United States you made on page 433. In each state, write an integer that represents how the number of members to the House of Representatives has changed for that state.

42. *Critical Thinking* Compare a number line to a thermometer. How are they alike? How are they different?

Mixed Review

43. **Standardized Test Practice** Juana is going to wrap a present shaped like a rectangular prism with dimensions 27 inches by 14 inches by 5 inches. What is the minimum amount of wrapping paper Juana will need? *(Lesson 10-6)*

   A 46 in²   **B** 583 in²   **C** 1,166 in²   **D** 1,890 in²

44. *Geometry* Name the polygon by the number of sides. *(Lesson 9-4)*

45. Express 1.35 as a percent. *(Lesson 8-5)*

46. Find $6\frac{2}{3} - 4\frac{3}{5}$ in simplest form. *(Lesson 6-5)*

47. Round 673.018 to the nearest tenth. *(Lesson 3-4)*

For **Extra Practice,** see page 585.

Comparing and Ordering Integers

What you'll learn

You'll learn to compare and order integers.

When am I ever going to use this?

Knowing how to compare integers can help you compare temperatures.

History books use timelines to show the order of events. This timeline shows the establishment of various cities.

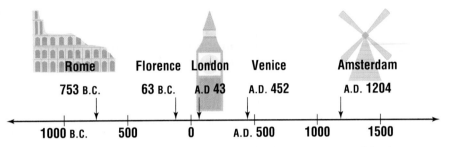

Rome 753 B.C.   Florence 63 B.C.   London A.D 43   Venice A.D. 452   Amsterdam A.D. 1204

1000 B.C.   500   0   A.D. 500   1000   1500

On a timeline, an event depicted to the left always occurred before an event to the right. So, Florence was established after Rome, but before London.

You can use a number line to compare numbers. On a number line, the number to the left is always less than the number to the right.

LOOK BACK
You can refer to Lesson 3-3 to review the symbols $<$ and $>$.

–7 –6 –5 –4 –3 –2 –1 0 1 2 3 4 5 6 7

Notice that -7 is to the left of -3. Therefore, $-7 < -3$. Also, $-3 > -7$ since -3 is to the right of -7.

Examples

Replace each ● with $<$, $>$, or $=$ to make a true sentence.

1 $-5 ● -2$

Graph -5 and -2 on a number line.

–7 –6 –5 –4 –3 –2 –1 0 1 2 3 4 5 6 7

-5 is to the left of -2, so $-5 < -2$.

2 $0 ● -5$

Graph 0 and -5 on a number line.

–7 –6 –5 –4 –3 –2 –1 0 1 2 3 4 5 6 7

0 is to the right of -5, so $0 > -5$.

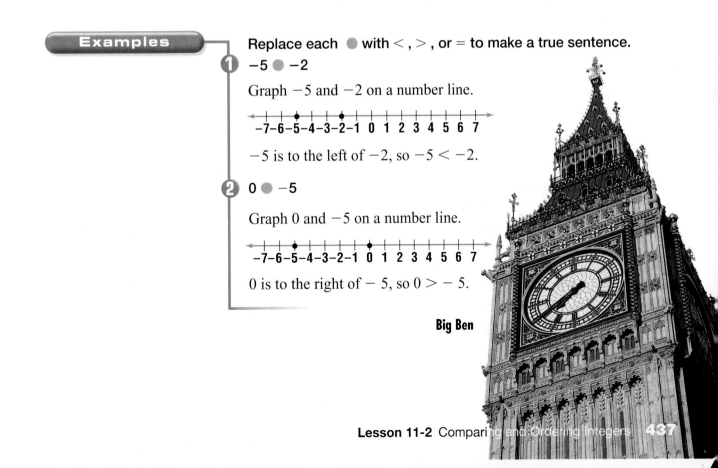

Big Ben

3 Order the integers −4, 3, 0, and −5 from least to greatest.

Graph each number on a number line first.

$$\overleftarrow{\underset{-7\,-6\,-5\,-4\,-3\,-2\,-1\ \ 0\ \ 1\ \ 2\ \ 3\ \ 4\ \ 5\ \ 6\ \ 7}{\vert\ \vert\ \bullet\ \bullet\ \vert\ \vert\ \vert\ \bullet\ \vert\ \vert\ \bullet\ \vert\ \vert\ \vert\ \vert}}\!\!\!\!\to$$

Then, write the integers as they appear on the number line from left to right. −5, −4, 0, and 3 are in order from least to greatest.

APPLICATION

Real World

LOOK BACK
You can refer to Lesson 2-7 to review median.

4 **Weather** The record cold temperatures for five states are recorded on the map. Find the median of these temperatures.

List the temperatures in order from least to greatest.

−80, −70, −52, −2, 12

The median is the middle number when the numbers are arranged in order. So the median of these low temperatures is −52.

RECORD COLD

Montana −70°
Alaska −80°
New York −52°
Hawaii 12°
Florida −2°

Source: *The USA Today Weather Almanac*

CHECK FOR UNDERSTANDING

Communicating Mathematics

Read and study the lesson to answer each question.

1. *Explain* how a number line can be used to order integers.

2. *Write a Problem* where integers would be compared or ordered.

3. *You Decide* Amy says that 5 is greater than 3, and therefore, −5 > −3. Cordelia disagrees. Who is correct? Explain your reasoning.

Guided Practice

Replace each ● with < , > , or = to make a true sentence.

4. +3 ● −1   5. −8 ● −4   6. 0 ● −2

7. Order −3, 5, 0, and −2 from least to greatest.

8. *Geography* Which is higher, 5 feet below sea level or 4 feet below sea level? Explain.

EXERCISES

Practice

Replace each ● with < , > , or = to make a true sentence.

9. +3 ● −4   10. −10 ● −100   11. −3 ● 0

12. +32 ● −32   13. 0 ● +5   14. −6 ● −7

15. −82 ● −85   16. −44 ● −33   17. −200 ● +123

18. Order 4, −5, 6, and −7 from least to greatest.

19. Order 0, 41, 3, −20, −10, and 10 from greatest to least.

20. Which is greater, −45 or −42?

21. Which is least, 0, −3, −17, or −8?

22. Is zero greater than, less than, or equal to negative ten?

Applications and Problem Solving

23. *Physical Science* The table shows the melting point of some common elements. As the temperature rises above the melting point, the element changes from a solid to a liquid.

  a. List the melting points from least to greatest.

  b. The average annual temperature in the interior of Antarctica is −71°F. Would mercury be a solid or a liquid at this temperature?

  c. Find the median of the melting points.

Element	Melting Point (°F)
Calcium	1,542
Gold	1,947
Helium	−458
Hydrogen	−435
Iron	2,795
Mercury	−38
Oxygen	−361
Silver	1,763
Tin	450

24. *Life Science* Some sea creatures live near the surface while others live in the depths of the ocean. Make a drawing showing the relative habitats of the following creatures.

- ribbon fish: 600 to 3,300 feet below the surface

- blue marlin: 0 to 600 feet below the surface

- brittle stars: 13,200 to 19,800 feet below the surface

- lantern fish: 3,300 to 13,200 feet below the surface

Brittle Star

25. *Critical Thinking* Why is any negative integer less than any positive integer?

Mixed Review

26. *Geography* New Orleans is 8 feet below sea level. Express this elevation as an integer. *(Lesson 11-1)*

27. *Standardized Test Practice* Fred had an equilateral triangle with an area of 3 square feet. He traced the triangle several times to make a hexagon. What is the area of the hexagon? *(Lesson 10-2)*

  A 8 ft²   **B** 12 ft²

  C 18 ft²   **D** 22 ft²

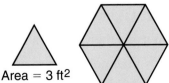

Area = 3 ft²

For **Extra Practice,** see page 585.

28. Find the GCF of 120 and 150. *(Lesson 5-3)*

11-3A Zero Pairs

A Preview of Lesson 11-3

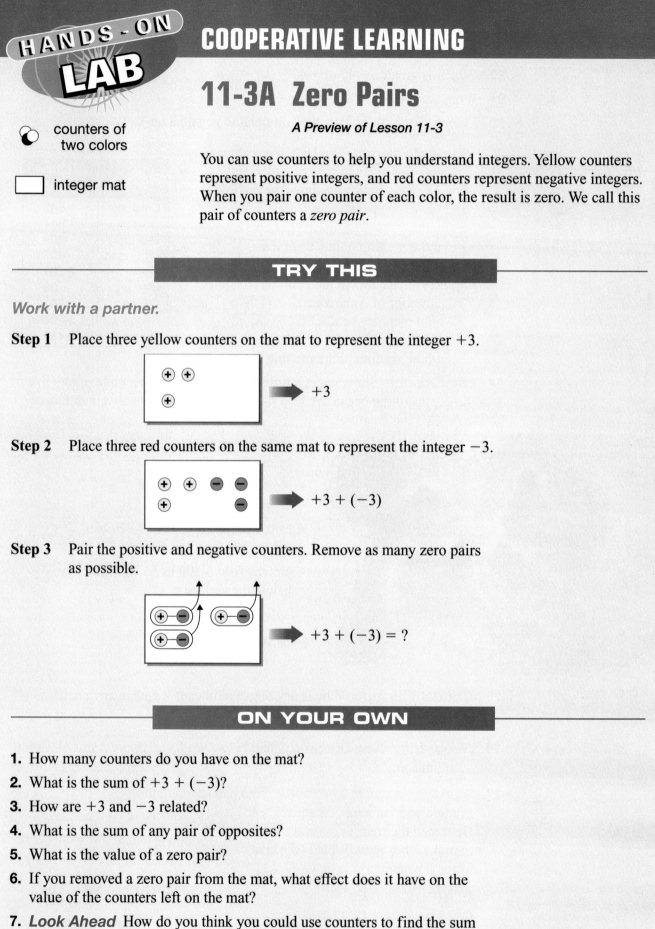

counters of
two colors

integer mat

You can use counters to help you understand integers. Yellow counters represent positive integers, and red counters represent negative integers. When you pair one counter of each color, the result is zero. We call this pair of counters a *zero pair*.

TRY THIS

Work with a partner.

Step 1 Place three yellow counters on the mat to represent the integer +3.

+3

Step 2 Place three red counters on the same mat to represent the integer −3.

+3 + (−3)

Step 3 Pair the positive and negative counters. Remove as many zero pairs as possible.

+3 + (−3) = ?

ON YOUR OWN

1. How many counters do you have on the mat?

2. What is the sum of +3 + (−3)?

3. How are +3 and −3 related?

4. What is the sum of any pair of opposites?

5. What is the value of a zero pair?

6. If you removed a zero pair from the mat, what effect does it have on the value of the counters left on the mat?

7. *Look Ahead* How do you think you could use counters to find the sum of +4 and −3?

Adding Integers

What you'll learn

You'll learn to add integers using models.

When am I ever going to use this?

Knowing how to add integers can help you keep score in games.

Word Wise

zero pair

Remember that yellow counters represent positive integers and red counters represent negative integers.

Monsa and Victor are playing a board game. In the game, each player rolls a number cube and moves a token. Some squares on the game board have further instructions.

- Monsa starts at 0 and rolls a 5.
- The fifth square tells him to roll again. He rolls a 6.
- His token lands on a square that tells him to move back 3 spaces.

How many spaces from the start is Monsa's token?

To find the location of Monsa's token, you will need to add integers. You can add integers using models.

Use yellow counters to represent positive integers and red counters to represent negative integers.

Step 1   Use 5 positive counters to represent Monsa's first roll ($+5$). Use 6 positive counters to represent Monsa's second roll ($+6$). Place all of the counters on a mat.

$$+5 + (+6) = +11$$

Step 2   Use 3 negative counters to represent the 3 spaces backward. Place the 3 negative counters on the mat with the 11 positive counters.

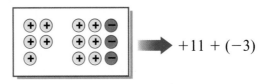

$$+11 + (-3)$$

Step 3   Pair the positive and negative counters. Remove as many **zero pairs** as possible since it does not change the value on the mat.

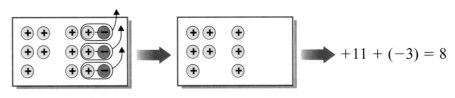

$$+11 + (-3) = 8$$

There are a total of 8 positive counters left on the mat. Monsa's token is 8 spaces from the starting point.

1 Use counters to find −4 + 3.

Step 1 Place 4 negative counters on the mat to represent −4. Place 3 positive counters on the same mat to represent adding 3.

Step 2 Pair the positive and negative counters. Remove as many zero pairs as possible.

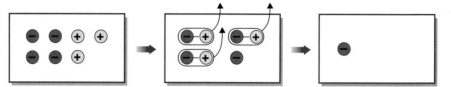

Step 3 Count the counters left on the mat. There is 1 negative counter. This represents −1. So, −4 + 3 = −1.

2 Use counters to find −2 + (−2).

Step 1 Place 2 negative counters on the mat to represent −2. Place 2 more negative counters on the same mat to represent adding −2.

Step 2 Since there are no positive counters, you cannot remove any zero pairs.

Step 3 Count the counters on the mat. There are 4 negative counters. This represents −4. So −2 + (−2) = −4.

You can also use a number line to add integers.

Example

APPLICATION

3 **Game Shows** On a popular TV game show, a contestant has 200 points and then loses 500 points. What is the contestant's score?

You need to find the sum of 200 + (−500). Consider a number line. Start at 0 and go 200 in the positive direction (right).

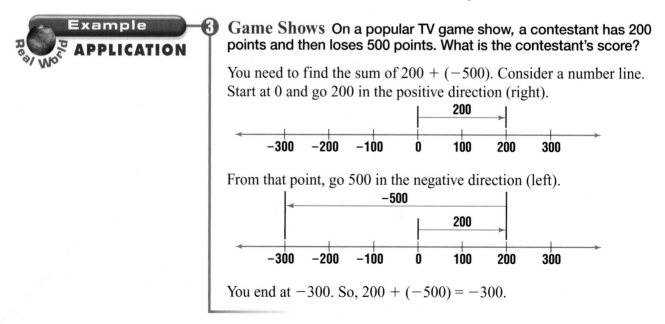

From that point, go 500 in the negative direction (left).

You end at −300. So, 200 + (−500) = −300.

Communicating Mathematics

Read and study the lesson to answer each question.

1. *Write* an addition sentence represented by the model.

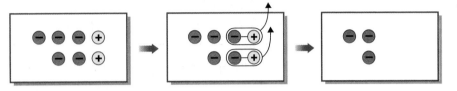

2. *Model* $-3 + 2$ and $2 + (-3)$ using counters. Compare and contrast these two problems.

Guided Practice

State whether each sum is *positive*, *negative*, or *zero*.

3. $-5 + 3$

4. $-3 + 7$

Find each sum. Use counters or a number line if necessary.

5. $3 + (-1)$

6. $-4 + (-8)$

7. $0 + (-2)$

8. $4 + (-6)$

9. Find the sum of -4, 8, and -12.

10. *Football* The Centerville Middle School football team has the ball on the 20-yard line. The team is heading towards the 50-yard line. On the first play, the team gains 6 yards. On the next play, the team loses 8 yards. Where is the ball after the second play?

EXERCISES

Practice

State whether each sum is *positive*, *negative*, or *zero*.

11. $3 + (-5)$

12. $-2 + (-7)$

13. $6 + 3$

14. $-4 + 6$

15. $-3 + (-3)$

16. $-10 + 10$

Find each sum. Use counters or a number line if necessary.

17. $-3 + (-6)$

18. $3 + (-8)$

19. $-12 + 12$

20. $3 + 0$

21. $-3 + 0$

22. $-5 + 4$

23. $6 + (-2)$

24. $-4 + 10$

25. $-4 + (-4)$

26. $-5 + 5$

27. $13 + (-3)$

28. $-11 + (-18)$

29. Find the sum of -18, 6, and 20.

30. What is -3 plus -6 plus 4?

31. *Algebra* Find the value of $a + b$ if $a = -9$ and $b = 6$.

Applications and Problem Solving

32. *Scuba Diving* A scuba diver dived 21 feet below sea level. Then the diver went up 12 feet.
    a. Write an addition statement representing the dive.
    b. What integer represents the diver's location with respect to sea level?

33. *Earth Science* At night, the average temperature on the surface of Saturn is $-150°C$. During the day, the temperature rises $27°C$. What is the average temperature on the planet's surface during the day?

34. **_Working on the_** CHAPTER **_Project_** Refer to the map of the United States you made on page 433. Write the integers for your state and each state that touches your state. Find the sum of the integers. Has your area of the country gained or lost members in the House of Representatives?

35. **_Critical Thinking_** Write an addition sentence that satisfies each statement.
   a. All addends are negative integers, and the sum is -5.
   b. One addend is zero, and the sum is -5.
   c. At least one addend is a positive integer, and the sum is -5.

Mixed Review

36. Which is greater, -66 or -75? _(Lesson 11-2)_

37. **_Geometry_** Draw a rectangular prism. _(Lesson 10-4)_

38. **_Algebra_** Evaluate ab if $a = \frac{3}{8}$ and $b = \frac{7}{15}$. _(Lesson 7-2)_

39. **_Standardized Test Practice_** Before sales tax, what is the total cost of three CDs selling for $13.98 each? _(Lesson 4-1)_

   A $13.98   **B** $20.97   **C** $27.96   **D** $41.94   **E** $48.93

For **Extra Practice**, see page 586.

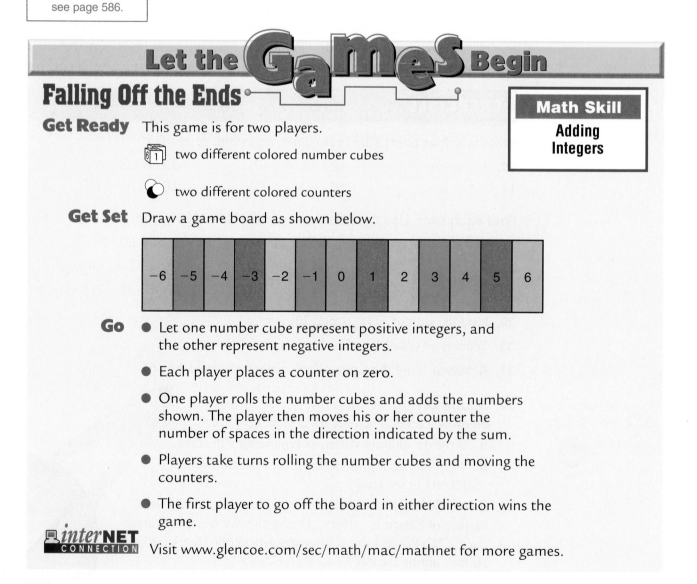

Let the Games Begin

Falling Off the Ends

Math Skill

Adding Integers

Get Ready   This game is for two players.

   two different colored number cubes

   two different colored counters

Get Set   Draw a game board as shown below.

| -6 | -5 | -4 | -3 | -2 | -1 | 0 | 1 | 2 | 3 | 4 | 5 | 6 |

Go
● Let one number cube represent positive integers, and the other represent negative integers.

● Each player places a counter on zero.

● One player rolls the number cubes and adds the numbers shown. The player then moves his or her counter the number of spaces in the direction indicated by the sum.

● Players take turns rolling the number cubes and moving the counters.

● The first player to go off the board in either direction wins the game.

interNET CONNECTION   Visit www.glencoe.com/sec/math/mac/mathnet for more games.

Subtracting Integers

What you'll learn

You'll learn to subtract integers using models.

When am I ever going to use this?

Knowing how to subtract integers can help you to compare elevations when studying geography.

A parade had 18 helium balloons. Five of the balloons were making their first appearance in the parade. How many of the balloons had been in the parade before?

To answer this question, you must find 18 − 5. You can model this subtraction problem using counters.

Step 1 Place 18 positive counters on a mat to represent the 18 balloons.

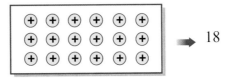

18

Step 2 Since subtraction is the opposite of addition, remove 5 of the positive counters from the mat to represent subtracting 5.

18 − 5

Step 3 Count the counters remaining on the mat.

18 − 5 = 13

There are 13 positive counters. This represents 13. So, 18 − 5 = 13. There were 13 balloons that had been in the parade previously.

You can also use counters to model subtraction problems involving negative integers.

Example ① **Use counters to find −5 − (−2).**

Step 1 Place 5 negative counters on the mat to represent −5.

Step 2 Remove 2 negative counters from the mat to represent subtracting −2.

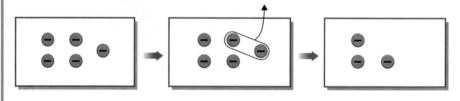

Step 3 Count the counters remaining on the mat. There are 3 negative counters. This represents −3. So, −5 − (−2) = −3.

Sometimes, you need to add zero pairs in order to subtract. When you add zero pairs, the value of the integers on the mat does not change.

Examples ② **Use counters to find −5 − 4.**

Step 1 Place 5 negative counters on the mat to represent −5.

Step 2 To subtract 4, you must remove 4 positive counters. But you cannot remove 4 positive counters because there are none on the mat. You must add 4 zero pairs to the mat. Then you can remove 4 positive counters.

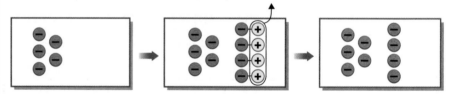

Step 3 Count the counters remaining on the mat. There are 9 negative counters. This represents −9. So, −5 − 4 = −9.

APPLICATION

③ **Weather** One morning when Melissa awoke, the temperature outside was −5°F. By noon, the temperature was 10°F. Find the change in temperature.

Explore You know the starting and ending temperatures. You want to know the change in the temperature.

Plan To find the change in temperature, subtract the starting temperature (−5) from the ending temperature (10). Use counters to find 10 − (−5).

Solve   Place 10 positive counters on the mat to represent $+10$. To subtract -5, you must remove 5 negative counters. But you cannot remove 5 negative counters because there are none on the mat. You must first add 5 zero pairs to the mat. Then you can remove 5 negative counters.

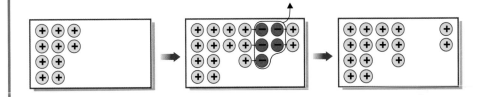

Count the counters remaining on the mat. There are 15 positive counters. This represents 15. So, $10 - (-5) = 15$. The change in temperature is 15°F.

Examine   Let a vertical number line represent a thermometer. Locate the starting and ending temperatures. The number line indicates the change in temperature is 15°F. The answer is correct.

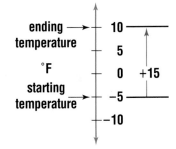

CHECK FOR UNDERSTANDING

Communicating Mathematics

Read and study the lesson to answer each question.

1. ***Write*** a subtraction sentence represented by the model.

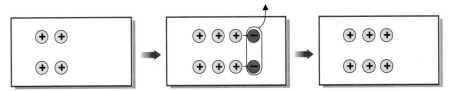

2. ***Explain*** when it is necessary to use zero pairs to model subtraction.

3. ***You Decide*** Andy says that you cannot subtract 8 from 5. Donnell disagrees. Who is correct? Explain.

Guided Practice

Find each difference. Use counters or a number line if necessary.

4. $3 - (-1)$ 5. $-4 - (-4)$ 6. $-2 - 5$ 7. $6 - 8$

8. ***Games*** Mai-Lin and Jamal are playing *Mother, May I*. They start by standing next to each other. Mai-Lin moves 4 steps forward and Jamal moves 2 steps backward. How many steps separate the two children?

Practice

Find each difference. Use counters or a number line if necessary.

9. $2 - 4$ **10.** $4 - 2$ **11.** $0 - 5$ **12.** $0 - (-5)$

13. $-3 - (-6)$ **14.** $3 - (-8)$ **15.** $-12 - 12$ **16.** $-7 - (-5)$

17. $6 - (-2)$ **18.** $-4 - 10$ **19.** $-7 - (-7)$ **20.** $5 - (-5)$

21. Use counters to find $3 + (-6) - (-3)$.

22. *Algebra* Find the value of $x - y$ if $x = 9$ and $y = 16$.

Applications and Problem Solving

23. *Geography* About one third of the Netherlands is below sea level. The map shows approximate elevations for parts of the country.

    a. What is the difference in elevation between Prins Alexander Polder and Leiden?

    b. What is the difference in elevation between Leiden and a location in the Dunes that is 15 feet above sea level?

24. *Communications* Javier lives in Santa Barbara, California. At 10:00 A.M., he called his grandmother in Orlando, Florida. She told him that it was 1:00 P.M. in Orlando. If we use the number 0 to represent time in Orlando, then what integer would represent the time in Santa Barbara?

25. *Critical Thinking* When you subtract a lesser number from a greater number, will the answer always be positive? Give examples to support your answer.

Mixed Review

26. *Algebra* Find the value of $x + y$ if $x = 7$ and $y = -12$. *(Lesson 11-3)*

27. Express the ratio, 2 student council representatives out of 28 students in the class, as a fraction in simplest form. *(Lesson 8-1)*

28. **Standardized Test Practice** There are $18\frac{2}{3}$ cups of jellybeans to be divided among a group of children. If each child gets $\frac{2}{3}$ cup of jellybeans, how many children are there? *(Lesson 7-6)*

    A 25

    B 26

    C 27

    D 28

    E Not Here

For **Extra Practice,** see page 586.

Multiplying Integers

What you'll learn

You'll learn to multiply integers using models.

When am I ever going to use this?

Knowing how to multiply integers can help you solve problems involving repeated addition with integers.

Cultural Kaleidoscope

A dragon head can weigh over 26 pounds. A dragon can be as long as 300 feet and require about 60 dancers to manipulate.

Making dragon heads for parades and celebrations is an ancient Chinese art form. It often takes 2 months to make one dragon head. At this rate, how long does it take to make 5 dragon heads?

To answer this question, you must multiply 5×2. Remember that multiplication is repeated addition. Therefore, 5×2 means $2 + 2 + 2 + 2 + 2$. You can model this multiplication problem using counters.

Step 1 5×2 means to *put in* 5 sets of 2 positive counters. Place these counters on the mat.

$\rightarrow 5 \times 2$

Step 2 Count the counters on the mat.

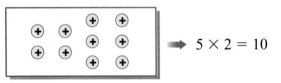

$\rightarrow 5 \times 2 = 10$

There are 10 positive counters. This represents 10. So, $5 \times 2 = 10$. It will take 10 months to make the dragon heads.

Example **1** **Use counters to find $4 \times (-2)$.**

Step 1 $4 \times (-2)$ means to *put in* 4 sets of 2 negative counters. Place these counters on the mat.

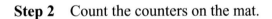

Step 2 Count the counters on the mat. There are 8 negative counters. This represents -8. So, $4 \times (-2) = -8$.

To multiply a positive integer times another number, you *put in* that many sets. To multiply a negative integer times another number, you do the opposite or *remove* that many sets.

Examples

2 Use counters to find −2 × 3.

Step 1 Since −2 is the opposite of 2, −2 × 3 means to *remove* 2 sets of 3 positive counters. But you cannot remove counters because there are none to remove. You must first add 2 sets of 3 zero pairs. Then you can remove 2 sets of 3 positive counters.

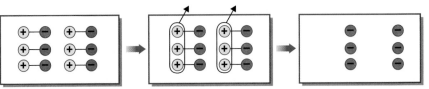

Step 2 Count the counters remaining on the mat. There are 6 negative counters. This represents −6. So, −2 × 3 = −6.

3 Use counters to find −4(−3).

Step 1 Since −4 is opposite of 4, −4(−3) means to *remove* 4 sets of 3 negative counters. But you cannot remove counters because there are none to remove. You must first add 4 sets of 3 zero pairs. Then you can remove 4 sets of 3 negative counters.

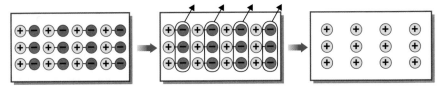

Step 2 Count the counters remaining on the mat. There are 12 positive counters. This represents 12. So, −4(−3) = 12.

CONNECTION **4** **Earth Science** **The temperature drops about 7°C for each kilometer above Earth. If the temperature at ground level is 0°C, find the temperature 3 kilometers above Earth.**

Explore You know the temperature drop per kilometer above Earth and the temperature at ground level. You want to know the temperature 3 kilometers above Earth.

Plan To find the temperature, multiply 3 times the amount of change per kilometer (−7).

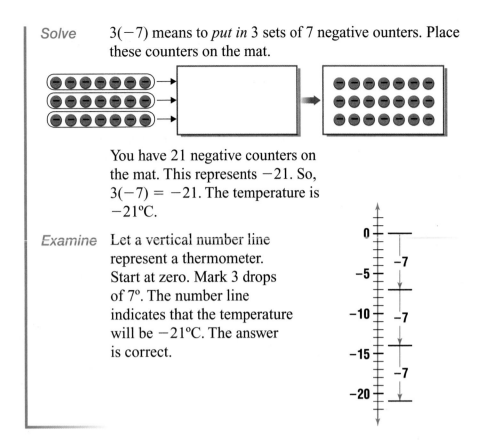

Solve $3(-7)$ means to *put in* 3 sets of 7 negative ounters. Place these counters on the mat.

You have 21 negative counters on the mat. This represents -21. So, $3(-7) = -21$. The temperature is $-21°C$.

Examine   Let a vertical number line represent a thermometer. Start at zero. Mark 3 drops of 7°. The number line indicates that the temperature will be $-21°C$. The answer is correct.

CHECK FOR UNDERSTANDING

Communicating Mathematics

Read and study the lesson to answer each question.

1. *Write* a multiplication sentence represented by the model.

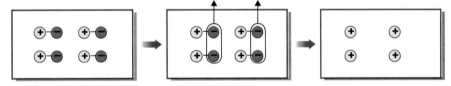

2. *Model* -2×6 and $6 \times (-2)$. Compare and contrast these two problems.

Guided Practice

Find each product. Use counters or a number line if necessary.

3. $3 \times (-1)$ **4.** $-4 \times (-8)$ **5.** $4(-4)$ **6.** $-2(5)$

7. Find the product of -3 and -7.

8. *Oceanography* A submarine is at the surface of the water. It starts to descend at a rate of 4 meters per second. How far below the surface will the submarine be after 5 seconds?

EXERCISES

Practice

Find each product. Use counters or a number line if necessary.

9. $-3 \times (-6)$ **10.** $3 \times (-8)$ **11.** 7×4 **12.** 3×0

13. $-3(0)$ **14.** $5(4)$ **15.** $6(-3)$ **16.** $9(-3)$

17. $8(-3)$ **18.** $3(-3)$ **19.** $5(-5)$ **20.** $7(-5)$

21. Find the product of 7 and −6.

22. Evaluate.

   a. −4(5 + (−9))

   b. 3(−4 − 7)

23. *Pet Care* Sam is a black Labrador retriever who weighs 80 pounds. Her owner puts her on a diet.

   a. If Sam loses 3 pounds each month, how much will she lose in 4 months?

   b. What will Sam weigh at the end of the 4 months?

24. *Time* Suppose you had a watch that loses 2 minutes each day.

   a. How many minutes will it lose in a week?

   b. How many minutes will it lose in the month of April?

25. *Critical Thinking* The product of 1 times any number is the number itself. What is the product of −1 times any number?

Mixed Review

26. *Algebra* Find the value of $s - t$ if $s = 6$ and $t = -5$. *(Lesson 11-4)*

27. *Geometry* Draw a line segment that is $2\frac{1}{2}$ inches long, and then bisect it using a straightedge and compass. *(Lesson 9-3)*

28. *Standardized Test Practice* If a can of green beans weighs 13 ounces, how many pounds will a case of 24 cans weigh? *(Lesson 7-7)*

   A 1.5 lb

   B 15 lb

   C 19.5 lb

   D 312 lb

   E Not Here

For **Extra Practice**, see page 586.

CHAPTER 11

Mid-Chapter Self Test

1. *Geography* Death Valley, California, has the lowest altitude in the United States. Its elevation is 282 feet below sea level. Express the elevation of Death Valley as an integer. *(Lesson 11-1)*

Replace each ● with < , > , or = to make a true sentence. *(Lesson 11-2)*

2. +3 ● −2   **3.** −7 ● −3   **4.** 0 ● −2

Find each sum, difference, or product. Use counters or a number line if necessary. *(Lessons 11-3, 11-4, and 11-5)*

5. −4 + (−6)   **6.** −7 + 12   **7.** −9 − (−5)

8. 6 − (−5)   **9.** −9 × 2   **10.** −8(−5)

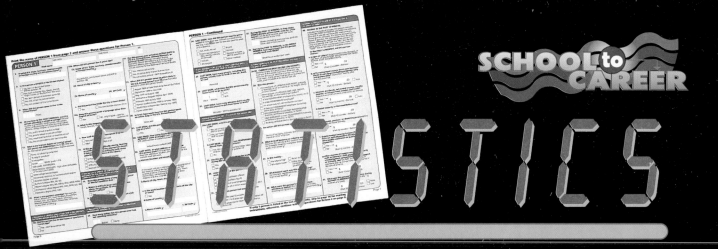

STATISTICS

Robert W. Cleveland
SURVEY STATISTICIAN

Robert Cleveland is a survey statistician for the Bureau of the Census. In the Bureau of the Census, census takers interview people and that information is given to the statisticians. Robert Cleveland works for the Income Statistics Branch of the Bureau of the Census. Each year, he works on the March Current Population Survey. This report includes information about people's jobs, income, and health.

To become a statistician, you will need at least a bachelor's degree. If you would like to become a statistician, you should take as many mathematics and computer courses as possible.

For more information:
American Statistical Association
1429 Duke Street
Alexandria, VA 22314

interNET
CONNECTION
www.glencoe.com/sec/
math/mac/mathnet

Someday, I'd like to study and interpret data like Robert Cleveland.

Your Turn
Find the results of a survey in a newspaper. Write a paragraph describing the data and explaining why it is important.

11-6A Work Backward

A Preview of Lesson 11-6

Callie and Nora want to get ready for the summer softball season that starts June 15. Before the season starts, they plan to spend 3 to 4 weeks hitting and fielding balls. Before hitting and fielding balls, they plan to spend 6 to 8 weeks running and exercising. The girls are trying to determine the latest and earliest dates to start getting ready for the season. Let's listen in!

Look at a calendar. Three weeks before June 15 is May 25 and 4 weeks before June 15 is May 18.

That means the latest we can begin hitting and fielding is May 25. The earliest we should begin is May 18.

Callie

Six weeks before May 25 is April 13. The latest we can start getting ready for the season is April 13.

Nora

Eight weeks before May 18 is March 23. We could start getting ready as early as March 23.

THINK ABOUT IT

Work with a partner.

1. **Discuss** how using a calendar and **working backward** helped Callie and Nora plan their preparation for the softball season.

2. **Explain** how Callie and Nora could solve their problem differently.

3. **Apply** the work backward strategy to solve this problem.

   Chris and Dani volunteer at the food bank at 9:00 A.M. on Saturday mornings. It takes 30 minutes to get from Dani's house to the food bank. Chris picks up Dani, but it takes him 15 minutes to get to Dani's house. If it takes Chris 45 minutes to get ready in the morning, what is the latest Chris should get out of bed?

For **Extra Practice,** see page 587.

ON YOUR OWN

4. The fourth step of the 4-step plan for problem solving tells you to *examine* your answer. *Tell* how you can check an answer to a problem solved by working backward.

5. *Write a Problem* in which an effective strategy to solve it would be to work backward.

6. *Look Ahead* You know $-4 \times (-3) = 12$. Explain how the work-backward strategy could be used to find $12 \div (-3)$.

MIXED PROBLEM SOLVING

STRATEGIES

Look for a pattern.
Solve a simpler problem.
Act it out.
Guess and check.
Draw a diagram.
Make a chart.
Work backward.

Solve. Use any strategy.

7. *Sales* Books Galore bookstore arranges its best sellers in the front window. In how many different orders can they arrange 4 best sellers?

8. *Education* A multiple choice test has 10 questions. A student receives $+3$ for each question answered correctly, -1 for each question answered incorrectly, and 0 for each question not answered.

   a. Joe answered 7 questions correctly and 2 questions incorrectly. He did not answer 1 question. What is his score on the test?

   b. Mary scored 23 points on the test. How many did she answer correctly? How many did she answer incorrectly?

9. *Geometry* The area of a square is 49 square feet.

   a. Find the length of each side of the square.

   b. Find the perimeter of the square.

10. *Life Science* A certain bacteria doubles its population every 12 hours. After 3 full days, there are 1,600 bacteria. How many bacteria were there at the beginning of the first day?

11. *Puzzles* In a magic square, each row, column, and diagonal have the same sum. Copy and complete the magic square.

-2	?	?
-3	-1	1
?	?	?

12. *Geography* The area of Rhode Island is 1,212 square miles. The area of Alaska is 591,004 square miles. About how many times larger is Alaska than Rhode Island?

13. *Money Matters* Gloria bought some watercolor paints and brushes. She spent $3.75 on brushes and 5 times that on paint. After paying for the items, she had $6.89 left. How much money did she have before she went shopping?

14. *Standardized Test Practice* Which problem does *not* have -8 as its answer?

   A $-5 + (-3)$

   B $2 - 10$

   C $-4(-2)$

   D $-6 - 2$

Dividing Integers

MUTUAL

What you'll learn

You'll learn to divide integers using models and patterns.

When am I ever going to use this?

Knowing how to divide integers can help you to find the mean of data with negative integers.

A radio station plans to play 15 minutes of music with no interruptions. How many 3-minute recordings can be played during this time?

To answer this question, you must divide 15 by 3. You can model this division problem using counters.

Step 1 Place 15 positive counters on the mat to represent 15.

➡ 15

Step 2 Separate the 15 counters into 3 equal-sized groups.

➡ $15 \div 3 = 5$

There are 3 groups of 5 positive counters each. So, $15 \div 3 = 5$. The station can play 5 recordings during the 15 minutes.

You can use counters to divide a negative integer by a positive integer.

Example

1 **Use counters to find $-20 \div 5$.**

Step 1 Place 20 negative counters on the mat to represent -20.

Step 2 Separate the 20 counters into 5 equal-sized groups.

There are 5 groups of 4 negative counters each. So, $-20 \div 5 = -4$.

You can also divide integers by *working backward*. For example, to find $56 \div 7$, think "what number times 7 equals 56?"

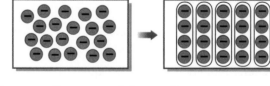

$8 \times 7 = 56$, so $56 \div 7 = 8$. *The quotient is positive.*

2 Find $-10 \div (-2)$.

$5 \times (-2) = -10$, so $-10 \div (-2) = 5$. *The quotient is positive.*

3 Find $18 \div (-6)$.

$-3 \times (-6) = 18$, so $18 \div (-6) = -3$. *The quotient is negative.*

4 Find $-28 \div 7$.

$-4 \times 7 = -28$, so $-28 \div 7 = -4$. *The quotient is negative.*

Do you notice any patterns in Examples 1–4? Notice that when you divide two positive integers or two negative integers, the quotient is positive. When you divide a negative integer and a positive integer, the quotient is negative.

Example

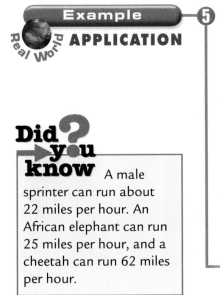
APPLICATION

5 **Sports** For each lap of a race, Angeleila was farther behind Paloma. Angeleila finished the race 12 meters behind Paloma. Both runners ran 4 laps. On average, how many meters did Angeleila fall behind each lap?

Represent Angeleila's final position with respect to Paloma as -12. You need to find $-12 \div 4$.

$-3 \times 4 = -12$, so $-12 \div 4 = -3$.

Angeleila lost 3 meters per lap.

Did you know A male sprinter can run about 22 miles per hour. An African elephant can run 25 miles per hour, and a cheetah can run 62 miles per hour.

CHECK FOR UNDERSTANDING

Communicating Mathematics

Read and study the lesson to answer each question.

1. *Write* a division sentence represented by the model.

2. *Describe* the quotient of two negative integers.

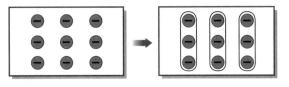

3a. *Write* a multiplication sentence with two negative factors. Then, write two related division sentences.

b. *Write* a multiplication sentence with one positive factor and one negative factor. Then, write two related division sentences.

Guided Practice

Find each quotient. Use counters or patterns if necessary.

4. $16 \div 2$ **5.** $18 \div (-9)$ **6.** $-27 \div 3$ **7.** $-81 \div (-9)$

8. *Oceanography* In an undersea exhibition, a self-contained module started at sea level. It descended to a depth of 24 meters. If this descent took 6 seconds, what was the rate of descent?

EXERCISES

Practice

Find each quotient. Use counters or patterns if necessary.

9. $12 \div 6$ **10.** $24 \div (-8)$ **11.** $-36 \div 9$ **12.** $-8 \div (-2)$
13. $-35 \div 7$ **14.** $32 \div (-4)$ **15.** $-15 \div (-5)$ **16.** $-12 \div (-2)$
17. $45 \div (-5)$ **18.** $-40 \div 8$ **19.** $36 \div (-6)$ **20.** $-42 \div (-7)$

21. Evaluate.

a. $\dfrac{(-3 + (-7))}{2}$ **b.** $\dfrac{(4 + (-6))(-1 + 7)}{-3}$

22. *Algebra* Find the value of $r \div n$ if $r = -12$ and $n = -6$.

Applications and Problem Solving

23. *Football* The Madison Middle School football team lost 12 yards in 3 plays. If the team lost an equal amount on each play, how many yards were lost on each play?

24. *Environment* The graph shows the change of the concentration of pollutants and small particles in the air. What is the mean percent change of these pollutants and particles?

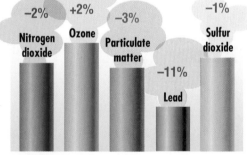

Change in Pollutants

-2% Nitrogen dioxide   +2% Ozone   -3% Particulate matter   -1% Sulfur dioxide   -11% Lead

Source: EPA

25. *Critical Thinking* Write four different division problems with a quotient of -7.

Mixed Review

26. **Standardized Test Practice** A diver is descending at the rate of 4 meters per minute. How far below the surface will the diver be in 8 minutes? *(Lesson 11-5)*

A 2 m   **B** 4 m   **C** 12 m   **D** 32 m

For **Extra Practice**, see page 587.

27. Estimate 31% of 15. *(Lesson 8-6)*

28. Find the value of ten plus five divided by five. *(Lesson 1-4)*

Integration: Geometry
The Coordinate System

What you'll learn

You'll learn to graph ordered pairs of numbers on a coordinate grid.

When am I ever going to use this?

Knowing how to locate points on a coordinate grid can help you locate places on maps.

Word Wise

coordinate system
coordinate grid
origin
x-axis
y-axis
quadrants
ordered pairs
x-coordinate
y-coordinate

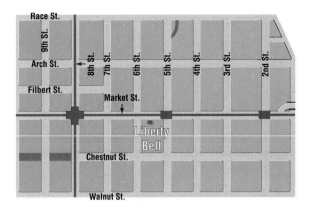

Patrick is studying the map of Philadelphia, Pennsylvania. He wants to take the subway to visit the Liberty Bell. He notices that there is a subway station located at the intersection of Market Street and 5th Street.

Similarly, points can be located on a coordinate system.

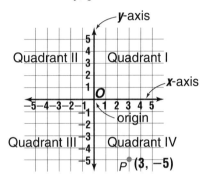

A **coordinate system**, or **coordinate grid**, consists of a horizontal number line and a vertical number line that intersect at their zero points. The point of intersection is called the **origin**. The horizontal line is called the **x-axis**, and the vertical line is called the **y-axis**.

The x-axis and y-axis divide the coordinate system into four **quadrants**. Point *P* is located in the fourth quadrant. It can be named by the **ordered pair** (3, −5). The first number in the ordered pair is the **x-coordinate**, and the second number is the **y-coordinate**.

Example

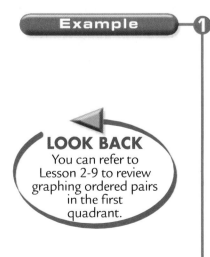

LOOK BACK
You can refer to Lesson 2-9 to review graphing ordered pairs in the first quadrant.

1 Name the ordered pair for point *A* and identify its quadrant.

- Start at 0. Move left along the x-axis until you are directly under point *A*. Since you moved four units to the left, the first coordinate of the ordered pair is −4.

- Now, move up parallel to the y-axis until you reach point *A*. Since you moved up 3 units, the second coordinate of the ordered pair is 3.

- The ordered pair for point *A* is (−4, 3). Point *A* is in the second quadrant.

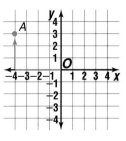

You can also graph a point on a coordinate grid. To graph a point means to place a dot at the point named by an ordered pair.

Examples

2 Graph $B(-3, -4)$.
- Start at 0. Move 3 units to the left on the x-axis.
- Then move 4 units down parallel to the y-axis to locate the point.
- Place a dot and label the dot B.

3 Graph $C(0, 1)$, $D(5, 1)$, $E(3, -2)$, and $F(-2, -2)$.

a. Draw $\overline{CD}$, $\overline{DE}$, $\overline{EF}$, and $\overline{FC}$.

b. Describe the figure formed.

a. Locate the points and draw the segments.

b. The figure formed looks like a parallelogram.

CHECK FOR UNDERSTANDING

Communicating Mathematics

Read and study the lesson to answer each question.

1. *Tell* how to graph the ordered pair $(-3, 6)$.

2. *Draw* a coordinate grid.

   a. *Identify* the portion(s) of the grid where the points are named by coordinates that are both negative. Color the portion(s) blue.

   b. *Identify* the portion(s) of the grid where the points are named by coordinates with one negative number and one positive number. Color the portion(s) red.

Guided Practice

Name the ordered pair for each point.

3. P

4. Q

5. R

6. S

Graph and label each point.

7. $T(1, -4)$

8. $U(-3, -2)$

9. $V(-4, 5)$

10. *Maps* Refer to the map at the beginning of the lesson. If the location of the station at Market Street and 8th Street is at $(0, 0)$, what are the coordinates of the station nearest the Liberty Bell?

Practice

Name the ordered pair for each point.

11. A 12. B 13. C

14. D 15. E 16. F

17. G 18. H 19. I

20. J 21. K 22. M

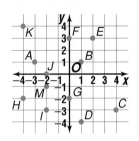

Graph and label each point.

23. $N(2, -2)$ 24. $P(5, 2)$ 25. $Q(-5, -5)$ 26. $R(0, 4)$

27. $S(-2, 3)$ 28. $T(-1, -5)$ 29. $V(-5, 0)$ 30. $W(1, 3)$

31. **a.** Graph $X(-1, -5)$, $Y(0, -2)$, and $Z(1, 1)$ on the same coordinate grid.

    b. Are points X, Y, and Z in the same line?

Applications and Problem Solving

32. *Geography* Longitude and latitude are used to locate places on a map.

    a. Find the longitude and latitude of the place where you live.

    b. Is there a place with a longitude of 0 and a latitude of 0? If so, where is it?

33. *Weather* Use the chart to form ordered pairs where the day is the *x*-coordinate and the temperature is the *y*-coordinate. Graph the point named by each ordered pair.

Day	1	2	3	4	5
Temperature (°F)	−5	−3	0	5	8

34. *Working on the* **CHAPTER Project** Refer to the map of the United States you made on page 433. For each state that had a change in the number of members to the House of Representatives, represent the data on a line plot showing positive and negative integers. What is the total change in the number of members in the House of Representatives?

35. *Critical Thinking* On a coordinate grid, graph and label each point.

    a. $A\left(2\frac{1}{2}, 3\frac{3}{4}\right)$ **b.** $B\left(4\frac{1}{4}, \frac{1}{2}\right)$ **c.** $C(0.5, 1.8)$ **d.** $D(-2.3, -1.5)$

Mixed Review

36. Solve $-24 \div 8 = t$. *(Lesson 11-6)*

37. **Standardized Test Practice** Is an angle that measures 92° *acute, right, obtuse,* or *straight*? *(Lesson 9-1)*

    A acute   **B** right

    C obtuse   **D** straight

38. Estimate $18\frac{3}{16} + 4\frac{1}{9}$. *(Lesson 6-2)*

39. *Statistics* Find Rita's mean golf score for the season if she played 9 times and her scores were 84, 88, 78, 79, 84, 84, 86, 83, and 81. *(Lesson 2-7)*

For **Extra Practice,** see page 587.

patty paper

scissors

notebook paper

11-8A Patty Paper Transformations

A Preview of Lesson 11-8

Coordinate grids can be used to graph *transformations*, or movements, of figures. This lab will help you understand two types of transformations, *translations* and *reflections*.

TRY THIS

Work with a partner.

1 To model one type of transformation, follow these steps.

- Draw triangle *A* and point *B* on a sheet of notebook paper. Line up the bottom of triangle *A* with a line on your notebook paper.

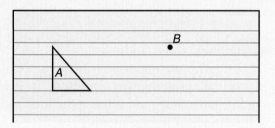

- Place a piece of patty paper on top of triangle *A*.
- Trace triangle *A* onto the patty paper.

- Slowly slide the traced triangle to point *B*. Be sure the bottom of the traced triangle slides along the line on your notebook paper. Stop when the top vertex of triangle *A* is on point *B*.

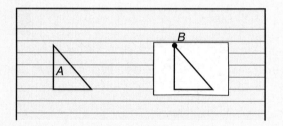

- Draw over the traced triangle. (You will need to press down fairly hard with your pen so that you can see the imprint on your notebook paper.) Then remove the patty paper.
- Trace over this imprint on your notebook paper with your pen. This movement is called a *translation*, or slide, of triangle *A*.

ON YOUR OWN

1. Describe what happened when you made the translation.
2. How does triangle *A* compare with the new triangle? Is it congruent or similar?

462   Chapter 11 Algebra: Investigating Integers

Work with a partner.

2 To model another type of transformation, follow these steps.

- Draw triangle *C* and line *m* on a sheet of notebook paper.

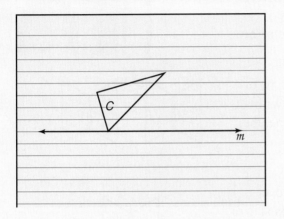

- Place a piece of patty paper on top of your drawing.
- Trace triangle *C* and line *m* onto the patty paper.

- Lift the patty paper up. Turn the patty paper upside down, toward your body. Lay the patty paper on your notebook paper so that line *m* on your patty paper is lined up with the line *m* on your notebook paper.

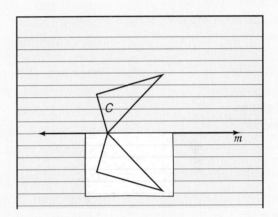

- Draw over the traced triangle, then remove the patty paper.
- Trace over this imprint on your notebook paper with your pen. This movement is called a *reflection*, or flip, of triangle *C*.

3. Describe what happened when you made a reflection.

4. How does triangle *C* compare with the new triangle?

5. *Look Ahead* Look at the clothing you and the other students are wearing. Can you find any patterns in the fabric that have used translations or reflections? Sketch these patterns.

Integration: Geometry
Graphing Transformations

What you'll learn

You'll learn to graph transformations on a coordinate grid.

When am I ever going to use this?

Knowing about transformations can help you do computer animations.

Word Wise

transformation
translation
reflection

The Zapotec Indians in Mexico weave colorful tepetes. Tepetes (tah-PAY-tays) are used as rugs and wall hangings.

A **transformation** is a movement of a figure. What transformations do you see in the design of the tepete?

In Hands-On Lab 11-8A, you learned about two kinds of transformations, a **translation** and a **reflection**.

When you slide a figure from one location to another without changing its size or shape, the new figure is called a *translation image*.

When you flip a figure over a line without changing its size or shape, the new figure is called a *reflection image*.

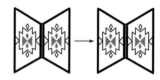

HANDS-ON MINI-LAB

Work with a partner. 🔲 grid paper   ✂ scissors

Try This

- On grid paper, draw a coordinate grid.
- Graph points $A(-2, 3)$, $B(-1, 1)$, and $C(-3, 1)$.
- Connect the points to make $\triangle ABC$.
- On a separate sheet of paper, trace $\triangle ABC$ and cut it out.
- Place the cut-out triangle on $\triangle ABC$.
- Slide the cut-out triangle 5 units right. Label the new vertices A', B', and C'. Then draw $\triangle A'B'C'$.

Study Hint

Reading Math The notation A' is read as A prime. A' is a point related to point A.

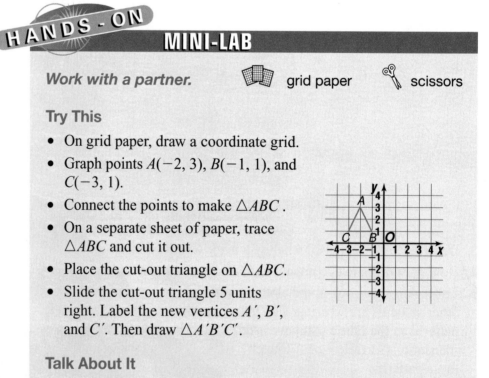

Talk About It

1. What are the coordinates of A', B', and C'?
2. What can you say about $\triangle ABC$ and $\triangle A'B'C'$?

Example **1** The vertices of rectangle *DEFG* are *D*(−2, −3), *E*(−2, 1), *F*(−4, 1) and *G*(−4, −3). On a coordinate grid, draw rectangle *DEFG* and its translation image that is 5 units right and 2 units up.

- On grid paper, draw a coordinate grid.

- Graph and label the vertices of rectangle *DEFG*.

- Draw rectangle *DEFG*.

- Translate each vertex 5 units to the right and 2 units up. The coordinates of the new vertices are *D′*(3, −1), *E′*(3, 3), *F′*(1, 3), and *G′*(1, −1).

- Label the new vertices *D′, E′, F′,* and *G′*.

- Then draw rectangle *D′E′F′G′*.

You can also graph reflections on a coordinate grid.

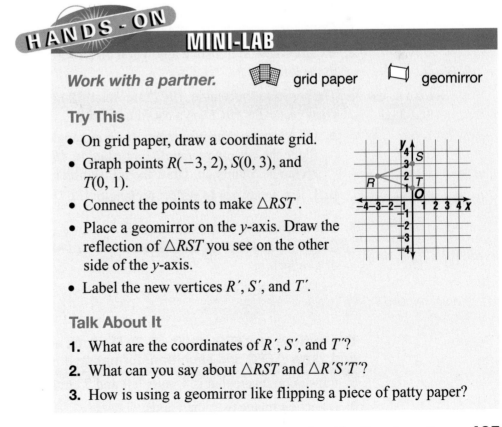

HANDS-ON MINI-LAB

Work with a partner. grid paper geomirror

Try This

- On grid paper, draw a coordinate grid.
- Graph points *R*(−3, 2), *S*(0, 3), and *T*(0, 1).
- Connect the points to make △*RST*.
- Place a geomirror on the *y*-axis. Draw the reflection of △*RST* you see on the other side of the *y*-axis.
- Label the new vertices *R′, S′,* and *T′*.

Talk About It

1. What are the coordinates of *R′, S′,* and *T′*?

2. What can you say about △*RST* and △*R′S′T′*?

3. How is using a geomirror like flipping a piece of patty paper?

Example — ② The vertices of parallelogram *WXYZ* are *W*(−2, 1), *X*(−1, 3), *Y*(4, 3), and *Z*(3, 1). On a coordinate grid, draw parallelogram *WXYZ* and its reflection image over the *x*-axis.

- On grid paper, draw a coordinate grid.

- Graph and label the vertices of parallelogram *WXYZ*.

- Draw parallelogram *WXYZ*.

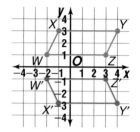

- Reflect the parallelogram by flipping it onto the other side of the *x*-axis. To be a mirror image, each new vertex must be the same distance from the *x*-axis as its corresponding vertex. The coordinates of the new vertices are *W′*(−2,−1), *X′*(−1,−3), *Y′*(4,−3), and *Z′*(3,−1).

- Label the vertices *W′*, *X′*, *Y′*, and *Z′*.

- Draw parallelogram *W′X′Y′Z′*.

CHECK FOR UNDERSTANDING

Communicating Mathematics

Read and study the lesson to answer each question.

1. *Compare and contrast* a translation and a reflection.

2. *Explain* how to translate a polygon on a coordinate grid 6 units to the left.

HANDS-ON MATH

3. The vertices of rectangle *ABCD* are *A*(1,−2), *B*(1,−4), C(4,−4), and D(4,−2). Draw rectangle *ABCD* on a coordinate grid.

   a. On a separate sheet of paper, trace rectangle *ABCD* and cut it out. Place the cut-out rectangle on rectangle *ABCD*. Slide the cut-out rectangle 5 units up. Draw the translation image.

   b. Use a geomirror to reflect rectangle *ABCD* over the *y*-axis.

Guided Practice

4. Tell whether the transformation of △*ABC* to △*A′B′C′* is a *translation* or a *reflection*.

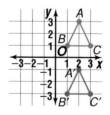

The vertices of △*EFG* are *E*(−3, 2), *F*(−3, 0), and *G*(0, 0). On a coordinate grid, draw △*EFG* and each transformation image.

5. translation image that is 4 units left and 5 units up

6. reflection image over the *y*-axis

7. **Nature** What type of transformation(s) do you see in the picture of the mountains and the lake? Explain your answer.

EXERCISES

Practice

Tell whether each transformation is a *translation* or a *reflection*.

8.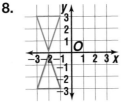

9.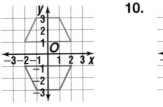

10.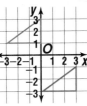

Family Activity

Find some examples of translations and reflections in your home. Make a sketch of each transformation. Label each transformation as a translation, a reflection, or both.

The vertices of △*MNP* are *M*(2, 1), *N*(2, 3), and *P*(4, 4). On a coordinate grid, draw △*MNP* and each transformation image.

11. translation image that is 7 units down

12. translation image that is 6 units left

13. translation image that is 3 units right and 4 units up

14. translation image that is 2 units left and 5 units down

15. reflection image over the *x*-axis

16. reflection image over the *y*-axis

Applications and Problem Solving

Real World

17. **Art** What type of transformation(s) do you see in the drawing by M. C. Escher? Explain your answer.

18. **Design** Draw a pattern for wrapping paper that uses both a translation and a reflection.

19. **Critical Thinking** The first coordinate of each vertex of a polygon is multiplied by −1. The image of the polygon having these new coordinates is drawn. Describe the relationship between the original polygon and its image.

Source: © M.C. Escher/Cordon Art–Baam–Holland Collection Haags Gemeentemuseum–The Hague

Mixed Review

20. **Geometry** Graph *A*(2, −3) and *B*(−3, 2). *(Lesson 11-7)*

21. **Standardized Test Practice** An architect made an 8-inch by 10-inch scale drawing of a house. If the scale on the drawing is 1 inch = 5 feet, what are the dimensions of the house? *(Lesson 8-3)*

  A 40 ft by 50 ft   **B** 35 ft by 45 ft

  C 20 ft by 25 ft   **D** 4 ft by 5 ft

For **Extra Practice**, see page 588.

22. Find the LCM of 15, 20, and 8. *(Lesson 5-7)*

*inter*NET
CONNECTION Chapter Review **For additional lesson-by-lesson review, visit:**
www.glencoe.com/sec/math/mac/mathnet

Vocabulary

After completing this chapter, you should be able to define each term, concept, or phrase and give an example or two of each.

Geometry
coordinate grid (p. 459)
coordinate system (p. 459)
ordered pair (p. 459)
origin (p. 459)
quadrant (p. 459)
reflection (p. 464)
transformation (p. 464)
translation (p. 464)
x-axis (p. 459)
x-coordinate (p. 459)
y-axis (p. 459)
y-coordinate (p. 459)

Number and Operations
integer (p. 434)
negative integer (p. 434)
opposite (p. 435)
positive integer (p. 434)
zero pair (p. 441)

Problem Solving
work backward (p. 454)

Understanding and Using the Vocabulary

State whether each sentence is *true* or *false*. If false, replace the underlined word or number to make a true sentence.

1. Integers that are greater than zero are called <u>negative</u> integers.
2. The opposite of the number 7 is $\frac{1}{7}$.
3. On the number line -21 is to the left of -8, so <u>$-21 < -8$</u>.
4. When adding two negative integers, the answer will <u>always</u> be negative.
5. When subtracting two negative integers, the answer will <u>never</u> be negative.
6. When dividing two negative integers, the answer will <u>always</u> be negative.
7. When multiplying two negative integers, the answer will <u>sometimes</u> be negative.
8. The <u>horizontal</u> line of a coordinate system is called the *x*-axis.
9. The first number in an ordered pair is the <u>*x*-coordinate</u>.
10. The *x*-axis and the *y*-axis intersect at the <u>quadrant</u>.
11. When you flip a figure, the transformation is called a <u>translation</u>.

In Your Own Words

12. ***Explain*** the difference between a translation and a reflection.

Objectives & Examples

Upon completing this chapter, you should be able to:

● identify, name, and graph integers
(Lesson 11-1)

Graph -3 on the number line.

$$-6\,-5\,-4\,-3\,-2\,-1\ 0\ 1\ 2\ 3\ 4\ 5\ 6$$

● compare and order integers
(Lesson 11-2)

Compare -7 and -9.

$$-10\,-9\,-8\,-7\,-6\,-5\,-4\,-3\,-2\,-1\ 0\ 1\ 2$$

-7 is to the right of -9, so $-7 > -9$.

● add integers using models
(Lesson 11-3)

Use counters to find $-4 + 6$.

$-4 + 6 = 2$

● subtract integers using models
(Lesson 11-4)

Use counters to find $-4 - 2$.

$-4 - 2 = -6$

Review Exercises

Use these exercises to review and prepare for the chapter test.

Draw a number line from -10 to 10. Graph each integer on the number line.

13. -9 **14.** 6 **15.** -8

16. Write an integer to describe a temperature drop of $13°$.

Replace each ● with $<$, $>$, or $=$ to make a true sentence.

17. $+4 ● +7$ **18.** $+1 ● -6$

19. $-7 ● -12$ **20.** $-9 ● +11$

21. Order $-15, 0, -1, 6, -7,$ and 10 from least to greatest.

Find each sum. Use counters or a number line if necessary.

22. $-3 + 5$ **23.** $7 + (-7)$

24. $-2 + (-5)$ **25.** $-7 + 1$

26. $4 + (-3)$ **27.** $-6 + 0$

Find each difference. Use counters or a number line if necessary.

28. $-2 - (-5)$ **29.** $-7 - 1$

30. $3 - (-4)$ **31.** $8 - (-3)$

32. $6 - 9$ **33.** $-4 - 7$

Objectives & Examples

Review Exercises

● multiply integers using models
(Lesson 11-5)

Use counters to find $2(-3)$.

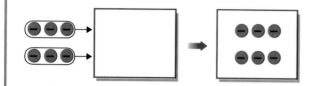

$$2(-3) = -6$$

Find each product. Use counters or a number line if necessary.

34. $6 \times (-2)$ **35.** $-3 \times (-5)$

36. -4×3 **37.** 5×0

38. $-5(2)$ **39.** $7(-1)$

● divide integers using models and patterns
(Lesson 11-6)

Find $14 \div (-2)$.

$-7 \times (-2) = 14$, so $14 \div (-2) = -7$.

Find each quotient. Use counters or patterns if necessary.

40. $35 \div (-7)$ **41.** $-36 \div 9$

42. $-40 \div 5$ **43.** $18 \div (-3)$

44. $-30 \div (-5)$ **45.** $-24 \div (-6)$

● graph ordered pairs of numbers on a coordinate grid *(Lesson 11-7)*

Graph $A(2, -1)$.
right 2, down 1

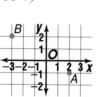

Graph $B(-3, 2)$.
left 3, up 2

Graph and label each point.

46. $M(-2, 2)$ **47.** $N(3, 0)$

48. $P(5, 6)$ **49.** $Q(-2, 4)$

50. $R(-2, -5)$ **51.** $S(7, -1)$

● graph transformations on a coordinate grid
(Lesson 11-8)

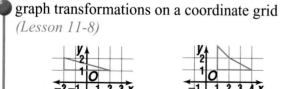

translation   reflection

The vertices of $\triangle ABC$ are $A(-1, 2)$, $B(0, 5)$, and $C(3, 1)$. On a coordinate grid, draw $\triangle ABC$ and each transformation image.

52. translation image that is 3 units left and 2 units down

53. reflection image over the x-axis

Applications & Problem Solving

54. Weather Low temperatures were recorded for one week in January. Order the temperatures from least to greatest. *(Lesson 11-2)*

S	M	T	W	T	F	S
−5°	0°	1°	−7°	−3°	2°	−2°

55. Football The Branson Middle School football team is 2 yards from its opponent's goal line. On the next two plays, its offense loses 17 yards and then gains 11 yards. How many yards is the team from the goal line? *(Lesson 11-3)*

56. Work Backward Pearl has money in her savings account. She made withdrawals of $100, $43, and $67. Her balance is now $245. How much did she have in her account before the three withdrawals? *(Lesson 11-6B)*

57. Crafts Diego wants to make a two-sided ornament out of fabric. He uses a pattern to cut one side of the ornament. Should he use a *reflection* or a *translation* of the pattern for the other side of the ornament? *(Lesson 11-8)*

Alternative Assessment

Open Ended

Suppose you are playing a game with a friend and you have decided to be the scorekeeper. During the first round, you scored 6 points and your friend scored 4 points. During the second round, you scored 3 points and your friend lost 3 points. During the third round, you lost 2 points and your friend scored 5 points. Finally, during the fourth round, you lost 1 point and your friend scored 5 points. Explain how to organize the information in order to keep score accurately.

Suppose on the fifth round, you scored 2 points and your friend lost 1 point. Find the total scores. Who is winning the game?

A practice test for Chapter 11 is provided on page 605.

Completing the CHAPTER Project

Use the following checklist to make sure your poster or brochure is complete.

☑ The map is clear and easy to read.

☑ The line plot is correct.

☑ The paragraph discussing the changes in the House of Representatives includes whether your part of the country has gained or lost representatives and how the total number of representatives has changed.

Add any finishing touches that you would like to make your poster or brochure attractive.

PORTFOLIO Select one exercise that involved graphing from this chapter that you found particularly challenging. Place it in your portfolio.

Section One: Multiple Choice

There are twelve multiple choice questions in this section. Choose the best answer. If a correct answer is *not here*, choose the letter for Not Here.

1. Which numbers are factors of 75?

A 3, 5, and 6

B 3, 5, and 9

C 3 and 5

D 5 and 10

2. If $n = -3$, what is the value of $8 + n$?

F -5

G -3

H 5

J 11

3. How many lines of symmetry does an equilateral triangle have?

A 0

B 1

C 2

D 3

4. What are the coordinates of point G on the graph?

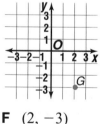

F $(2, -3)$

G $(3, -2)$

H $(-3, 2)$

J $(-2, -3)$

Please note that Questions 5–12 have five answer choices.

5. What integers are graphed?

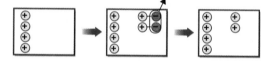

A $-5, -1, 2$

B $-7, -1, 2$

C $5, 1, 2$

D $-7, 1, 2$

E $-2, 1, 5$

6. Write the sentence represented by the model.

F $4 + (-4) = 0$

G $4 + (-2) = 2$

H $4 + (6) = 10$

J $4 - (-2) = 6$

K $4 - 2 = 2$

7. One boat is 5.05 meters long, and another is 4.8 meters long. How much space is needed to park the boats end to end?

A 9.85 meters

B 9.13 meters

C 10.3 meters

D 5.53 meters

E 0.25 meter

8. A round window in Kevin's house is 40 inches in diameter. Which is a good estimate for the area of the window?

F 60 square inches

G 120 square inches

H 600 square inches

J 1,200 square inches

K 4,800 square inches

9. Mr. Orta needs $\frac{7}{8}$ yard of material to upholster a chair seat and $\frac{5}{8}$ yard to upholster the back of the chair. How many yards of material will be needed to cover the chair?

 A 1 yard

 B $1\frac{1}{2}$ yards

 C $1\frac{3}{4}$ yards

 D 2 yards

 E Not Here

10. Bobbie charges $5.75 per hour for baby-sitting. If she works for 4 hours, how much will she earn?

 F $9.75

 G $20.75

 H $17.25

 J $23.00

 K Not Here

11. Barbara bought a 3-pound ham and $5\frac{1}{2}$ pounds of steak. How many more pounds of steak did she buy than ham?

 A 2 pounds

 B $2\frac{1}{2}$ pounds

 C $3\frac{1}{2}$ pounds

 D $8\frac{1}{2}$ pounds

 E Not Here

12. $-3(-13) =$

 F 16   G -16

 H 39   J -39

 K Not Here

Test Practice **For additional test practice questions, visit:**

www.glencoe.com/sec/math/mac/mathnet

Test-Taking Tip THE PRINCETON REVIEW

You could use educated guesses to help improve your scores. A wild guess involves pure chance, but eliminating possible answers as definitely wrong will allow you to increase your chances of getting a right answer.

Section Two: Free Response

This section contains seven questions for which you will provide short answers. Write your answers on your paper.

13. Find the quotient of 32 and -4.

14. Find the next two numbers in the sequence 64, 16, 4, 1, ...

15. Write the ordered pair represented by M.

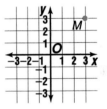

16. Find 98% of 6.

17. Draw the reflection of the letter E over a vertical line.

18. How can you find the surface area of a 3-inch by 4-inch by 5-inch rectangular prism?

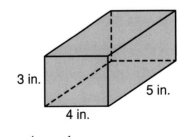

3 in.   5 in.   4 in.

19. Find $3\frac{4}{5} \times 7\frac{1}{2}$.

Algebra: Exploring Equations

What you'll learn in Chapter 12

- to use models to solve equations,
- to solve problems using an equation,
- to complete function tables, and
- to graph functions from function tables.

Thousands of NEW ITEMS

Baseball Card PRICE GUIDE NUMBER 1

CHAPTER Project

COLLECT A FORTUNE

In this project, you will start an imaginary collection of at least five items such as sport cards, coins, stamps, or dolls. Or you can use a collection you already have. You will need to research the history of the value of these items. You will prepare a table or some type of display showing the value of your collection to present to your class.

Getting Started

- Look at the table. Which card was worth the most in August, 1996? Which card was worth the most in June, 1997?

- How could you determine whether the value of Kristin's six-card collection increased or decreased in the 10 months between August, 1996 and June, 1997?

Kristin's Card Collection		
Player	Value Aug., 1996	Value June, 1997
John Stockton	$10	$7.50
Michael Jordan	$25	$20
Larry Bird	$18	$18
Charles Barkley	$50	$60
Shaquille O'Neal	$80	$30
Rudy Tomjanovich	$15	$15

Technology Tips

- Use a **spreadsheet** to find the value of a collection.

- Use **computer software** to make graphs.

 **inter NET CONNECTION**

Research For up-to-date information on card collecting, visit:

www.glencoe.com/sec/math/mac/mathnet

Working on the Project

You can use what you'll learn in Chapter 12 to help you make your presentation.

Page	Exercise
478	32
487	33
503	27
507	Alternative Assessment

Solving Addition Equations

What you'll learn

You'll learn to solve addition equations by using models.

When am I ever going to use this?

Knowing how to solve equations can help you determine golf scores.

Justin is playing a board game and needs to roll an 8 so he can land on Community Chest. If he rolls a 5 on the first number cube, what must he roll on the second number cube? *This problem will be solved in Example 1.*

You can use cups and counters to solve equations. A cup represents the unknown value, yellow counters represent positive integers, and red counters represent negative integers.

Example

Real World APPLICATION

1 **Games** Refer to the beginning of the lesson. What must Justin roll on the second number cube?

Explore   The number on the first number cube is 5. Justin wants the total of the two number cubes to be 8.

Plan   Let d represent the number tossed on the second number cube. Translate the problem into an equation using the variable d. Model the equation and solve.

Solve

number on the first number cube	plus	number on the second number cube	equals	sum of the two numbers
5	+	d	=	8

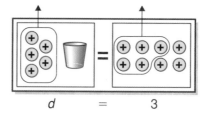

$$5 + d = 8$$

To get the cup by itself, subtract 5 from each side.

$$d = 3$$

The solution is 3. Justin must roll a 3 on the second number cube.

Examine   Check the solution by replacing the value of the variable in the original equation.

Check: $5 + d = 8$
$5 + 3 \stackrel{?}{=} 8$ *Replace d with 3.*
$8 = 8$ ✓

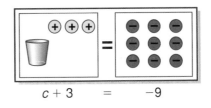

Example Use cups and counters to solve $c + 3 = -9$.

Use a cup to represent c. Add 3 positive counters on the left side of the mat to represent $+3$. Place 9 negative counters on the right side of the mat to represent -9.

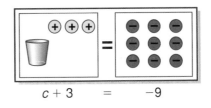

$$c + 3 \quad = \quad -9$$

To get the cup by itself, you need to remove 3 positive counters from each side. Since there are no positive counters on the right side of the mat, add 3 negative counters to each side to make 3 zero pairs on the left side of the mat. Then remove the zero pairs.

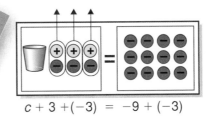

$$c + 3 + (-3) \quad = \quad -9 + (-3)$$

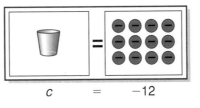

$$c \quad = \quad -12$$

Check: $c + 3 = -9$ *Replace c with –12.*

$$-12 + 3 \stackrel{?}{=} -9$$

$$-9 = -9 \quad \checkmark$$

The solution is -12.

CHECK FOR UNDERSTANDING

Communicating Mathematics

Read and study the lesson to answer each question.

1. **Explain** how the model represents $5 + w = -8$.

2. **Show** how to model the equation $y + 2 = 4$.

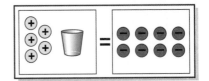

Lesson 12-1 Solving Addition Equations **477**

Solve each equation. Use cups and counters if necessary.

3. $x + 3 = 9$ **4.** $t + 25 = 15$ **5.** $4 + g = 11$

6. $r + 2 = -5$ **7.** $2 + c = -7$ **8.** $-12 + m = 8$

9. When n is added to 10, the result is 4. What is the value of n?

EXERCISES

Solve each equation. Use cups and counters if necessary.

10. $y + 7 = 18$ **11.** $x + 3 = -2$ **12.** $4 + g = 6$

13. $n + 4 = 3$ **14.** $-3 + m = -3$ **15.** $z + 3 = 5$

16. $2 + s = -1$ **17.** $2 + b = -4$ **18.** $x + (-3) = -1$

19. $3 + p = -17$ **20.** $d + 1 = -2$ **21.** $t + 9 = -34$

22. $-3 + c = -5$ **23.** $a + 25 = -100$ **24.** $-9 + z = -18$

25. $x + (-7) = -9$ **26.** $f + 6 = -8$ **27.** $-2 = m + 6$

28. If $8 + c = -12$, what is the value of c?

29. Find the value of n if $n + (-4) = 6$.

30. *Games* In the card game Clubs, it is possible to have a negative score. Suppose your friend had a score of -5 in the second hand. This made her total score after two hands equal -2. What was her score in the first hand?

31. *Sports* On the first day of the Masters golf tournament, Tiger Woods' score was -2. His score on the second day was added to the first day's score, and the total was -8. What was Tiger's score on the second day?

32. *Working on the* **CHAPTER Project**   Refer to the table on page 475. Enter the three columns of the table into a spreadsheet.

   a. Label column D *Change in 10 months*. Write a formula for cells D2, D3, D4, D5, D6, and D7 to find the amount of change in value of each basketball card.

	A	B	C	D
1	Player	Value August, 1996	Value June, 1997	Change in 10 months
2	John Stockton	$10	$7.50	
3	Michael Jordan	$25	$20	
4	Larry Bird	$18	$18	
5	Charles Barkley	$50	$60	
6	Shaquille O'Neal	$80	$30	
7	Rudy Tomjanovich	$15	$15	

   b. Write a formula for cell D8 that will find the total change in value of Kristin's six-card collection over the 10-month period. What was the total change in value?

33. *Critical Thinking* Replace the boxes with the numbers 2, 3, 7, 8, and 9 to make a true equation. Use each number exactly once.

$$\blacksquare\,\blacksquare + (-\blacksquare) = \blacksquare\,\blacksquare$$

Mixed Review **34. Standardized Test Practice** Which of the following illustrates a reflection of the letter L? *(Lesson 11-8)*

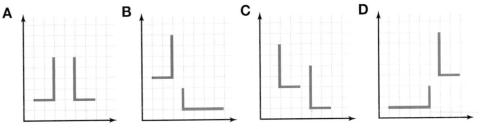

A   B   C   D

35. Find the product of -3 and -7. *(Lesson 11-5)*

36. *Geometry* Find the area of the triangle.
(Lesson 10-2)

42 m   41 m   18 m

37. *Vacations* Refer to the circle graph. How much greater is the percent of vacationers that travel and sightsee than the percent that go to a resort? *(Lesson 2-4)*

How People Spend Vacations

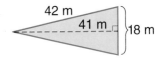

No vacation/don't know

Summer/winter resort

Visit family/friends

6%

17%

30%

21%

26%

Stay at home

Travel/sightsee

For **Extra Practice**, see page 588.

MATH IN THE MEDIA

x + 3 = 2x
x = 3

"SAY, WAIT A MINUTE! JUST YESTERDAY SHE SAID X WAS EQUAL TO TWO!"

1. What assumption is the student making about the value of x?

2. How would you check the teacher's solution?

Solving Subtraction Equations

What you'll learn

You'll learn to solve subtraction equations by using models.

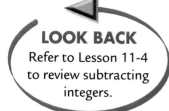

When am I ever going to use this?

Knowing how to solve equations can help you describe changes in temperature.

You probably have heard about the greenhouse effect on Earth. Did you know that it also affects other planets? Scientists have determined that, on Mars, the difference between the air temperature with the greenhouse effect and what it would be if it weren't for the greenhouse effect is 5° F. Suppose the temperature would be $-3°$ F on Mars if it weren't for the greenhouse effect. What is the air temperature with the greenhouse effect?

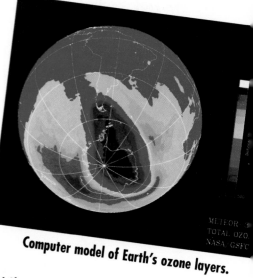

Computer model of Earth's ozone layers.

Example ①

CONNECTION

Earth Science Refer to the beginning of the lesson. What would the temperature be with the greenhouse effect?

Let g represent the temperature with the greenhouse effect. Translate the problem into an equation using the variable g. Then model the equation and solve.

temperature with the greenhouse effect	*minus*	*actual temperature*	*equals*	*difference*
g	$-$	-3	$=$	5

Rewrite as an addition equation. Remember that subtracting an integer is the same as adding its opposite.

$$g - (-3) = 5 \rightarrow g + 3 = 5$$

Use a cup to represent g. Add 3 positive counters on the left side of the mat to represent $+3$. Place 5 positive counters on the right side of the mat to represent $+5$.

To get the cup by itself, remove 3 positive counters from each side.

The solution is 2. The temperature with the greenhouse effect is 2° F.

Check the solution by replacing the value of the variable in the original equation.

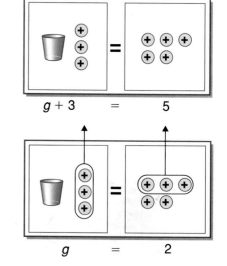

$g + 3 \quad = \quad 5$

$g \quad = \quad 2$

LOOK BACK

Refer to Lesson 11-4 to review subtracting integers.

Check: $g - (-3) = 5$

$2 - (-3) \stackrel{?}{=} 5$ *Replace g with 2.*

$2 + 3 \stackrel{?}{=} 5$ *Rewrite as an addition equation.*

$5 = 5 \checkmark$

2 **Use cups and counters to solve $h - (-4) = -3$.**

Rewrite as an addition equation.

$$h - (-4) = -3 \rightarrow h + 4 = -3$$

Use a cup to represent h. Add 4 positive counters beside the cup on the left side of the mat to represent $+4$. Place 3 negative counters on the right side of the mat to represent -3.

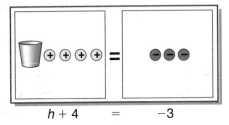

$$h + 4 \quad = \quad -3$$

To get the cup by itself, remove 4 positive counters from each side. Since there are no positive counters on the right side of the mat, add 4 negative counters to each side to make 4 zero pairs on the left side of the mat. Then remove the zero pairs.

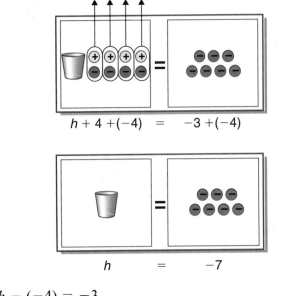

$$h + 4 + (-4) \quad = \quad -3 + (-4)$$

$$h \quad = \quad -7$$

Check: $h - (-4) = -3$

$-7 - (-4) \overset{?}{=} -3$ *Replace h with -7.*

$-7 + 4 \overset{?}{=} -3$ *Rewrite as an addition equation.*

$-3 = -3$ ✓

The solution is -7.

3 **Use cups and counters to solve $x - 6 = -2$.**

Rewrite as an addition equation.

$$x - 6 = -2 \rightarrow x + (-6) = -2$$

Use a cup to represent x. Add 6 negative counters on the left side of the mat to represent -6. Place 2 negative counters on the right side of the mat to represent -2.

(continued on the next page)

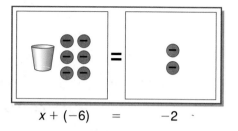

$$x + (-6) = -2$$

To get the cup by itself, remove 6 negative counters from each side. Since there are not enough negative counters on the right side of the mat, add 6 positive counters to each side to make 6 zero pairs on the left side of the mat. Then remove all zero pairs.

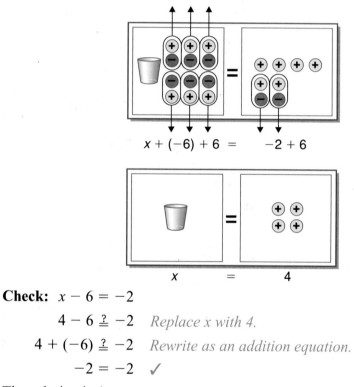

$$x + (-6) + 6 = -2 + 6$$

$$x = 4$$

Check: $x - 6 = -2$

$$4 - 6 \stackrel{?}{=} -2 \quad \textit{Replace x with 4.}$$

$$4 + (-6) \stackrel{?}{=} -2 \quad \textit{Rewrite as an addition equation.}$$

$$-2 = -2 \quad \checkmark$$

The solution is 4.

CHECK FOR UNDERSTANDING

Communicating Mathematics

Read and study the lesson to answer each question.

1. *Explain* how the model represents $y - 3 = 4$.

2. *Show* how to model the equation $x - 2 = 6$.

3. *Write* a few sentences explaining how you know whether to add or subtract counters from each side of the equation mat.

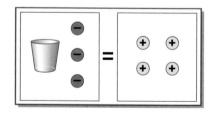

Guided Practice

Use cups and counters to solve each equation.

4. $x - 8 = 2$ 5. $a - 4 = -6$ 6. $g - 4 = 11$

7. $y - 2 = 5$ 8. $c - 2 = -7$ 9. $z - 7 = 9$

10. **Sports** After a loss of 5 yards, Ayani had a total of 16 yards. How many yards did he have before the 5-yard loss?

EXERCISES

Practice

Solve each equation. Use cups and counters if necessary.

11. $g - 4 = -6$
12. $z - (-3) = 5$
13. $x - 1 = -3$
14. $c - 5 = -5$
15. $b - 6 = -7$
16. $r - 7 = -15$
17. $z - 10 = -18$
18. $x - (-2) = -1$
19. $t - 2 = 3$
20. $h - (-5) = -2$
21. $y - 3 = 4$
22. $x - 9 = 4$
23. $d - 5 = -2$
24. $s - 8 = -1$
25. $v - 12 = -10$

26. Find the value of x if $x - (-8) = -14$.
27. If $t - (-13) = 4$, what is the value of t?

Applications and Problem Solving

28. **Earth Science** Refer to the graph.
   a. How did the temperature change between January 18 and January 19?
   b. The low temperature on January 21 was 5 degrees higher than the low on January 20. What was the low on January 21?

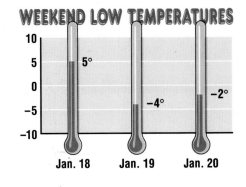

29. **Diving** A diver begins to go back to the boat from 140 feet below sea level. A few minutes later, the diver is 35 feet below sea level.
   a. Write an addition equation to represent the diver's movement.
   b. How many feet did the diver rise?

30. **Critical Thinking** Describe how you would solve $5 - x = -4$.

Mixed Review

31. **Standardized Test Practice** Sabra collected 6 silver dollars. Her friend Logan gave her some more, and then she had 15. To find out how many silver dollars she was given, Sabra wrote $s + 6 = 15$. What is the value of s? *(Lesson 12-1)*

   A 6   **B** 9   **C** 21   **D** 90

32. **Measurement** Find the area of a circle with a diameter of 6 meters to the nearest tenth. Use 3.14 for pi. *(Lesson 10-3)*

33. **Geometry** Use a protractor and draw an angle that measures 70°. *(Lesson 9-2)*

34. **Statistics** Refer to the circle graph of students' favorite sports. How much more of the students' votes were for swimming than tennis? *(Lesson 2-4)*

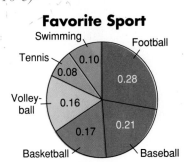

For **Extra Practice**, see page 588.

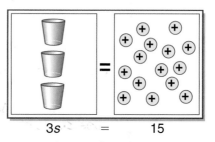

12-3 Solving Multiplication and Division Equations

What you'll learn

You'll learn to solve equations involving multiplication and division using models.

When am I ever going to use this?

Knowing how to solve equations involving multiplication and division can help you find your rate of pay.

If you love your dog, don't let it eat chocolate. An ingredient in chocolate, theobromine, can be poisonous to dogs. Just 2 ounces of chocolate could harm a 10-pound pup. Zina determines that half of her chocolate bar is enough to harm a 10-pound pup. How much does her chocolate bar weigh? *This problem will be solved in Example 3.*

Cups and counters can be used to solve multiplication and division equations.

Example ① **Use cups and counters to solve** $3s = 15$.

Use three cups to represent $3s$. Place them on the left side of the mat. Place 15 positive counters on the right side of the mat to represent 15.

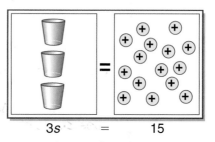

$$3s \qquad = \qquad 15$$

Since there are three cups, undo the multiplication by dividing each side by 3. Show division by 3 by forming 3 equal groups on each side of the mat.

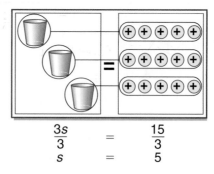

$$\frac{3s}{3} = \frac{15}{3}$$
$$s = 5$$

Check: $3s = 15$

$3(5) \overset{?}{=} 15$ *Replace s with 5.*

$15 = 15$ ✓

The solution is 5.

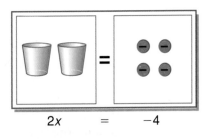

Examples

2 Use cups and counters to solve $2x = -4$.

Use two cups to represent $2x$. Place them on the left side of the mat. Place 4 negative counters on the right side of the mat to represent -4.

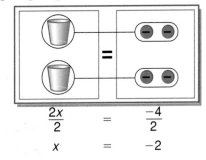

$$2x \qquad = \qquad -4$$

Undo the multiplication by dividing each side by 2. Show division by 2 by forming 2 equal groups on each side of the mat.

$$\frac{2x}{2} \qquad = \qquad \frac{-4}{2}$$
$$x \qquad = \qquad -2$$

Check: $\qquad 2x = -4$

$\qquad 2(-2) \stackrel{?}{=} -4 \qquad$ *Replace x with −2.*

$\qquad -4 = -4 \quad \checkmark$

The solution is -2.

CONNECTION

3 **Life Science** Refer to the beginning of the lesson. How much does Zina's chocolate bar weigh?

First, write an equation to describe the situation. Let b represent the weight of the chocolate bar.

one half of bar	*equals*	*2 ounces*
$\frac{1}{2}b$	$=$	2

Solve the equation $\frac{1}{2}b = 2$.

$$\frac{1}{2}b = 2$$

$$2\left(\frac{1}{2}b\right) = 2(2) \qquad \text{\textit{Undo the division by multiplying each side by 2.}}$$

$$b = 4$$

Check: $\frac{1}{2}b = 2$

$\qquad \frac{1}{2}(4) \stackrel{?}{=} 2 \qquad$ *Replace b with 4.*

$\qquad 2 = 2 \quad \checkmark$

The solution is 4. Zina's chocolate bar weighs 4 ounces.

Communicating Mathematics

Read and study the lesson to answer each question.

1. **Tell** what equation is represented by the model. Then solve.

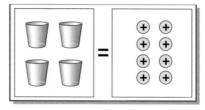

2. **Make** a model to represent the equation $3x = -9$. Then solve the equation.

3. **You Decide** Zack says that to solve the equation $2k = 50$ he would divide each side by 2. Aubrey says she would multiply each side by $\frac{1}{2}$. Who is correct? Explain.

Guided Practice

Complete each solution.

Family Activity

Use items you can find in your home to explain to a family member how you solve an equation involving multiplication or division.

4. $3x = 9$
$$\frac{3x}{\blacksquare} = \frac{9}{\blacksquare}$$
$$x = \blacksquare$$

5. $8g = 16$
$$\frac{8g}{\blacksquare} = \frac{16}{\blacksquare}$$
$$g = \blacksquare$$

6. $\frac{1}{4}a = -7$
$$\blacksquare\left(\frac{1}{4}a\right) = \blacksquare\left(-7\right)$$
$$a = \blacksquare$$

Solve each equation. Use cups and counters if necessary.

7. $5e = 25$

8. $2c = 6$

9. $\frac{1}{2}n = -5$

10. **Money Matters** Margie's average weekly earnings in 1997 were three times higher than in 1980. She earned $624 per week in 1997. How much did she earn per week in 1980?

Practice

Solve each equation. Use cups and counters if necessary.

11. $9a = 18$

12. $3x = 3$

13. $2c = -6$

14. $4k = -20$

15. $4x = -36$

16. $5e = -15$

17. $16x = 4$

18. $5h = 5$

19. $4m = 20$

20. $2x = -16$

21. $\frac{1}{8}y = 3$

22. $\frac{1}{3}x = 12$

23. $\frac{n}{2} = 19$

24. $53 = \frac{1}{2}y$

25. $3.5s = 7$

26. $24.8 = 1.24a$

27. $\frac{1}{4}w = -9$

28. $-12 = \frac{3}{8}g$

29. Solve the equation $3.1t = 25.42$.

30. Roni's mother Nikki is 3 times as old as Roni is. Nikki is 39. How old is Roni?

Applications and Problem Solving

31. **Money Matters** In 1995, Michael Jordan's income from endorsements was 10 times his salary for playing basketball. If his income from endorsements was about $40 million, about how much was his salary for playing basketball?

32. **Money Matters** The track team has 19 members. They stopped at a restaurant on the way home from a meet. If the total bill was $90.25, what was the average cost of each team member's meal?

33. *Working on the* CHAPTER Project Refer to the spreadsheet you made for Exercise 32 on page 478. Label column E *Change per month*. Write a formula for cells E2, E3, E4, E5, E6, and E7 to find the change in value per month for each card.

34. *Critical Thinking* Without solving, tell which equation, $\frac{1}{4}x = 13$ or $\frac{1}{8}x = 13$, has the greater solution. Explain.

Mixed Review **35.** *Algebra* Solve the equation $y - 11 = -8$. *(Lesson 12-2)*

36. If the figure at the right is a scale drawing with a scale 1 unit = 100 meters, what is the actual length of $\overline{AB}$? *(Lesson 8-3)*

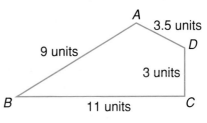

A
3.5 units
D
9 units
3 units
B
11 units
C

37. **Standardized Test Practice** Melika was making a quilt for her room. She had $9\frac{1}{2}$ yards of material. It took 7 yards for the quilt. How much material was not used for the quilt? *(Lesson 6-6)*

A $16\frac{1}{2}$ yd

B $9\frac{1}{2}$ yd

C $2\frac{1}{2}$ yd

D $1\frac{1}{2}$ yd

E Not Here

For **Extra Practice,** see page 589.

38. List all the common factors of 20 and 50. *(Lesson 5-3)*

CHAPTER 12

Mid-Chapter Self Test

Solve each equation. Use cups and counters if necessary. *(Lessons 12-1 and 12-2)*

1. $y + 6 = 12$ **2.** $x + 6 = -3$ **3.** $m + (-7) = 14$ **4.** $t - 8 = 14$

5. $k - (-10) = -5$ **6.** $3g = 18$ **7.** $2x = -4$ **8.** $\frac{1}{3}y = -9$

Solve.

9. *Transportation* Marta's car averages 24 miles per gallon. Her odometer shows that she has driven 72 miles. How many gallons of gasoline has her car used? *(Lesson 12-2)*

10. *Money Matters* Lorenzo spends one-third of his monthly allowance on snacks. If he spends $4 on snacks every month, how much money does he get for his allowance? *(Lesson 12-3)*

Solving Two-Step Equations

Lupita went bowling at Great Lanes Sport Center. Shoe rental was $1, and games were $2 each. If she spent $9 on shoe rental and games, how many games did she bowl?

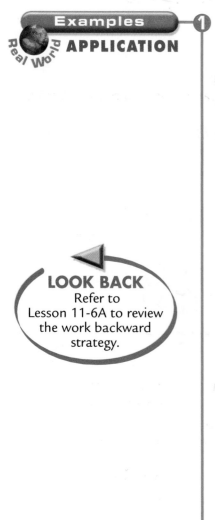

Examples

APPLICATION

① Sports Refer to the beginning of the lesson. How many games did Lupita bowl?

Let g equal the number of games bowled. Translate the problem into an equation using the variable g. Then model the equation and solve.

$$\underbrace{\$1 \text{ for shoe rental}}_{1} \quad \underbrace{plus}_{+} \quad \underbrace{\$2 \text{ per game}}_{2g} \quad \underbrace{is}_{=} \quad \underbrace{total\ cost}_{9}$$

The equation is $1 + 2g = 9$. This is a two-step equation because it involves two different operations, addition and multiplication. To solve this equation, you need to work backward using the reverse of the order of operations.

LOOK BACK
Refer to Lesson 11-6A to review the work backward strategy.

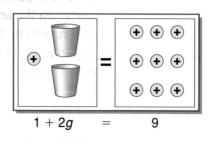

$$1 + 2g = 9$$

To get the cups by themselves, remove 1 positive counter from each side.

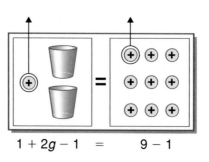

$$1 + 2g - 1 = 9 - 1$$

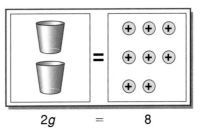

$$2g \quad = \quad 8$$

The Since there are 2 cups, undo the multiplication by dividing each side by 2. Form 2 equal groups on each side of the mat.

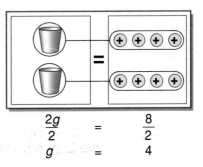

$$\frac{2g}{2} \quad = \quad \frac{8}{2}$$
$$g \quad = \quad 4$$

The solution is 4. Lupita bowled 4 games.

Check by replacing the value of the variable in the original equation.

Check: $\quad 1 + 2g = 9$

$\qquad 1 + 2(4) \stackrel{?}{=} 9 \quad$ *Replace g with 4.*

$\qquad\qquad 9 = 9 \quad \checkmark$

2 **Use cups and counters to solve $2x + 6 = -4$.**

Place two cups and 6 positive counters on the left side of the mat to represent $2x + 6$. Place 4 negative counters on the right side of the mat to represent -4.

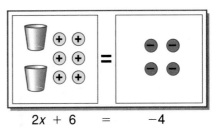

$$2x + 6 \quad = \quad -4$$

To get the cups by themselves, you need to remove 6 positive counters from each side. Since there are no positive counters on the right side of the mat, add 6 negative counters to each side to make 6 zero pairs on the left side of the mat. Then remove the zero pairs.

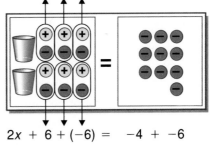

$$2x + 6 + (-6) = \quad -4 + -6$$

(continued on the next page)

Now the equation is $2x = -10$. Undo the multiplication by dividing each side by 2. Show division by forming 2 equal groups on each side of the mat.

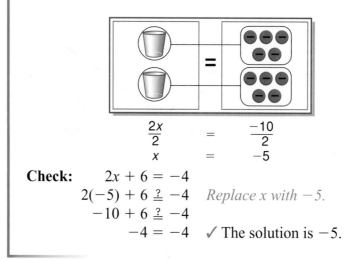

$$\frac{2x}{2} = \frac{-10}{2}$$
$$x = -5$$

Check: $2x + 6 = -4$

$2(-5) + 6 \overset{?}{=} -4$ *Replace x with -5.*

$-10 + 6 \overset{?}{=} -4$

$-4 = -4$ ✓ The solution is -5.

CHECK FOR UNDERSTANDING

Communicating Mathematics

Read and study the lesson to answer each question.

1. *Write* an equation for the model. Then solve.

2. *Make* a model to represent the equation $2x + 3 = -9$. Then solve the equation.

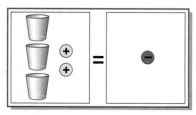

Guided Practice

Name the first step in solving each equation. Then solve.

3. $3x + 2 = 14$ 4. $2a - 3 = 11$ 5. $\frac{1}{2}y - 4 = 6$

6. Twice a number, n, plus 7 is -21. What is the value of n?

EXERCISES

Practice

Solve each equation.

7. $2t + 5 = 13$ 8. $\frac{1}{2}m + 3 = -5$ 9. $3h - 4 = 5$

10. $-11 = 4x - 3$ 11. $2x - (-34) = 16$ 12. $5r + 1 = 1$

13. $\frac{1}{4}t - 5 = -13$ 14. $-1 - 4y = 7$ 15. $-3y + 15 = 75$

16. Ten less than twice a number is sixteen. What is the number?

17. One half a number less five is eleven. Find the number.

Applications and Problem Solving

18. *Money Matters* Claudio ordered three novelty T-shirts from a catalog. The total price including shipping charges was $50. If the total shipping cost was $5, how much did each T-shirt cost?

19. *Geometry* The perimeter of a rectangle is 40 inches. Find its length if its width is 4 inches.

20. **Critical Thinking** Use what you know about solving two-step equations to solve the equation $\frac{1}{4}(k - 8) = -3$.

Mixed Review

21. **Algebra** Solve for q in the equation $-7q = -56$. *(Lesson 12-3)*

22. **Standardized Test Practice** Mr. Vega has 232 seedlings to be planted in flowerpots with 8 plants in each. How many flowerpots will he need to plant the seedlings? *(Lesson 11-6)*

    A 32　　　　**B** 29　　　　**C** 27　　　　**D** 19　　　　**E** Not Here

For **Extra Practice**, see page 589.

23. **Patterns** Find the next two numbers in the sequence 34, 36.5, 39, 41.5,... . *(Lesson 7-8)*

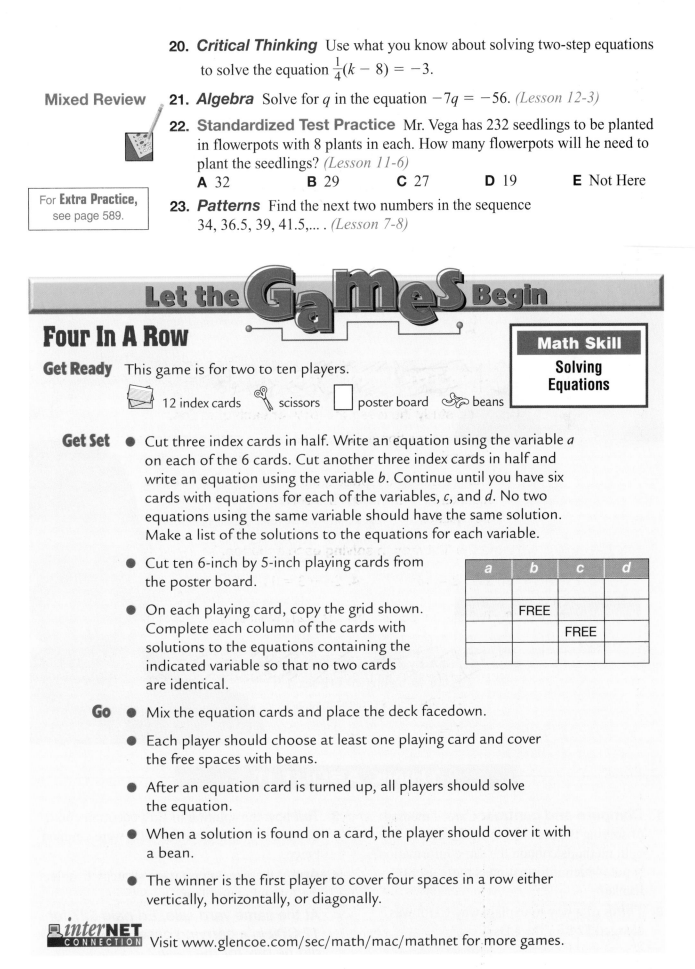

Let the Games Begin

Four In A Row

Math Skill
Solving Equations

Get Ready　This game is for two to ten players.

12 index cards　　scissors　　poster board　　beans

Get Set
- Cut three index cards in half. Write an equation using the variable *a* on each of the 6 cards. Cut another three index cards in half and write an equation using the variable *b*. Continue until you have six cards with equations for each of the variables, *c*, and *d*. No two equations using the same variable should have the same solution. Make a list of the solutions to the equations for each variable.

- Cut ten 6-inch by 5-inch playing cards from the poster board.

- On each playing card, copy the grid shown. Complete each column of the cards with solutions to the equations containing the indicated variable so that no two cards are identical.

a	b	c	d
	FREE		
		FREE	

Go
- Mix the equation cards and place the deck facedown.

- Each player should choose at least one playing card and cover the free spaces with beans.

- After an equation card is turned up, all players should solve the equation.

- When a solution is found on a card, the player should cover it with a bean.

- The winner is the first player to cover four spaces in a row either vertically, horizontally, or diagonally.

interNET CONNECTION　Visit www.glencoe.com/sec/math/mac/mathnet for more games.

12-4B Use an Equation

A Follow-Up of Lesson 12-4

Carol and Ed have just returned from a yard sale where Carol bought a personal CD player and some CDs. Carol bought 8 CDs but thinks she was charged for more. Let's listen in!

Those were good bargains, but I think I was charged for too many CDs.

How many CDs did you get and how much did you pay?

I paid a total of $39 for a $15 CD player and 8 CDs. I can find out if that is correct by adding $15 and $16.

Or we could let *n* represent the number of CDs and translate the problem into an equation. The total is $2 times the number of CDs plus $15. This can be written as 39 = 2n + 15.

Carol

Ed

But that's a two-step equation!

It's not difficult. First, subtract 15 from each side. Now the equation is 2n = 24. Dividing each side by 2 shows that you paid for 12 CDs. We better head back and get this straightened out.

SALE	
CD Player	$15
All CDs	$2

THINK ABOUT IT

1. *Compare and contrast* Carol's method of solving the problem to Ed's method. Do both methods contain the same information? If not, which information is more useful? Explain.

2. *Think* of a way to explain why Carol was charged $24 for the CDs.

3. *Tell* how the solution to Ed's equation could be used to find how many CDs were counted twice.

4. *Apply* the **use an equation** strategy to solve the following problem.

   At the same yard sale, Ed paid $27 for 12 CDs in a carrying case. How much did he pay for the case?

For **Extra Practice,** see page 589.

ON YOUR OWN

5. The third step of the 4-step plan for problem solving asks you to *solve.* **Tell** what other problem-solving strategy you could use to solve Carol's problem.

6. *Write a Problem* that can be solved by using an equation.

7. *Reflect Back* Explain how solving a two-step equation is similar to using the work backward strategy.

MIXED PROBLEM SOLVING

STRATEGIES

Look for a pattern.
Solve a simpler problem.
Act it out.
Guess and check.
Draw a diagram.
Make a chart.
Work backward.

Solve. Use any strategy.

8. *Money Matters* Jillisa bought a clock radio for $9 less than the regular price. If she paid $32, what was the regular price?

9. *School* The sixth grade class is planning a field trip. There are 589 students in the sixth grade. Each bus holds 48 people. About how many buses will they need?

10. *Design* A designer wants to arrange 12 glass bricks into a rectangular shape with the least perimeter possible. How many blocks will be in each row?

11. *Money Matters* Scott paid $2.50 in sales tax on a sweatshirt. The total cost was $42.49. What was the price of the sweatshirt before taxes?

12. *Geometry* A kite has two pairs of congruent sides. If two sides are 56 centimeters and 34 centimeters, what is the perimeter of the kite?

34 cm

56 cm

13. *Fashion* A catalog company offers 3 styles of ski sweaters each in 8 different colors. How many combinations of style and color are possible?

14. *Language Arts* Science fiction books were the most popular items at the Book Fair. On Monday, 86 science fiction books were sold. This is 8 more than twice the amount that were sold on Thursday. How many science fiction books were sold on Thursday?

15. *Travel* Kathy lives in Rockwood and works in Somerset. There is no direct route from Rockwood to Somerset, so Kathy goes through either Boulder Creek or Castleton. There is one road between Boulder Creek and Castleton. How many different ways can Kathy drive to work?

16. *Standardized Test Practice* The Swann family is going to a play. Ticket prices are shown below. Mr. Swann needs 2 adult tickets, 3 student tickets, and 1 child's ticket. Which number sentence could be used to find *T*, the cost in dollars of the tickets?

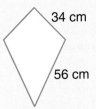

Ticket Prices	
Adult	$7.25
Student	$3.50
Child under 4	$1.75

A $T = (2 \times 7.25) + (3 \times 3.50) + 1.75$

B $T = 6 \times (7.25 + 3.50 + 1.75)$

C $T = (2 + 3) \times (7.25 + 3.50 + 1.75)$

D $T = 7.25 + 3.50 + 1.75 + 6$

E $T = (2 + 7.25) \times (3 + 3.50) \times (1.75)$

COOPERATIVE LEARNING

12-5A Function Machines

A Preview of Lesson 12-5

✂ scissors

🎞 tape

In this lab, you will use what you have learned about solving equations to help you work with function machines. A *function machine* takes a number called the *input*, performs one or more operations on it, and produces a result called the *output*.

TRY THIS

Work in groups of three.

Make a function machine for the rule $\boxed{+5}$.

• Take a sheet of paper and cut it in half lengthwise.

• On one of the halves, cut four slits — two on each side of the paper. Make each slit at least one inch wide.

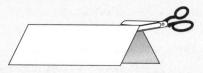

• From the other half of the paper, cut two narrow strips lengthwise. These strips should be able to pass through the slits you cut in the other half.

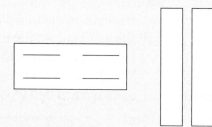

• On one of the narrow strips, write five consecutive numbers starting with 1. On the other narrow strip, write five consecutive numbers starting with 6. The numbers on both strips should align.

1	6
2	7
3	8
4	9
5	10

- Place the strips into the slits so that the numbers can be seen. Show 1 and 6. Once they appear, tape the ends of the strips together. When you pull the strips, they should move together. Mark the left-hand strip *input,* and the right-hand strip *output.* Write the function rule $\boxed{+5}$ between the input and output.

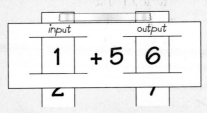

- Make a function table showing the input and the output.

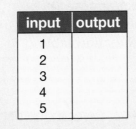

input	output
1	
2	
3	
4	
5	

ON YOUR OWN

1. What is the output when the input is 3?

2. What is the output when the input is 5?

3. Suppose you added more input numbers to the left strip. What would the output be if the input was 8?

4. What is the output if the input is *x*?

5. Make a function machine for the rule $\boxed{\times 3}$.

   a. What is the output when the input is 4?

   b. What is the output when the input is 6?

   c. Suppose you added more input numbers to the left strip. What would the output be if the input was 7?

   d. What was the input if the output was 33?

6. Make up your own function machine. Write pairs of inputs and outputs and have the other members of your group determine the rule.

7. Temperature is usually measured in Celsius (°C) or Fahrenheit (°F). The formula for changing from Celsius to Fahrenheit is $F = \frac{9}{5} \times C + 32$. This formula is like a function. The input is the Celsius temperature, and the output is the Fahrenheit temperature. Find the output for each input.

   a. 60°C   b. 15°C   c. 0°C   d. −10°C

8. *Look Ahead* Use the function machine to find the set of outputs that correspond to the set of inputs 1, 3, 5, and 7.

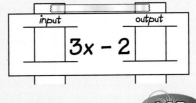

Functions

What you'll learn

You'll learn to complete function tables.

When am I ever going to use this?

Knowing how to complete function tables can help you determine the amount of profit a business will make for selling any number of items.

Word Wise

function
function table

A group of kangaroos is called a mob. The amount of plants that a mob of kangaroos eats depends on how many are in the mob. In other words, the amount of plants eaten is a **function** of the number of kangaroos.

A grown kangaroo can eat 14 pounds of grass and other plants per day. About how many pounds a day would be eaten by a mob of 3 grown kangaroos? 4 kangaroos? 5 kangaroos? This problem can be organized in a **function table**. *This problem will be solved in Example 1.*

Examples

CONNECTION

1 **Life Science** Refer to the beginning of the lesson. Determine the amount eaten by a mob of 3, 4, and 5 kangaroos.

To find the amount eaten by a mob of kangaroos in a day (output), you need to multiply the number in the mob (input) by 14.

Input	Function Rule	Output
Number in the Mob (n)	$14n$	Amount Eaten in a Day (lb)
3	14(3)	42
4	14(4)	56
5	14(5)	70

A mob of 3 eat 42 pounds of grass and plants, 4 eat 56 pounds, and 5 eat 70 pounds.

2 Find the rule for the function table.

Study the relationship between each input and output.

input (n)	output (■)
−2	2
0	4
1	5
3	7

input			output
−2	+ 4	→	2
0	+ 4	→	4
1	+ 4	→	5
3	+ 4	→	7

The output is 4 more than the input. So, the function rule is $n + 4$.

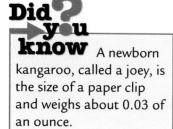

Did you know A newborn kangaroo, called a joey, is the size of a paper clip and weighs about 0.03 of an ounce.

Example **3**

CONNECTION

Life Science Insect-eating bats control insects without chemicals. The brown bat often eats 600 mosquitoes an hour. Make a function table showing the number of mosquitoes eaten in 2, 4, and 6 hours.

You can use the function rule 600h, where h is the number of hours.

Replace h in the rule 600h with the number of hours.

Replace h with 2.   *Replace h with 4.*   *Replace h with 6.*

$600h = 600 \cdot 2$ $600h = 600 \cdot 4$ $600h = 600 \cdot 6$
$\quad\quad = 1{,}200$ $\quad\quad = 2{,}400$ $\quad\quad = 3{,}600$

A brown bat can eat
1,200 mosquitoes in 2 hours,
2,400 mosquitoes in 4 hours,
and 3,600 mosquitoes in 6 hours.

input (h)	output (600h)
2	1,200
4	2,400
6	3,600

CHECK FOR UNDERSTANDING

Communicating Mathematics

Read and study the lesson to answer each question.

1. ***Make*** a function table for the function rule $y - 2$. Use inputs of -4, -1, 0, and 3.

2. ***Write*** the function rule for the table.

3. ***You Decide*** The output of a function table is 3 less than each input. Juanita says that the function rule is $x - 3$. Ruby says the rule is $3 - x$. Who is correct? Explain.

input (n)	output (■)
9	3
0	0
−3	−1
−6	−2

Guided Practice

Copy and complete each function table.

4.

input (n)	output (n+3)
−1	■
0	■
3	■

5.

input (n)	output ($\frac{1}{4}n$)
4	■
8	■
12	■

Find the rule for each function table.

6.

n	■
1	4
2	8
3	12

7.

n	■
0	0
2	1
4	2

8. If the input values are -2, 0, and 4 and the corresponding outputs are 3, 5, and 9, what is the function rule?

9. **Money Matters** Marc Wright of Windsor, Ontario, Canada, started his own greeting card business, called Kiddie Cards. Marc makes a profit of $0.75 for every card he sells.

   a. Write the function rule to represent Marc's profits.

   b. How much profit would Marc earn on a sale of 50 cards?

EXERCISES

Practice   **Copy and complete each function table.**

10.

input (n)	output (n + 4)
−3	■
0	■
3	■

11.

input (n)	output (3n)
0	■
3	■
4	■

12.

input (n)	output (5 − n)
5	■
6	■
7	■

13.

input (n)	output $\left(\frac{1}{8}n\right)$
−8	■
0	■
12	■

Find the rule for each function table.

14.

n	■
0	−3
3	0
6	3

15.

n	■
0	0
3	15
4	20

16.

n	■
−2	4
−1	2
3	−6

17.

n	■
−4	2
−2	1
0	0

18.

n	■
1	5
5	1
9	−3

19.

n	■
5	25
−5	25
0	0
−1	1

20. If a function rule is $2n + 1$, what is the output for an input of 2?

21. If a function rule is $6n − 4$, what is the output for an input of −3?

22. If the input values are 2 and 7 and the corresponding outputs are −6 and −21, what is the function rule?

23. If the input values are 5 and 10 and the corresponding outputs are 1 and 2, what is the function rule?

24. If the output values are −5 and −1 and the function rule is $n − 5$, what are the corresponding input values?

25. Spreadsheets In the spreadsheet, the number
of correct answers on a quiz is entered in
column A. The number of questions that
were on the quiz are entered in column B.
Column C computes the percent correct.
The formula in cell C1 is A1/B1*100.
The formula acts like the rule of a
function. What are the output values
for cells C2, C3, C4, and C5?

	A	B	C
1	9	10	90
2	10	10	
3	6	10	
4	8	10	
5	7	10	

26. Money Matters Ebony Hood of Washington, D.C., started
her own business selling scarves and fashion pins. Suppose she
sells a scarf for $2 and a pin for $4.

a. Write a function rule to represent the total cost of scarves (s)
and pins (p).

b. How much would 5 scarves and 3 pins cost?

27. Critical Thinking Find the rule for the
function table.

n	■
−2	−2
−1	0
2	6
3	8

Ebony Hood

Mixed Review

28. Geometry The area A of a
trapezoid can be found by
multiplying the height h and
one-half the sum of the bases b_1
and b_2. The formula is $A = \frac{1}{2}h(b_1 + b_2)$.
The area of the trapezoid at the right is
48 square centimeters. It is 6 centimeters
high, and the length of one base is 7
centimeters. What is the length of
the other base? *(Lesson 12-4)*

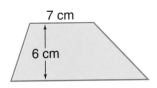

7 cm

6 cm

29. Write an integer to describe a temperature of twelve degrees below zero.
(Lesson 11-1)

30. Standardized Test Practice If △TUV is congruent to △PQR, then —
(Lesson 9-6)

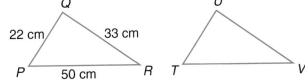

Q

22 cm   33 cm

P   50 cm   R

U

T   V

A the measure of side $\overline{TU}$ is equal to the measure of side $\overline{PR}$.

B side $\overline{TV}$ measures 50 centimeters.

C the perimeter of triangle TUV is 100 centimeters.

D the measure of angle P is equal to the measure of angle V.

E the measure of side $\overline{TV}$ is twice the measure of side $\overline{TU}$.

For **Extra Practice,**
see page 590.

Graphing Functions

What you'll learn

You'll learn to graph functions from function tables.

When am I ever going to use this?

Knowing how to graph functions can help you analyze data at a quick glance.

Getting an allowance of $2 a week is an example of a function. The equation $y = 2x$, where y is the amount received and x is the number of weeks, will give you the total allowance received after the number of weeks you choose. In the equation, $y = 2x$, x is the input, and y is the output. The function rule is $2x$.

We can use a coordinate system to graph an equation or function.

Example

Real World APPLICATION

Money Matters Refer to the beginning of the lesson. Make a function table for the rule 2x. Then graph the function.

Step 1 Record the input and output in a function table. We chose 0, 2, 4, and 6 for the input. List the input and output as ordered pairs.

input	function rule	output	ordered pairs
x	$2x$	y	(x, y)
0	2(0)	0	(0, 0)
2	2(2)	4	(2, 4)
4	2(4)	8	(4, 8)
6	2(6)	12	(6, 12)

Step 2 Graph the ordered pairs from the table in Step 1 on the coordinate plane.

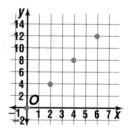

The x-coordinates represent the number of weeks from the start.

The y-coordinates represent the total allowance received after x weeks.

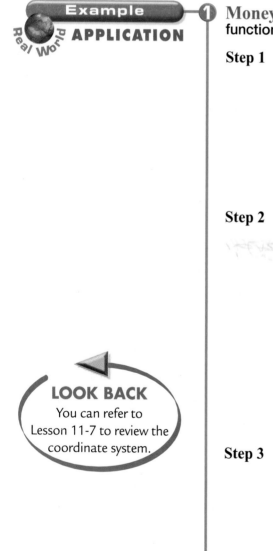

LOOK BACK
You can refer to Lesson 11-7 to review the coordinate system.

Step 3 The points appear to lie on a line. Draw the line that contains these points. The line is the graph of $y = 2x$.

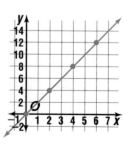

Example — 2

Make a function table for the graph. Then determine the rule.

(3, 1)
(−3, −1)
(−6, −2)

Use the ordered pairs to make a function table.

input (x)	output (y)	(x, y)
−6	−2	(−6, −2)
−3	−1	(−3, −1)
0	0	(0,0)
3	1	(3,1)

Study the input and output to determine a rule.

input   **output**

$-6 \quad \times \frac{1}{3} \rightarrow \quad -2$

$-3 \quad \times \frac{1}{3} \rightarrow \quad -1$

$0 \quad \times \frac{1}{3} \rightarrow \quad 0$

$3 \quad \times \frac{1}{3} \rightarrow \quad 1$

Each input is multiplied by $\frac{1}{3}$ to get the output.

The function rule is $\frac{1}{3}x$ or $\frac{x}{3}$.

CHECK FOR UNDERSTANDING

Communicating Mathematics

Read and study the lesson to answer each question.

1. *Explain* how you can use the function table to graph a function.

2. *Tell* why only those solutions graphed in the first quadrant in Example 1 make sense.

input	output
−1	2
0	0
1	−2
2	−4

Math Journal

3. *Write* a brief paragraph explaining how to graph a function when you know the function rule.

Guided Practice

Graph the functions represented by each function table.

4.

input	output
0	−3
3	0
6	3

5.

input	output
5	10
6	11
7	12

Copy and complete each function table. Then graph the function.

6.

input (n)	output (n − 5)
4	■
0	■
−2	■

7.

input (n)	output ($\frac{n}{2}$)
6.6	■
8.4	■
10.5	■

8. Make a function table for the rule $x - 6$ using 1, 4, and 5 as the input. Then graph the function.

9. Make a function table for the graph. Then determine the rule.

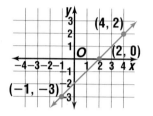

(4, 2)
(2, 0)
(−1, −3)

EXERCISES

Practice

Graph the functions represented by each function table.

10.

input	output
0	−5
2	−3
4	−1

11.

input	output
5	3
3	1
1	−1

12.

input	output
−2	4
0	0
2	−4

13.

input	output
0	0
4	1
8	2

14.

input	output
2	5
0	−1
−1	−4

15.

input	output
-4	0
0	1
4	2

Copy and complete each function table. Then graph the function.

16.

input (n)	output (2n + 4)
$5\frac{1}{2}$	■
$4\frac{1}{4}$	■
0	■

17.

input (n)	output (−3n)
3	■
0	■
−2	■

18.

input (n)	output (2n − 3)
−2	■
0	■
2	■
5	■

19.

input (n)	output ($\frac{1}{2}n + 1$)
−4	■
−2	■
0	■
2	■

20. Make a function table for the rule $a + 5$ using input values of -5, -3, and 0. Then graph the function.

21. Make a function table for the rule $3n - 3$ using input values of -2, 1, and 4. Then graph the function.

Make a function table for each graph. Then determine the rule.

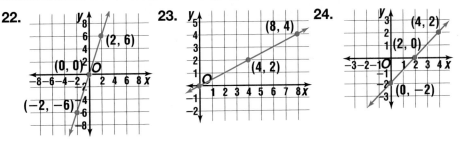

22.

23.

24.

Applications and Problem Solving

25. **Money Matters** During the summer, Jane earned $25 a week and was required to purchase a uniform for $20. No special clothing was required at Julie's summer job. Julie earned $20 a week.

  a. Write the function rule for each girl's wages.

  b. Graph each function on the same coordinate plane.

  c. What does the point of intersection of the two graphs represent?

26. **History** The line graph shows the number of cups of chocolate the Aztec emperor Montezuma drank each day.

  a. Make a function table for the graph.

  b. Determine the function rule.

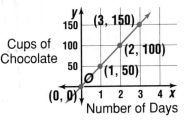

27. **Working on the CHAPTER Project** Refer to the table on page 475.

  You can graph a function to see how the value of a basketball card might change in the future. In June, 1997, the Charles Barkley card was worth $60. The value had been increasing $1 per month.

  a. Make a function table for the rule $60 + x$. Use inputs of 1 through 12 to predict the value of the card over the next 12 months.

  b. Graph the function using graphing software or a graphing calculator.

  c. What does the ordered pair $(5, 65)$ mean in this situation?

28. **Critical Thinking** Determine the rule for the line that passes through $A(-2, 6)$ and $B(3, -6)$.

Mixed Review

29. **Algebra** Complete the function table. *(Lesson 12-5)*

input (n)	output ($n + 7$)
−5	■
0	■
4	■

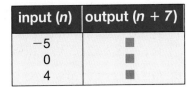

30. **Standardized Test Practice** If an airplane travels 438 miles per hour, how many miles will it travel in 5 hours? *(Lesson 8-2)*

  A 2,050 mi   **B** 2,090 mi   **C** 2,150 mi   **D** 2,190 mi   **E** Not Here

31. **Geometry** Find the circumference of a circle with a radius of 5 meters. Use 3.14 for pi. *(Lesson 7-4)*

For **Extra Practice**, see page 590.

32. Find the prime factorization of 120. *(Lesson 5-2)*

CHAPTER 12
Study Guide and Assessment

interNET
CONNECTION
Chapter Review For additional lesson-by-lesson review, visit:
www.glencoe.com/sec/math/mac/mathnet

Vocabulary

After completing this chapter, you should be able to define each term, concept, or phrase and give an example or two of each.

Patterns and Functions
function (p. 496)
function machine (p. 494)
function table (p. 496)
input (p. 494)
output (p. 494)

Problem Solving
use an equation (pp. 492–493)

Understanding and Using the Vocabulary

Choose the correct term, number, or equation to complete each sentence.

1. To solve an equation means to find a value for the (coordinate, variable) that makes the equation true.

2. The second step in solving $4b + 3 = 27$ is to (divide, multiply) each side of the equation by 4.

3. In an equation, when a number is (added to, subtracted from) the variable, add the number to each side to solve the equation.

4. A(n) (function, output) describes the relationship between two sets of numbers.

5. A(n) (input, function table) can be helpful in organizing information by using a mathematical rule to assign an output to a given input.

6. The solutions of a function table are found in the (output, input) column.

7. If the function rule is $6n$ and the input value is -3, then the output value is $(-18, -2)$.

8. On the graph of a function, every point has (coordinates, variables) that satisfy the function.

9. The diagram at the right models the equation $(2x + 3 = -9, x = 6)$.

In Your Own Words

10. **Tell** how to graph the function $y = 2x - 1$.

Objectives & Examples

Upon completing this chapter, you should be able to:

● solve addition equations using models
(Lesson 12-1)

Solve $m + (-2) = -5$.

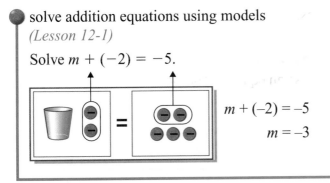

$$m + (-2) = -5$$
$$m = -3$$

● solve subtraction equations by using models
(Lesson 12-2)

Solve $t - 3 = 4$.

Rewrite as an addition equation.

$$t - 3 = 4 \rightarrow t + (-3) = 4$$

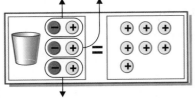

$$t + (-3) = 4$$

Since there are no negative counters on the right side, add 3 positive counters to each side to make 3 zero pairs on the left side. Then remove the zero pairs.

$$t + (-3) + 3 = 4 + 3$$
$$t = 7$$

● solve multiplication and division equations using models *(Lesson 12-3)*

Solve $3x = -9$.

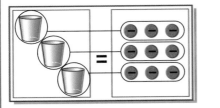

$$3x = -9$$
$$x = -3$$

Review Exercises

Use these exercises to review and prepare for the chapter test.

Solve each equation. Use cups and counters if necessary.

11. $x + 2 = 5$ **12.** $p + 7 = 4$

13. $6 + r = -7$ **14.** $-2 + w = -5$

15. $a + (-4) = -7$ **16.** $t + 2 = -6$

Solve each equation. Use cups and counters if necessary.

17. $x - 1 = 3$

18. $y - 3 = 11$

19. $f - 3 = -1$

20. $g - 4 = -5$

21. $h - 9 = -9$

22. $c - (-8) = -12$

Solve each equation. Use cups and counters if necessary.

23. $7q = 28$ **24.** $\frac{1}{3}d = 9$

25. $\frac{1}{5}k = -11$ **26.** $8x = 40$

27. $3m = -15$ **28.** $2.5g = 15$

29. $\frac{1}{4}y = -5$ **30.** $-4t = -4$

Objectives & Examples

Review Exercises

solve two-step equations using models
(Lesson 12-4)

Solve $2n + 1 = 7$.

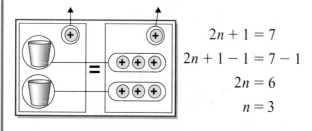

$$2n + 1 = 7$$
$$2n + 1 - 1 = 7 - 1$$
$$2n = 6$$
$$n = 3$$

Solve each equation. Use cups and counters if necessary.

31. $2r - 8 = 4$

32. $\frac{1}{5}k + 5 = 7$

33. $2x + 5 = -13$

34. $4n + 12.4 = -6.8$

---

complete function tables *(Lesson 12-5)*

Find the rule for the function table.

input (n)	output ($\blacksquare$)
−1	3
0	4
3	7

$$\left. \begin{array}{l} -1 + 4 = 3 \\ 0 + 4 = 4 \\ 3 + 4 = 7 \end{array} \right\} n + 4$$

Copy and complete each function table.

35.
n	$3n$
−1	$\blacksquare$
0	$\blacksquare$
2	$\blacksquare$

36.
n	$n - 2$
−1	$\blacksquare$
0	$\blacksquare$
5	$\blacksquare$

37. If a function rule is $2n + 1$, what is the output for $n = -3$?

Find the rule for each function table.

38.
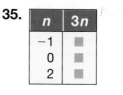

n	$\blacksquare$
−1	5
0	0
3	−15

39.
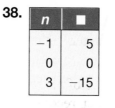

n	$\blacksquare$
−1	−5
0	−4
3	−1

---

graph functions from function tables
(Lesson 12-6)

Graph the function represented by the function table.

x	$x + 4$	output	ordered pair
−3	−3 + 4	1	(−3, 1)
0	0 + 4	4	(0, 4)
1	1 + 4	5	(1, 5)

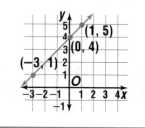

Copy and complete each function table. Then graph the function.

40.
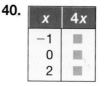

x	$4x$
−1	$\blacksquare$
0	$\blacksquare$
2	$\blacksquare$

41.
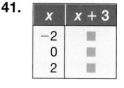

x	$x + 3$
−2	$\blacksquare$
0	$\blacksquare$
2	$\blacksquare$

Applications & Problem Solving

42. *Health* After dieting for 8 weeks and losing 17 pounds, Jeremy weighed 172 pounds. How much did he weigh before the diet? *(Lesson 12-1)*

43. *Use an Equation* Keith swam three times as many laps as Dan in the swim meet. Keith swam 24 laps. How many laps did Dan swim? *(Lesson 12-4B)*

44. *Money Matters* On her vacation, Mrs. Salgado planned to spend less than $120 each day on her hotel room and meals. She found a hotel room for $75 a night. How much can she spend on meals? *(Lesson 12-4)*

45. *Money Matters* Mr. Jackson got a new shipment of T-shirts to sell at his souvenir shop. The table shows the amount Mr. Jackson paid for each of the three types of T-shirts, and the selling price he marked on the shirts. Find the function rule he used. *(Lesson 12-5)*

Amount Paid	Selling Price
$4.00	$7.50
$5.50	$9.00
$6.00	$9.50

Alternative Assessment

Open Ended

Record the temperature of a cup of hot tap water to the nearest degree. Then record the temperature every minute for the next 10 minutes.

Graph the ordered pairs (time, temperature) on a coordinate plane.

Estimate the temperature after 30 minutes. Estimate again after 60 minutes.

Completing the CHAPTER Project

Use the following checklist to make sure your project is complete.

☑ You have included a list of the items in your collection.

☑ You have included your table or other display showing the value of at least five items in your collection.

☑ You may want to include pictures of the items in your collection.

Add any finishing touches that you would like to make your project attractive.

PORTFOLIO Select an item from this chapter that you feel shows your best work. Place it in your portfolio. Explain why you selected it.

A practice test for Chapter 12 is provided on page 606.

Section One: Multiple Choice

There are eleven multiple-choice questions in this section. Choose the best answer. If a correct answer is *not here*, mark the letter for **Not Here**.

1. Name the number of faces in a square pyramid.
   A 4
   B 5
   C 6
   D 7

2. What is the solution of the equation $5 + x = 3$?
   F -8
   G -2
   H 2
   J 8

3. Which is a reflection of the letter H?

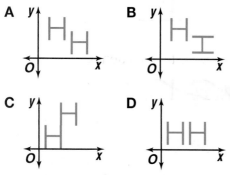

   A   **B**
   C   **D**

4. What numbers complete the function table?

n	$3n$
10	■
0	■
−5	■

   F 30, 0, −15
   G 30, 20, −10
   H 13, 3, −2
   J 13, 3, −10

5. Which ordered pair is represented by point M?

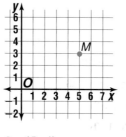

   A (5, 4)
   B (3, 5)
   C (2, 5)
   D (5, 3)

Please note that Questions 6–12 have five answer choices.

6. Diane had 10 coins in her pocket. There were 4 pennies, 1 nickel, 4 dimes, and 1 quarter. If she took out 1 coin at random, what is the probability it would be a dime?
   F $\frac{1}{10}$
   G $\frac{1}{4}$
   H $\frac{2}{5}$
   J $\frac{1}{2}$
   K Not Here

7. Sharon is knitting a sweater. She needs 8 skeins of red yarn that cost $1.89 each and 3 skeins of white yarn that cost $1.29 each. Which number sentence could be used to find S, the total cost in dollars?
   A $S = (1.89 \div 8) + (1.29 \div 3)$
   B $S = (8 \times 1.89) \times (3 \times 1.29)$
   C $S = (1.89 + 1.29) \times 11$
   D $S = (1.89 \div 8) \times (1.29 \div 3)$
   E $S = (8 \times 1.89) + (3 \times 1.29)$

8. Which best describes the two triangles?

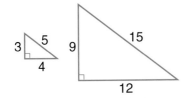

F The triangles are congruent.

G The triangles are reflections.

H The triangles are similar.

J The triangles are equilateral.

K The triangles are symmetrical.

9. Marty bought 5 packages of hamburger. If each package weighed 1.45 pounds, what was the weight of the 5 packages?

A 3.45 lb

B 6.45 lb

C 7.90 lb

D 7.25 lb

E Not Here

10. What is the perimeter of a triangle with sides of $3\frac{1}{2}$ cm, $2\frac{1}{2}$ cm, and $1\frac{3}{4}$ cm?

F 6 cm

G $6\frac{1}{2}$ cm

H $6\frac{3}{4}$ cm

J 7 cm

K Not Here

11. $0.0125 \div 1.7 =$

A 1   **B** 0.1

C 0.01   **D** 0.001

E Not Here

12. $-12 - (-10) =$

F -22 **G** -2

H 2   **J** 22

K Not Here

Test-Taking Tip   THE PRINCETON REVIEW

If you solve a problem and the answer you get is not one of the choices, check to see if your answer can be written in a different form. For example, if you solve a problem and get $\frac{x}{4}$ and it is not listed as a choice, try a different form such as $\frac{1}{4}x$. You may have the right answer, but in a different form.

Section Two: Free Response

This section contains six questions for which you will provide short answers. Write your answers on your paper.

13. Find the volume of the rectangular prism.

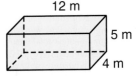

14. Complete the function table.

input (x)	output (x + 7)
−3	
1	
5	

15. In a survey, 17 out of 110 people said they liked boating. How many people would that be out of 100?

16. The fence around a pool measures 25 feet wide by 60 feet long. If the length is increased by 20 feet, what will the new dimensions be?

17. The base of a cone is a _____.

18. Make a function table for the rule $x - 2$ using 2, 5, and 8 as the input. Then graph the function.

GOING IN CIRCLES

What do these items have in common—a drum, a tambourine, and cymbals? You may have guessed they all have a circular shape. Circles are common in our daily lives.

What You'll Do

In this investigation, you will measure the circumference of common circular items. You will graph functions relating the radius, circumference, and area of circles. You will use your graphs to estimate the area of a circle.

Materials 🪱 string 📏 tape measure 🖩 calculator

🗒 grid paper ◑ circular objects

Procedure

1. Work in pairs. Measure the circumference of six different circular items to the nearest $\frac{1}{8}$ inch. To do this, wrap a string around the object and then measure the string or use a tape measure.

2. Make a table similar to the one below. Record the name and circumference of each item measured by you and your partner. Arrange the items from least to greatest circumference. If you use a spreadsheet, convert all fractions to decimals.

Name of Item	Radius (inches)	Circumference (inches)	Area (square inches)
pencil			
vegetable can			

3. Recall that the circumference of a circle can be found using the formula $C = 2\pi r$. Write and solve an equation to find the radius of each item. Record each radius in column 2 of your table.

4. Use the formula $A = \pi r^2$ to find the area of each circle. Record the area in column 4 of your table.

5. Draw three coordinate planes.
   a. For each item in the table, write the ordered pair, (radius, circumference). Then graph each ordered pair.

b. For each item in the table, write the ordered pair (radius, area). Then graph each ordered pair on the second coordinate plane.

c. For each item in the table, write the ordered pair (circumference, area). Then graph each ordered pair on the third coordinate plane.

6. Write a sentence or two describing each graph.

7. Measure the circumference of a new item. The circumference should be between the smallest and largest items you used for your table. Use the graph from Exercise 5c to estimate the area of the circle. How did your estimate compare to the actual area?

Making the Connection

Use the information about your circles as needed to help in these investigations.

Language Arts

Choose two circles from the investigation. Write directions for finding the side length of two squares that have approximately the same area as the circles.

Physical Education

In the sport of archery, athletes often use a target to improve their skill or for competitions. Research this sport and find the diameter of the target. Use your graphs to estimate the circumference and area of the target.

Go Further

• Use a graphing calculator to find equations for the three functions you graphed.

• Research the history of π.

• Find the area of each ring on an archery target.

*inter*NET Research **For current information on mathematics, visit:**
CONNECTION **www.glencoe.com/sec/math/mac/mathnet**

You may want to place your work on this investigation in your portfolio.

Using Probability

What you'll learn in Chapter 13

- to find and interpret the probability of an event or events,
- to solve problems by making a table,
- to predict actions of a larger group using a sample,
- to find probability using area models, and
- to find outcomes using lists, tree diagrams, and combinations.

CHAPTER Project

HERE TODAY, GONE TOMORROW

In order to survive in their environment, many animals are camouflaged, or hidden, from predators. In this project, you will use probability to explore how camouflage works. You will also research how one animal is camouflaged, make a poster about the animal, and write a report on how a probability experiment can be used to illustrate camouflage.

Getting Started

- You will need black paper, white paper, two sections of the classified ads from the newspaper, a hole punch, and poster board.
- Choose one animal that lives in the wild that you would like to research.
- Research the animal.
  - -Where does the animal live?
  - -What does the animal eat?
  - -Is survival of the animal aided by camouflage?
  - -Is the animal an endangered species? Find statistics that tell how many of the animals still exist. What is being done to help the animals survive?

Technology Tips

- Use an **electronic encyclopedia** or the **Internet** to research the animal.
- Use a **word processor** to write a report on how the probability experiment illustrates camouflage.
- Use **computer software** to design your animal poster.

*inter*NET CONNECTION Research For up-to-date information on animal species, visit:

www.glencoe.com/sec/math/mac/mathnet

Working on the Project

You can use what you'll learn in Chapter 13 to help you explore and understand camouflage.

Page	Exercise
518	27
529	23
543	Alternative Assessment

COOPERATIVE LEARNING

13-1A Fair and Unfair Games

A Preview of Lesson 13-1

⬡ 1 number cubes

In this lab, you will explore fair and unfair games. A *fair game* is a game in which players have an equal chance of winning. In an *unfair game*, players do not have an equal chance of winning.

TRY THIS

Work with a partner.

1 Rules of the game:
- Roll two number cubes.
- Add the two numbers that are face up.
- Player 1 gets one point if the sum is even. Player 2 gets one point if the sum is odd.
- Roll the number cubes 40 times.
- Record each sum in a chart like the one shown.

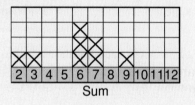

Sum

2 Rules of the game:
- Roll two number cubes.
- Multiply the two numbers that are face up.
- Player 1 gets one point if the product is even. Player 2 gets one point if the product is odd.
- Roll the number cubes 40 times.
- Record each product in a chart like the one shown.

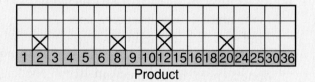

Product

ON YOUR OWN

1. Compare the number of even sums tossed to the number of odd sums tossed, and the number of even products tossed to the number of odd products tossed.

2. Which game do you think is fair? Which is unfair? Explain your reasoning.

3. Your charts should resemble a bar graph. Describe the shape of each of your graphs.

4. Which sum occurred most often? least often? Which product occurred most often? least often?

5. **Look Ahead** In Game 1, is getting a sum of 12 impossible, less likely than not getting 12, equally likely with not getting 12, more likely than not getting 12, or certain to occur? Explain your reasoning.

Theoretical Probability

What you'll learn

You'll learn to find and interpret the theoretical probability of an event.

When am I ever going to use this?

Knowing about probability can help you determine the chance of winning a game at a festival.

Word Wise

theoretical probability
outcomes
event
sample space
complementary events

LOOK BACK

You can refer to Lesson 5-4 to review ratios.

On a popular television game show, the contestants spin a wheel that is separated into equal sections marked 5¢, 10¢, 15¢, ... , $1. Each contestant gets two spins. The contestant closest to $1, without going over, wins a spot in the showcase. What is the **theoretical probability**, or chance, that the wheel will land on $1 the first time a contestant spins it?

There are twenty equally likely results, or **outcomes** on the wheel. The specific outcome or type of outcome is called an **event**. In this case, the outcomes are all of the different sections the wheel could stop on, and the event is landing on $1.

The set of all possible outcomes is called the **sample space**. The sample space for the wheel is {5¢, 10¢, 15¢, ..., $1}.

To find the probability of landing on $1, you can use a ratio.

Theoretical Probability	**Words:**	The theoretical probability of an event is the ratio of the number of ways the event can occur to the number of possible outcomes.
	Symbols:	$P(\text{event}) = \dfrac{\text{number of ways the event can occur}}{\text{number of possible outcomes}}$

In the sample space, there is one way to land on $1 and 20 possible outcomes. So, $P(\$1) = \frac{1}{20}$.

You can express the probability of an event as a fraction, decimal, or percent. The probability that an event will occur is a number from 0 to 1. You can interpret probabilities using a number line.

cannot occur	not too likely to occur	50-50 chance	very likely to occur	certain to occur
$P = 0$	$P = \frac{1}{4}$	$P = \frac{1}{2}$	$P = \frac{3}{4}$	$P = 1$
	$= 0.25$	$= 0.5$	$= 0.75$	
$= 0\%$	$= 25\%$	$= 50\%$	$= 75\%$	$= 100\%$

- A probability of 0 means that the event cannot occur.

- A probability of $\frac{1}{2}$ means that there is a 50-50 chance that the event will occur.

- A probability of 1 means that the event is certain to occur.

- The closer a probability is to 1, the more likely the event is to occur.

① A set of counters is numbered 1, 2, 3, ... , 10. Suppose you draw one counter without looking. Find the probability of choosing a number less than 3. Then tell how likely it is to choose a number less than 3.

There are two numbers less than 3: 1 and 2.

$\dfrac{2}{10}$ ← *number of ways to choose a number less than 3*
        ← *number of possible outcomes*

Therefore, $P(\text{less than } 3) = \dfrac{2}{10}$ or $\dfrac{1}{5}$.

Since the probability is close to $\dfrac{1}{4}$, it is not very likely to occur.

Study Hint

Reading Math The notation $P(\text{less than } 3)$ is read as the probability of a number less than 3 occurring.

---

② There are six equally likely outcomes on the spinner.

a. Find the probability of spinning blue.

b. Find the probability of spinning a color other than blue.

a. $\dfrac{2}{6}$ ← *number of ways to spin blue*
       ← *number of possible outcomes*

Therefore, $P(\text{blue}) = \dfrac{2}{6}$ or $\dfrac{1}{3}$.

b. $\dfrac{4}{6}$ ← *number of ways to spin a color other than blue*
       ← *number of possible outcomes*

Therefore, $P(\text{not blue}) = \dfrac{4}{6}$ or $\dfrac{2}{3}$.

In Example 2, either one or the other event must take place, but they cannot both happen at the same time. Notice that the sum of the probabilities of the two events is 1. These events are examples of **complementary events**.

Complementary Events	**Words:** Complementary events are two events in which either one or the other must take place, but they cannot both happen at the same time. The sum of their probabilities is 1.
	Symbols: $P(\text{event}_1) + P(\text{event}_2) = 1$

CONNECTION

③ Earth Science A meteorologist predicts a 30% chance of rain. What is the probability that it will not rain?

The probability of no rain is the complement of the probability of rain. To find the probability of no rain, you can use an equation.

$P(\text{rain}) + P(\text{no rain}) = 1$

$0.3 + P(\text{no rain}) = 1$ *Replace P(rain) with 30% or 0.3.*

$P(\text{no rain}) = 1 - 0.3$ *Subtract 0.3 from each side.*

$P(\text{no rain}) = 1.0 - 0.3$ *Rewrite 1 as 1.0.*

$P(\text{no rain}) = 0.7$

So, $P(\text{no rain}) = 0.7$ or 70%.

CHECK FOR UNDERSTANDING

Communicating Mathematics

Read and study the lesson to answer each question.

1. *Draw* a spinner that shows $P(\text{yellow}) = \frac{2}{5}$.

2. *Give two examples* of events in which the probability of each event occurring is 0.

Math Journal

3. *Write* a sentence describing the relationship between complementary events. Then give an example of complementary events.

Guided Practice

A number cube is marked with 1, 2, 3, 4, 5, and 6 on its faces. You roll the cube one time. Find the probability of each event. Then tell how likely the event is to happen.

4. $P(3)$ 5. $P(\text{greater than 2})$ 6. $P(\text{odd})$

7. $P(5 \text{ or } 6)$ 8. $P(\text{not } 6)$ 9. $P(7)$

10. *Industry* On a toy assembly line, 3% of the toys are defective. If an inspector selects a product at random, what is the probability the toy will pass inspection?

EXERCISES

Practice

Suppose you spin the spinner one time. Find the probability of each event. Then tell how likely the event is to happen.

11. $P(\text{red})$ 12. $P(\text{blue or red})$

13. $P(\text{green or red})$ 14. $P(\text{not yellow})$

A set of 26 counters is lettered a, b, c, ... , z. Suppose you choose one counter without looking. Find the probability of each event. Then tell how likely the event is to happen.

15. $P(\text{m, n, or p})$ 16. $P(26)$ 17. $P(\text{consonant})$

A set of 30 cards is numbered 1, 2, 3, ... , 30. Suppose you choose one card without looking. Find the probability of each event. Then describe the complementary event of each event, and find its probability.

18. $P(12)$　　　　**19.** $P(\text{odd})$　　　　**20.** $P(1 \text{ digit})$

21. $P(\text{integer})$　　　**22.** $P(\text{less than } 1)$　　　**23.** $P(\text{greater than } 18)$

24. When two number cubes are rolled, the probability of rolling doubles is $\frac{1}{6}$. What is the probability of not rolling doubles?

Applications and Problem Solving

25. *School* A multiple-choice test question has five possible answers. Suppose you guess at the answer.

　a. What is the probability of choosing the correct answer?

　b. Explain how you can use the problem-solving strategy *eliminating possibilities* to choose the correct answer.

　c. By eliminating possibilities, would the probability of choosing the correct answer increase or decrease? Explain.

26. *Advertising* A cereal company is having a contest. Each box of cereal contains one of the letters B, A, M, O. In every 20 boxes, there are four Bs, five As, ten Ms, and one O. What is the probability that a box of cereal will have a B?

27. *Working on the* **CHAPTER Project** Place one sheet of the classified ads on the floor. Use a hole punch to cut out 100 holes from each sheet of white paper, black paper, and classified ads. Scatter the circles on the spread-out sheet of classifieds.

　a. Randomly pick up as many paper circles as you can for 10 seconds. How many circles of each kind do you have?

　b. To find the probability of picking up a black circle, divide the number of black circles picked up by the total number of circles picked up. Find the probability of picking up each color.

　c. Write a paragraph that explains how probability relates to camouflage. Tell whether or not this activity is an example of experimental or theoretical probability.

For **Extra Practice,** see page 590.

28. *Critical Thinking* Write a problem in which the probability of an event occurring is 0.6.

Mixed Review

29. *Functions* Graph the function represented by the function table shown. *(Lesson 12-6)*

30. Find $-12 + (-6)$. *(Lesson 11-3)*

31. What is 32% of 148? *(Lesson 8-7)*

input	output
4	4
−2	−2
0	0

32. **Standardized Test Practice** Every class that sells 75 tickets to the school play earns an ice cream party. Kate's class has sold 47 tickets. How many more tickets must they sell to earn the ice cream party? *(Lesson 1-1)*

　A 23　　　　**B** 32　　　　**C** 28　　　　**D** 35　　　　**E** Not Here

BIOLOGICAL SCIENCE

Lyle Allard
Biological Science Technician

Lyle Allard is a biological science technician for the Fish Technology Center in Bozeman, Montana. At the Center, they are developing healthier fish foods. Mr. Allard takes care of experimental fish and watches their growth and behavior. He is a member of the Turtle Mountain Band of the Chippewa Tribe in North Dakota. One of his goals is to help Native American tribes improve their management of fish and wildlife.

To work in biological science, you will need a bachelor's degree in a biology-related field. Some careers in biological science are zoologist, ecologist, naturalist, park ranger, veterinarian, and zookeeper. A zoologist may use mathematics when recording or interpreting data about animals. If you like animals and being outdoors, then you may want to consider a career in biological science.

For more information:
U.S. Fish and Wildlife Service
1849 C Street
Washington, DC 20240

interNET CONNECTION
www.glencoe.com/sec/math/mac/mathnet

I love the outdoors and I think that it would be great to work with wildlife.

Your Turn

Interview a person who works with animals. Ask about the advantages and disadvantages of working with animals and how they use mathematics in their job. Then write a report that summarizes your interview.

13-2A Make a Table

A Preview of Lesson 13-2

Patsy and Adam are organizing their school's upcoming music festival. They are discussing how to conduct a survey to determine which songs the students would prefer to sing. Let's listen in!

Patsy

I have the list of songs to choose from. I think that we should ask all of the chorus members which of the songs they would like to sing.

There isn't enough time to ask everyone. Let's ask every fourth chorus member who comes into chorus practice tonight to choose six songs.

Adam

Song Survey		
Song	**Tally**	**Frequency**
Yesterday	IIII	4
You've Lost That Lovin' Feeling	IHI I	6
Georgia On My Mind	III	3
Bridge Over Troubled Water		0
A Whole New World	IHI III	8
Tomorrow	IHI	5
Just Around The River Bend	II	2
Yellow Rose Of Texas	IIII	4

OK. I'll make a frequency table like this to record the results of the survey. We'll choose the six songs with the most votes.

THINK ABOUT IT

Work with a partner.

1. *Analyze* the way Patsy and Adam conducted the survey. Do you think the survey reflected the opinions of the entire chorus? Why or why not?

2. *Describe* other types of information that can be concluded from the table.

3. *Tell* an advantage of organizing information in a table.

4. *Apply* the **make a table** strategy to solve the following problem.

   A list of test scores is shown.

71	89	65	77	79	98	84
86	70	97	93	80	91	72
100	75	73	86	99	77	68

   a. *Make a frequency table of the test scores.*

   b. *How many more students scored 71 to 80 than 91 to 100?*

For **Extra Practice,** see page 591.

ON YOUR OWN

5. The first step of the 4-step plan for problem solving asks you to *explore* the problem. **Tell** the advantages of exploring the set of data before you make a table.

6. *Write a Problem* that can be solved by making a table. Then ask a classmate to solve the problem by making a table.

7. *Explain* how the make a table strategy is used to help conduct the pet survey in Exercise 18 on page 525.

MIXED PROBLEM SOLVING

STRATEGIES

Look for a pattern.
Solve a simpler problem.
Act it out.
Guess and check.
Draw a diagram.
Make a chart.
Work backward.

Solve. Use any strategy.

8. *Number Sense* The difference between two whole numbers is 14. Their product is 1,800. Find the two numbers.

9. *Life Science* A snail at the bottom of a 10-foot hole crawls up 3 feet each day, but slips back 2 feet each night. How many days will it take the snail to reach the top of the hole and escape?

10. *School* The list shows the birth month of the students in Miss Miller's geography class.

June	October	May	April
April	May	October	June
July	August	April	May
March	June	September	July
July	April	December	March
June	October	January	June

   a. Make a frequency table of the students' birth months.

   b. How many of the students were born in April?

   c. How many more students were born in June than in August?

11. *Money Matters* Harris spent $5.69 on golf tees. About how much change should he expect to receive if he paid with a $20 bill?

12. *Sports* Coach Franco is deciding in which order the players will bat. She looks at the number of hits each player made in the previous game. Use the frequency table shown.

Number of Hits		
Player	**Tally**	**Frequency**
Parker	II	2
Martinez	III	3
Cruz	IIII	4
Plesich	II	2
Higgins	III	3
Reid	I	1
Hartley		0
Wilson	II	2

   a. Who had the most hits?

   b. Who had the least hits?

   c. Find the total number of hits in the previous game.

13. *Standardized Test Practice* There are 4 green, 2 purple, 1 orange, and 5 yellow marbles in a pouch. Forest chooses 1 marble at random, records its color, and replaces it. He repeats this process 25 times. Which color did Forest probably choose the greatest number of times?

   A orange

   B purple

   C green

   D yellow

Integration: Statistics
Making Predictions Using Samples

What you'll learn
You'll learn to predict the actions of a larger group using a sample.

When am I ever going to use this?
You can predict the number of left-handed students in your school by using a sample.

Word Wise
sample
population
random

The student council at Springfield Middle School is planning a year-end field trip to either Park A or Park B in San Antonio, Texas. Since the committee does not have time to survey every student to find their preference, they can survey a smaller group, or **sample**. The information from the sample will help them predict where the students would prefer to go on the field trip.

All of the students in the school make up the **population**. To predict the population's preference, the survey should represent the preferences of all of the students. Here are some suggestions.

- Survey every fifteenth student named on the school roster.

- Survey every tenth student who exits the school at the end of the day.

- Survey two students from each homeroom class.

Each of these methods is a way of making sure that the sample is **random**, or drawn by chance from the population.

Examples

1 Chase wanted to find out which place the students would prefer to go: Park A or Park B. He surveyed every twentieth student who entered the school at the beginning of the day. Is this a random sample?

Chase had no control over the order in which the students entered the school at the beginning of the day. By surveying every twentieth student, he avoided surveying an entire group of friends that might all prefer the same place. This is a random sample.

2 Chase wanted to find out which day students would prefer to go on the trip: Wednesday, Thursday, or Friday. He surveyed a small group of people standing in the hallway before school. Is this a random sample?

By surveying a group of students standing in the hallway, their preferences may be influenced by those in the group. This is *not* a random sample.

You can use the results of a survey to make predictions about the actions of the entire population.

Example

APPLICATION 3

School When Chase conducted the survey to find students' field trip preferences, he learned that 48 out of the 60 students surveyed preferred Park B.

a. What is the probability that any given student will want to go to Park B?

b. There are 395 students at Springfield Middle School. Predict how many students will want to go to Park B.

a. 48 out of 60, or $\frac{4}{5}$, prefer Park B. The probability is $\frac{4}{5}$, or 80%.

b. Use a proportion. Let s represent the number of students who will want to go to Park B.

$$\frac{48}{60} = \frac{s}{395}$$

$48 \times 395 = 60 \times s$ *Write the cross products.*

$18{,}960 = 60s$ *Multiply.*

$\frac{18{,}960}{60} = \frac{60s}{60}$ *Divide each side by 60.*

$316 = s$

Of the 395 students, about 316 will prefer to go to Park B.

LOOK BACK

You can refer to Lesson 8–2 to review proportions.

In the Mini-Lab, you will conduct a survey to predict the number of left-handed students in your school.

MINI-LAB

Work with a partner.

Try This

- Decide how you will choose a random sample to predict the number of left-handed students in your school.

- With your teacher's permission, conduct your survey.

Talk About It

1. Find the probability that a student selected at random from your sample is left-handed. Explain why this is an example of experimental probability.

2. Based on your sample, predict the number of left-handed students in your school.

3. Compare your results with the other groups. Explain why different groups may arrive at different conclusions.

Communicating
Mathematics

Read and study the lesson to answer each question.

1. *Tell* the difference between population and random sample.

2. *Explain* how you can use a random sample to predict the results of a school board election.

HANDS-ON
MATH

3. *Find* a newspaper article or advertisement that refers to a survey. Then write one or two sentences describing how the results of the survey can predict the actions of a larger group.

Guided Practice

Tell whether each of the following is a random sample. Explain your answer.

	Type of Survey	**Survey Location**
4.	favorite entertainment	Broadway show
5.	favorite flower	shopping center
6.	favorite holiday	fast-food restaurant

7. *Sports* In soccer, Karena scored 6 goals in her last 10 attempts.
   a. What is the probability of Karena scoring a goal on her next attempt?
   b. Suppose Karena attempts to score 15 goals. About how many goals will she make?

EXERCISES

Practice

Tell whether each of the following is a random sample. Explain your answer.

	Type of Survey	**Survey Location**
8.	favorite color	grocery store
9.	favorite hobby	model train store
10.	favorite musician	rock concert
11.	favorite sport	mall
12.	favorite season	public library
13.	favorite car	a specific car dealership
14.	favorite TV show	skating rink
15.	favorite candy bar	chocolate factory
16.	favorite food	Mexican restaurant

17. *Civics* Louisa conducted a survey to find out which political party people in her county preferred. She surveyed every tenth person standing in line at the Democratic headquarters. Is this a random sample? Explain.

18. *Life Science* Ashley conducted a survey of students' favorite pets from a random sample of 56 students at Park Middle School. The results are shown in the chart.
   a. What is the size of the sample?
   b. What is the probability that a student at Park Middle School prefers a gerbil?
   c. If there are 252 sixth graders, about how many will prefer gerbils?

Pet	Number
Dog	24
Cat	11
Gerbil	14
Bird	7

CONNECTION
For the latest ratings, visit:
www.glencoe.com/
sec/math/mac/mathnet

19. *Technology* A ratings company uses a sample of about 970 households in the United States to determine the rankings of TV shows. An estimated 38% of all TV sets in use were tuned to the top rated show. How many sets in the sample were tuned to the top rated show?

20. *Critical Thinking* A pre-election poll predicted that a certain candidate for mayor would receive 25% of the vote. The candidate received 15,248 votes. Estimate how many people voted in the mayoral election.

Mixed Review

21. **Standardized Test Practice** A bag contains 2 yellow counters, 1 red counter, 5 blue counters, and 3 green counters. Alec chooses 1 counter at random 20 times, records its color, and then replaces it in the bag. Which color did Alec probably choose the most number of times? *(Lesson 13-1)*
   A yellow   **B** red   **C** blue   **D** green

22. *Geometry* An angle measures 91.2°. Is it an obtuse or acute angle? *(Lesson 9-1)*

23. *Measurement* How many inches are in 5 yards? *(Lesson 5-6)*

For **Extra Practice**, see page 591.

CHAPTER 13

Mid-Chapter Self Test

A bag contains 4 blue marbles, 3 yellow marbles, 6 purple marbles, and 5 green marbles. One marble is chosen at random. Find the probability of each event. *(Lesson 13-1)*

1. *P*(yellow)

2. *P*(orange)

3. *P*(green or blue)

4. *P*(blue or purple)

5. *Technology* A quality control inspector found that 3 out of 50 computer keyboards were defective. *(Lesson 13-2)*
   a. What is the probability that a randomly-chosen computer keyboard will be defective?
   b. Suppose the company manufactured 15,000 computer keyboards. How many of the keyboards will be defective?

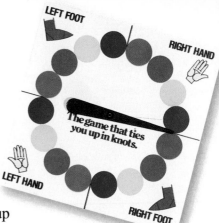

Integration: Geometry
Probability and Area

What you'll learn

You'll learn to find probability using area models.

When am I ever going to use this?

Knowing how to relate probability to the area can help you find the probability of a dart landing on a dartboard.

Have you ever played a popular floor game? If so, then you know that it is easy to get tied up in knots! The spinner shown is used to play the game. It has 16 equally likely outcomes. What is the probability of spinning left hand yellow?

There are 16 possible outcomes and 1 way the event can occur. Therefore, $P(\text{left hand, yellow}) = \frac{1}{16}$.

In Lesson 13-1, you learned to relate probability to spinners or number cubes. You can also relate probability to the areas of geometric shapes.

HANDS-ON MINI-LAB

Work with a partner.   dot paper   beans

Try This

- Draw two squares on dot paper similar to the ones shown.
- Drop 50 beans onto the paper, from about 8 inches above.
- Record the number of beans that land within the large square. Record the number of beans that land within the small square. Do not count those that land outside the squares.

Talk About It

1. Find the ratio $\dfrac{\text{number of beans in a small square}}{\text{number of beans in a large square}}$.

2. Find the ratio $\dfrac{\text{area of small square}}{\text{area of large square}}$.

3. Compare the two ratios.

4. Combine your results with another group and compare the ratios.

With a very large sample, the experimental probability should be very close to the theoretical probability.

The results of the Mini-Lab suggest the following conclusion.

$$\frac{\text{number landing in small square}}{\text{number landing in large square}} = \frac{\text{area of small square}}{\text{area of large square}}$$

1 The figure shown represents a dartboard.

a. Suppose you threw a dart randomly at the board and it hits the board. Find the probability of the dart landing in region C.

b. Suppose you threw the dart 500 times. How many times would you expect it to land in region C?

a. $P(\text{region C}) = \dfrac{\text{area of region C}}{\text{area of target}}$

$= \dfrac{8}{16} \text{ or } \dfrac{1}{2}$

b. Let c = times the dart lands in region C.

$\dfrac{c}{500} = \dfrac{1}{2}$ ← *area of region C*
            ← *area of dart board*

$c \times 2 = 500 \times 1$ *Find the cross products.*

$2c = 500$

$c = 250$ *Divide each side by 2.*

So, out of 500 times, the dart should land in region C about 250 times.

APPLICATION

Real World

2 **Games** At a county fair, a dart is thrown at the dartboard shown. It is equally likely that the dart will land anywhere on the dartboard. Find the probability that a dart will land in the red region. Then tell how likely the event is to happen.

$P(\text{red region}) = \dfrac{\text{area of red region}}{\text{area of target}}$

Area of red region $= \pi r^2$

$\approx 3.14 \times 1 \times 1$

$\approx 3.14 \text{ in}^2$

Area of target $= s^2$

$= 5 \times 5 \text{ or } 25 \text{ in}^2$

5 in.

1 in.

5 in.

$P(\text{red region}) \approx \dfrac{3.14}{25}$

$\approx \dfrac{3}{24} \text{ or } \dfrac{1}{8}$ *3 and 24 are compatible numbers.*

The probability of landing in the red region is about $\dfrac{1}{8}$ or 12.5%.

The event is not very likely to happen.

Cultural Kaleidoscope

During the 17th century, mathematicians Blaise Pascal and Pierre de Fermat studied the probability of games of chance.

CHECK FOR UNDERSTANDING

Communicating Mathematics

Read and study the lesson to answer each question.

1. *Tell* the probability of a randomly-thrown dart landing in region B of the figure shown.

2. *Draw* a dartboard in which the probability of landing in a shaded region is $\dfrac{3}{4}$.

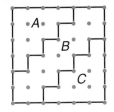

3. Use your dartboard from Exercise 2. Drop a bean 50 times onto the target. Then find the ratio that compares the number of times the bean lands in the shaded region to the number of times the bean lands in the target. How does this ratio compare to $\frac{3}{4}$, the theoretical probability of landing in the shaded region?

Guided Practice

Each figure represents a dartboard. It is equally likely that a dart will land anywhere on the dartboard. Find the probability of a randomly-thrown dart landing in the shaded region.

4.   **5.**

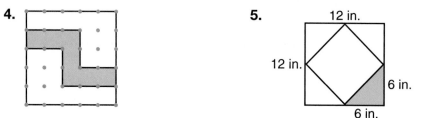

12 in.

12 in.

6 in.

6 in.

6. Suppose you threw a dart 150 times at the dartboard in Exercise 4. How many times would you expect it to land in the shaded region?

7. Suppose you threw a dart 150 times at the dartboard in Exercise 5. How many times would you expect it to land in the shaded region?

8. *Games* A randomly-thrown dart is thrown at the dartboard shown. It is equally likely that the dart will land anywhere on the dartboard. Find the probability that a dart will land in the shaded region of the figure shown. Then tell how likely the event is to happen.

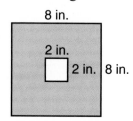

8 in.

2 in.

2 in. | 8 in.

EXERCISES

Practice

Each figure represents a dartboard. It is equally likely that a dart will land anywhere on the dartboard. Find the probability of a randomly-thrown dart landing in the shaded region.

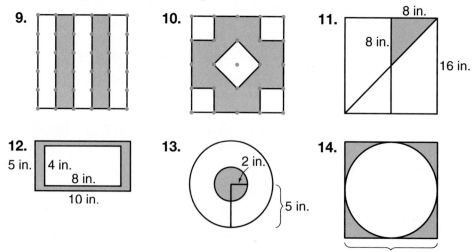

9.

10.

11.
8 in.
8 in.
16 in.

12.
5 in. | 4 in.
8 in.
10 in.

13.
2 in.
5 in.

14.
3 in.

Copy and complete.

	Dartboard	Times Dart is Thrown	Times Expected to Land in Shaded Region
15.	Exercise 9	200	
16.	Exercise 10	200	
17.	Exercise 11	250	
18.	Exercise 12	250	
19.	Exercise 13	275	
20.	Exercise 14	275	

Applications and Problem Solving

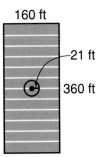

21. *Entertainment* A sky diver is the featured entertainer for the half-time show of the Johnstown High School homecoming football game. It is equally likely that the sky diver will land on any point of the field. Find the probability that the sky diver will land inside the circle.

160 ft
21 ft
360 ft

22. *Games* A dart is thrown at the dartboard shown. It is equally likely that the dart will land anywhere on the dartboard. What is the probability of the dart landing on the shaded region? Then tell how likely the event is to happen.

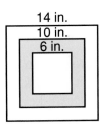

14 in.
10 in.
6 in.

23. *Working on the* CHAPTER Project Refer to page 518.
 a. Suppose you had randomly picked up 100 circles. How many circles of each color would you expect to pick up?
 b. Suppose you had randomly picked up 250 circles. How many circles of each color would you expect to pick up?
 c. Write a paragraph that explains how this experiment relates to camouflage.

24. *Critical Thinking* Suppose you made a dartboard from the tangram shown. If you throw 200 darts at the dartboard, how many would you expect to land in each region? Assume it is equally likely that the dart lands anywhere in the tangram.

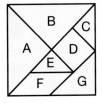

Mixed Review

25. *Statistics* Tell whether a survey on favorite jeans at a specific brand of jeans store is a random survey. Explain your answer. *(Lesson 13-2)*

26. **Standardized Test Practice** The area of a rectangle is 153 square feet, and the width is 9 feet. Use the formula $A = \ell w$ to find the length of the rectangle. *(Lesson 12-3)*

 A 17 ft **B** 144 ft **C** 162 ft **D** 1,377 ft

27. *Geometry* What is the area of a triangle whose base is 52 feet and whose height is 38 feet? *(Lesson 10-2)*

For **Extra Practice,** see page 591.

Lesson 13-3 Integration: Geometry Probability and Area **529**

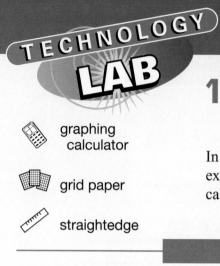
GRAPHING CALCULATORS

13-3B Probability

A Follow-Up of Lesson 13-3

In Lesson 13-3, you used area to find a probability. You can experiment with area and probability by using a graphing calculator.

graphing calculator

grid paper

straightedge

TRY THIS

Work with a partner.

- Use a straightedge to draw the figure shown on grid paper. Then shade the figure as shown.

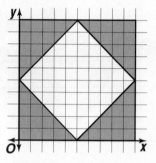

- Use the program to generate 50 random ordered pairs. When you run the program, it will ask you how many ordered pairs you want: type "50" and push the ENTER key. It will then ask you for a seed number. Type in any number and push the ENTER key. This helps the calculator pick random numbers. The calculator will now start giving you pairs of coordinates. The first number is x and the second number is y. Each time you push the ENTER key, it will give you another pair of numbers until it has given you 50 of them.

```
:Fix 1
:Disp "Number of Pairs"
:Input P
:Disp "Seed Number"
:Input N
:N → rand
:For (J,1,P,1)
:10 * rand → X
:Disp X
:10 * rand → Y
:Disp Y
:Disp " "
:Pause
:End
```

- Graph each pair on the grid paper. Keep a tally of how many points are in the shaded regions and how many points are in the unshaded region.

ON YOUR OWN

1. Write a ratio that compares the number of points graphed in the shaded regions to the total number of points graphed.
2. Use the formulas for the area of a triangle and the area of a square to find the ratio that compares the area of the shaded regions to the area of the square.
3. How do the ratios compare?
4. Generate and graph 50 more ordered pairs by changing the seed number. Add the results to your previous results. What fraction of the 100 points generated are in the shaded region? How does this fraction compare to the fraction of the square that is shaded?
5. Suppose you wanted to plot points on a grid in which the x- and y-coordinates went from 0 to 20, instead of 0 to 10. What change would you need to make in the program?

13-4 Finding Outcomes

What you'll learn

You'll learn to find outcomes using lists, tree diagrams, and combinations.

When am I ever going to use this?

Knowing how to find the number of possible outcomes can help you determine options when ordering a pizza.

Word Wise

tree diagram
combinations

Mr. Mason wrote the suffixes *-ness, -er,* and *-ly* on the chalkboard. He then asked his students to form words using the given suffixes and the words *quick, slow,* and *happy.* How many different words can be made using the suffixes and the words?

To solve the problem, you can determine the sample space by making a list.

> quickness   quicker   quickly
> slowness   slower   slowly
> happiness   happier   happily

There are 9 different words that can be formed.

You can also use a **tree diagram** to show the sample space.

Word	Suffix	Outcome
quick	-ness	quickness
	-er	quicker
	-ly	quickly
slow	-ness	slowness
	-er	slower
	-ly	slowly
happy	-ness	happiness
	-er	happier
	-ly	happily

Study Hint

Problem Solving
Use the draw-a-diagram strategy.

Example ①

At Pizza House, you can order a thin crust or deep dish pizza with cheese, pepperoni, or sausage. Draw a tree diagram that shows all of the ways you can order a one-topping pizza.

Pizza	Topping	Outcome
thin crust (T)	cheese (C)	TC
	pepperoni (P)	TP
	sausage (S)	TS
deep dish (D)	cheese (C)	DC
	pepperoni (P)	DP
	sausage (S)	DS

Study Hint

Reading Math The outcome TC means Thin crust Cheese pizza.

There are six ways to order a one-topping pizza.

2 Each spinner is spun once. Find the probability of spinning blue, green, and red in that order.

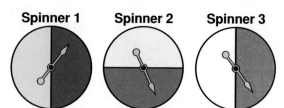

Find all of the possible outcomes.

Spinner 1	Spinner 2	Spinner 3	Outcome

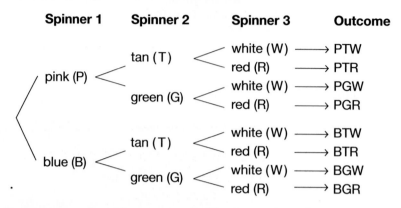

pink (P)
- tan (T)
  - white (W) ⟶ PTW
  - red (R) ⟶ PTR
- green (G)
  - white (W) ⟶ PGW
  - red (R) ⟶ PGR

blue (B)
- tan (T)
  - white (W) ⟶ BTW
  - red (R) ⟶ BTR
- green (G)
  - white (W) ⟶ BGW
  - red (R) ⟶ BGR

One outcome has blue, green, red in that order. There are eight possible outcomes. Therefore, $P(\text{BGR}) = \frac{1}{8}$.

Arrangements or listings where order is not important are called **combinations**. Combinations are another way to determine a sample space. To find combinations, you can make a list.

3 **Sports** At Teyas Middle School, there are five members on the gymnastics team. Coach Reyes must choose two of the members to be team captains. In how many ways can he choose two captains from five team members?

Let V, W, X, Y, and Z represent the team members. List all of the ways two team captains can be chosen.

VW	VX	VY	VZ	WV
WX	WY	WZ	XV	XW
XY	XZ	YV	YW	YX
YZ	ZV	ZW	ZX	ZY

Then count all of the *different* arrangements. Since order is not important, VW and WV are the same.

VW	VX	VY	VZ	WX
WY	WZ	XY	XZ	YZ

There are 10 ways Coach Reyes can choose two team captains.

Communicating Mathematics

Read and study the lesson to answer each question.

1. *Write a Problem* that can be solved by using the tree diagram shown.

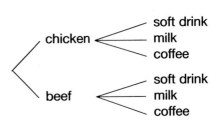

2. *Draw* a tree diagram that shows the results of tossing a nickel, a dime, and a quarter. Then find the probability of getting three tails.

3. *Write* about the advantages of using a tree diagram to count outcomes. Are there any disadvantages of tree diagrams?

Guided Practice

For each situation, make a list and draw a tree diagram to show the sample space.

4. a choice of a hot dog or hamburger and a choice of iced tea or lemonade

5. a choice of a blue or red shirt with blue, black, or red shorts

6. How many ways can a person choose two magazines from four magazines?

7. *School* A science quiz has one multiple-choice question with answer choices A, B, and C, and two true/false questions.

    a. Draw a tree diagram that shows all of the ways a student can answer the questions.

    b. What is the probability of answering all three questions correctly by guessing?

Family Activity

Make a diagram of your family tree going back to your great-grandparents. Explain how your diagram is like a tree diagram.

Practice

For each situation, make a list and draw a tree diagram to show the sample space.

8. a choice of lemon, apple, or pecan pie with milk or tea

9. a choice of a leather or nylon backpack in purple, green, black, or brown

10. rolling two number cubes

11. spinning each spinner once

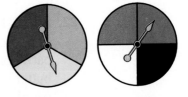

12. a choice of a portable or stationary basketball hoop with a graphite or acrylic backboard

13. a choice of a tower or spinner CD holder made from wood or plastic that holds 60, 100, or 250 CDs

14. How many different combinations of 2 recipes can a chef choose from 3 different recipes?

15. How many different ways can a person choose 3 kittens from a litter of 5 kittens?

16. How many different ways can a teacher choose 3 three-dimensional figures to put on a quiz from 4 different three-dimensional figures?

Applications and Problem Solving

17. *Food* At Otani Sushi Restaurant, customers can choose from the following six Nigiri-Sushi, or fish: Maguro (tuna), Ika (squid), Tako (octopus), Ebi (shrimp), Kani (crab), and Saba (mackerel). In how many ways can a customer choose two of the six Nigiri-Sushi?

18. *Fashion* Awenasa is buying a new navy sweater. The sweater will coordinate with her black, white, red, and yellow blouses and her blue, plaid, and black pants.

   a. How many different outfits will she have?

   b. Suppose Awenasa chooses one blouse and one pair of pants at random. What is the probability that she will choose the yellow blouse and the blue pants or the red blouse and the black pants?

19. *Critical Thinking* There are three 2-sided counters in a cup. One is red on one side and blue on the other, one is red and white, and the third is blue and white. Amiri shakes the cup and tosses out the chips. He receives one point if the chips are all different colors. Irene receives one point if two chips match. Use a tree diagram to help you determine whether or not this is a fair game. If this is not a fair game, how could you make it fair?

Mixed Review

20. *Probability* The figure shown represents a dartboard. It is equally likely that the dart will land anywhere on the dartboard. Find the probability of a randomly-thrown dart landing in the shaded region. *(Lesson 13-3)*

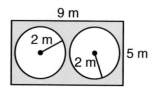

21. *Standardized Test Practice* Solve $n = -145 \div (-29)$. *(Lesson 11-6)*

   A -5 **B** 4 **C** 5 **D** -3 **E** Not Here

22. *Food* Refer to the graph. Suppose 2,000 people participated in the peanut butter survey. How many people can be expected to prefer creamy peanut butter? *(Lesson 8-6)*

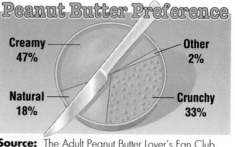

Peanut Butter Preference

Creamy 47%
Other 2%
Natural 18%
Crunchy 33%

Source: The Adult Peanut Butter Lover's Fan Club, Peanut Advisory Board

For **Extra Practice**, see page 592.

COOPERATIVE LEARNING

13-4B Simulations

A Follow-Up of Lesson 13-4

⬤ two-colored
counters

▢ cups

A simulation is an application of the Acting It Out problem-solving strategy.

A *simulation* is a way of acting out a problem. You can conduct a simulation by using manipulatives such as counters or number cubes. In this lab, you will conduct a simulation to explore the probability that in a family with three children, at least two of them are girls.

It is equally likely that a boy or a girl will be born. Similarly, it is equally likely that a two-colored counter will land on one color or the other. So, you can toss a two-colored counter to simulate outcomes.

TRY THIS

Work with a partner.

To explore the probability that at least two of the three children in a family are girls, follow these steps.

	Outcome		
Trial 1	B	B	G
Trial 2	G	G	G
Trial 3	G	G	B
Trial 4	B	B	G

Step 1 Place three counters in a cup and toss them onto your desk.

Step 2 Count the number of counters that land with the red side up. This represents the number of boys. Count the number of counters that land with the yellow side up. This represents the number of girls.

Step 3 Record the results in a table like the one shown.

Step 4 Repeat Steps 1-3 until you have 50 trials.

Suppose 23 of the 50 trials have at least two girls. The experimental probability that at least two of the three children in a family are girls is $\frac{23}{50}$.

You can refer to Lesson 5-4B for information on experimental probability.

ON YOUR OWN

1. Based on your results of the simulation, what is the experimental probability that, in a family with three children, at least two of them are girls?

2. Make a list showing all of the possible outcomes of the experiment. What is the theoretical probability that, in a family with three children, at least two of them are girls?

3. Compare the experimental probability with the theoretical probability.

4. **Reflect Back** Find the probability that in a family with four children, all four of them are girls. Then describe a simulation that you can use to explore this probability. Conduct the simulation. Record and explain your results.

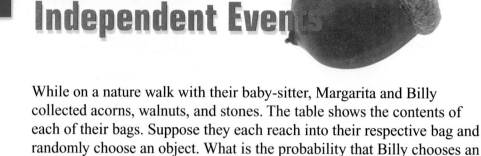

What you'll learn

You'll learn to find the probability of independent events.

When am I ever going to use this?

Knowing how to find the probability of independent events can help you determine the chance of rolling doubles in Monopoly.

Word Wise

independent event

While on a nature walk with their baby-sitter, Margarita and Billy collected acorns, walnuts, and stones. The table shows the contents of each of their bags. Suppose they each reach into their respective bag and randomly choose an object. What is the probability that Billy chooses an acorn and Margarita chooses a stone?

Name	Acorns	Walnuts	Stones
Billy	6	8	10
Margarita	12	12	16

The object that Billy chooses does not affect the object that Margarita chooses. They are called **independent events**.

To find the probability of two independent events, you multiply their probabilities.

Probability of Two Independent Events	**Words:**	The probability of two independent events, A and B, is the product of the probability of event A and the probability of event B.
	Symbols:	$P(A \text{ and } B) = P(A) \cdot P(B)$

$P(\text{acorn}) = \dfrac{6}{24}$ ← *number of ways Billy can choose an acorn*
← *number of possible outcomes: 6 + 10 + 8 = 24*

$P(\text{stone}) = \dfrac{16}{40}$ ← *number of ways Margarita can choose a stone*
← *number of possible outcomes: 12 + 16 + 12 = 40*

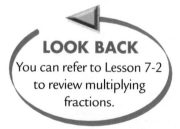

LOOK BACK

You can refer to Lesson 7-2 to review multiplying fractions.

Multiply to find the probability.

$$\frac{6}{24} \times \frac{16}{40} = \frac{\overset{1}{\cancel{6}}}{\underset{4}{\cancel{24}}} \times \frac{\overset{2}{\cancel{16}}}{\underset{5}{\cancel{40}}} \quad \textit{The GCF of 6 and 24 is 6.}$$
$$\textit{The GCF of 16 and 40 is 8.}$$
$$= \frac{2}{20} \text{ or } \frac{1}{10}$$

The probability that Billy chooses an acorn and Margarita chooses a stone is $\frac{1}{10}$. The event is not too likely to occur.

1 A number cube is rolled, and the spinner shown is spun. Find the probability of rolling 5 and landing on a consonant.

$$P(5) = \frac{1}{6} \qquad P(\text{consonant}) = \frac{3}{4}$$

$$P(5 \text{ and consonant}) = \frac{1}{6} \times \frac{3}{4}$$

$$= \frac{3}{24} \text{ or } \frac{1}{8}$$

The probability of rolling a 5 and landing on a consonant is $\frac{1}{8}$.

CONNECTION

2 **Life Science** Owls lay two to five rounded, white eggs. Suppose an owl lays two eggs. What is the probability that the first egg hatched is a female owl and the second egg hatched is a male owl?

Explore   You know the probability of female and the probability of male. You need to find the probability that the first egg hatched is a female owl and the second egg hatched is a male owl.

Plan   Multiply to find the probability.

Solve $P(\text{female owl}) = \frac{1}{2}$ $P(\text{male owl}) = \frac{1}{2}$

$$\frac{1}{2} \times \frac{1}{2} = \frac{1}{4}$$

So, P(female owl, then male owl) $= \frac{1}{4}$.

Examine   Use a tree diagram. One outcome has a female owl and then a male owl. There are four possible outcomes.

Therefore, the probability is $\frac{1}{4}$.

Did you know? The largest owls in the world are eagle owls, which grow to be about 710 millimeters long. The smallest owls are pygmy owls, which grow to be about 120 millimeters long. While eagle owls are found in Europe, Asia, and North Africa, pygmy owls are found in South and Central America.

First Egg	Second Egg	Outcome
	female ⟶	FF
female (F) <		
	male ⟶	FM
	female ⟶	MF
male (M) <		
	male ⟶	MM

Communicating Mathematics

Read and study the lesson to answer each question.

1. *Explain* how to find the probability of two independent events.

2. *Write* a probability problem involving two events that are independent. Then ask a classmate to solve your problem.

3. *You Decide* Two number cubes are rolled. Madeline says that the probability of the first number cube landing on 4 or 5 and the second number cube landing on 6 is $\frac{1}{9}$. Flora disagrees. She says that the probability is $\frac{1}{18}$. Who is correct? Explain.

Guided Practice

The two spinners shown are spun. Find the probability of each event.

4. P(green and 5)

5. P(orange and odd)

6. P(blue and prime)

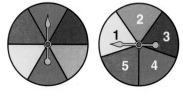

7. *School* A pencil box contains 4 lead pencils, 2 pens, 3 color pencils, and 6 markers. Another pencil box contains 3 pens, 2 markers, and 5 lead pencils.
   a. What is the probability of choosing a marker from the first pencil box?
   b. What is the probability of choosing a pen from the second pencil box?
   c. Find the probability of choosing a marker from the first box and then a pen from the second box.

Practice

A coin is tossed and a number cube is rolled. Find the probability of each event.

8. P(tails and 3)

9. P(heads and odd)

10. P(heads and 2 or 4)

11. P(heads and 7)

12. P(tails and prime)

13. P(heads or tails and composite)

The spinner shown is spun, and a card is chosen from the set of cards shown. Find the probability of each event.

14. P(2-digit and consonant)

15. P(even and vowel)

16. P(less than 14 and T)

 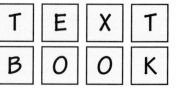

17. A red number cube and a green one are rolled. Find the probability of tossing a number greater than 4 on the green cube and a number other than 1 on the red one.

Applications and Problem Solving

18. *School* A quiz has one true/false question and one multiple-choice question with possible answer choices a, b, c, and d. If you guess each answer, what is the probability of answering both questions correctly?

19. *Sports* The probability that the Eagles will win on Saturday is 0.6. The probability that the Lions will win on Sunday is 0.8.
  a. What is the probability that both teams will win?
  b. What is the probability that both teams will lose?

20. *Critical Thinking* Five students have no absences for the first semester. They are able to draw from a bag that has 7 movie tickets and 3 gift certificates. Each person keeps what is drawn. Paul and Kikuyu are the first two students to draw. What is the probability that they both draw movie tickets?

Mixed Review

21. How many ways can a person choose 3 videos from a stack of 6 videos? *(Lesson 13-4)*

22. *Geometry* If a circle has a radius of 27 feet, what is its circumference? *(Lesson 7-4)*

23. **Standardized Test Practice** Nathan walks $\frac{3}{4}$ of a mile to school. James walks $\frac{5}{8}$ of a mile to school. How much farther does Nathan walk to school than James? *(Lesson 6-4)*

  A $\frac{1}{2}$ mile   **B** $\frac{3}{8}$ mile   **C** $\frac{1}{4}$ mile   **D** $\frac{1}{8}$ mile   **E** Not Here

For **Extra Practice,** see page 592.

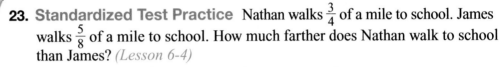

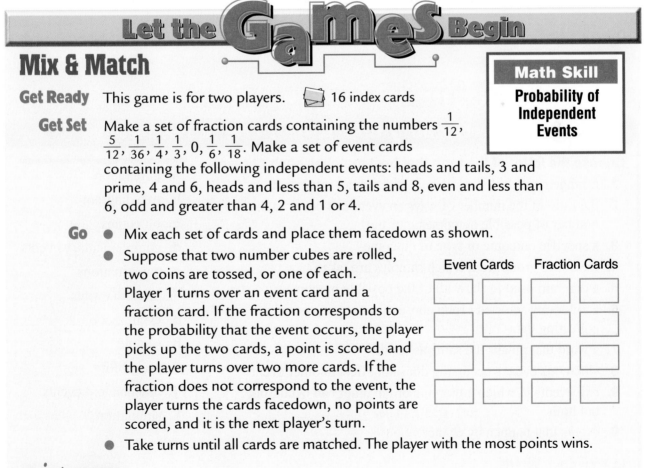

Mix & Match

Math Skill
Probability of Independent Events

Get Ready This game is for two players.   16 index cards

Get Set Make a set of fraction cards containing the numbers $\frac{1}{12}$, $\frac{5}{12}$, $\frac{1}{36}$, $\frac{1}{4}$, $\frac{1}{3}$, 0, $\frac{1}{6}$, $\frac{1}{18}$. Make a set of event cards containing the following independent events: heads and tails, 3 and prime, 4 and 6, heads and less than 5, tails and 8, even and less than 6, odd and greater than 4, 2 and 1 or 4.

Go ● Mix each set of cards and place them facedown as shown.

  ● Suppose that two number cubes are rolled, two coins are tossed, or one of each. Player 1 turns over an event card and a fraction card. If the fraction corresponds to the probability that the event occurs, the player picks up the two cards, a point is scored, and the player turns over two more cards. If the fraction does not correspond to the event, the player turns the cards facedown, no points are scored, and it is the next player's turn.

  Event Cards   Fraction Cards

  ● Take turns until all cards are matched. The player with the most points wins.

*inter*NET
CONNECTION Visit www.glencoe.com/sec/math/mac/mathnet for more games.

CHAPTER 13
Study Guide and Assessment

*inter*NET
CONNECTION
Chapter Review For additional lesson-by-lesson review, visit:
www.glencoe.com/sec/math/mac/mathnet

Vocabulary

After completing this chapter, you should be able to define each term, concept, or phrase and give an example or two of each.

Problem Solving
make a table (pp. 520-521)

Statistics and Probability
combinations (p. 532)
complementary events (p. 516)
event (p. 515)
fair game (p. 514)
independent events (p. 536)

outcomes (p. 515)
population (p. 522)
random (p. 522)
sample (p. 522)
sample space (p. 515)
simulation (p. 535)
theoretical probability (p. 515)
tree diagram (p. 531)
unfair game (p. 514)

Understanding and Using the Vocabulary

Choose the letter of the term that best matches each phrase.

1. arrangements or listings where order is not important
2. the ratio of the number of ways an event can occur to the number of possible outcomes
3. a specific outcome or type of outcome
4. the entire group from which samples are taken
5. a diagram used to show all of the possible outcomes
6. a randomly selected group chosen for the purpose of collecting data
7. a word that means the same as results
8. when one event's occurring does not affect another event
9. two events in which either one or the other can occur, but not both
10. events that happen by chance

a. population

b. tree diagram

c. outcomes

d. complementary events

e. combinations

 f. random events

g. event

h. sample

 i. probability

j. independent events

k. simulation

In Your Own Words

11. **Explain** how to find the probability of two independent events.

Objectives & Examples

Upon completing this chapter, you should be able to:

● find and interpret the theoretical probability of an event *(Lesson 13-1)*

There are six equally likely outcomes on the spinner. Find the probability of spinning blue.

$\dfrac{1}{6}$ ← *number of ways to spin blue*
     ← *number of possible outcomes*

So, $P(\text{blue}) = \dfrac{1}{6}$.

● predict the actions of a larger group using a sample *(Lesson 13-2)*

If 12 out of 50 people prefer to watch TV after 11 P.M., how many people out of 1,000 would prefer to watch TV after 11 P.M.?

Let p represent the number of people who would prefer to watch TV after 11 P.M.

$$\frac{12}{50} = \frac{p}{1,000}$$
$$12 \times 1,000 = 50 \times p$$
$$12,000 = 50p$$
$$\frac{12,000}{50} = \frac{50p}{50}$$
$$240 = p$$

Of the 1,000 people, 240 would prefer to watch TV after 11 P.M.

Review Exercises

Use these exercises to review and prepare for the chapter test.

Suppose you spin the spinner shown at the left once. Find the probability of each event. Then tell how likely the event is to happen.

12. $P(\text{green})$ **13.** $P(\text{red or white})$

14. $P(\text{pink})$ **15.** $P(\text{not blue})$

A bag contains a nickel, dime, and penny. Suppose you choose one coin without looking. Find the probability of each event. Then tell how likely the event is to happen.

16. $P(\text{nickel})$ **17.** $P(\text{dime or penny})$

Tell whether each of the following is a random survey. Explain your answer.

	Type of Survey	Survey Location
18.	favorite fast food	beach
19.	favorite sport	football game
20.	favorite pet	mall

21. The results of a survey showed that 14 out of 40 students at West Middle School are interested in publishing a school newspaper.

 a. What is the probability that a student at this school would be interested in publishing a school newspaper?

 b. If there are 420 students, how many would be interested in publishing a school newspaper?

Objectives & Examples

Review Exercises

● find probability using area models
(Lesson 13-3)

The figure shown represents a dartboard. Find the probability that a randomly-thrown dart lands in the shaded region.

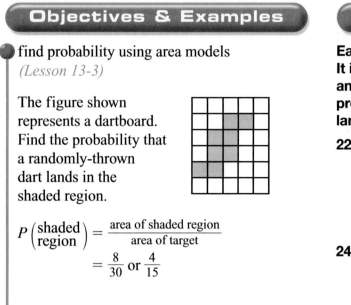

$$P\left(\begin{array}{c}\text{shaded} \\ \text{region}\end{array}\right) = \frac{\text{area of shaded region}}{\text{area of target}}$$

$$= \frac{8}{30} \text{ or } \frac{4}{15}$$

Each figure represents a dartboard. It is equally likely that a dart will land anywhere on the dartboard. Find the probability of a randomly-thrown dart landing in the shaded region.

22. **23.**

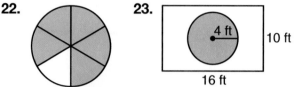

24. Suppose you threw a dart 150 times at the dartboard in Exercise 22. How many times would you expect it to land in the shaded region?

● find outcomes using lists, tree diagrams, and combinations *(Lesson 13-4)*

In how many ways can a person choose a 1-dip ice cream cone from two types of cones and three flavors of ice cream?

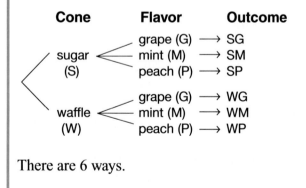

Cone	Flavor	Outcome
sugar (S)	grape (G) ⟶	SG
	mint (M) ⟶	SM
	peach (P) ⟶	SP
waffle (W)	grape (G) ⟶	WG
	mint (M) ⟶	WM
	peach (P) ⟶	WP

There are 6 ways.

For each situation, make a list and draw a tree diagram to show the sample space.

25. a choice of black or blue jeans in tapered leg, straight leg, or baggy style

26. a choice of soup or salad with beef, chicken, fish, or pasta

27. a choice of going to a basketball game, an amusement park, or a concert on a Friday or a Saturday

28. Tell how many different ways a person can choose four movies from a list of five movies.

● find the probability of independent events *(Lesson 13-5)*

Two number cubes are rolled.
Find P(odd and 4).

$P(\text{odd}) = \frac{1}{2}$ $P(4) = \frac{1}{6}$

$P(\text{odd and 4}) = \frac{1}{2} \times \frac{1}{6} \text{ or } \frac{1}{12}$

A coin is tossed and a number cube is rolled. Find the probability of each event.

29. P(heads and even)

30. P(heads and 1)

31. P(tails and 5 or 6)

32. P(tails and 2, 3, or 4)

Applications & Problem Solving

33. *Make a Table* The table shows the results of a survey on the number of hours teenagers use a computer on the weekend. *(Lesson 13-2A)*

Weekend Computer Use		
Hours	**Tally**	**Frequency**
0-2	卌 卌	10
3-4	卌 卌 I	11
5-6	卌	5
7 or more	卌 I	6

a. How many teenagers use a computer 5 or more hours during the weekend?

b. How many teenagers participated in the survey?

34. *Earth Science* The probability of rain on Saturday is 0.6. The probability of rain on Sunday is 0.3. What is the probability that it will rain on both days? *(Lesson 13-5)*

35. *School* The cheerleaders at Ross Middle School have three different skirts they can wear as part of their uniforms. One is purple, one is white, and one is yellow. They also have a choice of a yellow vest or a purple sweater to wear. Make a list and a tree diagram to show all of the possible uniform combinations. *(Lesson 13-4)*

36. *Party Planning* According to a survey of 30 students, 13 prefer chicken, and 17 prefer hot dogs. If 750 students will be attending the picnic, how many will prefer each selection? *(Lesson 13-2)*

Alternative Assessment

Open Ended

Suppose you are taking pictures to be used in the school yearbook. You can choose either color or black and white film. When the pictures are developed, you can choose a glossy or matte finish, and the pictures can be developed into 3×5, 4×6, 5×5, or 8×10 size prints. How can you determine the number of different ways you can develop a photo?

Draw a tree diagram to show all of the sample space. If the prints can also be cut with square or rounded corners, how many outcomes are possible?

A practice test for Chapter 13 is provided on page 607.

Completing the CHAPTER Project

Use the following checklist to make sure that your project is complete.

☑ The data about the animal is included on the poster.

☑ The paragraph that explains how probability relates to camouflage is included.

☑ Add any finishing touches that you would like to make your project complete.

 PORTFOLIO Select one of the assignments from this chapter and place it in your portfolio. Attach a note to it explaining why you selected it.

Section One: Multiple Choice

There are thirteen multiple-choice questions in this section. Choose the best answer. If a correct answer is *not here*, choose the letter for Not Here.

1. Find $6^3 \div 3^2 + 2$.

   A 6 **B** 216

   C 132 **D** 26

2. How many lines of symmetry does the figure have?

   F 2

   G 3

   H 4

   J 5

3. Which expression represents the volume of a cube with side r?

   A $6r$ **B** $6r^2$

   C r^3 **D** $3r^3$

4. What is the probability of a randomly-thrown dart landing in the shaded region of the figure?

   F 30%

   G 50%

   H 60%

   J 75%

5. Which two-dimensional figure best describes the base of a cone?

   A square

   B circle

   C triangle

   D rectangle

6. Which number is missing from this sequence?

   $\dots, 27, 81, \underline{?}, 729, 2{,}187, \dots$

   F 357 **G** 243

   H 261 **J** 539

Please note that Questions 7–13 have five answer choices.

7. Taye bought 3 bags of chips for $0.39 each and 2 cans of soft drink for $0.55 each. Which expression can be used to find the cost in dollars of the items?

   A $3 \times 0.39 \times 5 \times 0.55$

   B $3 \div 0.39 + 2 \div 0.55$

   C $3 \times 0.39 + 5 \times 0.55$

   D $3 \times 0.39 + 2 \times 0.55$

   E $3 \div 0.39 \times 2 \div 0.55$

8. In a survey, Akili found that 8 out of 40 students were the oldest child in their family. If 250 students had participated in the survey, how many would be expected to be the oldest child?

   F 120

   G 75

   H 60

   J 50

   K 85

9. What percent of the numbers from 1 to 25 contain the digit 3?

   A about 10%

   B about 20%

   C about 30%

   D about 40%

   E about 50%

10. In May, about 78 tools were rented each day at The Rent Center. Estimate the number of tools rented for the entire month of May.

    F 2,000

    G 2,400

    H 2,800

    J 3,200

    K 3,600

11. The perimeter of the triangle is $2\frac{3}{8}$ inches. Find the value of y.

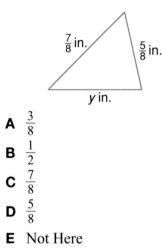

$\frac{7}{8}$ in. $\frac{5}{8}$ in.

y in.

A $\frac{3}{8}$

B $\frac{1}{2}$

C $\frac{7}{8}$

D $\frac{5}{8}$

E Not Here

12. Ms. Hayashi made a tablecloth for her kitchen. She bought $4\frac{3}{4}$ yards of material. She used $3\frac{1}{8}$ yards of material to make the tablecloth. How much material was not used to make the tablecloth?

F $7\frac{3}{8}$ yd   G $6\frac{1}{2}$ yd

H $7\frac{7}{8}$ yd   J $6\frac{3}{4}$ yd

K Not Here

13. What is the area of the octagon?

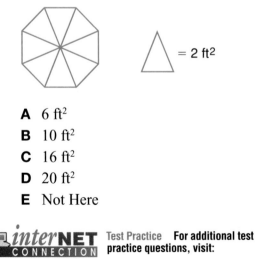

= 2 ft²

A 6 ft²

B 10 ft²

C 16 ft²

D 20 ft²

E Not Here

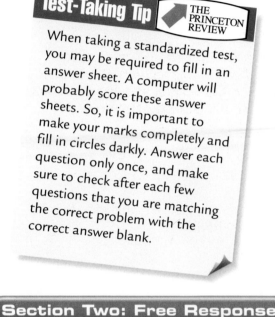

Test-Taking Tip   THE PRINCETON REVIEW

When taking a standardized test, you may be required to fill in an answer sheet. A computer will probably score these answer sheets. So, it is important to make your marks completely and fill in circles darkly. Answer each question only once, and make sure to check after each few questions that you are matching the correct problem with the correct answer blank.

Section Two: Free Response

This section contains three questions for which you will provide short answers. Write your answers on your paper.

14. A rectangular aquarium has sides measuring 28 inches long, 16 inches wide, and 12 inches high. The aquarium is filled with water to a height of 10 inches. What is the volume of the water?

15. A bag contains 3 yellow marbles, 5 red marbles, 6 blue marbles, and 16 green marbles.

    a. Suppose you choose one marble without looking, record its color, and then replace the marble. What is the probability of choosing a red marble?

    b. Suppose you choose one marble without looking, record its color, replace the marble, and then choose another marble. What is the probability of choosing a yellow marble and then a green marble?

16. How many different words can be made using the words *college*, *game*, *season*, and *election* and the prefixes *pre-* and *post-*?

inter NET CONNECTION   Test Practice **For additional test practice questions, visit:**

www.glencoe.com/sec/math/mac/mathnet

Student Handbook
Table of Contents

Basic Skills

Place Value

Write the place-value position for each digit in 721,056,938,504,550.

1. 4

2. 2

3. 6

4. 8

5. 3

6. 7

7. 1

8. 9

Write each number in words.

9. 47,900

10. 2,013

11. 540,006,000

12. 7,036,000,000

13. 263

14. 95,000,100,000,000

15. 7,261

16. 120,760

17. 102,000,016

18. 582

19. 8,000,070,000,600

20. 67,826

Write each number in standard form.

21. seventy-six million

22. nine trillion

23. six hundred thousand

24. forty-two

25. fifty-five trillion

26. three billion

27. seven hundred seventy-one

28. sixteen thousand

29. one hundred twenty-four million

30. six thousand nine hundred

31. eighty-eight billion

32. three hundred seventy trillion

Replace each ● with a number to make a true sentence.

33. 67,000 = ● hundreds

34. 39,000,000 = ● hundred thousands

35. 1,400,000,000,000 = ● millions

36. 760,000 = ● thousands

37. 4,200 = ● hundreds

38. 86,000,000 = ● thousands

39. ● = 17 millions

40. ● = 930 ten thousands

Basic Skills

Adding Whole Numbers

1. 40
$+\ 8$

2. 32
$+\ 5$

3. 63
$+\ 6$

4. 41
$+\ 8$

5. 53
$+\ 4$

6. 30
$+60$

7. 20
$+50$

8. 47
$+20$

9. 85
$+10$

10. 56
$+33$

11. 600
$+\ 50$

12. 506
$+\ 30$

13. 225
$+\ 40$

14. 704
$+\ 35$

15. 628
$+\ 71$

16. 500
$+200$

17. 320
$+430$

18. 405
$+503$

19. 342
$+127$

20. 315
$+583$

21. 27
$+\ 4$

22. 76
$+\ 9$

23. 59
$+\ 7$

24. 25
$+68$

25. 24
$+48$

26. 304
$+\ 57$

27. 845
$+\ 29$

28. 637
$+\ 36$

29. 304
$+509$

30. 228
$+534$

31. 83
$+56$

32. 94
$+72$

33. 62
$+85$

34. 380
$+270$

35. 761
$+187$

36. 684
$+\ 67$

37. 495
$+\ 48$

38. 347
$+\ 59$

39. 676
$+276$

40. 733
$+197$

41. 24
76
$+53$

42. 67
28
$+44$

43. 55
89
$+23$

44. 368
275
$+256$

45. 275
384
$+633$

46. 4,680
$+3,945$

47. 5,126
$+2,899$

48. 2,973
$+1,689$

49. 52,046
$+41,388$

50. 96,277
$+27,563$

Basic Skills

Adding Whole Numbers

1. 50
     + 9

2. 46
     + 3

3. 23
     + 2

4. 62
     + 5

5. 81
     + 4

6. 20
     +60

7. 40
     +30

8. 38
     +20

9. 61
     +10

10. 58
      +11

11. 100
      + 70

12. 207
      + 40

13. 563
      + 20

14. 716
      + 81

15. 334
      + 53

16. 300
      +500

17. 240
      +530

18. 621
      +347

19. 406
      +273

20. 748
      +111

21. 17
      + 5

22. 34
      + 8

23. 78
      + 4

24. 57
      +24

25. 22
      +39

26. 517
      + 64

27. 266
      + 29

28. 742
      + 38

29. 354
      + 37

30. 635
      + 46

31. 44
      +72

32. 35
      +83

33. 56
      +92

34. 580
      +340

35. 174
      +261

36. 674
      + 36

37. 398
      + 43

38. 264
      + 89

39. 593
      +327

40. 786
      +114

41. 62
       37
      +46

42. 28
       54
      +19

43. 71
       96
      +12

44. 265
       612
      +117

45. 422
       536
      +314

46. 3,276
      +4,563

47. 6,127
      +1,932

48. 2,985
      +1,316

49. 47,864
      +32,297

50. 93,760
      +42,163

EXTRA PRACTICE

Basic Skills

Subtracting Whole Numbers

1. 98 $-\ 5$	**2.** 87 $-\ 4$	**3.** 56 $-\ 3$	**4.** 45 $-\ 5$	**5.** 29 $-\ 7$
6. 60 -20	**7.** 80 -50	**8.** 56 -40	**9.** 90 -60	**10.** 78 -24
11. 798 $-\ 45$	**12.** 955 $-\ 23$	**13.** 354 $-\ 34$	**14.** 865 $-\ 52$	**15.** 697 $-\ 83$
16. 800 -500	**17.** 650 -300	**18.** 854 -630	**19.** 355 -103	**20.** 695 -132
21. 93 $-\ 7$	**22.** 47 $-\ 8$	**23.** 54 $-\ 5$	**24.** 78 -59	**25.** 60 -38
26. 760 $-\ 36$	**27.** 382 $-\ 67$	**28.** 630 $-\ 23$	**29.** 460 -248	**30.** 373 -126
31. 578 $-\ 93$	**32.** 247 $-\ 83$	**33.** 623 $-\ 93$	**34.** 738 -165	**35.** 954 -372
36. 232 -184	**37.** 540 -275	**38.** 727 -538	**39.** 660 -383	**40.** 840 -496
41. 315 -227	**42.** 712 -555	**43.** 408 -209	**44.** 705 -509	**45.** 400 -189
46. 6,791 $-\ 899$	**47.** 3,406 $-\ 408$	**48.** 5,690 $-\ 792$	**49.** 6,243 $-4,564$	**50.** 7,092 $-6,895$
51. 64,700 $-\ 3,792$	**52.** 41,905 $-\ 4,916$	**53.** 52,009 $-\ 7,314$	**54.** 80,490 $-60,495$	**55.** 68,418 $-39,529$

Basic Skills

Subtracting Whole Numbers

1. 67
    − 4

2. 25
    − 5

3. 78
    − 6

4. 93
    − 2

5. 56
    − 3

6. 40
    −10

7. 70
    −30

8. 62
    −40

9. 90
    −70

10. 86
    −32

11. 726
    − 14

12. 584
    − 32

13. 963
    − 51

14. 497
    − 83

15. 677
    − 21

16. 600
    −400

17. 730
    −300

18. 961
    −520

19. 874
    −352

20. 519
    −116

21. 43
    − 7

22. 72
    − 5

23. 95
    − 8

24. 86
    −27

25. 40
    −13

26. 461
    − 28

27. 532
    − 17

28. 670
    − 52

29. 982
    − 36

30. 621
    − 12

31. 726
    − 43

32. 942
    − 61

33. 527
    − 52

34. 438
    −229

35. 855
    −472

36. 860
    −472

37. 215
    −167

38. 742
    −563

39. 936
    −478

40. 487
    −199

41. 417
    −358

42. 324
    −137

43. 707
    −309

44. 602
    −408

45. 500
    −279

46. 7,213
    − 426

47. 2,146
    − 347

48. 6,290
    − 794

49. 7,418
    −2,439

50. 6,052
    −5,456

51. 64,205
    − 3,746

52. 88,644
    − 9,657

53. 30,716
    − 4,755

54. 64,658
    −23,659

55. 91,273
    −86,594

Basic Skills

Multiplying Whole Numbers

1. 40
× 5

2. 30
× 6

3. 20
× 8

4. 60
× 4

5. 50
× 7

6. 23
× 3

7. 44
× 2

8. 81
× 6

9. 72
× 3

10. 61
× 7

11. 721
× 4

12. 513
× 3

13. 234
× 2

14. 634
× 2

15. 831
× 3

16. 46
× 5

17. 53
× 7

18. 82
× 6

19. 27
× 4

20. 68
× 8

21. 704
× 6

22. 409
× 5

23. 806
× 8

24. 307
× 9

25. 208
× 7

26. 28
×10

27. 86
×10

28. 51
×10

29. 247
× 10

30. 4,328
× 10

31. 52
×20

32. 37
×50

33. 26
×40

34. 175
× 30

35. 1,469
× 80

36. 75
×19

37. 54
×27

38. 45
×81

39. 52
×64

40. 80
×76

41. 89
×45

42. 64
×37

43. 78
×62

44. 56
×82

45. 83
×59

46. 414
× 22

47. 321
× 43

48. 522
× 34

49. 613
× 32

50. 202
× 24

Basic Skills

Multiplying Whole Numbers

1. $\begin{array}{r} 30 \\ \times\ 7 \\ \hline \end{array}$

2. $\begin{array}{r} 70 \\ \times\ 4 \\ \hline \end{array}$

3. $\begin{array}{r} 50 \\ \times\ 8 \\ \hline \end{array}$

4. $\begin{array}{r} 80 \\ \times\ 3 \\ \hline \end{array}$

5. $\begin{array}{r} 90 \\ \times\ 6 \\ \hline \end{array}$

6. $\begin{array}{r} 71 \\ \times\ 8 \\ \hline \end{array}$

7. $\begin{array}{r} 64 \\ \times\ 2 \\ \hline \end{array}$

8. $\begin{array}{r} 92 \\ \times\ 3 \\ \hline \end{array}$

9. $\begin{array}{r} 53 \\ \times\ 2 \\ \hline \end{array}$

10. $\begin{array}{r} 81 \\ \times\ 6 \\ \hline \end{array}$

11. $\begin{array}{r} 624 \\ \times\ 2 \\ \hline \end{array}$

12. $\begin{array}{r} 434 \\ \times\ 2 \\ \hline \end{array}$

13. $\begin{array}{r} 712 \\ \times\ 3 \\ \hline \end{array}$

14. $\begin{array}{r} 221 \\ \times\ 4 \\ \hline \end{array}$

15. $\begin{array}{r} 511 \\ \times\ 7 \\ \hline \end{array}$

16. $\begin{array}{r} 27 \\ \times\ 5 \\ \hline \end{array}$

17. $\begin{array}{r} 36 \\ \times\ 2 \\ \hline \end{array}$

18. $\begin{array}{r} 54 \\ \times\ 4 \\ \hline \end{array}$

19. $\begin{array}{r} 92 \\ \times\ 6 \\ \hline \end{array}$

20. $\begin{array}{r} 75 \\ \times\ 3 \\ \hline \end{array}$

21. $\begin{array}{r} 906 \\ \times\ 5 \\ \hline \end{array}$

22. $\begin{array}{r} 702 \\ \times\ 7 \\ \hline \end{array}$

23. $\begin{array}{r} 503 \\ \times\ 9 \\ \hline \end{array}$

24. $\begin{array}{r} 807 \\ \times\ 8 \\ \hline \end{array}$

25. $\begin{array}{r} 209 \\ \times\ 3 \\ \hline \end{array}$

26. $\begin{array}{r} 92 \\ \times 10 \\ \hline \end{array}$

27. $\begin{array}{r} 87 \\ \times 10 \\ \hline \end{array}$

28. $\begin{array}{r} 43 \\ \times 10 \\ \hline \end{array}$

29. $\begin{array}{r} 761 \\ \times\ 10 \\ \hline \end{array}$

30. $\begin{array}{r} 5,276 \\ \times\ 10 \\ \hline \end{array}$

31. $\begin{array}{r} 76 \\ \times 40 \\ \hline \end{array}$

32. $\begin{array}{r} 19 \\ \times 50 \\ \hline \end{array}$

33. $\begin{array}{r} 51 \\ \times 20 \\ \hline \end{array}$

34. $\begin{array}{r} 247 \\ \times\ 30 \\ \hline \end{array}$

35. $\begin{array}{r} 1,236 \\ \times\ 80 \\ \hline \end{array}$

36. $\begin{array}{r} 92 \\ \times 16 \\ \hline \end{array}$

37. $\begin{array}{r} 74 \\ \times 23 \\ \hline \end{array}$

38. $\begin{array}{r} 56 \\ \times 47 \\ \hline \end{array}$

39. $\begin{array}{r} 81 \\ \times 32 \\ \hline \end{array}$

40. $\begin{array}{r} 45 \\ \times 72 \\ \hline \end{array}$

41. $\begin{array}{r} 61 \\ \times 37 \\ \hline \end{array}$

42. $\begin{array}{r} 72 \\ \times 59 \\ \hline \end{array}$

43. $\begin{array}{r} 12 \\ \times 86 \\ \hline \end{array}$

44. $\begin{array}{r} 93 \\ \times 72 \\ \hline \end{array}$

45. $\begin{array}{r} 26 \\ \times 41 \\ \hline \end{array}$

46. $\begin{array}{r} 723 \\ \times\ 46 \\ \hline \end{array}$

47. $\begin{array}{r} 812 \\ \times\ 51 \\ \hline \end{array}$

48. $\begin{array}{r} 245 \\ \times\ 67 \\ \hline \end{array}$

49. $\begin{array}{r} 123 \\ \times\ 94 \\ \hline \end{array}$

50. $\begin{array}{r} 679 \\ \times\ 77 \\ \hline \end{array}$

Basic Skills

Dividing Whole Numbers

1. $3\overline{)72}$ **2.** $4\overline{)96}$ **3.** $2\overline{)78}$ **4.** $3\overline{)84}$ **5.** $3\overline{)57}$

6. $6\overline{)918}$ **7.** $8\overline{)976}$ **8.** $5\overline{)965}$ **9.** $7\overline{)903}$ **10.** $4\overline{)752}$

11. $12\overline{)60}$ **12.** $17\overline{)51}$ **13.** $25\overline{)75}$ **14.** $15\overline{)90}$ **15.** $24\overline{)72}$

16. $34\overline{)204}$ **17.** $18\overline{)126}$ **18.** $27\overline{)135}$ **19.** $46\overline{)184}$ **20.** $53\overline{)424}$

21. $24\overline{)240}$ **22.** $32\overline{)320}$ **23.** $25\overline{)500}$ **24.** $17\overline{)510}$ **25.** $15\overline{)600}$

26. $6\overline{)384}$ **27.** $23\overline{)483}$ **28.** $34\overline{)612}$ **29.** $14\overline{)546}$ **30.** $48\overline{)720}$

31. $31\overline{)1,953}$ **32.** $99\overline{)1,881}$ **33.** $47\overline{)1,927}$ **34.** $26\overline{)1,742}$ **35.** $19\overline{)1,045}$

36. $18\overline{)3,672}$ **37.** $23\overline{)9,223}$ **38.** $32\overline{)9,824}$ **39.** $15\overline{)7,545}$ **40.** $27\overline{)8,154}$

41. $8\overline{)91}$ **42.** $6\overline{)87}$ **43.** $5\overline{)99}$ **44.** $7\overline{)87}$ **45.** $6\overline{)80}$

46. $8\overline{)685}$ **47.** $7\overline{)538}$ **48.** $4\overline{)273}$ **49.** $6\overline{)580}$ **50.** $5\overline{)387}$

51. $12\overline{)75}$ **52.** $23\overline{)97}$ **53.** $18\overline{)99}$ **54.** $33\overline{)75}$ **55.** $27\overline{)56}$

56. $16\overline{)134}$ **57.** $37\overline{)299}$ **58.** $53\overline{)483}$ **59.** $29\overline{)210}$ **60.** $62\overline{)439}$

61. $49\overline{)29,670}$ **62.** $84\overline{)25,880}$ **63.** $32\overline{)38,693}$ **64.** $26\overline{)80,311}$ **65.** $46\overline{)92,330}$

66. $100\overline{)706}$ **67.** $100\overline{)842}$ **68.** $200\overline{)900}$ **69.** $500\overline{)705}$ **70.** $300\overline{)602}$

71. $400\overline{)1,632}$ **72.** $300\overline{)2,205}$ **73.** $600\overline{)8,407}$ **74.** $200\overline{)9,820}$ **75.** $500\overline{)7,513}$

Basic Skills

Dividing Whole Numbers

1. $5\overline{)95}$ **2.** $2\overline{)58}$ **3.** $4\overline{)92}$ **4.** $3\overline{)78}$ **5.** $8\overline{)88}$

6. $4\overline{)848}$ **7.** $6\overline{)984}$ **8.** $3\overline{)351}$ **9.** $6\overline{)762}$ **10.** $2\overline{)276}$

11. $32\overline{)96}$ **12.** $19\overline{)95}$ **13.** $13\overline{)78}$ **14.** $21\overline{)63}$ **15.** $17\overline{)68}$

16. $32\overline{)256}$ **17.** $64\overline{)192}$ **18.** $97\overline{)679}$ **19.** $72\overline{)360}$ **20.** $16\overline{)144}$

21. $31\overline{)620}$ **22.** $14\overline{)980}$ **23.** $22\overline{)220}$ **24.** $29\overline{)870}$ **25.** $12\overline{)600}$

26. $8\overline{)576}$ **27.** $23\overline{)644}$ **28.** $56\overline{)952}$ **29.** $13\overline{)481}$ **30.** $46\overline{)690}$

31. $27\overline{)2,592}$ **32.** $68\overline{)3,060}$ **33.** $85\overline{)1,020}$ **34.** $53\overline{)4,081}$ **35.** $39\overline{)3,198}$

36. $19\overline{)4,902}$ **37.** $59\overline{)9,558}$ **38.** $23\overline{)5,175}$ **39.** $18\overline{)7,614}$ **40.** $12\overline{)6,264}$

41. $3\overline{)97}$ **42.** $4\overline{)65}$ **43.** $5\overline{)87}$ **44.** $8\overline{)93}$ **45.** $7\overline{)74}$

46. $5\overline{)273}$ **47.** $9\overline{)858}$ **48.** $6\overline{)412}$ **49.** $3\overline{)136}$ **50.** $7\overline{)529}$

51. $13\overline{)68}$ **52.** $16\overline{)97}$ **53.** $14\overline{)33}$ **54.** $24\overline{)54}$ **55.** $18\overline{)71}$

56. $34\overline{)162}$ **57.** $19\overline{)135}$ **58.** $51\overline{)426}$ **59.** $68\overline{)345}$ **60.** $46\overline{)234}$

61. $65\overline{)58,974}$ **62.** $47\overline{)32,569}$ **63.** $12\overline{)14,586}$ **64.** $52\overline{)74,853}$ **65.** $29\overline{)62,185}$

66. $100\overline{)902}$ **67.** $100\overline{)764}$ **68.** $200\overline{)536}$ **69.** $300\overline{)847}$ **70.** $500\overline{)604}$

71. $300\overline{)1,475}$ **72.** $400\overline{)3,206}$ **73.** $200\overline{)2,568}$ **74.** $600\overline{)9,612}$ **75.** $500\overline{)8,520}$

EXTRA PRACTICE

Extra Practice

Lesson 1-1 *(Pages 4–7)*
Use the four-step plan to solve each problem.

1. On a map of Ohio, each inch represents approximately 9 miles. Chad is planning to travel from Cleveland to Cincinnati. If the distance on the map from Cleveland to Cincinnati is about 23 inches, how far will he travel?

2. Sylvia has $102. If she made three purchases of $13, $37, and $29, how much money does she have left?

3. It took Jason 56 hours to paint a house. It only took Nolan 43 hours and 16 minutes to paint an identical house. How many more hours did it take Jason to paint the house than it took Nolan?

4. A cassette tape holds sixty minutes of music. If Logan has already taped eight songs that are each five minutes long on the tape, how many more minutes of music can she still tape on the cassette?

5. The Cornells want to buy a car that costs $4,260. They plan to make a down payment of $1,500 and pay the rest in twelve equal payments. What will be the amount of each payment?

Lesson 1-2 *(Pages 8–11)*
Find the next three numbers or shapes for each pattern.

1.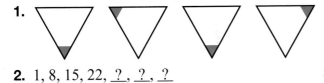

2. 1, 8, 15, 22, _?_, _?_, _?_

3. 1, 2, 6, 24, _?_, _?_, _?_

Solve by using patterns.

4. Sarah is conditioning for track. On the first day, she ran 2 laps. The second day she ran 3 laps. The third day she ran 5 laps, and on the fourth day she ran 8 laps. If this pattern continues, how many laps will she run on the seventh day?

5. The times that a new movie is showing are 9:30 A.M., 11:12 A.M., 12:54 P.M., and 2:36 P.M. What are the next three times this movie will be shown?

Lesson 1-3 *(Pages 12–15)*
Round each number to the underlined place-value position.

1. 4̲6
2. 1,2̲49
3. 9̲,499
4. 2,96̲0
5. 6̲,001
6. 16̲3
7. 6̲58
8. 6̲,710
9. 12,6̲50
10. 18̲,305
11. 156,9̲99
12. 960,7̲15

Estimate. State whether the answer shown is reasonable.

13. $61 \times 5 = 234$
14. $889 - 43 = 846$
15. $\$2.94 + \$6.13 + \$9.25 = \$18.32$
16. $415 \times 4 = 1,660$
17. $385 \div 5 = 65$
18. $2,107 - 182 = 1,625$
19. $108 + 496 + 229 = 833$
20. $5,627 \div 331 = 17$

Lesson 1-4 (Pages 16–19)

Find the value of each expression.

1. $14 - 5 + 7$
2. $12 + 10 - 5 - 6$
3. $50 - 6 + 12 + 4$
4. $12 - 2 \times 3$
5. $16 + 4 \times 5$
6. $5 + 3 \times 4 - 7$
7. $2 \times 3 + 9 \times 2$
8. $6 \times 8 + 4 \div 2$
9. $7 \times 6 - 14$
10. $8 + 12 \times 4 \div 8$
11. $13 - 6 \times 2 + 1$
12. $80 \div 10 \times 8$
13. $1 + 2 + 3 + 4$
14. $1 \times 2 \times 3 \times 4$
15. $6 + 6 \times 6$
16. $14 - 2 \times 7 + 0$
17. $156 - 6 \times 0$
18. $30 - 14 \times 2 + 8$

Lesson 1-5 (Pages 22–25)

Evaluate each expression if $m = 2$ and $n = 4$.

1. $m + m$
2. $n - m$
3. mn
4. $2m$
5. $2n$
6. $2n + 2m$
7. $m \times 0$
8. $64 \div n$
9. $12 - m$
10. $2mn$

Evaluate each expression if $a = 3$, $b = 4$, and $c = 12$.

11. $a + b$
12. $c - a$
13. $a + b + c$
14. $b - a$
15. $c - a \times b$
16. $a + 2 \times b$
17. $b + c \div 2$
18. ab
19. $a + 3b$
20. $a + c \div 6$
21. $25 + c \div b$
22. abc
23. $144 - abc$
24. $c \div a + 10$
25. $2b - a$
26. $2ab$

Lesson 1-6 (Pages 28–31)

Write each product using exponents.

1. $2 \cdot 2 \cdot 2 \cdot 2 \cdot 2$
2. $6 \cdot 6 \cdot 6 \cdot 7 \cdot 7$
3. $9 \; 9 \cdot 9 \cdot 9 \cdot 9 \cdot 9 \cdot 10$
4. $k \cdot k \cdot k \cdot t \cdot t \cdot t$
5. $14 \cdot 14 \cdot 6$
6. $3 \cdot 3 \cdot 3 \cdot 3 \cdot y \cdot y$

Write each power as a product.

7. 13^4
8. 9^6
9. $2^3 \cdot 3^2$
10. x^5
11. 169^3
12. $13,410^2$

Evaluate each expression.

13. 5^6
14. 17^3
15. 2^{12}
16. $3^5 \cdot 2^3$
17. $6^4 \cdot 3$
18. $2^2 \cdot 3^2 \cdot 4^2$
19. 176^2
20. $6 \cdot 4^3$

Lesson 1-7A *(Pages 32–33)*
Solve.

1. Find an even number between 70 and 80 that is divisible by 2 and 9.

2. Last week the Tri-River Animal Shelter sent a total of 34 cats and dogs to new homes. There were 8 more cats than dogs. How many of each were adopted?

3. Admission to the Cincinnati Zoo is $8 for adults, $5 for children, and $3 for senior citizens. Eleven members of the Ruiz family paid a total of $55 for admission. If 6 children were in the group, how many adults and senior citizens were in the group?

4. Sylvia's soccer team played a total of 18 matches. Her team won twice as many matches as they lost. How many matches did they win?

5. Jordan makes $5 per hour mowing lawns and $7 per hour painting houses during the summer. This week, Jordan made $102. If he worked twice as many hours mowing lawns as he did painting houses, how many hours did he work at each job?

Lesson 1-7 *(Pages 34–37)*
Tell whether the equation is *true* or *false* by replacing the variable with the given value.

1. $q - 7 = 7; q = 28$
2. $g - 3 = 10; g = 13$
3. $r - 3 = 4; r = 7$
4. $t + 3 = 21; t = 24$

Identify the solution to each equation from the list given.

5. $7 + a = 10; 3, 13, 17$
6. $14 + m = 24; 7, 10, 34$
7. $j \div 3 = 2; 4, 6, 8$
8. $20 = 24 - n; 2, 3, 4$

Solve each equation mentally.

9. $b + 7 = 12$
10. $s + 10 = 23$
11. $4x = 36$
12. $6 = t \div 5$
13. $b - 3 = 12$
14. $w \div 2 = 8$

Lesson 2-1 *(Pages 46–49)*
Make a frequency table for each set of data.

1. To the nearest mile, how many miles did members of the track team run during practice yesterday?
   1 4 3 5 2 3 3 2 1 4 3 2 3 1 5 3 4 2 3 1 5 2

2. What were the high temperatures in Indiana cities on March 13?
   52 57 48 53 52 49 48 52 51
   47 51 49 57 53 48 52 52 49

Use the frequency table to the right to answer Exercises 3–4.

3. Describe the data shown in the table.

4. Which flavor should the ice cream shop stock the most?

ICE CREAM FLAVORS SOLD IN JULY		
Flavor	Tally	Frequency
Vanilla	卌 卌 卌 卌 IIII	24
Chocolate	卌 卌 卌 III	18
Strawberry	卌 卌 II	12
Chocolate Chip	卌 卌 卌 I	16
Peach	卌 III	8
Butter Pecan	卌 卌 I	11

Lesson 2-2 *(Pages 50–53)*

Choose an appropriate scale and an interval for each set of data.

1. 2, 7, 13, 3, 4, 12, 9

2. 11, 15, 13, 18, 19, 20

3. 56, 85, 23, 78, 42, 63

4. 10, 25, 88, 64, 99, 37

5. 165, 167, 169, 164, 170, 166, 167, 165, 169

6. 132, 865, 465, 672, 318, 940, 573, 689

7. 1,450; 7,896; 5,638; 7,142; 4,287; 8,612

Lesson 2-3 *(Pages 54–57)*

Make a bar graph for each set of data.

1. a vertical bar graph

Favorite Subject	
Subject	*Frequency*
Math	4
Science	6
History	2
English	8
Phys. Ed.	12

2. a horizontal bar graph

Final Grades	
Subject	*Score*
Math	88
Science	82
History	92
English	94

Make a line graph for each set of data.

3.

Test	Score
1	62
2	75
3	81
4	83
5	78
6	92

4.

Day	Absences
Mon.	3
Tues.	6
Wed.	2
Thur.	1
Fri.	8

Lesson 2-3B *(Pages 58–59)*

The double bar graph shows the quarterly sales (in millions of dollars) of two companies. Use the graph to answer the following questions.

1. In which quarter did Company A have its lowest sales?

2. About how much higher are the sales for Company A than Company B in the 3rd quarter?

3. The sales of Company A in the 2nd quarter are about equal to the sales of Company B in which quarter?

4. In which quarter is there the least difference between the sales of Company A and the sales of Company B?

Lesson 2-4 (Pages 60–63)

The circle graph shows the favorite subject of students at Midland Middle School.

1. The percents are 5%, 12%, 20%, 25%, and 38%. Match each percent with the appropriate section of the graph.

2. Suppose math and history were combined. Would the combination be preferred by $\frac{1}{4}$ of the students?

3. Which two subjects together are preferred by the same percent as English?

4. Which two subjects together are preferred by half of the students?

Students' Favorite Subjects

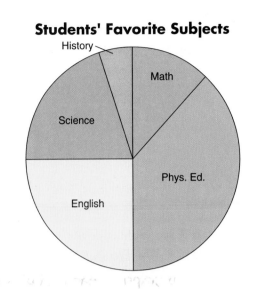

Lesson 2-5 (Pages 64–67)

Use the following graph to solve each problem.

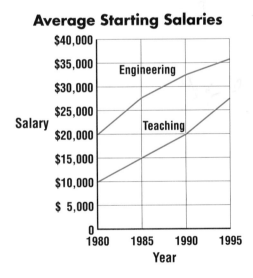

1. Give the expected starting salary in 1988 for:
   a. an engineer
   b. a teacher

2. How much more did an engineer make than a teacher in 1980?

3. Which of these two professions do you think will start with a higher salary in 2000?

4. How much less was the difference in salaries in 1995 than in 1980?

Lesson 2-6 (Pages 68–70)

Make a stem-and-leaf plot for each set of data.

1. 23, 15, 39, 68, 57, 42, 51, 52, 41, 18, 29

2. 5, 14, 39, 28, 14, 6, 7, 18, 13, 28, 9, 14

3. 189, 182, 196, 184, 197, 183, 196, 194, 184

4. 71, 82, 84, 95, 76, 92, 83, 74, 81, 75, 96

Lesson 2-7 *(Pages 71–75)*

Find the mean, median, mode, and range for each set of data.

1. 1, 5, 9, 1, 2, 4, 8, 2
2. 2, 5, 8, 9, 7, 6, 3, 5
3. 1, 2, 4, 2, 2, 1, 2
4. 12, 13, 15, 12, 12, 11
5. 256, 265, 247, 256
6. 957, 562, 462, 848, 721
7. 46, 54, 66, 54, 46, 66
8. 81, 82, 83, 84, 85, 86, 87

Lesson 2-8 *(Pages 78–81)*

Tell whether the mean, median, or mode would be best to describe each set of data. Explain each answer.

1. 627, 452, 573, 602, 498
2. Favorite bagel flavor
3. $42,360; $51,862; $47,650; $23,400; $52,961

The graphs below display the same information.

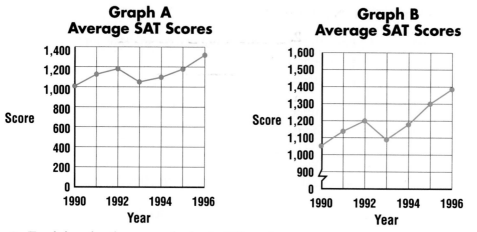

4. Explain why these graphs look different.
5. If Mr. Roush wishes to show that SAT scores have improved greatly since he became principal in 1993, which graph should he use?

Lesson 2-9 *(Pages 82–85)*

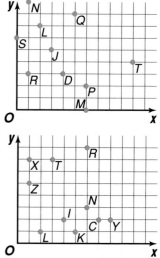

Use the grid at the right to name the point for each ordered pair.

1. (6, 2)
2. (3, 5)
3. (1, 3)
4. (6, 0)
5. (5, 8)
6. (2, 7)
7. (0, 6)
8. (4, 3)
9. (10, 4)
10. (1, 9)

Use the grid at the right to name the ordered pair for each point.

11. *L*
12. *I*
13. *C*
14. *R*
15. *T*
16. *X*
17. *Z*
18. *K*
19. *Y*
20. *N*

Lesson 3-1 *(Pages 95–98)*
Write each fraction or mixed number as a decimal.

1. $\dfrac{4}{10}$
2. $\dfrac{66}{100}$
3. $\dfrac{73}{100}$
4. $\dfrac{5}{100}$
5. $\dfrac{79}{100}$
6. $\dfrac{94}{100}$
7. $6\dfrac{85}{1,000}$
8. $\dfrac{875}{10,000}$
9. $\dfrac{1,264}{1,000}$
10. $\dfrac{527}{10,000}$

Write each expression as a decimal.

11. two hundredths
12. sixteen hundredths
13. four tenths
14. two and twenty-seven hundredths
15. nine and twelve hundredths
16. fifty-six and nine tenths
17. twenty-seven thousandths
18. one hundred ten-thousandths
19. two ten-thousandths
20. twenty and six hundred thousandths
21. two thousand, four hundred seventy-five and six tenths
22. twelve thousand, ninety-seven and sixty-two thousandths

Lesson 3-2 *(Pages 102–104)*
Use a centimeter ruler to measure each line segment.

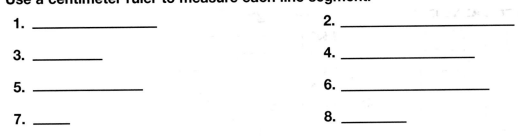

1. _____
2. _____
3. _____
4. _____
5. _____
6. _____
7. ____
8. _____

Lesson 3-3 *(Pages 105–108)*
State the greater number in each group.

1. 0.112 or 0.121
2. 0.9985 or 0.998
3. 0.556 or 0.519
4. 1.19 or 11.9
5. 0.6, 6.0 or 0.06
6. 0.0009 or 0.001

Order each set of decimals from least to greatest.

7.	8.	9.	10.
415.65	0.0256	1.2356	50.12
451.65	0.2056	1.2355	5.012
451.66	0.0255	1.25	50.22
451.56	0.0009	1.2335	5.901
415.56	0.2560	1.2353	50.02

Order each set of decimals from greatest to least.

11.	12.	13.	14.
13.664	26.6987	1.00065	2.014
13.446	26.9687	1.00100	2.010
1.3666	26.9666	1.00165	22.00
1.6333	26.9688	1.00056	22.14

Lesson 3-4 *(Pages 109–111)*

Round each number to the underlined place-value position.

1. 5.<u>6</u>4
2. 12.3<u>7</u>6
3. 0.05<u>3</u>62
4. <u>6</u>.17

5. 15.<u>2</u>98
6. 0.002<u>6</u>325
7. 758.9<u>9</u>9
8. <u>4</u>.25

9. 32.65<u>8</u>3
10. <u>0</u>.025
11. 1.00<u>4</u>9
12. 9.<u>2</u>5

13. 67.4<u>9</u>2
14. 25.<u>1</u>9
15. 26.<u>9</u>6
16. 4.00<u>0</u>98

Lesson 3-5 *(Pages 112–115)*

Estimate using rounding.

1. 0.245
   +0.256

2. 2.45698
   −1.26589

3. 0.5962
   +1.2598

4. 17.985
   − 9.001

5. 0.256 + 0.6589

6. 1.2568 − 0.1569

7. 12.999 + 5.048

Estimate using clustering.

8. 4.5 + 4.95 + 5.2 + 5.49
9. 2.25 + 1.69 + 2.1 + 2.369
10. $12.15 + $11. 63 + $12 + $11.89
11. 0.569 + 1.005 + 1.265 + 0.765

Lesson 3-6A *(Pages 116–117)*

Solve.

1. Jaleel bought a t-shirt at a concert. He paid for the $17.49 shirt with a $20 bill. Should he expect about $4 or about $2 in change?

2. There are 3,261 seats in the Northmore High School stadium. What is a reasonable number of rows in the stadium if each row holds about 55 people?

3. Jayson wants to spend his allowance on CD ROM games for his computer. He has saved $80. About how many games can Jayson buy if each CD ROM costs $29.95?

4. Lucas needs to buy school supplies. He found a backpack for $29.95, a notebook for $8.49, and a package of pens and pencils for $3.70. Is $30 or $50 needed to pay for these supplies?

5. When Jamilah added 0.00276 and 0.0149 the calculator showed 0.1766. Is this answer reasonable?

Lesson 3-6 *(Pages 118–121)*
Add or subtract.

1. $0.46 + 0.72$

2. $13.7 + 2.6$

3. $17.9 + 7.41$

4. $19.2 + 7.36$

5. $0.5113 + 0.62148$

6. $12.56 - 10.21$

7. $0.2154 - 0.1526$

8. $2.3125 + 1.02$

9. $1.025 - 0.58697$

10. $14.526 - 12.654$

11. $2.3568 + 5$

12. $20 - 5.98671$

13. $15.256 + 0.236$

14. $3.7 + 1.5 + 0.2$

15. $0.23 + 1.2 + 0.36$

16. $0.896352 - 0.25639$

17. $25.6 - 2.3$

18. $13.5 - 2.8456$

19. $1.265 + 1.654$

20. $24.56 - 24.32$

21. $0.256 - 0.255$

Lesson 4-1 *(Pages 133–136)*
Multiply.

1. $\begin{array}{r} 0.2 \\ \times\, 65 \\ \hline \end{array}$

2. $\begin{array}{r} 0.73 \\ \times\, 12 \\ \hline \end{array}$

3. $\begin{array}{r} 0.65 \\ \times\, 27 \\ \hline \end{array}$

4. $\begin{array}{r} 9.6 \\ \times\, 13 \\ \hline \end{array}$

5. $\begin{array}{r} 12.15 \\ \times\, 6 \\ \hline \end{array}$

6. $\begin{array}{r} 0.91 \\ \times\, 16 \\ \hline \end{array}$

7. $\begin{array}{r} 0.265 \\ \times\, 7 \\ \hline \end{array}$

8. $\begin{array}{r} 2.612 \\ \times\, 14 \\ \hline \end{array}$

9. $\begin{array}{r} 0.003 \\ \times\, 55 \\ \hline \end{array}$

10. $\begin{array}{r} 0.67 \\ \times\, 21 \\ \hline \end{array}$

Solve each equation.

11. $r = 19 \times 0.111$

12. $1.65 \times 72 = a$

13. $9.6 \times 101 = q$

14. $24 \times 1.201 = d$

15. $610 \times 7.5 = j$

16. $z = 0.001 \times 6$

17. $x = 510 \times 0.0135$

18. $b = 9.2 \times 17$

19. $14.1235 \times 4 = m$

Lesson 4-2 *(Pages 137–139)*
Find each product mentally. Use the distributive property.

1. 5×18

2. 9×27

3. 8×83

4. 7×21

5. 3×47

6. 2×106

7. 6×34

8. 56×3

9. 27×8

10. 5×3.4

11. 6×40.7

12. 1.5×30

13. 0.9×71

14. 30×2.08

15. 16×7

16. 33×4

17. 0.6×12

18. 80×7.9

Lesson 4-3 (Pages 141–143)

Multiply.

1. $9.6 \cdot 10.5$

2. $3.2 \cdot 0.1$

3. $1.5 \cdot 9.6$

4. $5.42 \cdot 0.21$

5. $7.42 \cdot 0.2$

6. $0.001 \cdot 0.02$

7. $0.6 \cdot 542$

8. $6.7 \cdot 5.8$

9. $3.24 \cdot 6.7$

Solve each equation.

10. $9.8 \cdot 4.62 = s$

11. $7.32 \cdot 9.7 = v$

12. $t = 0.008 \cdot 0.007$

13. $a = 0.001 \cdot 56$

14. $c = 4.5 \cdot 0.2$

15. $9.6 \cdot 2.3 = h$

16. $5.63 \cdot 8.1 = q$

17. $10.35 \cdot 9.1 = u$

18. $t = 28.2 \cdot 3.9$

19. $g = 102.13 \cdot 1.221$

20. $n = 2.02 \times 1.25$

21. $z = 8.37 \times 89.6$

Lesson 4-4 (Pages 145–148)

Find the perimeter of each figure.

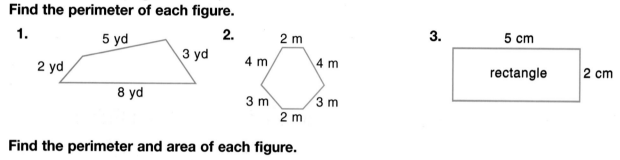

1. 5 yd, 3 yd, 2 yd, 8 yd

2. 2 m, 4 m, 4 m, 3 m, 3 m, 2 m

3. 5 cm, rectangle, 2 cm

Find the perimeter and area of each figure.

4. $19 \text{ cm} \times 3 \text{ cm}$

5. $5 \text{ in.} \times 3 \text{ in.}$

6. 9 ft square

7. 4.3 m square

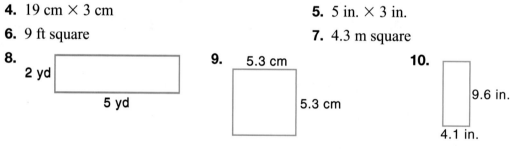

8. 2 yd, 5 yd

9. 5.3 cm, 5.3 cm

10. 9.6 in., 4.1 in.

Lesson 4-5A (Pages 150–151)

Solve.

1. Genaro wants to carpet a den that measures 18 feet by 24 feet. In the center of the room is a tile hearth for his stove which he does not want to carpet. If the hearth measures 6 feet by 6 feet, how much carpet does he need?

2. Jamika and Jordan are carving pumpkins for a Halloween party. Jamika can carve 3 pumpkins in 2 hours. Jordan can carve 2 pumpkins in 1 hour. If they work the same amount of time, how long will it take to carve 14 pumpkins?

3. Lonny has 24 feet of fence to make a pen for his rabbit. What dimensions should he make the pen so the area is the greatest possible?

4. How many cuts must be made to divide a 6-foot submarine sandwich equally among 10 people?

Lesson 4-5 *(Pages 152–155)*

Find each quotient.

1. $6\overline{)1.26}$ 2. $8\overline{)23.2}$ 3. $6\overline{)89.22}$ 4. $15\overline{)54.75}$

5. $13\overline{)128.31}$ 6. $9\overline{)2.583}$ 7. $47\overline{)11.28}$ 8. $26\overline{)32.5}$

9. $37.1 \div 14$ 10. $5.88 \div 4$ 11. $3.7 \div 5$

12. $41.4 \div 18$ 13. $9.87 \div 3$ 14. $8.45 \div 25$

Round each quotient to the nearest tenth.

15. $26.5 \div 4$ 16. $46.25 \div 8$ 17. $19.38 \div 9$

18. $8.5 \div 2$ 19. $90.88 \div 14$ 20. $23.1 \div 4$

Round each quotient to the nearest hundredth.

21. $19.5 \div 27$ 22. $26.5 \div 19$ 23. $46.23 \div 25$

24. $46.25 \div 25$ 25. $4.26 \div 9$ 26. $18.74 \div 19$

Lesson 4-6 *(Pages 157–159)*

Find each quotient.

1. $0.5\overline{)18.45}$ 2. $0.08\overline{)5.2}$ 3. $2.6\overline{)0.65}$ 4. $1.3\overline{)12.831}$

5. $0.87\overline{)5.133}$ 6. $2.54\overline{)24.13}$ 7. $3.7\overline{)35.89}$ 8. $26\overline{)32.5}$

9. $5.88 \div 0.4$ 10. $3.7 \div 0.5$ 11. $6.72 \div 2.4$

12. $9.87 \div 0.3$ 13. $8.45 \div 2.5$ 14. $90.88 \div 14.2$

15. $33.6 \div 8.4$ 16. $25.389 \div 4.03$ 17. $85.92 \div 4.8$

18. $63.18 \div 16.2$ 19. $18.49 \div 4.3$ 20. $9.363 \div 0.003$

21. $1.02 \div 0.3$ 22. $6.4 \div 0.8$ 23. $7.2 \div 0.9$

Lesson 4-7 *(Pages 161–163)*

Find each quotient to the nearest hundredth.

1. $9\overline{)0.36}$ 2. $13\overline{)39.39}$ 3. $45\overline{)0.585}$ 4. $8\overline{)0.24}$

5. $6\overline{)0.312}$ 6. $7\overline{)0.161}$ 7. $7\overline{)7.21}$ 8. $3\overline{)9.18}$

9. $\$0.72 \div 12$ 10. $0.36 \div 9$ 11. $0.56 \div 14$

12. $32.2 \div 8$ 13. $0.3869 \div 5.3$ 14. $0.39 \div 7.8$

Solve each equation.

15. $0.0426 \div 71 = q$ 16. $f = 0.1185 \div 7.9$ 17. $j = 0.84 \div 12$

18. $m = 4.544 \div 64$ 19. $v = 0.384 \div 9.6$ 20. $0.2262 \div 8.7 = d$

Lesson 4-8 *(Pages 164–166)*

Write the unit that you would use to measure each of the following. Then estimate the mass or capacity.

1. a bag of sugar

2. a pitcher of fruit punch

3. the mass of a dime

4. the amount of water in an ice cube

5. a vitamin

6. a pencil

7. the mass of a puppy

8. a bottle of perfume

9. a grain of sand

10. the mass of a car

Lesson 4-9 *(Pages 167–169)*

Complete.

1. 400 mm = _____ cm

2. 4 kg = _____ g

3. 660 cm = _____ m

4. 0.3 L = _____ mL

5. 30 mm = _____ cm

6. 84.5 g = _____ kg

7. _____ m = 54 cm

8. _____ L = 563 mL

9. _____ mg = 21 g

10. 4 L = _____ mL

11. 61.2 mg = _____ g

12. 4,497 mL = _____ L

13. _____ mm = 45 cm

14. 632 mL = _____ L

15. 61 g = _____ mg

16. _____ mg = 0.51 kg

17. 0.63 L = _____ mL

18. 18 km = _____ cm

Lesson 5-1 *(Pages 178–180)*

Determine whether the first number is divisible by the second number.

1. 89; 3

2. 64; 2

3. 125; 5

4. 156; 4

5. 216; 9

6. 330; 10

7. 225; 3

8. 524; 6

State whether each number is divisible by 2, 3, 5, 6, 9, and 10.

9. 1,986

10. 2,052

11. 110

12. 315

13. 405

14. 918

15. 243

16. 735

17. 1,233

18. 5,103

19. 8,001

20. 9,270

Lesson 5-2 *(Pages 182–184)*
Tell whether each number is *prime*, *composite*, or *neither*.

  1. 20 **2.** 65 **3.** 37 **4.** 26 **5.** 54

  6. 155 **7.** 201 **8.** 0 **9.** 49 **10.** 17

Find the prime factorization of each number.

  11. 72 **12.** 2,648 **13.** 32 **14.** 86 **15.** 120

  16. 576 **17.** 68 **18.** 240 **19.** 24 **20.** 70

  21. 102 **22.** 121 **23.** 164 **24.** 225 **25.** 54

Lesson 5-3A *(Pages 186–187)*
Solve.

  1. Mychal needs to go to the bank, the post office, and the bicycle shop. In how many different orders can she do this?

  2. How many different four digit numbers can be made from the digits 2, 5, 7, and 9?

  3. Ai-lien usually buys her lunch from the a la carte line in the cafeteria. Today, she has a choice of chicken, fish, or spaghetti; soup or salad; and fresh fruit or yogurt. How many possible choices does she have?

  4. A shoe company offers 8 different styles of running shoes in 4 different colors. How many combinations of style and color are possible?

Lesson 5-3 *(Pages 188–190)*
Find the GCF of each set of numbers by making a list.

  1. 8, 18 **2.** 6, 9 **3.** 4, 12 **4.** 18, 24

  5. 25, 30 **6.** 36, 54 **7.** 64, 32 **8.** 16, 32, 56

Find the GCF of each set of numbers by using prime factorization.

  9. 6, 15 **10.** 8, 24 **11.** 14, 22 **12.** 12, 27

  13. 17, 51 **14.** 48, 60 **15.** 54, 72 **16.** 14, 28, 42

Find the GCF of each set of numbers using either method.

  17. 5, 20 **18.** 9, 21 **19.** 10, 25 **20.** 12, 28

  21. 16, 24 **22.** 42, 48 **23.** 60, 75 **24.** 27, 45, 63

Lesson 5-4 *(Pages 193–196)*

Replace each ▪ with a number so that the fractions are equivalent.

1. $\frac{12}{16} = \frac{\blacksquare}{4}$

2. $\frac{7}{8} = \frac{\blacksquare}{32}$

3. $\frac{3}{4} = \frac{75}{\blacksquare}$

4. $\frac{8}{16} = \frac{\blacksquare}{2}$

5. $\frac{6}{18} = \frac{1}{\blacksquare}$

6. $\frac{27}{36} = \frac{3}{\blacksquare}$

7. $\frac{1}{4} = \frac{16}{\blacksquare}$

8. $\frac{9}{18} = \frac{\blacksquare}{2}$

State whether each fraction or ratio is in simplest form. If not, write each fraction or ratio in simplest form.

9. $\frac{50}{100}$

10. $\frac{24}{40}$

11. 2:5

12. 8 out of 24

13. 20 to 27

14. $\frac{4}{10}$

15. $\frac{3}{5}$

16. 14:19

17. 9:12

18. $\frac{6}{8}$

19. $\frac{15}{18}$

20. 9 out of 20

Lesson 5-5 *(Pages 198–201)*

Express each mixed number as an improper fraction.

1. $3\frac{1}{16}$

2. $2\frac{3}{4}$

3. $1\frac{3}{8}$

4. $1\frac{5}{12}$

5. $7\frac{3}{5}$

6. $6\frac{5}{8}$

7. $3\frac{1}{3}$

8. $1\frac{7}{9}$

9. $2\frac{3}{16}$

10. $1\frac{2}{3}$

Express each improper fraction as a mixed number.

11. $\frac{33}{10}$

12. $\frac{103}{25}$

13. $\frac{22}{5}$

14. $\frac{13}{2}$

15. $\frac{29}{6}$

16. $\frac{101}{100}$

17. $\frac{21}{8}$

18. $\frac{19}{6}$

19. $\frac{23}{5}$

20. $\frac{99}{50}$

Lesson 5-6 *(Pages 202–205)*

Draw a line segment of each length.

1. $2\frac{1}{4}$ inches

2. $1\frac{3}{8}$ inches

3. $\frac{3}{4}$ inch

4. $1\frac{1}{2}$ inches

5. $3\frac{1}{8}$ inches

6. $2\frac{1}{4}$ inches

Find the length of each line segment to the nearest half, fourth, or eighth inch.

7. ▬▬▬▬▬▬

8. ▬▬▬▬▬▬

9. ▬▬▬▬

10. ▬▬▬▬▬▬

Lesson 5-7 *(Pages 206–209)*

Determine whether the first number is a multiple of the second number.

1. 30; 7   **2.** 42; 14   **3.** 10; 8   **4.** 30; 10

5. 13; 7   **6.** 84; 28   **7.** 150; 25   **8.** 21; 14

Find the LCM for each set of numbers.

9. 5, 15   **10.** 13, 39   **11.** 16, 24   **12.** 18, 20

13. 8, 12   **14.** 12, 15   **15.** 9, 27   **16.** 5, 6

17. 12, 18, 3   **18.** 12, 35, 10   **19.** 21, 14, 6

20. 3, 6, 9   **21.** 6, 10, 15   **22.** 15, 75, 25

Lesson 5-8 *(Pages 210–213)*

Find the LCD for each pair of fractions.

1. $\dfrac{2}{5}$ $\dfrac{4}{15}$ **2.** $\dfrac{2}{28}$ $\dfrac{15}{42}$ **3.** $\dfrac{8}{16}$ $\dfrac{14}{24}$ **4.** $\dfrac{9}{25}$ $\dfrac{21}{30}$

Replace each ⬤ with <, >, or = to make a true sentence.

5. $\dfrac{1}{2}$ ⬤ $\dfrac{1}{3}$ **6.** $\dfrac{2}{3}$ ⬤ $\dfrac{3}{4}$ **7.** $\dfrac{5}{9}$ ⬤ $\dfrac{4}{5}$ **8.** $\dfrac{3}{6}$ ⬤ $\dfrac{6}{12}$

9. $\dfrac{12}{23}$ ⬤ $\dfrac{15}{19}$ **10.** $\dfrac{9}{27}$ ⬤ $\dfrac{13}{39}$ **11.** $\dfrac{7}{8}$ ⬤ $\dfrac{9}{13}$ **12.** $\dfrac{5}{9}$ ⬤ $\dfrac{7}{8}$

13. $\dfrac{25}{100}$ ⬤ $\dfrac{3}{8}$ **14.** $\dfrac{6}{7}$ ⬤ $\dfrac{8}{15}$ **15.** $\dfrac{5}{9}$ ⬤ $\dfrac{19}{23}$ **16.** $\dfrac{120}{567}$ ⬤ $\dfrac{1}{2}$

17. $\dfrac{5}{7}$ ⬤ $\dfrac{2}{3}$ **18.** $\dfrac{9}{36}$ ⬤ $\dfrac{7}{28}$ **19.** $\dfrac{2}{5}$ ⬤ $\dfrac{2}{6}$ **20.** $\dfrac{5}{9}$ ⬤ $\dfrac{12}{13}$

Lesson 5-9 *(Pages 214–216)*

Express each decimal as a fraction or mixed number in simplest form.

1. 0.5   **2.** 0.8   **3.** 0.32   **4.** 0.875

5. 0.54   **6.** 0.38   **7.** 0.744   **8.** 0.101

9. 0.303   **10.** 0.486   **11.** 0.626   **12.** 0.448

13. 0.074   **14.** 0.008   **15.** 9.36   **16.** 10.18

17. 0.06   **18.** 0.75   **19.** 0.48   **20.** 0.9

21. 0.005   **22.** 0.4   **23.** 1.875   **24.** 5.08

Lesson 5-10 *(Pages 217–219)*

Write each repeating decimal using bar notation.

1. 0.757575...   2. 0.444444...   3. 2.875875...   4. 0.333333...

5. 6.404040...   6. 5.272727...   7. 0.8521685216...   8. 0.833333...

Express each fraction or mixed number as a decimal. Use bar notation to show a repeating decimal.

9. $\frac{3}{16}$ 10. $\frac{8}{33}$ 11. $\frac{7}{12}$ 12. $\frac{14}{25}$

13. $\frac{7}{10}$ 14. $\frac{5}{8}$ 15. $\frac{11}{15}$ 16. $\frac{8}{9}$

17. $\frac{15}{16}$ 18. $\frac{1}{12}$ 19. $\frac{7}{20}$ 20. $\frac{5}{18}$

Lesson 6-1 *(Pages 228–231)*

Round each number to the nearest half.

1. $\frac{11}{12}$ 2. $\frac{5}{8}$ 3. $\frac{2}{5}$ 4. $\frac{1}{10}$ 5. $\frac{1}{6}$ 6. $\frac{2}{3}$

7. $\frac{9}{10}$ 8. $\frac{1}{8}$ 9. $\frac{4}{9}$ 10. $1\frac{1}{8}$ 11. $\frac{12}{11}$ 12. $2\frac{4}{5}$

13. $\frac{7}{9}$ 14. $7\frac{1}{10}$ 15. $10\frac{2}{3}$ 16. $\frac{1}{3}$ 17. $\frac{7}{16}$ 18. $\frac{5}{7}$

Lesson 6-2 *(Pages 232–234)*

Estimate.

1. $14\frac{1}{10} - 6\frac{4}{5}$ 2. $8\frac{1}{3} + 2\frac{1}{6}$ 3. $4\frac{7}{8} + 7\frac{3}{4}$ 4. $11\frac{11}{12} - 5\frac{1}{4}$

5. $3\frac{2}{5} - 1\frac{1}{4}$ 6. $4\frac{2}{5} + \frac{5}{6}$ 7. $4\frac{7}{12} - 1\frac{3}{4}$ 8. $4\frac{2}{3} + 10\frac{3}{8}$

9. $7\frac{7}{15} - 3\frac{1}{12}$ 10. $2\frac{1}{20} + 1\frac{1}{3}$ 11. $18\frac{1}{4} - 12\frac{3}{5}$ 12. $12\frac{5}{9} + 8\frac{5}{8}$

13. $8\frac{2}{3} - 5\frac{1}{2}$ 14. $3\frac{1}{8} - 2\frac{3}{5}$ 15. $9\frac{2}{7} - \frac{1}{3}$ 16. $11\frac{7}{8} - \frac{5}{6}$

17. $2\frac{2}{5} + 2\frac{1}{4}$ 18. $8\frac{1}{2} - 7\frac{4}{5}$ 19. $8\frac{3}{4} + 4\frac{2}{3}$ 20. $1\frac{1}{8} + 7\frac{1}{10}$

Lesson 6-2B *(Pages 236–237)*
Solve.

1. Ben wants a part-time job. There are three 10 hour per week jobs available. One pays $45 per week. One pays $4.75 per hour. One pays $3.25 per hour plus tips which average $20 per week. If Ben wants to take the job that pays the most, for which job should he apply?

2. Micaela gave the clerk $50 to pay for a $21.79 hat and a $17.33 belt. Should she expect $10, $20, or $40 change?

3. Jackie used her calculator to multiply 678 and 34. Should she expect the product to be about 2,305, 23,050 or 230,500?

4. If Carl blinks 15 times per minute, about how many times does he blink in one day?
   a. 216 times
   b. 2,160 times
   c. 21,600 times
   d. 216,000 times

Lesson 6-3 *(Pages 238–241)*
Add or subtract. Write the answer in simplest form.

1. $\frac{2}{5} + \frac{2}{5}$
2. $\frac{5}{8} + \frac{3}{8}$
3. $\frac{9}{11} - \frac{3}{11}$
4. $\frac{3}{14} + \frac{5}{14}$

5. $\frac{7}{8} - \frac{3}{8}$
6. $\frac{3}{4} - \frac{1}{4}$
7. $\frac{15}{27} - \frac{7}{27}$
8. $\frac{1}{36} + \frac{5}{36}$

9. $\frac{2}{9} - \frac{1}{9}$
10. $\frac{7}{8} + \frac{5}{8}$
11. $\frac{9}{16} - \frac{5}{16}$
12. $\frac{6}{8} + \frac{4}{8}$

13. $\frac{1}{2} + \frac{1}{2}$
14. $\frac{1}{3} - \frac{1}{3}$
15. $\frac{8}{9} + \frac{7}{9}$
16. $\frac{5}{6} - \frac{3}{6}$

17. $\frac{3}{9} + \frac{8}{9}$
18. $\frac{8}{40} + \frac{12}{40}$
19. $\frac{56}{90} - \frac{26}{90}$
20. $\frac{2}{9} + \frac{8}{9}$

Lesson 6-4 *(Pages 243–245)*
Add or subtract. Write the answer in simplest form.

1. $\frac{1}{3} + \frac{1}{2}$
2. $\frac{2}{9} + \frac{1}{3}$
3. $\frac{1}{2} + \frac{3}{4}$
4. $\frac{1}{4} + \frac{3}{12}$

5. $\frac{5}{9} - \frac{1}{3}$
6. $\frac{5}{8} - \frac{2}{5}$
7. $\frac{3}{4} - \frac{1}{2}$
8. $\frac{7}{8} - \frac{3}{16}$

9. $\frac{9}{16} + \frac{13}{24}$
10. $\frac{8}{15} + \frac{2}{3}$
11. $\frac{5}{14} + \frac{11}{28}$
12. $\frac{11}{12} + \frac{7}{8}$

13. $\frac{2}{3} - \frac{1}{6}$
14. $\frac{9}{16} - \frac{1}{2}$
15. $\frac{5}{8} - \frac{11}{20}$
16. $\frac{14}{15} - \frac{2}{9}$

17. $\frac{9}{20} + \frac{2}{15}$
18. $\frac{5}{6} + \frac{4}{5}$
19. $\frac{23}{25} - \frac{27}{50}$
20. $\frac{19}{25} - \frac{1}{2}$

Lesson 6-5 *(Pages 246–249)*

Add or subtract. Write the answer in simplest form.

1. $5\frac{1}{2} + 3\frac{1}{4}$

2. $2\frac{2}{3} + 4\frac{1}{9}$

3. $7\frac{4}{5} + 9\frac{3}{10}$

4. $9\frac{4}{7} - 3\frac{5}{14}$

5. $13\frac{1}{5} - 10$

6. $3\frac{3}{4} + 5\frac{5}{8}$

7. $3\frac{2}{5} + 7\frac{6}{15}$

8. $10\frac{2}{3} + 5\frac{6}{7}$

9. $15\frac{6}{9} - 13\frac{5}{12}$

10. $13\frac{7}{12} - 9\frac{1}{4}$

11. $5\frac{2}{3} - 3\frac{1}{2}$

12. $17\frac{2}{9} + 12\frac{1}{3}$

13. $6\frac{5}{12} + 12\frac{5}{8}$

14. $8\frac{3}{5} - 2\frac{1}{5}$

15. $23\frac{2}{3} - 4\frac{1}{2}$

Lesson 6-6 *(Pages 250–253)*

Subtract. Write the answer in simplest form.

1. $11\frac{2}{3} - 8\frac{11}{12}$

2. $3\frac{4}{7} - 1\frac{2}{3}$

3. $7\frac{1}{8} - 4\frac{1}{3}$

4. $18\frac{1}{9} - 12\frac{2}{5}$

5. $12\frac{3}{10} - 8\frac{3}{4}$

6. $43 - 5\frac{1}{5}$

7. $8\frac{1}{5} - 4\frac{1}{4}$

8. $14\frac{1}{6} - 3\frac{2}{3}$

9. $25\frac{4}{7} - 21$

10. $17\frac{3}{9} - 4\frac{3}{5}$

11. $18\frac{1}{9} - 1\frac{3}{7}$

12. $16\frac{1}{4} - 7\frac{1}{5}$

13. $18\frac{1}{5} - 6\frac{1}{4}$

14. $4 - 1\frac{2}{3}$

15. $26 - 4\frac{1}{9}$

Lesson 6-7 *(Pages 254–257)*

Complete.

1. 2 h 10 min = 1 h _____ min

2. 3 h 65 min = _____ h 5 min

3. 3 min 14 s = 2 min _____ s

4. 1 h 15 min 10 s = _____ min 10 s

Add or subtract. Rename if necessary.

5.　　6 h 14 min
　　　−2 h　8 min

6.　　5 h 35 min 25 s
　　　+　　45 min 35 s

7.　　5 h　4 min 45 s
　　　−2 h 40 min　5 s

8.　　15 h 16 min
　　　− 8 h 35 min 16 s

9.　　9 h 20 min 10 s
　　　+1 h 39 min 55 s

10.　　2 h 40 min 20 s
　　　+3 h　5 min 50 s

Find the elapsed time.

11. 10:30 A.M. to 6:00 P.M.

12. 8:45 P.M. to 1:30 A.M.

Lesson 7-1 *(Pages 268–270)*

Round each fraction to 0, $\frac{1}{2}$, or 1.

1. $\frac{3}{4}$
2. $\frac{5}{8}$
3. $\frac{3}{25}$
4. $\frac{1}{20}$
5. $\frac{5}{11}$
6. $\frac{11}{18}$
7. $\frac{1}{3}$

Round each mixed number to the nearest whole number.

8. $6\frac{4}{9}$
9. $19\frac{3}{4}$
10. $2\frac{1}{15}$
11. $17\frac{2}{7}$

12. $8\frac{9}{16}$
13. $1\frac{3}{5}$
14. $15\frac{41}{45}$

Estimate each product.

15. $\frac{2}{3} \times \frac{4}{5}$
16. $\frac{1}{6} \times \frac{2}{5}$
17. $\frac{4}{9} \times \frac{3}{7}$
18. $\frac{5}{12} \times \frac{6}{11}$

19. $\frac{3}{8} \times \frac{8}{9}$
20. $\frac{3}{5} \times \frac{5}{12}$
21. $\frac{2}{5} \times \frac{5}{8}$
22. $5\frac{3}{7} \times \frac{4}{5}$

Lesson 7-2 *(Pages 273–276)*

Find each product. Write in simplest form.

1. $\frac{5}{6} \times \frac{15}{16}$
2. $\frac{6}{14} \times \frac{12}{18}$
3. $\frac{2}{3} \times \frac{3}{13}$
4. $\frac{4}{9} \times \frac{1}{6}$

5. $\frac{3}{4} \times \frac{5}{6}$
6. $\frac{9}{10} \times \frac{3}{4}$
7. $\frac{8}{9} \times \frac{2}{3}$
8. $\frac{6}{7} \times \frac{4}{5}$

9. $\frac{8}{11} \times \frac{11}{12}$
10. $\frac{5}{6} \times \frac{3}{5}$
11. $\frac{6}{7} \times \frac{7}{21}$
12. $\frac{8}{9} \times \frac{9}{10}$

Solve each equation. Write the solution in simplest form.

13. $q = \frac{7}{11} \times \frac{12}{15}$
14. $r = \frac{7}{9} \times \frac{5}{7}$
15. $\frac{8}{13} \times \frac{2}{11} = a$

16. $m = \frac{4}{7} \times \frac{2}{9}$
17. $d = \frac{4}{9} \times \frac{24}{25}$
18. $\frac{1}{9} \times \frac{6}{13} = s$

19. $j = \frac{4}{7} \times 6$
20. $\frac{7}{10} \times 5 = p$
21. $10 \times 3\frac{1}{5} = k$

Lesson 7-3 *(Pages 277–279)*

Express each mixed number as an improper fraction.

1. $2\frac{1}{3}$
2. $2\frac{1}{2}$
3. $2\frac{2}{3}$
4. $3\frac{1}{2}$
5. $1\frac{1}{4}$

Find each product. Write in simplest form.

6. $3\frac{5}{8} \times 4\frac{1}{2}$
7. $\frac{4}{5} \times 2\frac{3}{4}$
8. $6\frac{1}{8} \times 5\frac{1}{7}$
9. $2\frac{2}{3} \times 2\frac{1}{4}$

10. $6\frac{2}{3} \times 7\frac{3}{5}$
11. $7\frac{1}{5} \times 2\frac{4}{7}$
12. $8\frac{3}{4} \times 2\frac{2}{5}$
13. $4\frac{1}{3} \times 2\frac{1}{7}$

Solve each equation. Write the solution in simplest form.

14. $4\frac{3}{5} \times 2\frac{1}{2} = c$
15. $z = 5\frac{5}{6} \times 4\frac{2}{7}$

16. $6\frac{8}{9} \times 3\frac{5}{6} = f$
17. $2\frac{1}{9} \times 1\frac{1}{2} = x$

18. $4\frac{7}{15} \times 3\frac{3}{4} = h$
19. $k = 5\frac{7}{9} \times 6\frac{3}{8}$

Lesson 7-4 *(Pages 280–283)*

Find the circumference of each circle shown or described to the nearest tenth. Use $\frac{22}{7}$ or 3.14 for π.

1.

9 cm

2.

2.1 yd

3.

0.6 m

4.

10 in.

5. $d = 5.6$ m   **6.** $r = 3.21$ yd   **7.** $r = 0.5$ in.

8. $d = 4$ m   **9.** $r = 16$ cm   **10.** $d = 9.1$ m

11. $r = 0.1$ yd   **12.** $d = 65.7$ m   **13.** $r = 1$ cm

Lesson 7-5 *(Pages 285–288)*

Find the reciprocal of each number.

1. $\frac{12}{13}$ **2.** $\frac{7}{11}$ **3.** 5 **4.** $\frac{1}{4}$ **5.** $\frac{7}{9}$ **6.** $\frac{9}{2}$ **7.** $\frac{1}{5}$

Find each quotient. Write in simplest form.

8. $\frac{2}{3} \div \frac{1}{2}$ **9.** $\frac{3}{5} \div \frac{2}{5}$ **10.** $\frac{7}{10} \div \frac{3}{8}$ **11.** $\frac{5}{9} \div \frac{2}{3}$

12. $4 \div \frac{2}{3}$ **13.** $8 \div \frac{4}{5}$ **14.** $9 \div \frac{5}{9}$ **15.** $\frac{2}{7} \div 7$

Solve each equation. Write the solution in simplest form.

16. $\frac{1}{14} \div 7 = b$ **17.** $\frac{2}{13} \div \frac{5}{26} = y$ **18.** $g = \frac{4}{7} \div \frac{6}{7}$

19. $w = \frac{7}{8} \div \frac{1}{3}$ **20.** $15 \div \frac{3}{5} = n$ **21.** $\frac{9}{14} \div \frac{3}{4} = v$

22. $u = \frac{8}{9} \div \frac{5}{6}$ **23.** $j = \frac{4}{9} \div 36$ **24.** $d = \frac{15}{16} \div \frac{5}{8}$

Lesson 7-6 *(Pages 289–291)*

Write each mixed number as an improper fraction. Then write its reciprocal.

1. $4\frac{1}{5}$ **2.** $3\frac{2}{9}$ **3.** $2\frac{3}{8}$ **4.** $5\frac{3}{5}$

Find each quotient. Write in simplest form.

5. $\frac{3}{5} \div 1\frac{2}{3}$ **6.** $2\frac{1}{2} \div 1\frac{1}{4}$ **7.** $7 \div 4\frac{9}{10}$ **8.** $1\frac{3}{7} \div 10$

9. $3\frac{3}{5} \div \frac{4}{5}$ **10.** $8\frac{2}{5} \div 4\frac{1}{2}$ **11.** $6\frac{1}{3} \div 2\frac{1}{2}$ **12.** $5\frac{1}{4} \div 2\frac{1}{3}$

Solve each equation. Write the solution in simplest form.

13. $n = 4\frac{1}{8} \div 3\frac{2}{3}$ **14.** $p = 2\frac{5}{8} \div \frac{1}{2}$ **15.** $1\frac{5}{6} \div 3\frac{2}{3} = z$

16. $k = 21 \div 5\frac{1}{4}$ **17.** $b = 12 \div 3\frac{3}{5}$ **18.** $q = 18 \div 2\frac{1}{4}$

Lesson 7-7 *(Pages 292–294)*
Complete.

1. 3 gal = <u>?</u> pt
2. 24 pt = <u>?</u> gal
3. 20 lb = <u>?</u> oz
4. 2 gal = <u>?</u> fl oz
5. 20 pt = <u>?</u> qt
6. 18 qt = <u>?</u> pt
7. 2,000 lb = <u>?</u> T
8. 3 T = <u>?</u> lb
9. 6 lb = <u>?</u> oz
10. 9 lb = <u>?</u> oz
11. 15 qt = <u>?</u> gal
12. 4 pt = <u>?</u> c
13. 4 gal = <u>?</u> qt
14. 4 qt = <u>?</u> fl oz
15. 12 pt = <u>?</u> c
16. 10 pt = <u>?</u> qt
17. 24 fl oz = <u>?</u> c
18. 1.5 pt = <u>?</u> c
19. $\frac{1}{4}$ lb = <u>?</u> oz
20. 5 T = <u>?</u> lb
21. 2 lb = <u>?</u> oz

Lesson 7-8A *(Pages 296–297)*
Solve.

1. What are the next three numbers in the pattern? 6, 7, 9, 12, 16, . . .

2. Kip and Celia began working for the same company in 1997. Celia earned $19,000 per year, and Kip earned $16,500 per year. Each year Kip received a $1,500 raise and Celia received a $1,000 raise.
   a. In what year will they earn the same amount of money?
   b. What will be their annual salary in that year?

3. What is the missing measurement in the pattern?
   . . . $\frac{1}{2}$ in., $\frac{1}{4}$ in., ____ in., $\frac{1}{16}$ in., . . .

4. Daria eats half of a pizza at 12:00 P.M. At 2:00 P.M., she eats half of what was left of her pizza from lunch. At 4:00 P.M., she eats half of what was left from the pizza. If Daria continues eating at this rate for six more hours, how much of the pizza has she eaten?

5. At Snowfalls Middle School, the bell rings at 8:05, 8:55, 9:00, and 9:50 each morning. If this pattern continues, when would the next three bells ring?

Lesson 7-8 *(Pages 298–300)*
Find the next two numbers in each sequence.

1. 14, 21, 28, 35, . . .
2. 36, 42, 48, 54, . . .
3. 3, 9, 27, 81, . . .
4. 2, 6, 10, 14, . . .
5. 1,600, 800, 400, 200, . . .
6. 192, 96, 48, 24, . . .
7. 15, $14\frac{1}{3}$, $13\frac{2}{3}$, 13, . . .
8. 11, $11\frac{1}{2}$, 12, $12\frac{1}{2}$, . . .
9. $\frac{1}{5}$, 2, 20, 200, . . .
10. 36, 6, 1, $\frac{1}{6}$, . . .

Find the missing number in each sequence.

11. 5, 10, <u>?</u>, 40, . . .
12. <u>?</u>, 193, 293, 393, . . .
13. 8, 32, 128, <u>?</u>, . . .
14. 11, <u>?</u>, 19, 23, . . .
15. 9, $8\frac{3}{4}$, <u>?</u>, $8\frac{1}{4}$, . . .
16. 64, <u>?</u>, 16, 8, . . .
17. <u>?</u>, 19, 26, 33, . . .
18. $\frac{1}{2}$, $1\frac{1}{2}$, $4\frac{1}{2}$, <u>?</u>, . . .
19. $\frac{1}{81}$, <u>?</u>, 1, 9, . . .
20. 12, <u>?</u>, 432, 2,592, . . .

Lesson 8-1 *(Pages 312–315)*

Express each ratio as a fraction in simplest form.

1. 21 sugar cookies out of an assortment of 75 cookies.

2. 10 girls in a class of 25 students.

3. 34 non-smoking tables in a restaurant with 50 tables.

4. 7 striped ties out of 21 ties.

Express each ratio as a rate.

5. $2.00 for 5 cans of tomato soup

6. $200.00 for 40 hours of work

7. 540 parts produced in 18 hours

Lesson 8-2 *(Pages 317–320)*

Use cross products to determine whether each pair of ratios forms a proportion.

1. $\dfrac{3}{10}, \dfrac{7}{25}$

2. $\dfrac{5}{12}, \dfrac{3}{8}$

3. $\dfrac{12}{16}, \dfrac{9}{12}$

4. $\dfrac{5}{4}, \dfrac{125}{100}$

5. $\dfrac{4}{5}, \dfrac{80}{100}$

Solve each proportion.

6. $\dfrac{15}{21} = \dfrac{5}{b}$

7. $\dfrac{22}{25} = \dfrac{n}{100}$

8. $\dfrac{24}{48} = \dfrac{h}{50}$

9. $\dfrac{9}{27} = \dfrac{y}{42}$

10. $\dfrac{4}{7} = \dfrac{16}{x}$

11. $\dfrac{4}{6} = \dfrac{a}{9}$

12. $\dfrac{6}{14} = \dfrac{21}{m}$

13. $\dfrac{3}{7} = \dfrac{21}{d}$

14. $\dfrac{4}{10} = \dfrac{18}{e}$

15. $\dfrac{9}{10} = \dfrac{27}{f}$

Lesson 8-3A *(Pages 322–323)*

Solve.

1. Four cars were crossing a bridge. The red car was behind the yellow car. The green car was last. The blue car was not first. In what order did the cars cross the bridge?

2. The streets in Sandra's town are arranged in square blocks. Sandra left school and walked 5 blocks west and 3 blocks south to Marcia's house. She then walked 1 block south and 3 blocks east to the park. Finally she walked 2 blocks east and 2 blocks north to get home. How far is Sandra's home from the school?

3. Nida arranged five friends in a line for a photograph. Benito stood between Emma and Toma. Sam stood between Toma and Clara. Clara was on the left. In what order were the five people arranged for the photograph?

4. How many different rectangles can be made with 36 one-inch tiles?

5. In how many different ways can Amelia, Bob, Cornelia, and Damien seat themselves in four chairs at a round table?

Lesson 8-3 *(Pages 324–327)*

The map at the right has a scale of $\frac{1}{4}$ in. = 5 km. Use a ruler to measure each map distance to the nearest $\frac{1}{4}$ inch. Then find the actual distances.

1. Bryan to Napoleon
2. Stryker to Evansport
3. Ney to Bryan
4. Defiance to Napoleon
5. Ridgeville to Bryan
6. Brunersburg to Ney

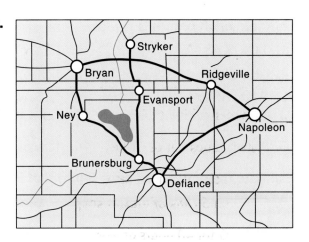

Lesson 8-4 *(Pages 330–333)*

Express each percent as a fraction in simplest form.

1. 13%	**2.** 25%	**3.** 8%	**4.** 105%	**5.** 60%
6. 70%	**7.** 80%	**8.** 45%	**9.** 20%	**10.** 14%
11. 75%	**12.** 120%	**13.** 5%	**14.** 2%	**15.** 450%

Express each fraction as a percent.

16. $\frac{77}{100}$	**17.** $\frac{3}{4}$	**18.** $\frac{17}{20}$	**19.** $\frac{3}{25}$	**20.** $\frac{3}{10}$
21. $\frac{27}{50}$	**22.** $\frac{2}{5}$	**23.** $\frac{3}{50}$	**24.** $\frac{9}{20}$	**25.** $\frac{8}{5}$
26. $\frac{1}{4}$	**27.** $\frac{1}{5}$	**28.** $\frac{19}{20}$	**29.** $\frac{7}{10}$	**30.** $\frac{11}{25}$

Lesson 8-5 *(Pages 334–336)*

Express each percent as a decimal.

1. 5%	**2.** 22%	**3.** 50%	**4.** 420%	**5.** 75%
6. 1%	**7.** 100%	**8.** 3.7%	**9.** 0.9%	**10.** 9%
11. 90%	**12.** 900%	**13.** 78%	**14.** 62.5%	**15.** 15%

Express each decimal as a percent.

16. 0.02	**17.** 0.2	**18.** 0.002	**19.** 1.02	**20.** 0.66
21. 0.11	**22.** 0.354	**23.** 0.31	**24.** 0.09	**25.** 5.2
26. 2.22	**27.** 0.008	**28.** 0.275	**29.** 0.3	**30.** 6.0

Lesson 8-6 (Pages 337–339)

Estimate each percent.

1. 11% of 48
2. 1.9% of 50
3. 29% of 500
4. 41% of 50
5. 32% of 300
6. 411% of 50
7. 149% of 60
8. 4.1% of 50
9. 62% of 200
10. 58% of 100
11. 52% of 400
12. 68% of 30
13. 9% of 25
14. 48% of 1000
15. 98% of 725

Estimate the percent of each figure that is shaded.

16.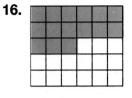
17.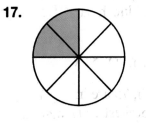
18.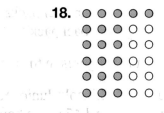

Lesson 8-7 (Pages 340–343)

Find the percent of each number.

1. 38% of 150
2. 20% of 75
3. 0.2% of 500
4. 25% of 70
5. 10% of 90
6. 16% of 30
7. 39% of 40
8. 250% of 100
9. 6% of 86
10. 12.5% of 160
11. 9% of 29
12. 3% of 46
13. $66\frac{2}{3}$% of 60
14. 89% of 47
15. 435% of 30
16. 25% of 48
17. 5% of 420
18. 55% of 134
19. 28% of 4
20. 14% of 40
21. 14% of 14
22. 90% of 140
23. 40% of 45
24. 0.5% of 200

Lesson 9-1 (Pages 352–355)

Use a protractor to find the measure of each angle. Then classify the angle as *acute*, *right*, or *obtuse*.

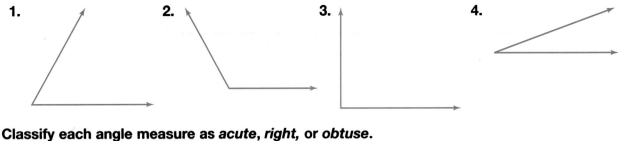

1.
2.
3.
4.

Classify each angle measure as *acute*, *right*, or *obtuse*.

5. 86°
6. 101°
7. 90°
8. 145.6°

Classify each pair of angles as *complementary*, *supplementary*, or *neither*.

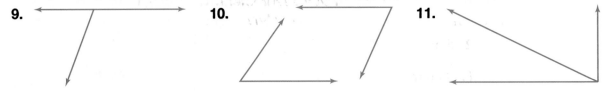

9.
10.
11.

Lesson 9-1B (Pages 356–357)
Solve.

1. At Morgon Middle School, 37 of the eighth grade students are involved in a club activity, 63 play sports, and 21 are involved in both sports and clubs.
   a. How many students participate in only clubs?
   b. How many students participate in only sports?
   c. What is the total number of students who participate in extracurricular activities?

2. In a class of 26 students going on a field trip, 8 have backpacks, 13 have lunch bags, and 6 have both backpacks and lunch bags. How many have neither a lunch bag nor a backpack?

3. What are the next two figures in the pattern? ↑ ↖ ← ↙

4. Of the 320 students at Lincoln Junior High, 75 are in the orchestra, 120 are in the marching band, and 45 are in both orchestra and marching band. How many students are in neither orchestra nor marching band?

Lesson 9-2 (Pages 358–361)
Use a protractor to draw angles having the following measurements.

1. 165°   **2.** 20°   **3.** 90°   **4.** 41°

Estimate the measure of each angle.

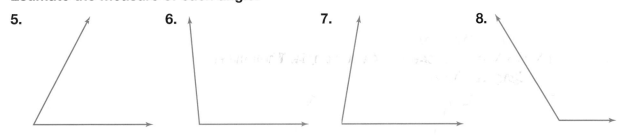

5.   **6.**   **7.**   **8.**

Lesson 9-3 (Pages 364–367)
Draw the angle or line segment with the given measurement. Then use a straightedge and a compass to bisect each angle or line segment.

1. 3 in.   **2.** 5 cm   **3.** 110°   **4.** 48 mm

5. 70°   **6.** 33 mm   **7.** 25°   **8.** 150°

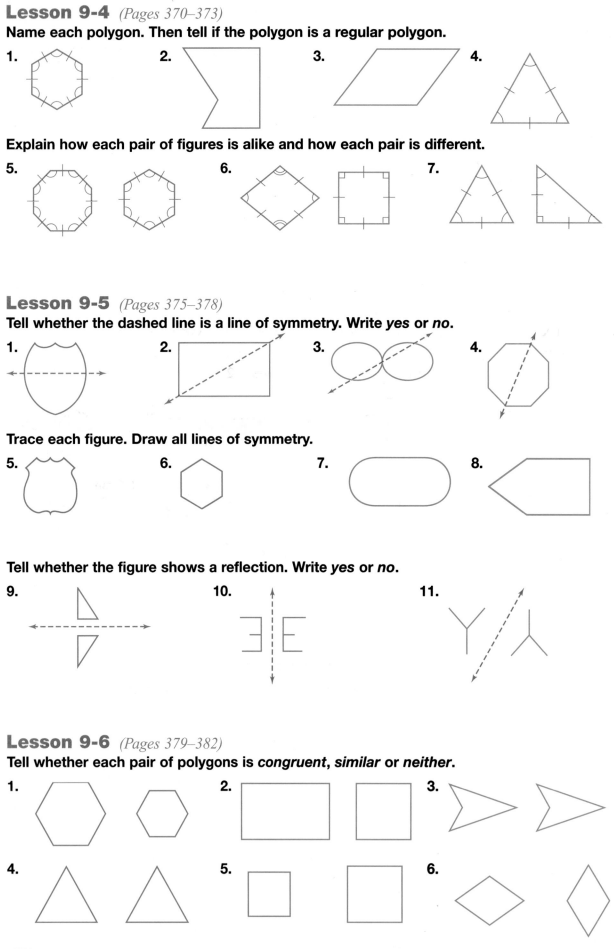

Lesson 9-4 *(Pages 370–373)*

Name each polygon. Then tell if the polygon is a regular polygon.

1. **2.** **3.** **4.**

Explain how each pair of figures is alike and how each pair is different.

5. **6.** **7.**

Lesson 9-5 *(Pages 375–378)*

Tell whether the dashed line is a line of symmetry. Write *yes* or *no*.

1. **2.** **3.** **4.**

Trace each figure. Draw all lines of symmetry.

5. **6.** **7.** **8.**

Tell whether the figure shows a reflection. Write *yes* or *no*.

9. **10.** **11.**

Lesson 9-6 *(Pages 379–382)*

Tell whether each pair of polygons is *congruent*, *similar* or *neither*.

1. **2.** **3.**

4. **5.** **6.**

EXTRA PRACTICE

582 Extra Practice

Lesson 10-1 (Pages 398–401)

Find the area of each parallelogram to the nearest tenth.

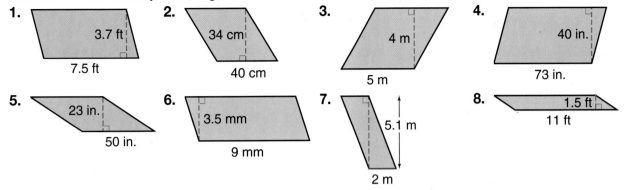

1. 3.7 ft, 7.5 ft

2. 34 cm, 40 cm

3. 4 m, 5 m

4. 40 in., 73 in.

5. 23 in., 50 in.

6. 3.5 mm, 9 mm

7. 5.1 m, 2 m

8. 1.5 ft, 11 ft

Lesson 10-2 (Pages 402–405)

Find the area of each triangle. Round decimal answers to the nearest tenth.

1. base, 6 ft
   height, 3 ft

2. base, 4.2 in.
   height, 6.8 in.

3. base, 9.1 m
   height, 7.2 m

4. base, 13.2 cm
   height, 16.2 cm

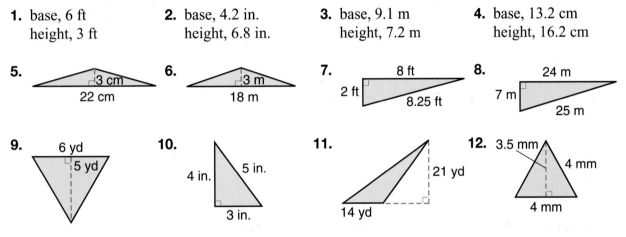

5. 3 cm, 22 cm

6. 3 m, 18 m

7. 8 ft, 2 ft, 8.25 ft

8. 24 m, 7 m, 25 m

9. 6 yd, 5 yd

10. 4 in., 5 in., 3 in.

11. 21 yd, 14 yd

12. 3.5 mm, 4 mm, 4 mm

Lesson 10-3 (Pages 406–409)

Find the area of each circle to the nearest tenth. Use 3.14 for π.

1. radius, 4 m

2. diameter, 6 in.

3. radius, 16 m

4. diameter, 11 in.

5. radius, 9 cm

6. diameter, 24 mm

7. 7 m

8. 12 cm

9. 10 in.

10. 8 m

11. 7 in.

12. 20 in.

13. 8.5 mm

14. 22.4 m

Lesson 10-4 *(Pages 412–414)*
Name each figure.

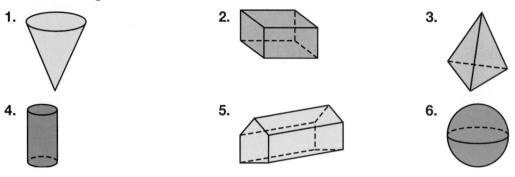

1.

2.

3.

4.

5.

6.

State the number of faces, edges, and vertices in each figure.

7. square prism

8. triangular pyramid

9. cone

10. pentagonal pyramid

Lesson 10-5A *(Pages 416–417)*
Solve.

1. Ten 3-inch cubes fit in a carton. What are the possible dimensions of the carton?

2. Eve is making a pyramid-shaped display of boxes of soda. Each box is a cube. The bottom layer of the pyramid has 8 boxes. If there is one less box in each layer and there are 6 layers in the pyramid, how many boxes will Eve need to make the display?

3. A construction worker wants to arrange 20 cement bricks into the shape of a rectangle with the smallest perimeter possible. How many bricks will be in each row?

4. A rectangular prism can be formed by using exactly 10 cubes. Find the length, width, and height of the prism.

Lesson 10-5 *(Pages 418–420)*
Find the volume of each rectangular prism to the nearest tenth.

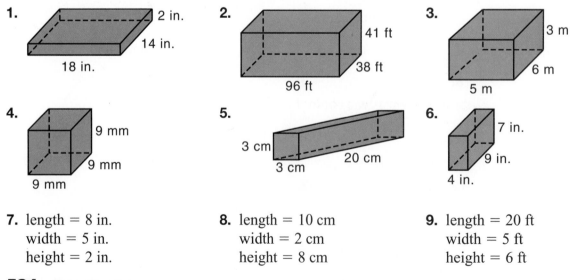

1. 2 in. 14 in. 18 in.

2. 41 ft 38 ft 96 ft

3. 3 m 6 m 5 m

4. 9 mm 9 mm 9 mm

5. 3 cm 3 cm 20 cm

6. 7 in. 9 in. 4 in.

7. length = 8 in.
width = 5 in.
height = 2 in.

8. length = 10 cm
width = 2 cm
height = 8 cm

9. length = 20 ft
width = 5 ft
height = 6 ft

Lesson 10-6 *(Pages 421–424)*

Find the surface area of each rectangular prism to the nearest tenth.

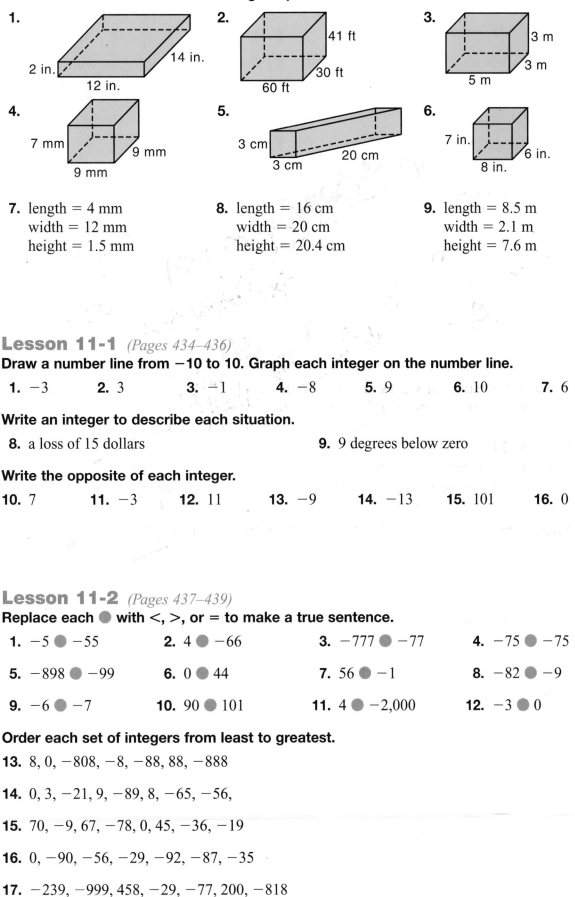

1. 2 in. 14 in. 12 in.

2. 41 ft 30 ft 60 ft

3. 3 m 3 m 5 m

4. 7 mm 9 mm 9 mm

5. 3 cm 3 cm 20 cm

6. 7 in. 6 in. 8 in.

7. length = 4 mm
width = 12 mm
height = 1.5 mm

8. length = 16 cm
width = 20 cm
height = 20.4 cm

9. length = 8.5 m
width = 2.1 m
height = 7.6 m

Lesson 11-1 *(Pages 434–436)*

Draw a number line from −10 to 10. Graph each integer on the number line.

1. −3   **2.** 3   **3.** −1   **4.** −8   **5.** 9   **6.** 10   **7.** 6

Write an integer to describe each situation.

8. a loss of 15 dollars   **9.** 9 degrees below zero

Write the opposite of each integer.

10. 7   **11.** −3   **12.** 11   **13.** −9   **14.** −13   **15.** 101   **16.** 0

Lesson 11-2 *(Pages 437–439)*

Replace each ● with <, >, or = to make a true sentence.

1. −5 ● −55   **2.** 4 ● −66   **3.** −777 ● −77   **4.** −75 ● −75

5. −898 ● −99   **6.** 0 ● 44   **7.** 56 ● −1   **8.** −82 ● −9

9. −6 ● −7   **10.** 90 ● 101   **11.** 4 ● −2,000   **12.** −3 ● 0

Order each set of integers from least to greatest.

13. 8, 0, −808, −8, −88, 88, −888

14. 0, 3, −21, 9, −89, 8, −65, −56,

15. 70, −9, 67, −78, 0, 45, −36, −19

16. 0, −90, −56, −29, −92, −87, −35

17. −239, −999, 458, −29, −77, 200, −818

Lesson 11-3 *(Pages 441–444)*
Find each sum. Use counters or a number line if necessary.

1. $-4 + (-7)$
2. $-1 + 0$
3. $7 + (-13)$
4. $-20 + 2$
5. $4 + (-6)$
6. $-12 + 9$
7. $-12 + (-10)$
8. $5 + (-15)$
9. $17 + 9$
10. $18 + (-18)$
11. $-4 + (-4)$
12. $0 + (-9)$
13. $-12 + (-9)$
14. $-8 + 7$
15. $3 + (-6)$
16. $-9 + 16$
17. $-5 + (-3)$
18. $-5 + 5$
19. $-3 + (-3)$
20. $-11 + 6$
21. $-10 + 6$
22. $-5 + (-9)$
23. $18 + (-20)$
24. $-4 + (-4)$
25. $2 + (-4)$
26. $-3 + (-11)$
27. $-17 + 9$
28. $-6 + 10$
29. $-6 + (-12)$
30. $8 + 8$
31. $-9 + 0$
32. $4 + (-5)$

Lesson 11-4 *(Pages 445–448)*
Find each difference. Use counters or a number line if necessary.

1. $7 - (-4)$
2. $-4 - (-9)$
3. $13 - (-3)$
4. $2 - (-5)$
5. $-9 - 5$
6. $-11 - (-18)$
7. $-4 - (-7)$
8. $-6 - (-6)$
9. $-6 - 6$
10. $17 - 9$
11. $-12 - (-9)$
12. $0 - (-4)$
13. $-7 - 0$
14. $-12 - (-10)$
15. $-2 - (-1)$
16. $3 - (-5)$
17. $5 - (-1)$
18. $-5 - (-6)$
19. $9 - (-1)$
20. $1 - 9$
21. $-5 - 1$
22. $-1 - 4$
23. $0 - (-7)$
24. $8 - 13$
25. $-4 - (-6)$
26. $9 - 9$
27. $-7 - (-7)$
28. $7 - 5$
29. $8 - (-5)$
30. $5 - 8$
31. $1 - 6$
32. $-8 - (-8)$

Lesson 11-5 *(Pages 449–452)*
Find each product. Use counters or a number line if necessary.

1. $3 \times (-5)$
2. -5×1
3. $-8 \times (-4)$
4. $6 \times (-3)$
5. -3×2
6. $-1 \times (-4)$
7. $8 \times (-2)$
8. $-5 \times (-7)$
9. $3 \times (-9)$
10. -9×4
11. $-4 \times (-5)$
12. $5 \times (-2)$
13. $-8(3)$
14. $-9(-1)$
15. $7(-3)$
16. $2(3)$
17. $-6(0)$
18. $-5(-1)$
19. $5(-5)$
20. $-2(-3)$
21. $8(-4)$
22. $-2(4)$
23. $-4(-4)$
24. $2(9)$
25. $-2(-12)$
26. $7 \times (-4)$
27. $-5 \times (-9)$
28. -2×11
29. $4(-2)$
30. $4(-4)$
31. $-3(-11)$
32. $-3(3)$

Lesson 11-6A *(Pages 454–455)*
Solve.

1. Morgan and Stephen want to get ready for the fall football season that starts August 29. Before the season starts, they plan to spend 3 to 4 weeks practicing plays. Before practicing plays, they plan to spend 6 to 8 weeks running and exercising. What are the earliest and latest dates that Morgan and Stephen can begin getting ready for the football season?

2. Daniella and Wayne must arrive at school no later than 7:25 A.M. weekday mornings. It takes 20 minutes to get from Wayne's house to the school. Daniella picks up Wayne, but it takes her 15 minutes to get to Wayne's house. If it takes Daniella 55 minutes to get ready in the morning, what is the latest she should get out of bed?

3. Melina bought some school supplies. She spent $4.35 on paper, pens, and pencils and 3 times that on notebooks. After paying for the items, she had $7.26 left. How much money did she have before she went shopping?

4. In the trivia bowl, each finalist must answer four questions correctly. Each question is worth twice as much as the question before it. The fourth question is worth $6,000. How much is the first question worth?

5. Connor, Petra, Daria, and Lyle each handed in term papers of the Industrial Revolution. Petra's paper was 4 pages shorter than Connor's. Daria's paper was 7 pages longer than Petra's. Lyle's paper was half as long as Daria's. Lyle's paper was 14 pages long. How long was Connor's paper?

Lesson 11-6 *(Pages 456–458)*
Find each quotient. Use counters or patterns if necessary.

1. $12 \div (-6)$	**2.** $-7 \div (-1)$	**3.** $-4 \div 4$	**4.** $6 \div (-6)$
5. $0 \div (-4)$	**6.** $45 \div (-9)$	**7.** $15 \div (-5)$	**8.** $-6 \div 2$
9. $-28 \div (-7)$	**10.** $20 \div (-2)$	**11.** $-40 \div (-8)$	**12.** $12 \div (-4)$
13. $-18 \div 6$	**14.** $9 \div (-1)$	**15.** $-30 \div 6$	**16.** $-54 \div (-9)$
17. $28 \div (-7)$	**18.** $-24 \div 8$	**19.** $24 \div (-4)$	**20.** $-14 \div 7$
21. $9 \div 3$	**22.** $-18 \div (-6)$	**23.** $-9 \div (-1)$	**24.** $18 \div (-9)$
25. $-25 \div (-5)$	**26.** $15 \div (-3)$	**27.** $-36 \div 9$	**28.** $-4 \div 2$
29. $-40 \div 8$	**30.** $-32 \div 4$	**31.** $-27 \div (-9)$	**32.** $-8 \div 8$

Lesson 11-7 *(Pages 459–461)*
Name the ordered pair for each point.

1. M	**2.** A	**3.** D	**4.** E
5. P	**6.** Q	**7.** B	**8.** C
9. F	**10.** G	**11.** N	**12.** R
13. K	**14.** H		

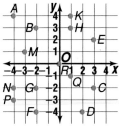

Graph and label each point.

15. $S(4, -1)$	**16.** $T(-3, -2)$	**17.** $W(2, 1)$	**18.** $Y(-5, 3)$
19. $Z(-1, -3)$	**20.** $U(3, -3)$	**21.** $V(1, 2)$	**22.** $X(-1, 4)$

Lesson 11-8 *(Pages 464–467)*

The vertices of △QRS are Q(1, −2), R(3, 4), and S(−2, 2). On a coordinate grid, draw △QRS and each transformation image.

1. translation image that is 5 units down

2. translation image that is 4 units left

3. translation image that is 2 units right and 3 units up

4. translation image that is 1 unit left and 8 units down

5. reflection image over the x-axis

Lesson 12-1 *(Pages 476–479)*

Solve each equation. Use cups and counters if necessary.

1. $x + 4 = 14$ 2. $b + (−10) = 0$ 3. $−2 + w = −5$

4. $k + (−3) = −5$ 5. $−4 + h = 6$ 6. $−7 + d = −3$

7. $m + 11 = 9$ 8. $f + (−9) = −19$ 9. $p + 66 = 22$

10. $−34 + t = 41$ 11. $e + 56 = −24$ 12. $−29 + a = −54$

13. $17 + m = −33$ 14. $b + (−44) = −34$ 15. $w + (−39) = 55$

Lesson 12-2 *(Pages 480–483)*

Solve each equation. Use cups and counters if necessary.

1. $y − (−7) = 2$ 2. $a − 10 = −22$ 3. $g − (−1) = 9$

4. $c − 8 = 5$ 5. $z − (−2) = 7$ 6. $n − 1 = −87$

7. $j − 15 = −22$ 8. $x − 12 = 45$ 9. $y − 65 = −79$

10. $q − 16 = −31$ 11. $q − (−6) = 12$ 12. $j − 18 = −34$

13. $k − (−2) = −8$ 14. $r − 76 = 41$ 15. $n − 63 = −81$

Lesson 12-3 *(Pages 484–487)*
Solve each equation. Use cups and counters if necessary.

1. $5x = 30$
2. $18w = 2$
3. $\frac{1}{2}a = 7$
4. $2d = -28$
5. $\frac{1}{4}c = -3$
6. $11n = 77$
7. $\frac{1}{3}z = 15$
8. $9y = -63$
9. $6m = -54$
10. $5f = -75$
11. $20p = 5$
12. $\frac{1}{4}x = 16$
13. $4t = -24$
14. $7b = 21$
15. $19h = 0$
16. $22d = -66$
17. $\frac{1}{3}y = 11$
18. $3m = -78$
19. $8x = -2$
20. $9c = -72$
21. $\frac{1}{2}p = 35$
22. $\frac{1}{5}k = 20$
23. $33y = 99$
24. $6z = -5$

Lesson 12-4 *(Pages 488–491)*
Solve each equation.

1. $3x + 7 = 13$
2. $\frac{1}{2}r + 6 = -3$
3. $2h - 5 = 7$
4. $-10 = 5x + 5$
5. $2x - (-16) = 26$
6. $6r + 2 = 2$
7. $\frac{1}{5}t - 6 = -12$
8. $-2 - 3y = -11$
9. $-4y + 16 = 64$

Lesson 12-4B *(Pages 492–493)*
Solve.

1. Eric bought a personal CD player for $12 less than the regular price. If he paid $54, what was the regular price?

2. Lisa paid $2.58 in sales tax on an Ohio State Buckeyes sweatshirt. The total cost was $45.57. What was the price of the sweatshirt before taxes?

3. School T-shirts are the most popular item sold in the school store. On Monday, 49 T-shirts were sold. This is 3 more than twice the amount that were sold on Tuesday. How many T-shirts were sold on Tuesday?

4. Jayson made a long-distance phone call to his grandmother. The first 5 minutes cost $3, and each minute after that cost $0.50. How many minutes did they talk on the phone if the cost of the call was $12?

5. Paige went bowling at Champion Bowling Center. Shoe rental was $1.75 and games were $2.25 each. If she spent $8.50 on shoe rental and games, how many games did she play?

Lesson 12-5 *(Pages 496–499)*

Copy and complete each function table.

1.

input (*n*)	output (*n* − 4)
5	
2	
−1	

2.

input (*n*)	output (3*n*)
1	
0	
−2	

Find the rule for each function table.

3.

n	■
−1	4
0	5
3	8

4.

n	■
−6	−3
0	0
8	4

Lesson 12-6 *(Pages 500–503)*

Copy and complete each function table. Then graph the function.

1.

input (*x*)	output (*x* + 1)
2	■
0	■
−3	■

2.

input (*x*)	output (2*x*)
2	■
0	■
−3	■

3.

input (*x*)	output (*x* − 3)
4	■
0	■
−1	■

4.

input (*x*)	output $\left(\frac{x}{5}\right)$
10	■
0	■
−5	■

5.

input (*x*)	output (−3*x*)
2	■
−1	■
−2	■

6.

input (*x*)	output (2*x* − 3)
2	■
0	■
−1	■

Lesson 13-1 *(Pages 515–518)*

A set of 30 tickets are placed in a bag. There are 6 baseball tickets, 4 hockey tickets, 4 basketball tickets, 2 football tickets, 3 symphony tickets, 2 opera tickets, 4 ballet tickets, and 5 theater tickets. One ticket is drawn without looking. Find the probability of each event. Then tell how likely the event is to happen.

1. *P*(basketball)
2. *P*(sports event)
3. *P*(opera or ballet)
4. *P*(soccer)
5. *P*(not symphony)
6. *P*(theater)
7. *P*(basketball or hockey)
8. *P*(not a sports event)
9. *P*(not opera)
10. *P*(baseball)
11. *P*(football)
12. *P*(not soccer)
13. *P*(opera)
14. *P*(not theater)
15. *P*(symphony)
16. *P*(soccer or football)
17. *P*(opera or theater)
18. *P*(hockey)

Lesson 13-2A *(Pages 520–521)*

Solve.

1. Sean made this frequency table to show the time it took his classmates to run 100 meters.

100 Meter Times (in seconds)		
Time	Tally	Frequency
13.0 – 13.9	III	3
14.0 – 14.9	IIII I	6
15.0 – 15.9	IIII IIII I	11
16.0 – 16.9	IIII IIII III	13
17.0 – 17.9	IIII	4
18.0 – 18.9	II	2

   a. How many students ran 100 meters in less than 16 seconds?
   b. How many students took 17 seconds or longer to run 100 meters?
   c. How many students did Sean survey?
   d. In what interval did the greatest number of times occur?
   e. How many students ran 100 meters in 15.0 - 16.9 seconds?

2. The list shows the birth month of the students in Mrs. Barr's geometry class.
   a. Make a frequency table of the students birth month.
   b. How many of the students were born in September?
   c. Which months were not represented?

   September October July
   June February October
   October July October
   October October July
   February October February
   March March December
   March June March
   September August September

Lesson 13-2 *(Pages 522–525)*

Tell whether each of the following is a random sample. Explain your answer.

	Type of Survey	Survey Location
1.	favorite television show	department store
2.	favorite baseball team	a specific baseball park
3.	favorite movie	shopping center
4.	favorite cookie	park
5.	favorite sport	soccer game
6.	favorite ice cream flavor	school
7.	favorite entertainer	Garth Brooks concert
8.	favorite fruit	apple orchard
9.	favorite food	county fair
10.	favorite vacation	popular theme park

Lesson 13-3 *(Pages 526–529)*

Each figure represents a dartboard. It is equally likely that a dart will land anywhere on the dartboard. Find the probability of a randomly-thrown dart landing in the shaded region.

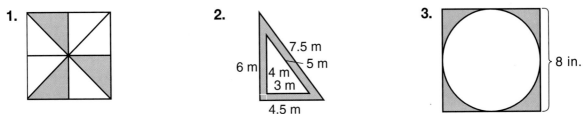

1.

2. 7.5 m, 5 m, 6 m, 4 m, 3 m, 4.5 m

3. 8 in.

Lesson 13-4 *(Pages 531–534)*

For each situation, make a list and draw a tree diagram to show the sample space.

1. tossing a quarter and rolling a number cube

2. spinning each spinner once

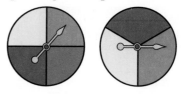

3. a choice of a red, blue, or green sweater with a white, black, or tan skirt.

4. a choice of a chicken, ham, turkey, or bologna sandwich with coffee, milk, juice, or soda

5. How many ways can a person choose to watch two television shows from four television shows?

Lesson 13-5 *(Pages 536–539)*

The spinner shown is spun and a card is chosen from the set of cards shown. Find the probability of each event.

1. P(15 and C)

2. P(even and H)

3. P(odd and a vowel)

4. P(multiple of 5 and O)

5. P(multiple of 10 and consonant)

6. P(composite and L)

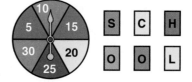

A bag contains 3 quarters and 5 dimes. Another bag contains 12 pennies and 8 nickels. One coin is randomly-chosen from each bag. Find the probability of each event.

7. P(dime and nickel)

8. P(quarter)

9. P(quarter and penny)

10. P(not a nickel)

11. P(dime and penny)

12. P(not a dime and nickel)

Mixed Problem Solving

Solve using any strategy.

1. *Money Matters* The Martin family bought tickets to the Science Museum. Admission is $8 for adults and $5 for children under 12. They spent $49 for admission. How many adult and children tickets did the Martin family purchase?

2. How many 3-digit numbers can you make using the even digits from 1 through 9, without repeating the same digit twice in any number?

3. *Life Science* Your heart at rest beats approximately 72 times per minute. At this rate, about how many times per day does your heart beat?
   a. 103,680
   b. 10,368
   c. 1,368
   d. 1,036,800

4. *Sports* Mrs. Zimmer sent her grandson an autographed souvenir baseball from the Baseball Hall of Fame in Cooperstown, N.Y. The baseball arrived in a cube-shaped box. Each side of the box measured 4 inches. Make a pattern for this box.

5. *Money Matters* Marcus belongs to a teen book club. He purchases a new paperback at $3.95 every other week for a full year. At this rate, how much will he pay over a 3-year period?

6. The bus leaves downtown for the mall at 7:35 A.M., 8:10 A.M., 8:45 A.M., and 9:20 A.M. If the bus continues to run on this schedule, what time does the bus leave between 10:00 A.M. and 11:00 A.M.?

7. *Civics* The American flag has 50 white stars on a blue background. The stars are arranged in alternating rows of 5 and 6. How many rows are there on the American flag?

8. Cadet Girl Scout Troop 548 held their annual cookie sale. Bethany sold half as many boxes of cookies as Olivia. Julie sold 15 more boxes than Bethany. Kara sold 9 fewer boxes than Alaina. Olivia sold three times as many boxes as Kara. How many boxes of cookies did Troop 548 sell if Julie sold 39 boxes?

Mixed Problem Solving

Solve using any strategy.

1. *Sports* The Chicago Cubs play 81 games each year. If 2,225,000 people attend the games, what is a reasonable estimate of the number of people that attend each game, 28,000 or 280,000?

2. *Money Matters* The Changs had their washing machine repaired. They were charged $45 for the first 15 minutes and $12 for every 15 minutes after that. The Chang's bill, excluding parts and tax, was $93. How many minutes did the repair work take?

3. *School* Of the 100 6th grade students at the John Glenn Middle School, 30 are taking French, 40 are taking Spanish, and 15 are taking both French and Spanish. How many students are taking either French or Spanish?

4. *School* Jared has to read a mystery novel for English class. If the novel contains 168 pages, how many pages must he read per day to finish the book in 7 days?

5. *Geometry* The Parks Commission is preparing a triangular section of Evening Shade Park in order to plant flowers. The base of the triangular region measures 18.4 feet, and the height is 12.5 feet. What is the area of the triangular flower garden?

6. The graph shows the types of footwear men say they wear in their home.

   a. Which two types of footwear are worn for about the same frequency?

   b. Which two types of footwear make up about 60% of the type of footwear worn?

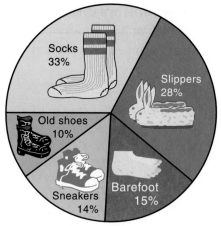

Happy Feet

Source: L.B. Evans

7. *Sports* Vicki wants to try out for the basketball team on Saturday. Tryouts begin at 9:30 A.M. sharp. She wants to arrive 20 minutes early to practice her foul shots. It takes her 35 minutes to eat breakfast and get ready. She plans to walk 10 minutes to her friend Madison's house. Vicki and Madison will then walk the remaining 15 minutes to the tryouts. What time should Vicki plan to get up on Saturday morning?

CHAPTER 1

Test

Use the four-step plan to solve.

1. *Geography* Amad and Marisela went hiking and canoeing down the Danube River, Europe's second largest river. They traveled from its source in the Black Forest 1,760 miles to its end at the Black Sea. If they averaged 11 miles each day, how long was the trip?

Find the next three numbers or draw the next three shapes for each pattern.

2. 192, 96, 48, 24, _?_, _?_, _?_

3.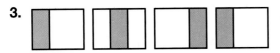

Round each number to the underlined place-value position.

4. 7$\underline{5}$4

5. $\underline{4}$,839

Estimate. State whether the answer shown is reasonable.

6. $198 \div 5 = 50$

7. $1,492 + 1,941 = 3,300$

8. *Earth Science* Each year, 16,000,000 thunderstorms occur throughout the world. About how many thunderstorms occur each week?

Find the value of each expression.

9. $48 - 24 \div 3 \times 6$

10. $63 \div 7 - 2 \times 3$

Evaluate each expression if $m = 2$, $n = 7$, and $p = 21$.

11. $p - m$

12. $m + p \div n$

13. Write $5 \cdot 5 \cdot 5 \cdot 5$ using exponents.

14. Write d^6 as a product.

Evaluate each expression.

15. 10^4

16. 3 cubed

Identify the solution to each equation from the given list.

17. $32 \div x = 2$; 16, 30, 64

18. $y + 7 = 23$; 5, 16, 7

Solve each equation mentally.

19. $5z = 55$

20. $15 - w = 9$

CHAPTER TEST

1. The test scores of the students in Mrs. Wimberly's history class are shown below. Make a frequency table of the data.

72	95	87	77	88	79	92	97	100
82	75	93	92	75	77	71	98	99
85	84	77	80	91	91	85	83	89

Geography The lengths, in miles, of the Great Lakes are Lake Superior, 350; Lake Michigan, 307; Lake Huron, 206; Lake Erie, 241; and Lake Ontario, 193.

2. What scale would you use for the vertical axis of a vertical bar graph?

3. Make a bar graph for this data.

4. Find the median length of the lakes.

Refer to the circle graph for Exercises 5–6.

5. Which color is the most popular?

6. Which three colors together are favored by about half the students?

Students' Favorite Colors

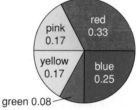

green 0.08

Jobs David mows lawns, rakes leaves, shovels snow, and does other outdoor jobs. Last year, his earnings were: January–$55, February–$70, March–$35, April–$23, May–$38, June–$55, July–$74, August–$78, September–$69, October–$55, November–$38, December–$58.

7. Find David's mean monthly earnings.

8. Find the mode.

9. Find the median.

10. David wants to buy a new bike that costs $179. If his earnings this year are the same as last year, when will he have enough to buy the bike?

11. Make a line graph for this data.

12. Change your graph to make it misleading.

Make a stem-and-leaf plot for each set of data.

13. 86, 85, 92, 73, 75, 96, 84, 92, 74, 87

14. 46¢, 59¢, 42¢, 69¢, 55¢, 48¢, 66¢, 43¢, 85¢

Use the grid to name the point for each ordered pair.

15. (3, 1) 16. (4, 0) 17. (2, 5)

Use the grid to name the ordered pair for each point.

18. *A* 19. *B* 20. *D*

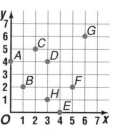

Test

Write each fraction or mixed number as a decimal.

1. $\dfrac{341}{10,000}$

2. $7\dfrac{13}{1,000}$

3. $\dfrac{9}{100}$

4. Write two hundred and six thousandths as a decimal.

Use a centimeter ruler to measure each line segment.

5. _____ **6.** _____

Use a centimeter ruler to measure one side of each figure.

7.  **8.**

9. *School* Jennifer's semester math average is 76.7. Her twin sister, Julie, is close with an average of 76.35. Whose average is higher?

10. Order 17.4, 1.747, 1.8, 17.36, and 17.09 from least to greatest.

11. Order 0.89, 0.98, 0.889, 0.982, and 0.88 from greatest to least.

12. *Money Matters* Sonia and four friends ate at a restaurant. Dividing the bill, each owed $7.146. What should each have paid?

Round each number to the underlined place-value position.

13. 901.2̲63

14. 0.99̲9

15. 2.4̲45

16. 7̲8.1

Estimate using any strategy.

17. 7.14 + 7 + 6.7 + 6.9

18. 45.9 − 6.12

19. 5.34 + 6.33 + 1.9

20. *Earth Science* The table shows the five most common elements in Earth's crust along with the percent of the crust each element represents. About how much more is the percent that is silicon than the percent that is aluminum?

Element	Percent
Oxygen	46.60
Silicon	27.72
Aluminum	8.13
Iron	5.00
Calcium	3.63

21. *Money Matters* At breakfast, Phil ordered juice for $0.89, scrambled eggs and toast for $3.69, and a glass of milk for $0.59. Did he spend about $6 or $4?

Add or subtract.

22.
$$\begin{array}{r} 9.04 \\ +12.8 \\ \hline \end{array}$$

23. 54.29 − 3.8

24. 54 + 1.8

25.
$$\begin{array}{r} 71.34 \\ -43.78 \\ \hline \end{array}$$

Multiply.

1. 0.81
   × 22

2. 22.38 × 803

3. 30.98
   × 5.6

Find each product mentally. Use the distributive property.

4. 6 × 17

5. 7 × 9.3

6. 21.8 × 30

Solve each equation.

7. $x = 47.3 \times 5.6$

8. $20.86 \cdot 4.11 = m$

9. *Algebra* Evaluate $b(a + c)$ if $a = 7.8$, $b = 5.05$, and $c = 0.02$.

Find the perimeter of each figure.

10.

24 m

25 m   7 m

11. a square with
    sides of length
    5.4 inches

12.   17.1 in.

10.5 in.

Find the area of each figure.

13. rectangle: ℓ, 10.5 mi; w, 17.1 mi

14. square: s, 2.5 cm

Find each quotient.

15. $19.36 \div 44$

16. $21.6 \overline{)49.68}$

Solve each equation.

17. $x = 58.305 \div 11.5$

18. $1{,}274.7 \div 2.1 = t$

19. *Algebra* Evaluate $c \div d$ if $c = 108.9$ and $d = 3.3$.

Write the unit that you would use to measure each of the following. Then estimate the mass or capacity.

20. a bottle of windshield-washer fluid

21. a potato

Complete.

22. $739 \text{ mL} = \underline{\ ?\ } \text{ L}$

23. $\underline{\ ?\ } \text{ mg} = 2.1 \text{ kg}$

24. *Transportation* Danielle drove 11.5 kilometers each day for 5 days. How far did she travel in all?

25. Find the area of the figure.

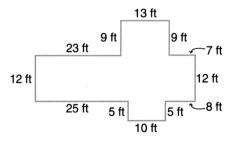

State whether each number is divisible by 2, 3, 5, 6, 9, or 10.

1. 75   **2.** 864

3. Games In one card game, all 52 cards are to be dealt out to the players. If there are six players, will all players have the same number of cards? Explain.

4. Which of the numbers 21, 31, 54, or 75 is prime?

Find the prime factorization of each number.

5. 72   **6.** 108   **7.** 58

8. List all of the factors of 20 from least to greatest.

Find the GCF of each set of numbers.

9. 27, 45   **10.** 18, 48   **11.** 21, 38, 42

State whether each fraction is in simplest form. If not, write each fraction or ratio in simplest form.

12. $\dfrac{40}{100}$ **13.** $\dfrac{6}{25}$ **14.** $\dfrac{45}{50}$

Express each mixed number as an improper fraction.

15. $2\dfrac{5}{7}$ **16.** $4\dfrac{1}{3}$ **17.** $12\dfrac{4}{5}$

Express each improper fraction as a mixed number.

18. $\dfrac{7}{3}$ **19.** $\dfrac{16}{5}$ **20.** $\dfrac{21}{4}$

21. Draw a line segment that is $1\dfrac{7}{8}$ inches long.

Find the LCM for each set of numbers.

22. 6, 16   **23.** 12, 18   **24.** 3, 9, 15

Replace each ⬤ with <, >, or = to make a true sentence.

25. $\dfrac{5}{6}$ ⬤ $\dfrac{7}{9}$ **26.** $\dfrac{6}{11}$ ⬤ $\dfrac{4}{9}$ **27.** $\dfrac{8}{12}$ ⬤ $\dfrac{12}{18}$

Express each decimal as a fraction or mixed number in simplest form.

28. 0.05   **29.** 5.1   **30.** 0.42

Express each fraction or mixed number as a decimal. Use bar notation to show a repeating decimal.

31. $6\dfrac{5}{6}$ **32.** $\dfrac{7}{16}$

33. *Food* The Pretzel Factory sells salted, unsalted, cinnamon, herb, and chocolate pretzels in two sizes, regular and jumbo. How many different ways can a person buy a pretzel?

CHAPTER 6 Test

Round each number to the nearest half unit.

1. $\dfrac{7}{15}$

2. $7\dfrac{2}{11}$

3. $\dfrac{11}{14}$

4. *Decorating* One wall space in Eric's room is $24\dfrac{3}{8}$ inches wide. When looking for a poster to fit in that space, should he round up or down?

Estimate.

5. $4\dfrac{7}{10} - 2\dfrac{1}{5}$

6. $\dfrac{7}{8} + \dfrac{1}{6}$

7. $\dfrac{15}{16} - \dfrac{4}{9}$

Add or subtract. Write the answer in simplest form.

8. $\dfrac{3}{17} + \dfrac{16}{17}$

9. $\dfrac{7}{8} + \dfrac{1}{6}$

10. $6\dfrac{3}{10} - 2\dfrac{9}{10}$

11. $4\dfrac{2}{9} + 5\dfrac{2}{9}$

12. $12 - 7\dfrac{3}{8}$

13. $3\dfrac{1}{8} - 1\dfrac{1}{16}$

14. $\dfrac{15}{16} - \dfrac{7}{16}$

15. $\dfrac{2}{15} + \dfrac{3}{10} - \dfrac{1}{5}$

16. $5\dfrac{7}{8} + 1\dfrac{5}{6}$

Solve each equation. Write the solution in simplest form.

17. $m = \dfrac{15}{16} - \dfrac{11}{16}$

18. $y = \dfrac{4}{5} + \dfrac{3}{5}$

19. $9\dfrac{4}{5} = 7\dfrac{3}{5} + t$

20. $d - 8\dfrac{3}{8} = 7\dfrac{5}{24}$

21. *Gardening* Mr. Hannon had $\dfrac{1}{2}$ ton of dirt for his garden delivered on Tuesday. On Thursday, another $\dfrac{1}{4}$ ton was delivered. How much dirt was delivered in all?

22. *Tickets* Emeka got to TicketCentral at 4:52 A.M. to wait in line for concert tickets. If TicketCentral opened at 7:30 A.M., how long did he wait?

Add or subtract.

23.
    10 h 12 min
  − 4 h 50 min

24.
    3 h 4 min 10 s
  +9 h 58 min 28 s

25.
    6 h 55 min
  −5 h 49 min

Estimate each product.

1. $6\frac{7}{8} \times 8\frac{1}{6}$

2. $37 \times \frac{4}{9}$

3. $\frac{9}{10} \times \frac{5}{8}$

Find the product. Write in simplest form.

4. $1\frac{4}{5} \times 2\frac{2}{3}$

5. $\frac{9}{10} \times \frac{5}{8}$

6. $3\frac{1}{5} \times 1\frac{1}{4}$

7. *Cooking* Kahlil's chocolate chip pie recipe calls for $1\frac{1}{2}$ cups of sugar. How much sugar will he need to make half of the recipe?

8. *Civics* In Homeland, U.S.A., $\frac{3}{4}$ of the adult residents are registered to vote. In the last May primary election, $\frac{2}{3}$ of those registered voted. What portion of the adult residents voted?

Solve each equation. Write the solution in simplest form.

9. $4 \times \frac{3}{5} = z$

10. $r = \frac{1}{8} \div \frac{3}{4}$

11. $k = 3\frac{2}{3} \times 2\frac{1}{3}$

12. $g = 6 \div 1\frac{4}{5}$

13. $\frac{3}{8} \times \frac{2}{5} = f$

14. $m = 5\frac{3}{4} \div 1\frac{1}{2}$

Find the circumference of each circle shown or described. Use $\frac{22}{7}$ or 3.14 for π. Round decimal answers to the nearest tenth.

15.

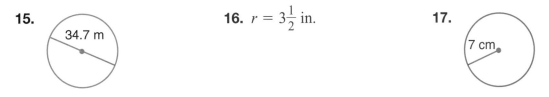

34.7 m

16. $r = 3\frac{1}{2}$ in.

17.

7 cm

Find each quotient. Write in simplest form.

18. $\frac{3}{4} \div \frac{4}{7}$

19. $2\frac{2}{3} \div 1\frac{5}{9}$

20. $\frac{2}{3} \div \frac{8}{9}$

Complete.

21. $2\frac{1}{2}$ qt = __?__ c

22. 9,000 lb = __?__ T

23. 8 c = __?__ fl oz

Find the next two numbers in each sequence.

24. $4, 6\frac{1}{2}, 9, 11\frac{1}{2}, \ldots$

25. $810, 270, 90, 30, \ldots$

Test

Express each ratio as a fraction in simplest form.

1. 2 notebooks out of 6 notebooks are blue.
2. 56 calculators out of 100 are graphing calculators.

Express each ratio as a rate.

3. 384 kilometers in 4 hours
4. 5 videos for $79.95

5. *Civics* In an election, 2,000 registered voters voted for Marilyn Williams, and 500 registered voters voted for Emilio Cardona. What ratio represents the portion of the voters that voted for Emilio Cardona?

Solve each proportion.

6. $\dfrac{4}{6} = \dfrac{x}{15}$

7. $\dfrac{m}{12} = \dfrac{12}{16}$

8. $\dfrac{10}{p} = \dfrac{4}{14}$

9. *Mechanics* The machine part shown has a scale of 1 unit = 1.25 centimeters. Find the actual length of the side labeled *b*.

a: 12 units

b: 20 units

10. *Geography* Bartolome is planning to travel from Dallas to Houston. His map has a scale of 1 inch is approximately 30 miles. If the distance on the map from Dallas to Houston is $7\frac{1}{4}$ inches, what is the approximate distance between the two cities?

Express each fraction as a percent.

11. $\dfrac{4}{5}$

12. $\dfrac{9}{100}$

13. $\dfrac{13}{25}$

Express each percent as a fraction in simplest form and as a decimal.

14. 80%

15. 7%

16. 0.1%

Express each decimal as a percent.

17. 0.012

18. 0.92

19. 3.1

Estimate each percent.

20. 9.5% of 51

21. 49% of 26

22. 308% of 9

Find the percent of each number.

23. 60% of 35

24. 49% of 26

25. 2% of 50

CHAPTER TEST

Test

Use a protractor to find the measure of each angle. Then classify the angle as *acute, right,* or *obtuse.*

1.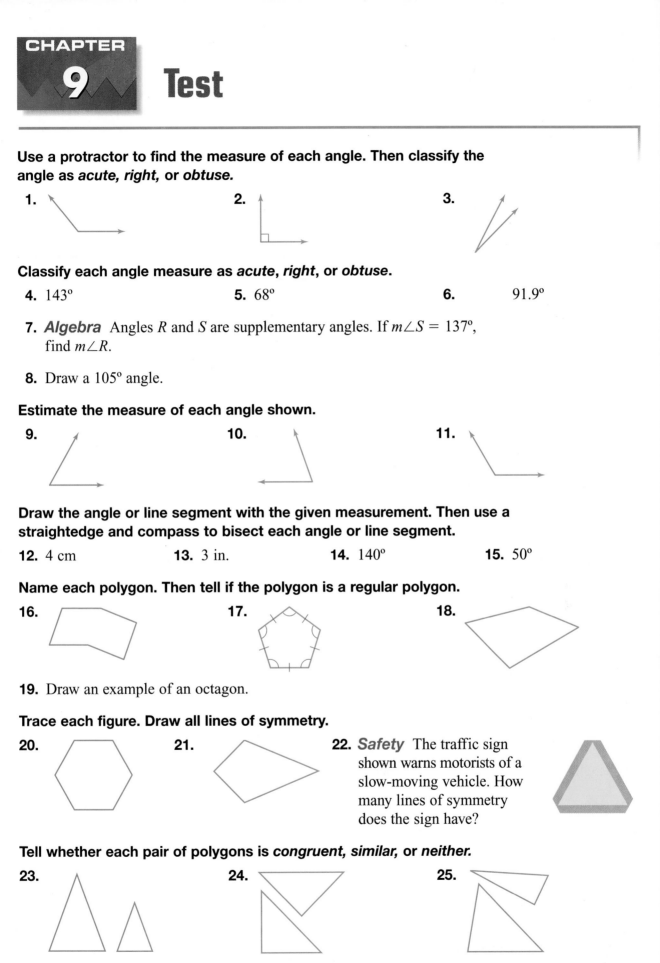

2.

3.

Classify each angle measure as *acute, right,* or *obtuse.*

4. 143°

5. 68°

6. 91.9°

7. *Algebra* Angles R and S are supplementary angles. If $m\angle S = 137°$, find $m\angle R$.

8. Draw a 105° angle.

Estimate the measure of each angle shown.

9.

10.

11.

Draw the angle or line segment with the given measurement. Then use a straightedge and compass to bisect each angle or line segment.

12. 4 cm

13. 3 in.

14. 140°

15. 50°

Name each polygon. Then tell if the polygon is a regular polygon.

16.

17.

18.

19. Draw an example of an octagon.

Trace each figure. Draw all lines of symmetry.

20.

21.

22. *Safety* The traffic sign shown warns motorists of a slow-moving vehicle. How many lines of symmetry does the sign have?

Tell whether each pair of polygons is *congruent, similar,* or *neither.*

23.

24.

25.

CHAPTER 10 Test

Find the area of each figure to the nearest tenth.

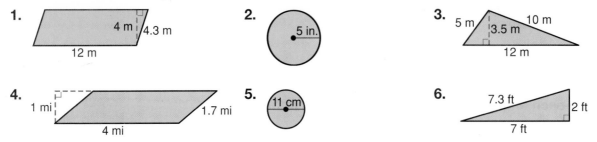

1. 4 m, 4.3 m, 12 m

2. 5 in.

3. 5 m, 10 m, 3.5 m, 12 m

4. 1 mi, 1.7 mi, 4 mi

5. 11 cm

6. 7.3 ft, 2 ft, 7 ft

7. **Traffic Signs** A triangular yield sign has a base of 32 inches and a height of 30 inches. Find the area of the sign.

8. **Gardening** Keisha plans to plant bulbs in a circular flower bed that has a radius of 2 meters. If she will plant 40 bulbs per square meter, how many bulbs should she buy?

Name each figure.

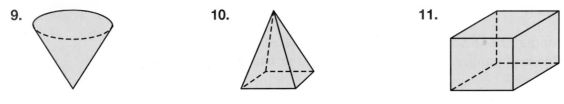

9.

10.

11.

12. State the number of faces, edges, and vertices in a triangular prism.

Find the volume of each rectangular prism in Exercises 13–16 to the nearest tenth.

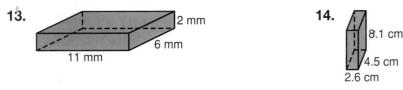

13. 2 mm, 6 mm, 11 mm

14. 8.1 cm, 4.5 cm, 2.6 cm

15. length, 11 inches; width, 8 inches; height, $5\frac{1}{2}$ inches

16. length, 4.5 meters; width, 2 meters; height, 5 meters

17. Find the surface area of the rectangular prism in Exercise 13.

18. Find the surface area of the rectangular prism in Exercise 14.

19. Find the surface area of a rectangular prism whose length is 20 yards, width is 30 yards, and height is 10 yards.

20. **Pools** A rectangular diving pool is 20 feet by 15 feet by 8 feet. How much water is required to fill the pool?

1. Write an integer to describe a loss of 12 dollars.

2. Write the integer that is the opposite of -5.

3. Graph -3 on a number line.

Replace each ● with >, <, or = to make a true sentence.

4. 8 ● -12 5. -77 ● -777 6. -127 ● -9

7. Order $-10, 0, -12, 3, -1$, and 11 from least to greatest.

Find each sum or difference. Use counters or a number line if necessary.

8. $-5 + (-7)$ 9. $6 - (-19)$ 10. $-13 + 10$ 11. $-4 - (-9)$

12. *Weather* The temperature at 6:00 A.M. was $-5°$F. What was the temperature at 8:00 A.M. if it was 7 degrees warmer?

Find each product or quotient. Use counters, a number line, or patterns if necessary.

13. $21 \div (-7)$ 14. $6 \times (-7)$ 15. $-60 \div (-6)$ 16. $-8(3)$

17. *Health* While Jane was recovering from the flu, her temperature dropped $1°$F each hour for 3 hours. What was the total drop?

Name the ordered pair for each point.

18. K
19. R
20. M

On grid paper, draw a coordinate grid. Then graph and label each point.

21. $A(5, -1)$ 22. $B(-2, -3)$ 23. $C(-4, 5)$

The vertices of △ABC are A(2, 3), B(−4, 1), and C(−1, 0). On a coordinate grid, draw △ABC and each transformation image.

24. a translation 3 units down   25. a reflection over the *x*-axis

Solve each equation. Use cups and counters if necessary.

1. $x + (-3) = -1$

2. $r - 2 = 5$

3. $w + 4 = -3$

4. $b - 4 = -1$

5. $\frac{1}{10}p = 2$

6. $5m = 30$

7. $\frac{1}{4}w = -8$

8. $4x + 1 = -15$

9. $10 = \frac{1}{3}n - 2$

10. What number should replace the cup on the equation mat at the right?

11. *Earth Science* The temperature at 10:00 P.M. was half what it was at noon. If the temperature at 10:00 P.M. was 26°, what was it at noon?

12. *Sports* Rodney was sacked for a 12-yard loss. On the next play, he had a gain of 5 yards. What was his net yardage on the two plays?

13. *Money Matters* Jeannette bought a sweater. The sale price was $38, which was $15 less than the original price. Find the original price.

Copy and complete each function table.

14.
input (n)	output (n + 5)
−5	■
0	■
1	■

15.
input (n)	output (6n)
−3	■
−1	■
1	■

Find the rule for each function table.

16.
input (n)	output (■)
−2	−5
0	−3
3	0

17.
input (n)	output (■)
−2	−14
0	0
3	21

18. *Age* Bret is 12 years old, and his father is 35 years old. When Bret is 20, his father will be 43 years old. Write a function rule for this relationship.

19. *Patterns* The triangles below were formed using toothpicks.

  a. Write the function rule to find the number of toothpicks for any number of triangles.

  b. How many toothpicks will be needed to make 8 triangles?

20. Graph the function given by the rule $\frac{1}{2}n$.

A set of 20 cards has two cards numbered 10, three cards numbered 15, three cards numbered 20, and one card for each of the numbers 1 through 12. Suppose you choose one card without looking. Find the probability of each event.

1. $P(8)$

2. $P(3 \text{ or } 10)$

3. $P(\text{multiple of } 5)$

4. $P(15)$

5. $P(\text{a multiple of } 3)$

6. $P(\text{prime})$

Tell whether each of the following is a random sample.

Type of Survey	Survey Location
7. favorite brand of tennis shoes	library
8. favorite subject	science fair
9. favorite TV show	department store

10. *Entertainment* According to a survey, 52 out of 100 people prefer listening to country music. Suppose 375 people participated in the survey. How many people can be expected to prefer listening to country music?

Each figure represents a dartboard. It is equally likely that a dart will land anywhere on the dartboard. Find the probability of a randomly-thrown dart landing in the shaded region.

11.

12.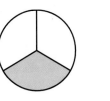

13.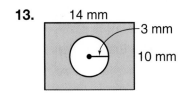

14. Suppose you throw a dart 135 times at the dartboard in Exercise 12. How many times would you expect it to land in the shaded region?

15. Suppose you throw a dart 150 times at the dartboard in Exercise 13. How many times would you expect it to land in the shaded region?

16. *Food* At The Snack Shack, a person has a choice of iced tea or soda and nachos, popcorn, or a chocolate bar.

   a. Make a list and draw a tree diagram to show the sample space.

   b. How many possible outcomes are there?

   c. What is the probability the next customer who orders a beverage and a snack will choose an iced tea and nachos?

A coin is tossed, and the spinner shown is spun.
Find the probability of each event.

17. $P(\text{tails and blue})$

18. $P(\text{heads and not green})$

19. $P(\text{not heads and red})$

20. $P(\text{heads or tails and yellow})$

Getting Acquainted with the Graphing Calculator

W hen some students first see a graphing calculator, they think, "Oh, no! Do we *have* to use one?", while others may think, "All right! We get to use these neat calculators!" There are as many thoughts and feelings about graphing calculators as there are students, but one thing is for sure: a graphing calculator *can* help you learn mathematics. Keep reading for answers to some frequently asked questions.

What is it?

So what is a graphing calculator? Very simply, it is a calculator that draws graphs. This means that it will do all of the things that a "regular" calculator will do, *plus* it will draw graphs of equations.

What does it do?

A graphing calculator can do more than just calculate and draw graphs. For example, you can program it and work with data to make statistical graphs and computations. If you need to generate random numbers, you can do that on the graphing calculator. If you need to find the absolute value of a number, you can do that too. It's really a very powerful tool, so powerful that it is often called a pocket computer.

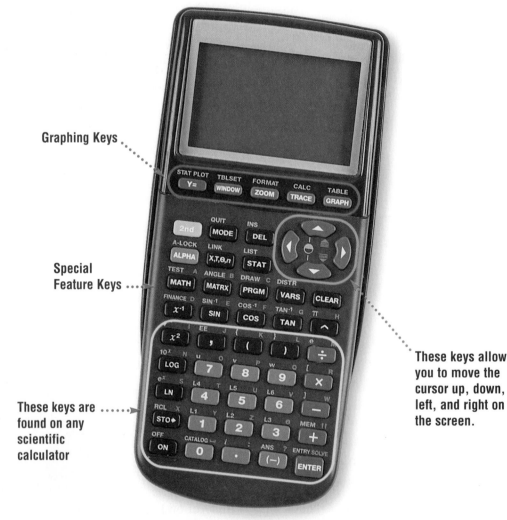

Graphing Keys

Special Feature Keys

These keys allow you to move the cursor up, down, left, and right on the screen.

These keys are found on any scientific calculator

Basic Keystrokes

- The yellow commands written above the calculator keys are accessed with the [2nd] key, which is also yellow. Similarly, the green characters above the keys are accessed with the [ALPHA] key, which is also green. In this text, commands that are accessed by the [2nd] and [ALPHA] keys are shown in brackets. For example, [2nd] [QUIT] means to press the [2nd] key followed by the key below the yellow [QUIT] command.

- [2nd] [ENTRY] copies the previous calculation so you can edit and use it again.

- [2nd] [QUIT] will return you to the home (or text) screen.

- Negative numbers are entered using the [(-)] key, not the minus sign, [-].

- [2nd] [OFF] turns the calculator off.

Order of Operations

As with any scientific calculator, the graphing calculator observes the order of operations.

Example	Keystrokes	Display
4 + 13	4 [+] 13 [ENTER]	4 + 13 *17*
5³	5 [∧] 3 [ENTER]	5 ^ 3 *125*
4 (9 + 18)	4 [(] 9 [+] 18 [)] [ENTER]	4(9 + 18) *108*
√24	[2nd] [√] 24 [ENTER]	√ (24 *4.8989 79486*

Programming

Programming features allow you to write and execute a series of commands for tasks that may be too complex or cumbersome to perform otherwise. Each program is given a name. Commands begin with a colon (:), which the calculator enters automatically, followed by an expression or an instruction.

When you press [PRGM], you see three menus: EXEC, EDIT, and NEW. EXEC allows you to execute a stored program, EDIT allows you to edit or change a program, and NEW allows you to create a program.

- To begin entering a new program, press [PRGM] [▶] [▶] [ENTER] .

- You do not need to type each letter using the [ALPHA] key. Any command that contains lowercase letters should be entered by choosing it from a menu. Check your user's guide to find any commands that are unfamiliar.

- After a program is entered, press [2nd] [QUIT] to exit the program mode and return to the home screen.

- To execute a program, press [PRGM] . Then use the down arrow key to locate the program name and press [ENTER] twice, or press the number or letter next to the program name followed by [ENTER] .

- If you wish to edit a program, press [PRGM] [▶] and choose the program from the menu.

- To immediately re-execute a program after it is run, press [ENTER] when Done appears on the screen.

- To stop a program during execution, press [ON] or [2nd] [QUIT].

While a graphing calculator cannot do everything, it can make some things easier and help your understanding of math. To prepare for whatever lies ahead, you should try to learn as much as you can. Who knows? Maybe one day you will be designing the next satellite or building the next skyscraper with the help of a graphing calculator!

Getting Acquainted with Spreadsheets

What do you think of when people talk about computers? Maybe you think of computer games or using a word processor to write a school paper. But a computer is a powerful tool that can be used for many things.

One of the most common computer applications is a spreadsheet program. Here are answers to some of the questions you may have if you're new to using spreadsheets.

What is it?

You have probably seen tables of numbers in newspapers and magazines. Similar to those tables, a spreadsheet is a table that you can use to organize information. But a spreadsheet is more than just a table. You can also use a spreadsheet to perform calculations or make graphs.

Why use a spreadsheet?

The advantage a spreadsheet has over a simple calculator is that when a number is changed, the entire spreadsheet is recalculated and the new results are displayed. So with a spreadsheet, you can see patterns in data and investigate what happens if one or more of the numbers is changed.

How do I use a spreadsheet?

A spreadsheet is organized into boxes called *cells*. The cells are named by a letter, that identifies the column, and a number, that identifies the row. In the spreadsheet below, cell C4 is highlighted.

	A	B	C
1	Width	Length	Area
2	3	4	12
3	2	10	20
4	5	12	60
5	8	14	112

To enter information in a spreadsheet, simply move the cursor to the cell you want to access and click the mouse. Then type in the information and press Enter.

How do I enter formulas?

If you want to use the spreadsheet as a calculator, begin by choosing the cell where you want the result to appear.

- For a simple calculation, type = followed by the formula. For example, in the spreadsheet above, the formula in cell C2 is entered as "=A2*B2." *Notice that * is the symbol for multiplication in a spreadsheet.*

- Sometimes you will want similar formulas in more than one cell. First type the formula in one cell. Then select the cell and click the copy button. Finally select the cells where you want to copy the formula and click the paste button.

- Often it is useful to find the sum or average of a row or column of numbers. The spreadsheet allows you to choose from several functions like this instead of entering the formula manually. To enter a function, click the cell where you want the result to appear. Then click on the = button above the cells. A list of formulas will appear to the left. Click the down arrow button and choose your function. The spreadsheet will enter a range, which you may alter. For example, to find the average of row 2 of the spreadsheet below, the function chooses to find the average of cells B2, C2, and D2.

	A	B	C	D	E
1	Student	Test 1	Test 2	Test 3	Average
2	Kathy	88	85	91	88
3	Ben	86	89	92	89
4	Carmen	92	86	92	90
5	Anthony	80	88	87	85

The formula for cell E2 is =(B2+C2+D2)/3.

Spreadsheet software is one of the most common tools used in business today. You should try to learn as much as you can to prepare for your future. Who knows? Maybe you'll use what you're learning today as a company president tomorrow!

Selected Answers

Chapter 1
Problem Solving, Numbers, and Algebra

Pages 6–7   Lesson 1-1
1. Explore–Decide what facts you know and what you need to find out.   Plan–Make a plan for solving the problem and estimate the answer. Solve–Solve the problem. Make a new plan if necessary.   Examine–Check to see if your answer makes sense.   **5.** about 15 million   **7.** 180 miles   **9.** $40   **11.** 65°F   **13.** 84,480 pennies   **15.** Yes; Sample answer: a team can score 7 field goals and 1 safety or 3 touchdowns, 2 two-point conversions, and 1 one-point conversion.

Pages 10–11   Lesson 1-2
3.

Number of Folds	1	2	3	4
Number of Thicknesses	2	4	8	16

Sample answer: The number of thicknesses doubles.
5. 8, 4, 2   **7.** 3, 4, 3   **9.** 512, 2,048, 8,192   **11.** 17, 22, 28   **13.** 21, 31, 43   **15.** 5:10 P.M.   **17.** 256   **19.** C

Pages 14–15   Lesson 1-3
1. On the number line, 957 is closer to 1,000 than 900.   **5.** 400   **7.** 300   **9.** 10,000   **11.** $900 - 400 = 500$; reasonable   **13.** $500 \div 50 = 10$; not reasonable   **15.** 100   **17.** 50   **19.** 440   **21.** 5,500   **23.** 3,700   **25.** 5,600   **27.** 10,000   **29.** 2,000   **31.** 9,000   **33.** $80 \times 3 = 240$; reasonable   **35.** $100 - 20 = 80$; not reasonable   **37.** $5,000 + 5,000 = 10,000$; reasonable   **39.** $800 - 200 = 600$; reasonable   **41.** $5.00 + 35.00 = 40.00$; not reasonable   **43.** Sample answer: $200 + 60 + 600 + 300 = 1,160$; about 1,160 lbs   **45a.** $75 + 36 + 21 + 6 + 2 + 2 + 10 = 152$ **47.** 11, 16, 22   **49.** 10:22 A.M., 11:02 A.M., 11:06 A.M.   **51.** C

Pages 18–19   Lesson 1-4
1. Multiply and divide in order from left to right. Then add and subtract in order from left to right.   **3.** +; 14

5. ×; 33   **7a.** $12 \times \$15 + 28 \times \$6$ **7b.** $348   **9.** 117   **11.** 43   **13.** 15   **15.** 23   **17.** 9   **19.** 55   **21a.** $2 \times 15 + 1 \times 14 + 3 \times 18$ **21b.** 98 g   **23a.** Sample answer: $3 \times \$29.95 + 2 \times \$16.95 + \$12.95 = \$136.70$ **25.** 9,000   **27.** B

Page 19   Mid-Chapter Self Test
1. $7   **3.** 37, 45, 53   **5.** 150   **7.** $18,000 \div 300 = 60$ **9.** 9

Pages 20–21   Lesson 1-5A
1. the unknown value   **3.** 13   **5.** the unknown value   **7.** 5 bags

Pages 24–25   Lesson 1-5
1. $5 \times r$ or $5 \cdot r$ **3.** Nicholas; $2 + 14 \div 2 = 9$ **5.** 4   **7.** 18   **9.** 16   **11.** 15   **13.** 10   **15.** 22   **17.** 105   **19.** 60   **21.** 35   **23.** 22   **25.** 9   **27.** 18   **29.** 54   **31.** 9   **33.** 24   **35.** 30   **37.** 9   **39.** 12 inches   **41.** yes; 30   **43.** 21   **45.** A

Page 26   Lesson 1-5B
1. 3   **3.** 28   **5.** 14   **7.** 20   **9.** Yes, unless the scientific calculator does not have parentheses keys.

Pages 29–31   Lesson 1-6
1. An exponent is a number that tells how many times a number, called the base, is used as a factor. A power is a number that is expressed using exponents.   **3.** Tamika; $3^2 + 2^4 \cdot 6 = 105$.   **5.** m^5 **7.** $c \cdot c \cdot c$ **9.** $h \cdot h \cdot h \cdot h \cdot h \cdot h$ **11.** 100,000   **13.** about 1,000,000,000 people   **15.** 3^6 **17.** $3^2 \cdot 4^4$ **19.** r^4 **21.** $6^3 \cdot 1^3$ **23.** $14 \cdot 14$ **25.** $7 \cdot 7 \cdot 7 \cdot 7 \cdot 7$ **27.** $16 \cdot 16 \cdot 16 \cdot 16 \cdot 16 \cdot 16$ **29.** $x \cdot x \cdot x \cdot y \cdot y \cdot y \cdot y$ **31.** 1,000   **33.** 32   **35.** 9   **37.** 1,027   **39.** 103   **41.** 1; 1; 10; 1   **43.** 5,832 in³   **45.** 18   **47.** $400 + 700 + 200 = 1,300$; reasonable   **49.** 379 students

Pages 32–33   Lesson 1-7A
3. The piece that is beside the top left-hand puzzle piece. It will connect to the corner piece.

5. Sample answer: When making a plan to solve a problem, you can use the guess-and-check strategy to guess an answer and compare the answer to the problem. If your answer does not work, then you know to make another guess.   **7.** Sample answer: You could guess values for p to find what number divided by 5 is 45.   **9.** 15 dogs, 9 cats   **11.** There are 8 bones in the wrist, 14 in the fingers, and 5 in the palm.   **13.** 67,740 feet per minute   **15.** B

Pages 35–37   Lesson 1-7

1. Sample answer: Both algebraic expressions and equations can contain variables, operations, and numbers. Whereas an equation contains an equals sign, an algebraic expression does not contain an equals sign.   **3.** false   **5.** 7   **7.** 4   **9.** 9   **11.** 10   **13.** true   **15.** true   **17.** false   **19.** 15   **21.** 7   **23.** 13   **25.** 7   **27.** 44   **29.** 26   **31.** 7   **33.** 3   **35.** 225   **37.** 2   **39.** 16   **41a.** $19 + g = 65$ **41b.** 46 grams   **43.** C   **45.** 12   **47.** about $70

Pages 38–41   Study Guide and Assessment

1. multiply and divide   **3.** 4   **5.** 3 and 4   **7.** 2^5 **9.** 10   **11.** 11   **13.** $39.92   **15.** 531 votes

17.

19. 48, 56, 64   **21.** 30   **23.** 650   **25.** 2   **27.** 2   **29.** 7   **31.** 16   **33.** 7   **35.** 21   **37.** 44   **39.** 4,096   **41.** 125   **43.** t^5 **45.** 3   **47.** 63   **49.** 9   **51.** 6   **53.** 13 years 10 months   **55.** about 63,000

Pages 42–43   Standardized Test Practice

1. B   **3.** B   **5.** D   **7.** E   **9.** B   **11.** A   **13.** 10 ft   **15.** 21   **17.** 27

Chapter 2
Statistics: Graphing Data

Pages 48–49   Lesson 2-1

1. Sample answer: to make it easier to study the data.

5a.

Age	Tally	Frequency
10	I	1
11	I	1
12	IIII	5
13	III	3
14	II	2

5b. 12

7.

Days Absent	Tally	Frequency
0	IIII	5
1	IIII I	6
2	II	2
3	III	3
4		0
5	II	2

9a. Sample answer: All the shoes sold were between sizes 9 and 11.   **9b.** 77 pairs of shoes   **9c.** 10   **11a.** Sample answer: Based on the data in the table, they should keep potato chips and corn chips well stocked.   **11b.** Sample answer: There are too few to tell.   **13.** A

Pages 51–53   Lesson 2-2

1. Sample answer: The scale must include all of the data. Since the shortest distance is 12 and the longest distance is 63, a scale from 10 to 70 is appropriate.   **5.** 2; 10 is too large.

7a.

Shelby's Video Game Scores		
Score	Tally	Frequency
60,000–69,900	I	1
50,000–59,900	I	1
40,000–49,900	IIII	4
30,000–39,900	IIII	5
20,000–29,900	IIII	4
10,000–19,900	II	2
0–9,900	I	1

7b. Most of her scores were between 30,000 and 39,900.   **9.** a   **11.** a   **13.** b

15.

Amount	Tally	Frequency
81–100	I	1
61–80		0
41–60	IIII	4
21–40	I	1
0–20	II	2

17a. No; since Texas is a much larger state, the distances will be larger. If she used the same scale the frequency table would not include all the distances.   **17b.** Sample answer: Yes; however, the table will have a large number of rows.   **19a.** Sample answer: 5,000–20,999

19b.

Elevation	Tally	Frequency
19,000–20,999	I	1
17,000–18,999		0
15,000–16,999		0
13,000–14,999	⊮ III	8
11,000–12,999	IIII	4
9,000–10,999		0
7,000–8,999	II	2
5,000–6,999	⊮ II	7

21.

Number of Hits	Tally	Frequency
0	I	1
1	II	2
2	III	3
3	II	2
4	I	1

23. D

Pages 55–57   Lesson 2-3

1. Call Them   **3.** Sample answer: If the vertical scale is much higher than the highest value, it makes the graph flatter. Changing the interval does not affect either type of graph.   **5.** 0–70; 10

7.

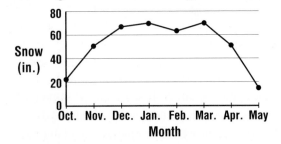

Favorite Soft Drinks

9.

Average Snowfall for Brighton, Utah

11. Sample answer: 0 to 300   **13.** Table A; It shows the change and direction of change over time.

15.

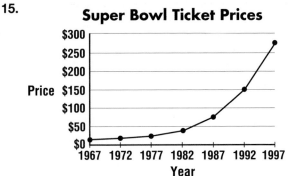

Super Bowl Ticket Prices

19. B   **21.** 4

Pages 58–59   Lesson 2-3B

3. Sample answer: Mickey's; It is the second favorite and the healthiest.   **7.** $30
9. 1,380 miles   **11.** C

Pages 62–63   Lesson 2-4

1. Sample answer: When you want to represent data that looks at part to whole relationships.
5a. 13%–Special; 15%–egg noodles; 31%–short pasta; 41%–long pasta   **5b.** special and egg noodles
5c. 72%   **5d.** special and egg noodles   **5e.** either long pasta and special (54%) or short pasta and egg noodles (46%)   **7a.** LP album, 7-12 in. singles and music videos   **7b.** cassette single–5%; cassette album–25%; CDs–68%

9.

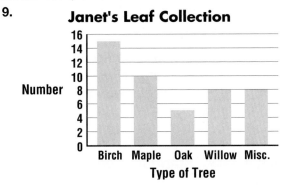

Janet's Leaf Collection

Pages 65–67   Lesson 2-5

1. Sample answer: If the extension aligns with the segment between 1980 to 1990, world population could be about 7 billion.   **5a.** Sample answer: It will continue to bounce higher.   **5b.** Sample answer: It will bounce about as high as it did when dropped from 110 cm.   **5c.** Sample answer: Ball A might be a Super Ball.   **5d.** Sample answer: No; at some point both balls will begin to level off. If you dropped Ball A from 100 feet it will not bounce 100 feet.   **7.** The point where they cross indicates the times were the same.   **9.** D

Page 67 Mid-Chapter Self Test

1.

Height	Tally	Frequency
54	l	1
55	l	1
56	l	1
57	llll	4
58	l	1
59	l	1
60	ll	2
61		0
62		0
63	l	1

3.

Favorite Ice Cream Flavor

5a. the 5th week   **5b.** the 11th week   **5c.** Sample answer: No; the horizontal scale doesn't extend far enough.

Pages 69–70   Lesson 2-6

1. when there are many numbers in a set of data

5.

Stem	Leaf
2	0 4 7
3	4 5 6 6 8
4	3 5 7
5	3 4 8 8
6	
7	8 2|7 = 27

7. 1, 2, 3, 4, 5
9. 0, 1, 2, 3, 4

11.

Stem	Leaf
9	2 4 9 9
10	
11	
12	4 4
13	0 2 3
14	0
15	
16	2 7 12|4 = 124

13.

Stem	Leaf
0	9
1	
2	7 8 9
3	4 4 4
4	5 6
5	3 6 6 7 7 7 8

5|3 = 53°

15a.

Stem	Leaf
0	5 5 6 7 8 9 9 9 9 9
1	0 1 1 3 4 6 7 8 8 9 9
2	2 2 3 6 7 7 9
3	7
4	1 1|0 = 10

15b. 9 days   **17.** Sample answer: 400

Pages 73–74   Lesson 2-7

1. The mean because the numbers do not differ greatly.   **5.** 10, 26, 27, 29, 36, 37, 57, 83, 88; 36; no mode; ≈ 43.7; 78   **7.** 44; 44; no mode; 22   **9.** 15; 15; 14 and 16; 4   **11.** 31; 31; 31; 26   **13.** The mean would be lower.   **15.** yes; Sample answer: The total would be higher so the mean would also be higher.   **17.** Sample answer: The mode because she would want to know what size they sold the most shakes.

19.

Stem	Leaf
0	0 2 2 4 5 7 7 9
1	0 2 3 3 8 9
2	0 3 2|0 = 20

Page 77   Lesson 2-7B

1.

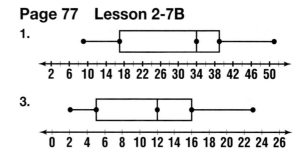

3.

Pages 79–81   Lesson 2-8

1. Sample answer: Using different scales and not having a title.   **3.** Sample answer: If the range is very high the mean could be misleading. If the set of data has a gap between two groups of numbers that are all about the same, the mean could be misleading. For example, 1, 2, 3, 2, and 8, 7, 6, 9, 8.   **5.** mode– want to find the most frequent   **7a.** Graph B is misleading because the distance between 0 and 25 is the same as the distance between 25 and 30.
7b. Graph A because it shows a general rise in price.   **9.** $1\frac{1}{2}$; median   **11.** $1962; mean   **13.** There may be a few very low priced CDs, thus giving a lower average.   **15a.** mean: 134 ÷ 24 ≈ 5.6 or 6 pets; median: 3 pets; mode: 2 pets

15b. Sample answer: The median best represents the data. The student with 24 pets has a big effect on the mean.   **17.** A   **19.** 25, 36, 49

Pages 84–85   Lesson 2-9

1. From the origin, go 7 units left, then up 9 units.
3. Sample answer: Raul; to graph (8, 3) you move 8 units right and 3 units up and to graph (3, 8) you move 3 units right and 8 units up.   **5.** *F*   **7.** (8, 0)
9. (5, 6)   **11.** *C*   **13.** *K*   **15.** *Q*   **17.** *L*   **19.** *M*
21. (7, 4)   **23.** (3, 7)   **25.** (0, 9)   **27.** (0, 0)
29. (9, 8)   **35.** C   **37.** 7

Pages 86–89   Study Guide and Assessment

1. f   **3.** a   **5.** c   **7.** i   **9.** j   **11.** Sample answer: A bar graph represents data using rectangles to show the frequency of responses. A line graph uses dots at the frequency points connected by segments to show how data changes over time.

13.

Favorite Color	Tally	Frequency
blue	IIII	4
red	IIII	4
pink	II	2
orange	I	1
yellow	I	1
green	II	2
brown	I	1

15. scale = 0 to 50, interval 5   **17.** scale = 0 to 160, interval 20
19.

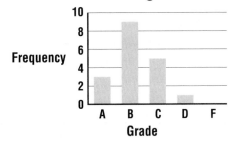

Alma's Grades During Junior High

21. less   **23.** 22   **25.** 25   **27.** median   **29.** mean
31. *A*   **33.** (3, 6)
35.

Stem	Leaf
4	0 5
5	0
6	5 9
7	5 5
8	5
9	0 2 5

9|0 = 90

Pages 90–91   Standardized Test Practice

1. D   **3.** B   **5.** B   **7.** C   **9.** C   **11.** E   **13.** 7 + *r* = 22; 15   **15.** 25

Chapter 3
Adding and Subtracting Decimals

Page 94   Lesson 3-1A

1. thirty hundredths
3. 9 tenths;

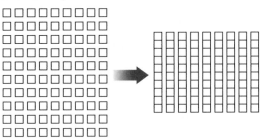

Pages 97–98   Lesson 3-1

1. 0.05, $\frac{5}{100}$, five hundredths   **3.** Mercedes is correct because the last digit, 8, is in the thousandths place.   **5.** 0.43   **7.** 4.0029   **9.** 0.0702   **11.** 0.3
13. 0.07   **15.** 19.53   **17.** 0.009   **19.** 47.047
21. 172.0001   **23.** 20.9   **25.** 3.003   **27.** 0.0801
29. 0.01   **31.** greatest: 0.750; least: 0.057   **33.** B
35. $19

Page 101   Lesson 3-2A

3. Students should notice that the measures are related to powers of 10 and start to notice how decimals are used.

Pages 103–104   Lesson 3-2

5. cm; about 2.5 cm   **7.** 3.2 cm or 32 mm   **9.** cm; about 90 cm   **11.** m; about 5 m   **13.** 1.3 cm or 13 mm   **15.** 3.8 cm or 38 mm   **17.** 3 cm or 30 mm
19. 2 cm or 20 mm   **21.** 300 cm   **25.** C   **27.** 10

Pages 107–108   Lesson 3-3

1.

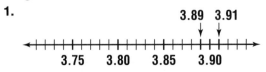

3. 1.63 > 1.54   **5.** 0.06   **7.** 13.05, 13.507, 13.84, 13.9   **9.** 389.225, 388.404, 388.246, 385.867,

385.841   **11.** 0.331   **13.** 1.018   **15.** 47.553
17. 0.7   **19.** 14.95, 15, 15.01, 15.8   **21.** 0.0316,
0.0306, 0.025, 0.0249, 0.0208   **23.** 397.877, 379.9,
379.88, 379.8778, 378.87   **25.** 0.603   **27a.** 2.1,
2.9, 4.2, 5.1, 5.8, 7.0, 9.0, 13.0, 14.0   **27b.** 5.8
29. 5.7 cm or 57 mm   **31.** Sample answer: 95, 91,
93, 89; no number occurs more often than the others

Pages 110–111   Lesson 3-4

1. Sample answer: Since the digit to the right of the
ones place is greater than 5, add one to the ones
place. $10.79 rounded to the nearest dollar is $11.
3. Sharon is correct. Carlos rounded to the nearest
hundred.   **5.** 8.20   **7.** 20   **9.** 18   **11.** 20.5
13. 49.8   **15.** 19.78   **17.** 100.0   **19.** 7.000
21. $2.00   **23.** 102   **25.** Sample answer: Using
the exact data in this case would be most appropriate.
27. No, because no numbers differ greatly.

Page 111   Mid-Chapter Self Test

1. 0.7   **3.** 0.681   **5.** 3 cm   **7.** 638.178   **9.** 3.40

Pages 114–115   Lesson 3-5

1. Sample answer: Round both amounts to the
nearest dollar. $17 − $4 = $13   **5.** 20 + 30 = 50
7. 11 + 11 + 11 + 11 = 44   **9.** 20 + 20 + 20 =
60   **11.** Sample answer: Round 0.38 to 0.4 and
round 0.21 to 0.2. 0.4 + 0.2 = 0.6. Since 0.6 > 0.5
both samples cannot be stored in the 0.5-liter
container.   **13.** 9 − 5 = 4   **15.** 32 + 17 = 49
17. $30 − $20 = $10   **19.** $58 − $27 = $31   **21.** 4
+ 4 + 4 = 12   **23.** 1 + 1 + 1 + 1 = 4   **25.** $54
+ $54 + $54 = $162   **27.** $65 − $40 = $25
29. $200 − $110 = $90   **31.** 102°F − 99°F = 3°F
33. about 1 million square feet   **35a.** Sample
answer: An estimate for Mandy's family is $9 + $9
+ $9 + $4 or $31. Since all costs were rounded up,
her whole family can go on the tour for less than
$35.   **35b.** Sample answer: An estimate for each
meal is $5 for breakfast, $7 for lunch, and $12 for
dinner. The estimate for one person would be $5 +
$7 + $12 or $24. The estimate for a family of 4
would be $24 + $24 + $24 + $24 or $96. This
could be rounded up to $100.   **37.** 0.270   **39.** 21
41. 8

Pages 116–117   Lesson 3-6A

3. Sample answer: If you round to the nearest
hundred, the 1986 and 1987 amounts or the 1987

and 1988 amounts total 200 million. If you round to
the nearest ten the 1986 and 1988 amounts total 200
million.   **9a.** They sold 12 subscriptions during
week three.   **9b.** Since each symbol means 4
subscriptions, a half-magazine means 2
subscriptions.   **9c.** They sold 16 subscriptions
during week two.   **11.** Sample answer: By drawing
a line graph and extending it to match the segment
between 1993 and 1994, over 1.6 million CDs will
be distributed in 2000.   **13.** B

Pages 119–121   Lesson 3-6

1. Sample answer: Annex two zeros. Then line up
the decimal points, rename and subtract.   **5.** 2.24
7. 0.75   **9.** 6.592   **11.** 6.4   **13.** 0.24   **15.** $11.77
17. 1.747   **19.** 2.02   **21.** 13.253   **23.** 151.575
25. $12.10   **27.** 26.42   **29.** 2.1   **31.** 10.72
33a. 2.6 pounds   **33b.** 18 pounds   **35.** 0.2 −
0.1876543 = 0.0123457   **37.** B   **39.** 3

Pages 122–125   Study Guide and Assessment

1. false, less   **3.** false, one hundredth   **5.** false,
centimeters   **7.** false, 600.012   **9.** true   **11.** 0.08
13. 14.017   **15.** 0.2   **17.** 0.053   **19.** eight ten-
thousandths   **21.** 6 cm   **23.** 2 cm   **25.** 11.6
27. 0.0289, 0.0319, 0.032, 0.31   **29.** 6.75, 6.39,
6.32, 6.02   **31.** 7   **33.** 13.6   **35.** 5 − 1 = 4
37. 6 + 6 + 6 + 6 + 6 = 30   **39.** 12.912
41. 77.1   **43.** 7320.4, 7321.5, 7321.539, 7342.98,
7346.24, 7346.4   **45.** about $210

Pages 126–127   Standardized Test Practice

1. A   **3.** C   **5.** A   **7.** D   **9.** B   **11.** D   **13.** divide
120 by 8   **15.** 241 miles   **17.** 68   **19.** mode

Chapter 4 Multiplying and Dividing Decimals

Page 132   Lesson 4-1A

1.

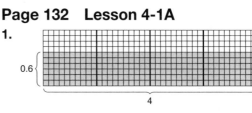

3.

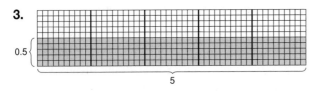

5. 3.2

Pages 134–136   Lesson 4-1

1. Sample answer: Round 40.32 to 40 and round 251 to 250. You can find 40×250 since 4 and 25 are compatible numbers.   **3.** Sample answer: Yes; by rounding 34.78 to 30 and rounding 452 to 500, the product of 30 and 500 is 15,000.   **5.** 239.4   **7.** 1.6   **9.** 0.32   **11.** 66,593.12 miles per hour   **13.** 1.12   **15.** 30.78   **17.** 64.2   **19.** 2,606.5   **21.** 9.425   **23.** 0.252   **25.** 3,940.2   **27.** 21,286.734   **29.** Sample answer: Since 0.5 is less than 5, 0.5×5 will be less than 25.   **31.** $29.25   **33.** Sample answer: 5×9.178 **35.** Sample answer: 80   **37a.** Sample answer: 800 points   **37b.** Sample answer: Use clustering to total the hundreds and use rounding to estimate to the nearest ten.   **37c.** Sample answer: You would round to the nearest ten because rounding to the nearest hundred they would all be the same.

Pages 138–139   Lesson 4-2

1. First find 8^2. Then add 64 and 6. Finally, multiply 70(9).   **5.** $80 + 64 = 144$ **7.** $30 + 2 = 32$ **9.** $100 + 25 = 125$ **11.** $3 \times 20 + 3 \times 7$ **13.** 84   **15.** 216   **17.** 48.6   **19.** 124.2   **21.** 92.7   **23.** 630   **25.** $15.75   **27a.** 0.32   **27b.** 0.05   **27c.** 0.54   **27d.** 7.3125   **29.** 35.9 in.

Page 140   Lesson 4-3A

1. 0.6×1.9

3.

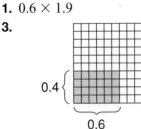

5.

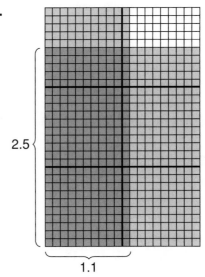

Pages 142–143   Lesson 4-3

1. Sample answer: The total decimal places of the factors are 5. 0.40563   **3.** 4.08   **5.** 27.9258   **7.** 0.0078   **9.** 69.608   **11.** 0.49   **13.** 13.0968   **15.** 16,688.0428   **17.** 9.49   **19.** 0.0186   **21.** 0.00212   **23.** 5.82788   **25.** 108.8472   **27.** 0.00696   **29.** 5.41   **31.** Sample answer: 0.1×0.6 **33.** D   **35.** 216

Page 144   Lesson 4-3B

1. E4   **3.** $28.79, $37.49, $25.41, $12.00, $12.34, $17.95   **5.** No; you would still multiply.

Pages 147–148   Lesson 4-4

5. 7.5 in.   **7.** $P = 121.6$ cm; $A = 924.16$ cm^2 **9.** 43 m   **11.** 27 m   **13.** $P = 6$ in.; $A = 2.25$ in.^{2} **15.** $P = 12.8$ m; $A = 7.68$ m^2 **17.** $P = 46.4$ in.; $A = 135.56$ in.^{2} **19.** 15 in^2; 20 in.   **21a.** perimeter   **21b.** 880 ft^2

23.

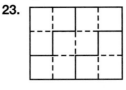

25. C

Page 149   Lesson 4-4B

1a.

Length	Width	Area
7	1	7
6	2	12
5	3	15
4	4	16

1b.

Length	Width	Area
8	1	8
7	2	14
6	3	18
5	4	20

1c.

Length	Width	Area
9	1	9
8	2	16
7	3	21
6	4	24
5	5	25

3. Sample answer: The longer the length and shorter the width, the lesser the area of the rectangular shape.

Pages 150–151 Lesson 4-5A

1. Sample answer: She thought she was dividing the length into 2-foot sections.   **3.** 17 cuts   **5.** Sample answer: This strategy allows you to break a difficult problem into manageable parts.   **7.** 28 feet
9a. $40 million   **9b.** $50 million   **9c.** $3 million
11. His competition average was 17 pins higher than his previous average.   **13.** D

Pages 153–155 Lesson 4-5

1. Sample answer: The estimate tells you the answer is close to 1.   **3.** Sample answer: Estimate 80 ÷ 40 = 2; The estimate shows that the answer is reasonable.   **5.** 0.415   **7.** 7.9   **9.** 2.6   **11.** 0.25
13. 21.2   **15.** 9.23   **17.** 2.7   **19.** 3.3   **21.** 22.42
23. 9.79   **25.** 5.5   **27.** 4.53   **29.** $14.67
31. $0.19, $0.07, $0.06   **33.** Sample answer: 0.9984 ÷ 8   **35.** about 27 cm   **37.** B

Page 155 Mid-Chapter Self Test

1. $20,000   **3.** 420   **5.** 0.0204   **7.** 9.8 cm, 0.94 cm^2 **9.** 9.63

Page 156 Lesson 4-6A

1. 2   **3.** 30   **5.** 3

Pages 158–159 Lesson 4-6

1. Sample answer: 1.84 ÷ 0.8.   **3.** 19   **5.** 694
7. 54.3   **9.** 2.2   **11.** 112   **13.** 0.4   **15.** 0.9
17. 2,967   **19.** 0.6   **21.** 8.37   **23.** 107   **25.** 3.9
27. 3.41   **29.** 0.84   **31.** 19 tanks   **33.** Sample answer: 0.896 ÷ 0.35 = 2.56   **35.** two hundred four and two thousand three hundred ninety-eight ten-thousandths   **37.** 7

Pages 162–163 Lesson 4-7

1. Sample answer: 4.53 ÷ 15   **3.** Sample answer: Multiply the divisor and dividend by 10. The problem becomes 1,561.4 ÷ 148. Estimate: 1,500 ÷ 150 = 10.   **5.** 4.03   **7.** 5.02   **9.** 17.05   **11.** 10.2
13. 301   **15.** 5.21   **17.** 10.94   **19.** 20.26
21. 10.42   **23.** 2.00   **25.** 101   **27.** 0.807
29. 0.0032   **31.** 6.08 m   **33.** never   **35.** D
37. Sample answer: 0 to 120

Pages 165–166 Lesson 4-8

1. kilogram   **3.** kilogram; 500 kg   **5.** liter; 2 L
7. Sample answer: The fruit is heavier, so the loaf has to be smaller to have the same mass.   **9.** milliliter, 200 mL   **11.** kilogram, 100 kg   **13.** kilogram, 2 kg   **15.** milligram, 1 mg   **17.** gram, 100 g
19. milligram, 500 mg   **23.** 1.765 g   **25a.** 4,389 g
25b. more than 4 kg   **27.** 201.5   **29.** C

Pages 168–169 Lesson 4-9

1. divide by 1,000   **3.** Sample answer: Parker and Dinh are both incorrect. Parker multiplied by 10 and Dinh divided by 10. They should have multiplied by 1,000.   **5.** multiply; 7   **7.** multiply; 5,020
9. divide; 0.15   **11.** 21   **13.** 2,500   **15.** 5.24
17. 81.7   **19.** 1.953   **21.** 3,290   **23.** 5,250
25. 10   **27.** 0.0067   **29.** 58 cm   **31.** 0.213 L
33. about 5.6 cans   **35.** liters   **37.** 77.65

Pages 170–173 Study Guide and Assessment

1. true   **3.** false, perimeter   **5.** true   **7.** true
9. true   **11.** false, multiply   **13.** Sample answer: Round 6.9 to 7 and 88 to 90. 7 × 90 = 630
15. 488.05   **17.** 47.46   **19.** 96,819.2   **21.** 322.8
23. 34   **25.** 0.204   **27.** 62.0984   **29.** 539.76
31. 27.3 m   **33.** 41 mi; 104.16 mi^2 **35.** 30.4 ft, 57.76 sq ft   **37.** 6.37   **39.** 0.52175   **41.** 19.21
43. 141   **45.** 5.9   **47.** 0.01   **49.** 0.08   **51.** 5.436
53. kilogram, 5 kg   **55.** milliliter, 15 mL
57. 0.001   **59.** 5.2 kilometers   **61.** $319.52

Pages 174–175 Standardized Test Practice

1. D   **3.** C   **5.** C   **7.** C   **9.** B   **11.** C   **13.** 88, 93, 95   **15.** liters   **17.** 0.0235   **19.** Sample answer: The 539-gram can of soup is a better buy. It costs 0.135¢ per gram while the 306-gram can costs 0.137¢ per gram.

Pages 179–180   Lesson 5-1

1. Sample answer:

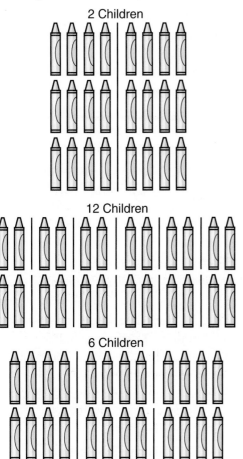

2 Children

12 Children

6 Children

3. 2: if the ones digit is divisible by 2. 3: if the sum of the digits is divisible by 3. 5: if the ones digit is 0 or 5. 6: if the number is divisible by 2 and 3. 9: if the sum of the digits is divisible by 9. 10: if the ones digit is zero.   **5.** no   **7.** none   **9.** 2, 3, 6   **11.** yes   **13.** yes   **15.** no   **17.** yes   **19.** yes   **21.** 2, 3, 6, 9   **23.** none   **25.** 3, 9   **27.** 2, 5, 10   **29.** yes   **31.** no   **33.** Sample answer: 90   **35.** Yes; each chaperone will have 15 students.   **37.** 50   **39.** 275.856

Page 181   Lesson 5-2A

1. 2 ▢▢   3 ▢▢▢   4 ▢▢▢▢

5 ▢▢▢▢▢

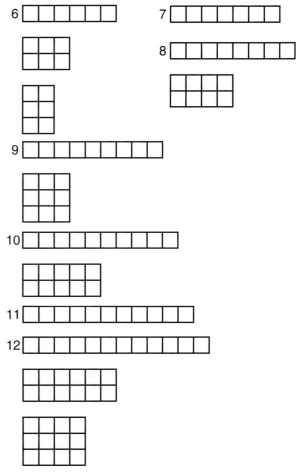

3. 4, 6, 8, 9, 10, 12   **5.** 12, 14, 15, 16, 18, 20, 21, 22, 24; these numbers have several factors.

Pages 183–184   Lesson 5-2

1.

1 × 10

2 × 5

So, 10 is a composite number.   **3.** Both are correct. The factors are arranged in a different order.
5. neither   **7.** composite   **9.** 3×5^2 **11.** $2^3 \times 13$
13. neither   **15.** prime   **17.** composite
19. composite   **21.** composite   **23.** composite
25. $2 \times 3 \times 7$ **27.** 5×13 **29.** prime   **31.** 2×3^2
33. $2 \times 3 \times 17$ **35.** $2^3 \times 3 \times 5$ **37.** 83   **39.** 2;
Two is the only even number that has exactly 2 factors, 1 and itself.   **41.** 3 and 5, 5 and 7, 11 and 13, 17 and 19, 29 and 31, 41 and 43, 59 and 61, 71 and 73   **43.** yes   **45.** 4.36   **47.** $17 \times 17 \times 17 \times 17 \times 17 \times 17 \times 17$

Pages 186–187   Lesson 5-3A

3. 12 possible options   **7.** 30   **9.** 6   **11.** $1,680
13. C

Pages 189–190   Lesson 5-3

1. Sample answer: Find the prime factorization of 12 and 18. Multiply all common factors of both numbers to find the greatest common factor.   **3.** 5
5. 7   **7.** 8   **9.** 8 ft   **11.** 2   **13.** 15   **15.** 11
17. 17   **19.** 12   **21.** 6   **23.** 3   **25.** 17   **27.** 19
29. 3   **31a.** 8 in. by 8 in.   **31b.** 18   **33.** Sample answer: 8 and 9   **35.** 24.9 yd^2 **37.** 9.016

Pages 191–192   Lesson 5-4A

1. 5 out of 6, 5 to 6, or 5:6; $\frac{5}{6}$ **3.** 5 out of 9, 5 to 9, or 5:9; $\frac{5}{9}$ **5.** $\frac{3}{4}, \frac{6}{8}$
7.

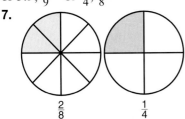

$$\frac{2}{8} \qquad \frac{1}{4}$$

Pages 195–196   Lesson 5-4

1a. $\frac{8}{12}$ **1b.** 4   **1c.** $\frac{2}{3}$ **1d.**
3. 9   **5.** 3   **7.** no; 1 to 9
9. yes   **11.** 4   **13.** 9
15. 24   **17.** 6   **19.** 20   **21.** no; $\frac{5}{19}$ **23.** no; 3 out of 4   **25.** no; $\frac{3}{20}$ **27.** no; 3:5   **29.** no; $\frac{21}{25}$ **31.** no; 1 out of 4   **33.** Sample answer: $\frac{4}{6}$ **35a.** Gwynn, $\frac{2}{5}$; Ripken, $\frac{1}{3}$; Gonzalez, $\frac{11}{36}$; Williams, $\frac{8}{21}$; Bonds, $\frac{1}{3}$; Rodriguez, $\frac{5}{14}$ **35b.** Ripken and Bonds   **37.** $\frac{36}{48}$
39. C   **41.** 26.98, 27.025, 27.13, 27.131, 27.9

Page 197   Lesson 5-4B

1. $\frac{7}{48}$ **3.** $\frac{17}{48}$

Pages 200–201   Lesson 5-5

1. Sample answer: a fraction with a numerator greater than or equal to the denominator.
3.
5. $\frac{7}{5}$ **7.** $2\frac{3}{4}$ **9.** 4   **11.** $\frac{16}{5}$ **13.** $\frac{9}{8}$ **15.** $\frac{16}{9}$
17. $\frac{26}{3}$ **19.** $\frac{38}{5}$ **21.** $6\frac{1}{3}$ **23.** $1\frac{7}{8}$ **25.** $7\frac{1}{4}$
27. $2\frac{1}{9}$ **29.** $3\frac{4}{7}$ **31.** $\frac{55}{8}$ **33.** Kentucky Derby–$\frac{5}{4}$; Preakness Stakes–$\frac{19}{16}$; Belmont Stakes–$\frac{3}{2}$ **35.** Sample answer: If the numerator is less than the denominator, then the fraction is less than one. If the numerator is the same as the denominator, then the fraction is equal to one. If the numerator is greater than the denominator, then

the fraction is greater than one.
37. $2^2 \times 3 \times 17$ **39.** 0.13

Page 201   Mid-Chapter Self Test

1. 3, 5   **3.** 2, 5, 10   **5.** $2^3 \times 11$ **7.** 18   **9.** 4:25

Pages 203–205   Lesson 5-6

1. Divide 24 by 12.   **3.** A cubit is about 18 in. A span is about 8 to 9 in.   **5.** 3
7. ▬▬▬▬▬▬▬▬▬▬
9. $\frac{7}{8}$ in.   **11.** 5   **13.** 108   **15.** 10,560
17. ▬▬▬▬▬   **19.** ▬▬
21. ▬▬▬▬▬▬   **23.** $\frac{1}{2}$ in.
25. $1\frac{3}{8}$ in.   **27.** $\frac{1}{4}$ in.   **29.** $1\frac{1}{2}$ ft   **31.** 1 in., $\frac{1}{2}$ in., $\frac{1}{4}$ in., $\frac{1}{8}$ in.   **33a.** 1 in.   **33b.** $1\frac{3}{4}$ in.   **33c.** $2\frac{1}{8}$ in.
33d. $3\frac{7}{8}$ in.   **33e.** 3 in.   **33f.** $\frac{3}{4}$ in.   **35.** $\frac{43}{8}$ **37.** C

Pages 208–209   Lesson 5-7

1. The least common multiple is the least number, other than zero, that is a multiple of two or more numbers.   **3.** 30   **5.** yes   **7.** 9   **9.** 22   **11.** 4 packages of hot dogs, 5 packages of hot dog buns, 2 packages of plates   **13.** yes   **15.** yes   **17.** no
19. no   **21.** 12   **23.** 21   **25.** 80   **27.** 208
29. 36   **31.** 280   **33.** 18   **35.** 12   **37.** Sample answer: 18 and 24, 36 and 72   **39.** $9\frac{2}{5}$
41. Sample answer: $6 + 234 = 240$

Pages 212–213   Lesson 5-8

1. Use the LCM of the denominators to rename fractions so they will have the same denominators. Then compare the numerators.   **3.** The LCD of $\frac{2}{5}$ and $\frac{4}{9}$ is 45. So, rename the fractions with a denominator of 45. $\frac{2}{5} = \frac{18}{45}$ and $\frac{4}{9} = \frac{20}{45}$. Since 18 < 20, $\frac{2}{5} < \frac{4}{9}$.   **5.** 20   **7.** =   **9.** kitchen   **11.** 12
13. 45   **15.** >   **17.** =   **19.** >   **21.** <   **23.** =
25. $\frac{2}{7}$ **27.** $\frac{3}{5}, \frac{3}{7}, \frac{2}{5}, \frac{1}{6}$ **29.** $\frac{16}{90}$ and $\frac{8}{45}$; Since $\frac{16}{90} = \frac{8}{45}$, the cost is the same.   **31.** 300   **33.** $\frac{25}{28}$ **35.** 23

Pages 215–216   Lesson 5-9

1. Sample answer: Write the decimal as a fraction with a denominator of 100. Then simplify by using the GCF.   **3.** Ann; The decimal is written to the hundredths place. So, the denominator will be 100.
5. $\frac{9}{20}$ **7.** $4\frac{3}{8}$ **9.** $3\frac{1}{5}$ **11.** $8\frac{13}{20}$ **13.** $5\frac{16}{25}$
15. $2\frac{2}{5}$ **17.** $13\frac{9}{1,000}$ **19.** $7\frac{89}{100}$ **21.** 0.38; $\frac{19}{50}$
23. $\frac{1}{2}$ mi; $\frac{2}{5}$ mi; $\frac{3}{10}$ mi

25. krypton: $83\frac{4}{5}$; selenium: $78\frac{24}{25}$; sulfur: $32\frac{3}{50}$; carbon: $12\frac{11}{1,000}$ **27.** $\frac{3}{7}$ **29.** D

31.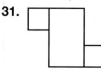

Pages 218–219 Lesson 5-10

1. Divide the numerator by the denominator. **3.** $0.\overline{4}$
5. 0.375 **7.** $0.58\overline{3}$ **9.** $0.\overline{7}$ **11.** $2.\overline{45}$ **13.** $0.8\overline{31}$
15. 0.75 **17.** $0.\overline{2}$ **19.** $0.\overline{45}$ **21.** $0.91\overline{6}$ **23.** $4.\overline{09}$
25. $0.1\overline{6}$ oz **29.** $29\frac{3}{4}$ **31.** D

Pages 220–223 Study Guide and Assessment

1. g **3.** e **5.** b **7.** f **9.** 3 **11.** 3, 9
13. prime **15.** neither **17.** 2×3^3 **19.** $2^2 \times 31$
21. 6 **23.** 2 **25.** 20 **27.** no; $\frac{5}{6}$ **29.** no; 7 to 8
31. $\frac{18}{5}$ **33.** $3\frac{4}{5}$ **35.** ▬▬▬
37. $1\frac{3}{4}$ inches **39.** 75 **41.** 84 **43.** 20 **45.** >
47. $\frac{4}{5}$ **49.** $3\frac{13}{20}$ **51.** 0.625 **53.** 24 ways
55. $30.40

Pages 224–225 Standardized Test Practice

1. A **3.** B **5.** A **7.** B **9.** A **11.** C **13.** D
15. 129 **17.** $20.49 **19.** $5.63

Chapter 6
Adding and Subtracting Fractions

Pages 230–231 Lesson 6-1

1.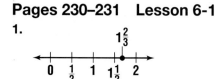

3. Sample answer: If the numerator is close in value to the denominator, round up. If the numerator is about half the denominator, round to $\frac{1}{2}$. If the numerator is much smaller than the denominator, round down. **5.** 4 **7.** $\frac{1}{2}$ **9.** down
11. $2\frac{1}{2}$ cups **13.** 1 **15.** $6\frac{1}{2}$ **17.** $\frac{1}{2}$ **19.** 12
21. 4 **23.** $10\frac{1}{2}$ **25.** $5\frac{1}{2}$ **27.** down **29.** up
31. up **33a.** $\frac{23}{28}, \frac{2}{7}, \frac{5}{28}, \frac{1}{2}, \frac{11}{14}, \frac{17}{28}, \frac{5}{7}, \frac{3}{4}$
33b. salads, turkey **33c.** popcorn, chicken, frozen yogurt **35.** Sample answer: $5\frac{3}{8}, 5\frac{7}{16}, 5\frac{3}{5}$
37. sixteen ten-thousandths **39.** D

Pages 233–234 Lesson 6-2

1.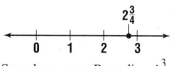

3. Sample answer: Rounding $4\frac{3}{8}$ to 4 is correct, but since the measure was rounded down, there would not be enough material to complete the picture frame. Nina should buy 20 inches of framing material. **5–9.** Sample answers are given.
5. $7 + 4 = 11$ **7.** $2\frac{1}{2} - \frac{1}{2} = 2$ **9.** $6 + 10 = 16$
11–27. Sample answers are given. **11.** $1 + \frac{1}{2} = 1\frac{1}{2}$
13. $8 - 1 = 7$ **15.** $1 + \frac{1}{2} = 1\frac{1}{2}$ **17.** $1 + 6 = 7$
19. $9 + 0 = 9$ **21.** $22 + 5 = 27$ **23.** 5 min − 1 min = 4 min **25.** 20 cups − 11 cups = 9 cups
27. $7 + 7 + 7 + 7 + 7 = 35$ **29.** 64 in. − 62 in. = 2 in. **31.** Sample answer: $\frac{5}{8}$ and $\frac{7}{16}$ **33.** $\frac{17}{100}$
35. 400,000 cm

Pages 236–237 Lesson 6-2B

1. Sample answer: Franco and Cesar eliminated the routes that they could not take to see which route they could. **3.** C **5.** Sample answer: Is $15,840 \div 18$ equal to 80, 88, 880, or 8,800? **7.** Viho is 8 years old. His father is 32 and his grandfather is 64.
9. 50¢ **11.** about 7 cups **13.** 12 **15.** D

Pages 240–241 Lesson 6-3

1. Sample answer: Add numerators and keep the denominators the same.

3.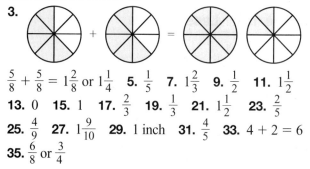

$\frac{5}{8} + \frac{5}{8} = 1\frac{2}{8}$ or $1\frac{1}{4}$ **5.** $\frac{1}{5}$ **7.** $1\frac{2}{3}$ **9.** $\frac{1}{2}$ **11.** $1\frac{1}{2}$
13. 0 **15.** 1 **17.** $\frac{2}{3}$ **19.** $\frac{1}{3}$ **21.** $1\frac{1}{2}$ **23.** $\frac{2}{5}$
25. $\frac{4}{9}$ **27.** $1\frac{9}{10}$ **29.** 1 inch **31.** $\frac{4}{5}$ **33.** $4 + 2 = 6$
35. $\frac{6}{8}$ or $\frac{3}{4}$

Page 242 Lesson 6-4A

1. Sample answer: you rename the units so that the answer makes sense. **5.** Sample answer: find a common denominator.

Pages 244–245 Lesson 6-4

1. Sample answer: You rename fractions with unlike denominators to get a common unit name.
3. $\frac{7}{16}$ **5.** $\frac{1}{12}$ **7.** $\frac{1}{4}$ **9.** $\frac{7}{8}$ **11.** $\frac{1}{16}$ **13.** $1\frac{1}{10}$
15. $\frac{8}{15}$ **17.** $\frac{7}{24}$ **19.** $\frac{7}{20}$ **21.** $\frac{17}{36}$ **23.** $\frac{3}{8}$

25. $\frac{5}{16}$ inch **27.** $\frac{3}{4}$ **29.** $\frac{1}{12}$ **31.** $1\frac{3}{10}$ **33.** $1\frac{1}{2}$
35. 30 feet **37.** Sample answer: $\frac{2}{8}$ or $\frac{1}{4}$ in. **39.** D

Pages 248–249 Lesson 6-5

1. Sample answer: Find the LCM of the denominators. **3.** Sample answer:

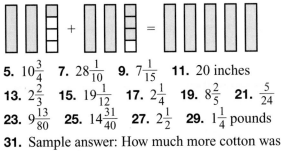

5. $10\frac{3}{4}$ **7.** $28\frac{1}{10}$ **9.** $7\frac{1}{15}$ **11.** 20 inches
13. $2\frac{2}{3}$ **15.** $19\frac{1}{12}$ **17.** $2\frac{1}{4}$ **19.** $8\frac{2}{5}$ **21.** $\frac{5}{24}$
23. $9\frac{13}{80}$ **25.** $14\frac{31}{40}$ **27.** $2\frac{1}{2}$ **29.** $1\frac{1}{4}$ pounds
31. Sample answer: How much more cotton was planted in 1996 than was estimated to be planted in 1997? **33.** D **35.** 0.8 kg

Page 249 Mid-Chapter Self Test

1. Sample answer: To avoid ruining the chili, you should round down to one-half teaspoon of salt. You can always add more. **3.** $22 - 15 = 7$ **5.** $\frac{3}{10}$
7. $\frac{7}{30}$ **9.** $20\frac{1}{6}$

Pages 251–253 Lesson 6-6

1. Sample answer:

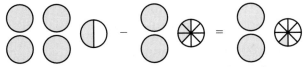

3. Sample answer: When Gail borrowed one from 10 she just carried it next to the 1 in $\frac{1}{5}$ and made it $\frac{11}{5}$. **5.** 30 **7.** $2\frac{5}{6}$ **9.** $\frac{7}{8}$ **11.** 2 **13.** 7 **15.** 29
17. $5\frac{1}{2}$ **19.** $3\frac{7}{10}$ **21.** $4\frac{7}{12}$ **23.** $7\frac{5}{12}$ **25.** $9\frac{3}{4}$
27. $7\frac{34}{35}$ **29.** $\frac{2}{3}$ **31.** $1\frac{23}{30}$ **33.** $1\frac{2}{3}$ **35a.** softball; $2\frac{5}{8}$ to $3\frac{1}{8}$ in. larger **35b.** softball; 1 to 2 ounces heavier **37b.** $1\frac{3}{4}$ in. **39.** $2\frac{17}{24}$ **41.** 1,010.5
43. 11

Pages 255–257 Lesson 6-7

1. 11 minutes **3.** Sample answer: Adding and subtracting measures of time is similar to adding and subtracting mixed numbers because you have to rename. **5.** 10 **7.** 10 min 45 s **9.** 3 h 50 min 40 s
11. 6 h 45 min **13.** $28 **15.** 68 **17.** 6; 27
19. 27; 23 **21.** 10 h 38 min **23.** 16 h 1 min
25. 20 min **27.** 16 min 55 s **29.** 16 h 15 min 20 s
31. 10 h 26 min 20 s **33.** 3 h 50 min 40 s

35. 59 min **37.** 10 h 30 min **39.** 21 h 20 min 10 s
41. 45 min **43.** 10:29 A.M. **45.** $3\frac{3}{4}$ in. **47.** B
49. 100,000,000 miles

Pages 258–261 Study Guide and Assessment

1. 5 **3.** $\frac{4}{5}$ **5.** LCD **7.** $5\frac{1}{2}$ **9.** 80 **11.** $\frac{1}{2}$
13. $11\frac{1}{2}$ **15.** down **17.** up **19.** $8 - 4 = 4$
21. $0 + 5 = 5$ **23.** $1\frac{4}{15}$ **25.** $\frac{3}{4}$ **27.** $\frac{1}{2}$ **29.** $1\frac{5}{18}$
31. $5\frac{2}{7}$ **33.** $1\frac{3}{10}$ **35.** $4\frac{7}{8}$ **37.** $5\frac{1}{3}$ **39.** $5\frac{7}{8}$
41. 68 **43.** 7 h 36 min **45.** 59 min 52 s
47. $8.50 **49.** $1\frac{1}{6}$ miles

Pages 262–263 Standardized Test Practice

1. B **3.** D **5.** C **7.** C **9.** E **11.** A
13. when both factors are less than one **15.** $\frac{5}{9}$
17. $1\frac{1}{16}$ **19.** 14 h 10 min

Chapter 7 Multiplying and Dividing Fractions

Pages 269–270 Lesson 7-1

1. Round $3\frac{1}{2}$ to 4. Round $9\frac{1}{8}$ to 9. Since $4 \times 9 = 36$, $3\frac{1}{2} \times 9\frac{1}{8} \approx 36$. **3.** Both Juan and Odina are correct. If you round $8\frac{1}{2}$ to 9 and $6\frac{1}{4}$ to 6, the product is 9×6, or 54. If you round $8\frac{1}{2}$ to 8 and $6\frac{1}{4}$ to 6, the product is 8×6, or 48. **5.** $\frac{1}{2}$ **7–9.** Sample answers are given. **7.** $\frac{1}{8} \times 16 = 2$ **9.** $3 \times 9 = 27$ **11.** $\frac{1}{2}$
13. $\frac{1}{2}$ **15.** 6 **17.** 3 or 4 **19–29.** Sample answers are given. **19.** $\frac{1}{4} \times 32 = 8$ **21.** $\frac{1}{7} \times 21 = 3$
23. $\frac{1}{6} \times 42 = 7$ **25.** $\frac{4}{9} \times 45 = 20$ **27.** $8 \times 3 = 24$
29. $15 \times 7 = 105$ **31.** about $\frac{2}{3} \times 6$, or about 4 million **33.** 1 h 3 min 51 s **35.** 0.498

Page 271–272 Lesson 7-2A

1. 6 square units **3.** $\frac{1}{6}$ **5.** $\frac{1}{12}$ **7.** To multiply fractions, multiply the numerators and then multiply the denominators.

Pages 275–276 Lesson 7-2

1. The rectangle is separated into thirds. Then the rectangle is separated into halves. The shaded portion shows overlap between $\frac{1}{3}$ of the rectangle and $\frac{1}{2}$ of the rectangle. This area represents the product, $\frac{1}{6}$. **3.** To find the product of fractions, you need to multiply the numerators and then multiply the denominators. If necessary, simplify the answer. **5.** $\frac{3}{20}$ **7.** $\frac{2}{15}$ **9.** $\frac{1}{3}$ **11.** about 105 pounds **13.** $\frac{8}{27}$ **15.** $\frac{1}{9}$ **17.** $\frac{2}{15}$ **19.** $\frac{2}{3}$ **21.** $\frac{10}{27}$ **23.** $\frac{2}{27}$ **25.** $6\frac{2}{3}$ **27.** $\frac{1}{12}$ **29.** $\frac{4}{9}$ **31.** $10\frac{1}{2}$ **33.** $\frac{4}{15}$ **35a.** $\frac{9}{64}$ **35b.** $\frac{16}{25}$ **37.** $\frac{3}{4} \times 36 = 27$ **39.** 2, 3, 6

Pages 278–279   Lesson 7-3

1. To multiply mixed numbers, express each mixed number as an improper fraction and then multiply as with fractions. **3.** $\frac{19}{3}$ **5.** $\frac{23}{4}$ **7.** $3\frac{3}{5}$ **9.** 38 **11.** $1\frac{1}{9}$ **13a.** $\frac{3}{4}$ **13b.** $\frac{3}{8}$ **15.** $\frac{23}{3}$ **17.** $\frac{14}{3}$ **19.** $\frac{33}{5}$ **21.** $\frac{35}{4}$ **23.** $6\frac{2}{3}$ **25.** $10\frac{1}{2}$ **27.** $4\frac{7}{8}$ **29.** 42 **31.** 21 **33.** $8\frac{1}{7}$ **35.** $7\frac{1}{3}$ **37.** 30 **39.** $7\frac{1}{2}$ **41.** $8\frac{1}{4}$ sq ft **43.** more; $2\frac{2}{3} \times 4\frac{1}{2} \approx 3 \times 4$ or 12 **45.** $3\frac{5}{12}$ **47.**

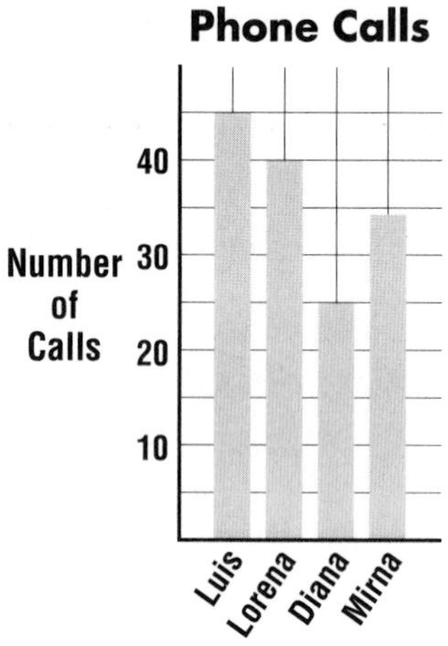

Phone Calls

(bar graph: Number of Calls on y-axis labeled 10, 20, 30, 40; x-axis labels: Luis, Lorena, Diana, Mirna)

Pages 282–283   Lesson 7-4

1. Sample answer: Multiply 2 by π by 3. **5.** $\pi = \frac{22}{7}$; $7\frac{6}{7}$ yd **7.** 6.0 m **9.** 12.6 in. **11.** $\pi = \frac{22}{7}$; $7\frac{1}{3}$ in. **13.** 37.7 yd **15.** 25.1 ft **17.** 9.7 yd **19.** 52.8 cm **21.** $\pi = \frac{22}{7}$; $15\frac{5}{7}$ ft **23.** The circumference is twice as long. **25.** $\frac{1}{2}$

Page 283   Mid-Chapter Self Test

1–3. Sample answers are given. **1.** $16 \times \frac{3}{4} = 12$ **3.** $3 \times \frac{1}{2} = 1\frac{1}{2}$ **5.** $\frac{1}{4}$ **7.** $3\frac{3}{8}$ **9.** 42.7 cm

Page 284   Lesson 7-5A

1.

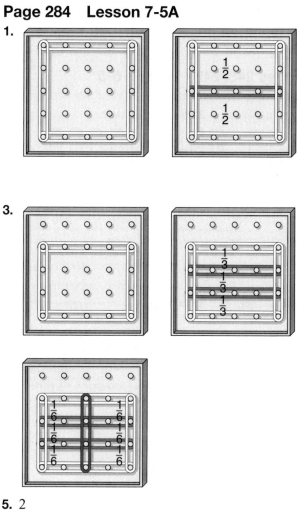

3.

5. 2

Pages 287–288   Lesson 7-5

1. Multiply $\frac{1}{2}$ by $\frac{3}{2}$. **3.** 2 **5.** $\frac{1}{8}$ **7.** $\frac{1}{6}$ **9.** $1\frac{1}{8}$ **11.** 20 **13.** $1\frac{2}{3}$ **15.** $\frac{1}{5}$ **17.** 3 **19.** 7 **21.** 1 **23.** $\frac{5}{9}$ **25.** $\frac{4}{5}$ **27.** $\frac{1}{10}$ **29.** $\frac{1}{3}$ **31.** $\frac{5}{16}$ **33.** $\frac{8}{9}$ **35.** $\frac{2}{3}$ **37.** $\frac{9}{10}$ **39.** $\frac{2}{9}$ **41.** $3\frac{1}{2}$ **43.** 8 quarters **45a.** $\frac{11}{10}$ **45b.** $\frac{11}{10}$ **47.** $0.\overline{7}$ **49.** about 4 inches

Pages 290–291   Lesson 7-6

1. First, write $4\frac{5}{8}$ as the improper fraction $\frac{37}{8}$. Then invert the numerator and the denominator. The reciprocal of $4\frac{5}{8}$ is $\frac{8}{37}$. **3.** Change each mixed number to an improper fraction. Multiply the first fraction by the reciprocal of the second fraction. Divide 8 and 12 by the GCF, 4. Multiply $\frac{3}{1}$ and $\frac{3}{2}$ to find the product $\frac{9}{2}$, or $4\frac{1}{2}$. **5.** $\frac{23}{3}$; $\frac{3}{23}$ **7.** $3\frac{1}{5}$ **9.** $4\frac{1}{2}$ **11.** $1\frac{1}{7}$ **13.** 36 slices **15.** $\frac{32}{5}$; $\frac{5}{32}$ **17.** $\frac{14}{5}$; $\frac{5}{14}$ **19.** $\frac{15}{4}$; $\frac{4}{15}$ **21.** $\frac{37}{4}$; $\frac{4}{37}$ **23.** $7\frac{3}{5}$ **25.** 12 **27.** $\frac{4}{5}$ **29.** $4\frac{3}{8}$ **31.** $\frac{2}{3}$ **33.** 26 **35.** $\frac{2}{3}$

37. $8\frac{1}{3}$ **39a.** $\frac{7}{10}, 5\frac{11}{12}, 9\frac{5}{8}$ **39b.** $\frac{1}{10}, \frac{7}{12}, 6\frac{3}{8}$

39c. $\frac{3}{25}, 8\frac{2}{3}, 13$ **39d.** $1\frac{1}{3}, 1\frac{7}{32}, 4\frac{12}{13}$ **41.** 6

photographs **43.** Greater than; $\frac{2}{3}$ is less than $\frac{3}{4}$.

45. $\frac{10}{21}$ **47.** B **49.** two tenths; seven hundredths;

four thousandths

Page 293–294   Lesson 7-7

1. fluid ounce, cup, pint, quart, gallon, ounce,
pound, and ton **3.** 8 **5.** 4 **7.** 2 **9.** about
13,000 pounds **11.** 6 **13.** 144 **15.** 2 **17.** 40
19. 4,500 **21.** 10 **23.** $1\frac{1}{2}$ **25.** 40 **27.** 500
pounds **29.** 16 **31.** 3, 5, 9 **33.** E

Page 295   Lesson 7-7B

1.

Planet/ Moon	Amount of water	Weight compared to Earth	Weight compared to Jupiter
Sun	28 cups	more	more
Mercury	$\frac{1}{3}$ cups	less	less
Venus	$\frac{9}{10}$ cups	less	less
Moon	$\frac{1}{6}$ cups	less	less
Mars	$\frac{3}{8}$ cups	less	less
Jupiter	3 cups	more	

3. 66 pounds **5.** 6 ounces

Pages 296–297   Lesson 7-8A

3.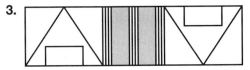

5. Sample answer: Once you identify the pattern,
then you can continue the pattern to solve the
problem. **9a.** $\frac{31}{32}$ **9b.** Richard will not finish the
sandwich, but the amount will be so small that it will
be like he did eat it all. **11.** 1,476 sq cm **13.** C

Pages 299–300   Lesson 7-8

1. Sample answer: A sequence is a list of numbers in
a specific order. **3.** Joshua; In the sequence, $2\frac{1}{2}$ is
added to each number. The next number is $12\frac{1}{2} + 2\frac{1}{2}$,
or 15. **5.** 17, 10 **7.** 125 **9.** 75, 90 **11.** $\frac{3}{4}, \frac{3}{8}$
13. 432, 2,592 **15.** 33 **17.** 1 **19.** 400 **21.** x^5
23. $\frac{1}{16}$ (sixteenth note), $\frac{1}{32}$ (thirty-second note), and
$\frac{1}{64}$ (sixty-fourth note) **25a.** $\frac{1}{2}, \frac{1}{4}, \frac{1}{8}, \frac{1}{16}, \frac{1}{32}, \frac{1}{64}$,
$\frac{1}{128}, \frac{1}{256}, \frac{1}{512}, \frac{1}{1,024}$ **25b.** 1 **27.** A

Page 301   Lesson 7-8B

1. Multiply the value in cell B5 by the value in
cell B2. **3.** 8.000 m, 3.200 m, 1.280 m, 0.512 m,
0.2048 m, 0.08192 m

Pages 302–305   Study Guide and Assessment

1. k **3.** i **5.** h **7.** f **9.** d **11–16.** Sample
answers are given. **11.** $10 \times 3 = 30$
13. $1 \times 13 = 13$ **15.** $\frac{2}{3} \times 18 = 12$ **17.** $\frac{7}{10}$
19. $7\frac{1}{2}$ **21.** $4\frac{1}{2}$ **23.** $13\frac{1}{2}$ **25.** 57.1 ft
27. $\pi = \frac{22}{7}$; $5\frac{1}{2}$ yd **29.** $1\frac{1}{9}$ **31.** $\frac{2}{3}$ **33.** $\frac{1}{3}$
35. $\frac{21}{40}$ **37.** 11,000 **39.** 48 **41.** 176, 156
43. 2 **45.** 57 students **47.** 144 one-cup servings

Pages 306–307   Standardized Test Practice

1. C **3.** B **5.** B **7.** B **9.** D **11.** A
13. $5.05 **15.** $\frac{1}{2}$ yd **17.**

Chapter 8
Exploring Ratio, Proportion, and Percent

Page 311   Lesson 8-1A

1a. $\frac{1}{2}$ square unit **1b.** Triangle C is $\frac{1}{2}$ the size of
triangle B. **3a.** $\frac{1}{2}$ **3b.** $\frac{1}{4}$ **3c.** 1 **3d.** $\frac{2}{1}$ **3e.** $\frac{1}{2}$
3f. $\frac{1}{2}$

Pages 313–315   Lesson 8-1

1. A ratio is a comparison of two numbers by
division. A rate is a ratio of two measurements that
have different units. **3.** Sample answer: Rates would
be useful when shopping in the grocery store. They
allow you to determine better buys. **5.** Sample
answer: 15 to 34, $\frac{15}{34}$, 15 out of 34 **7.** $\frac{2}{5}$ **9.** $0.65
per soft drink **11–15.** Sample answers are given.
11. 9 out of 16, 9 to 16, 9:16 **13.** 18 out of 29, 18
to 29, 18:29 **15.** 23 out of 25, 23 to 25, 23:25
17. $\frac{1}{3}$ **19.** $\frac{7}{9}$ **21.** $\frac{4}{9}$ **23.** 79 kilometers per hour
25. $8.75 per ticket **27.** $0.11 per egg **29.** Sample
answer: 13 out of 24; 13 to 24; 13:24 **31a.** $\frac{2}{25}$
31b. $\frac{6}{25}$ **33.** 25 miles **35a.** $\frac{1}{144}$ **35b.** $\frac{1}{64}$

37. $4\frac{5}{8}$ **39a.** 6.6 billion lb **39b.** 34.1 billion lb

Page 316   Lesson 8-1B

1. $\frac{1}{4}$; $\frac{3}{4}$ **3.** Sample answer: The results are about the same. **5.** 0.625

Pages 319–320   Lesson 8-2

1. A proportion is an equation stating that two ratios are equivalent. **5.** no **7.** 72 **9.** 120 dentists **11.** yes **13.** no **15.** no **17.** 8 **19.** 11 **21.** 100 **23.** 100 **25.** 6 pairs **27.** 1,404 **29.** 15 **31.** $7\frac{1}{6}$ **33.** D

Page 321   Lesson 8-2B

1. Multiply the contents of cell B2 by 2. **3.** 6 cups of mayonnaise, 24 T of mustard, 24 T of relish, 12 T of vinegar, 3 t of salt, $1\frac{1}{2}$ t of pepper, 60 cups of potatoes, 3 cups of parsley

Pages 322–323   Lesson 8-3A

1.

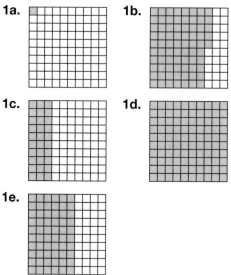

3. 15 handshakes **5.** A diagram will help you understand and picture the information more clearly.

7.

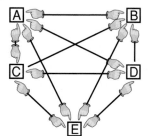

$\frac{1}{4}$ = 28 in.
$\frac{1}{4}$ = 28 in.
$\frac{1}{4}$ = 28 in.
$\frac{1}{2}$ of $\frac{1}{4}$ = 14 in.

9. Reynelda, Juliet, Keela, Pedro, Daniel
11. 417.5 kg **13.** 6 cuts

Pages 325–327   Lesson 8-3

1. An architect would make a scale drawing or a model in order to show a building exactly as it looks, but smaller. **3a.** $2\frac{1}{4}$ feet **3b.** $3\frac{1}{2}$ feet **3c.** $3\frac{1}{2}$ feet **5a.** $1\frac{2}{3}$ yards **5b.** $1\frac{1}{3}$ yards **5c.** 2 yards **7a.** $14\frac{1}{4}$ miles **7b.** $23\frac{3}{4}$ miles **7c.** $16\frac{5}{8}$ miles **7d.** $45\frac{1}{8}$ miles **7e.** $8\frac{5}{16}$ miles **9.** 20 feet **11a.** length: 5.18 m; width: 2.16 m; height: 1.58 m **11b.** length: 4.93 m; width: 1.92 m; height: 1.47 m **13.** 5.4

Page 327   Mid-Chapter Self Test

1. Sample answer: 11 to 19, $\frac{11}{19}$, 11:19 **3.** $\frac{3}{7}$ **5.** 60 mph **7.** 96 **9.** 2.8

Page 329   Lesson 8-4A

1a. **1b.** **1c.** **1d.** **1e.**

3. Sample answer: Grid C from the first set is the same as grid A from the second set. Grid A from the first set is the same as grid B from the second set. Grid B from the first set is the same as grid C from the second set. Grid D from the first set is the same as grid D from the second set. Grid E from the first set is the same as grid E from the second set.

Pages 332–333   Lesson 8-4

1a. 25% **1b.** 47% **1c.** $33\frac{1}{3}$% **1d.** 80% **1e.** 50% **1f.** $66\frac{2}{3}$% **5.** $\frac{11}{20}$ **7.** 34% **9.** 125% **11.** $\frac{7}{50}$ **13.** $\frac{1}{100}$ **15.** $1\frac{3}{10}$ **17.** $1\frac{1}{20}$ **19.** $\frac{17}{100}$

21. **23.**

25. 63% **27.** 150% **29.** 74% **31.** 76% **33.** 45% **35a.** 48% **35b.** 9% **37a.** $\frac{1}{144}$, 0.01, 1% **37b.** $\frac{1}{64}$, 0.02, 2% **39.** 55 miles **41.** $1\frac{1}{4}$ **43.** 30

Pages 335–336   Lesson 8-5

1. **3.** Betty is correct. The decimal 3.78 is equivalent to 378%. Therefore, it is more than 100%. **5.** 0.81 **7.** 0.008 **9.** 90% **11.** 2% **13.** 0.32

15. 0.01 **17.** 0.06 **19.** 0.003 **21.** 0.39 **23.** 0.04
25. 96% **27.** 36.4% **29.** 71.6% **31.** 7%
33. 8% **35.** 20% **37.** 1.8% **39a.** 0.0321, 0.12, 0.09, 26.5 **39b.** $\frac{7}{100}$, $1\frac{129}{1,000}$, $\frac{3}{100}$, $\frac{47}{1}$ **39c.** 4%, 53.8%, 60%, 37.5% **41a.** 18% **41b.** 82%
43. $1\frac{14}{25}$ **45.** D **47.** 5.4

Pages 338–339 Lesson 8-6

1. 75% = $\frac{3}{4}$. Round 1,976 to 2,000. $\frac{3}{4} \times 2,000 = 1,500$. So, 75% of 1,976 is about 1,500. **5.** Sample answer: $\frac{1}{4} \times 200 = 50$ **7–17.** Sample answers are given. **7.** 50% **9.** $\frac{1}{5} \times 35 = \$7$ **11.** $\frac{1}{3} \times 60 = 20$ **13.** $\frac{1}{2} \times 50 = 25$ **15.** $\frac{2}{5} \times 100 = 40$ **17.** $1 \times 300 = 300$ **19.** 75% **21.** $66\frac{2}{3}$% **23.** 100% **25–27.** Sample answers are given. **25.** $\frac{3}{8} \times 48 = 18$ **27.** $\frac{1}{100} \times 10 = \frac{1}{10}$ **29.** 50% × 5,000 = 2,500; 20% × 5,000 = 1,000; 2,500 − 1,000 = 1,500
31. c; because a and b equal 11 while c equals 0.11.
33. 23.04 m² **35.** 38

Pages 342–343 Lesson 8-7

1. Sample answer: **Method 1:** Change the percent to a fraction. First, change the percent to a fraction. Then multiply the number by the fraction.
Method 2: Change the percent to a decimal. First, change the percent to a decimal. Then multiply the number by the decimal. **Method 4:** Use a calculator. First, enter the percent by entering the number ⎡2nd⎤ [%]. Then press ⎡X⎤ and the number by which you are multiplying. Press ⎡=⎤.
3. Kosey is correct. 125% = 1.25. 1.25 × 150 = 187.5. So, 125% of 150 is 187.5 **5.** 6.72 **7.** 5.88 **9.** 18 **11.** 92 **13.** 6 **15.** 98.98 **17.** 181.25 **19.** 137 **21.** 219 **23.** 6.4 **25a.** $45.00 **25b.** $33.75 **27a.** 20% **27b.** 84 **27c.** 300 **29.** Sample answer: $\frac{3}{8} \times 64 = 24$

Pages 344–347 Study Guide and Assessment

1. false; division **3.** true **5.** false; proportion **7.** true **9.** true **11.** true **13.** false; 34.6% **15.** Sample answer: Express the percent as a fraction with a denominator of 100 and then simplify. **17.** $\frac{11}{20}$ **19.** 15.75 pounds per week **21.** yes **23.** 25 **25.** 30 **27.** 1.5 units **29.** $\frac{3}{100}$ **31.** $1\frac{1}{2}$ **33.** 60% **35.** 55% **37.** 5% **39.** 0.38 **41.** 0.66 **43.** 130% **45.** 59.1% **47.** Sample answer: $\frac{1}{4} \times 80 = 20$ **49.** Sample answer: $\frac{2}{5} \times 240 = 96$ **51.** 16.02 **53.** 17 **55.** 0.162 **57.** 50 mph **59.** 8 cans

Pages 348–349 Standardized Test Practice

1. C **3.** A **5.** D **7.** D **9.** D **11.** C **13.** D **15.** 60% **17.** $1\frac{3}{8}$ inches **19.** 21

Chapter 9
Geometry: Investigating Patterns

Pages 353–355 Lesson 9-1

1. Sample answer: **Step 1:** Place the center of the protractor on the vertex of the angle with the straightedge along one side. **Step 2:** Using the scale that begins with 0° on the side of the angle, read the angle measure where the other side crosses the same scale. **3.** 50°; acute **5.** obtuse **7.** acute **9.** 74° **11.** 65°; acute **13.** 135°; obtuse **15.** 15°; acute **17.** right **19.** obtuse **21.** acute **23.** acute **25.** neither **27.** acute **29.** 18° **31.** Sample answer: A right angle is formed by the two pieces. The angles on the inside corner are obtuse and the angles on the outside corner are acute. **33a.** cut into the hill and flatten the road **33b.** larger **35.** $0.76 = \frac{19}{25}$; $0.15 = \frac{3}{20}$; $0.09 = \frac{9}{100}$ **37.** 52

Pages 356–357 Lesson 9-1B

1. Sample answer: Jesse's and Estella's thinking does make sense because 22 + 33 = 55, the number of students who play sports. **3.** 82 students **5.** You found that 5 students have only lunch bags, 10 have only backpacks, and 9 have neither a lunch bag nor a backpack. 5 + 10 + 9 = 24. So, there are 24 students going on the trip. **7.** A right angle measures 90°. The measure of the angle given is 90.4°. Since 90.4° is greater than 90°. It is only logical that the angle is an obtuse angle. **9.** 456
11a. Sample answer:

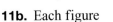

11b. Each figure has one more side than the previous figure, so the next figure must have six sides. **13.** C

Pages 360–361 Lesson 9-2

3. The corner of a sheet of paper is a 90° angle. A 45° angle is half of a 90° angle. So, to show a 45° angle, you can fold the corner in half.

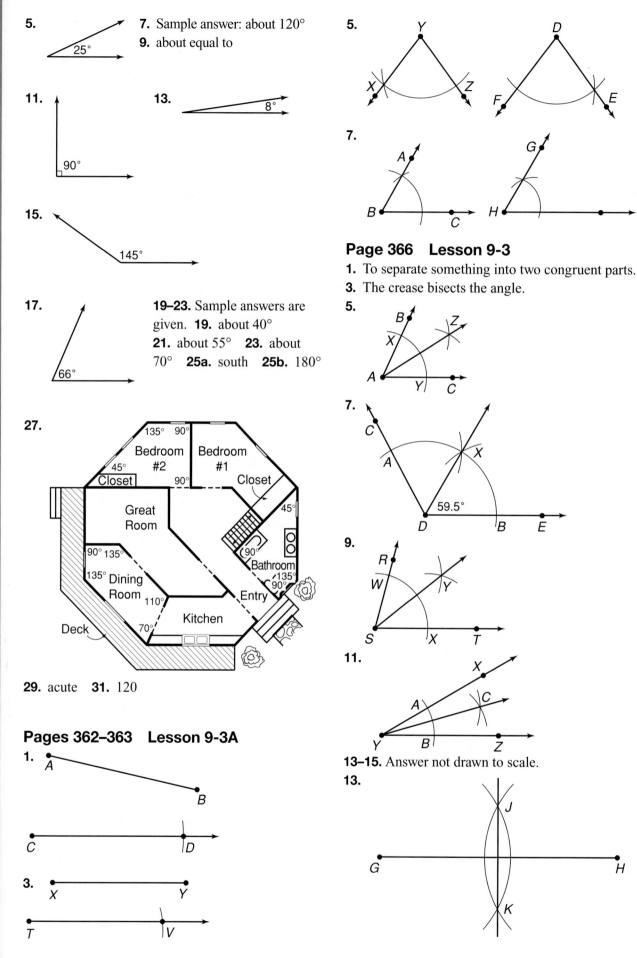

5.

25°

7. Sample answer: about 120°
9. about equal to

11.

90°

13.

8°

15.

145°

17.

66°

19–23. Sample answers are given. **19.** about 40°
21. about 55° **23.** about 70° **25a.** south **25b.** 180°

27.

135° 90°

Bedroom #2

Bedroom #1

45°

Closet

90°

Closet

Great Room

45°

90° 135°

90°

135° Dining Room

110°

Bathroom
135°
90°

Entry

70°

Kitchen

Deck

29. acute **31.** 120

Pages 362–363   Lesson 9-3A

1.
A

B

C   D

3.
X   Y

T   V

5.
Y

X   Z

D

F   E

7.
A

B   C

G

H

Page 366   Lesson 9-3

1. To separate something into two congruent parts.
3. The crease bisects the angle.

5.
B   Z
X
A   Y   C

7.
C
A   X

D 59.5° B   E

9.
R
W   Y

S   X   T

11.
X
A   C
Y   B   Z

13–15. Answer not drawn to scale.

13.
J

G   H

K

15.

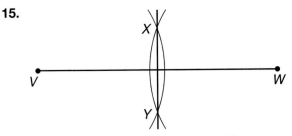

17. The diagonals bisect the corners of the square.

21. $\frac{37}{50}$

23.

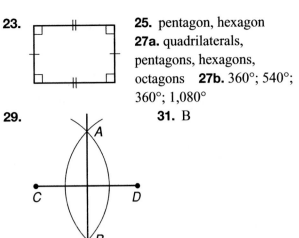

25. pentagon, hexagon
27a. quadrilaterals, pentagons, hexagons, octagons   **27b.** 360°; 540°; 360°; 1,080°

29.

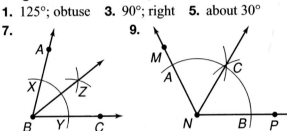

31. B

Pages 368–369   Lesson 9-4A

1. acute   **3.** right   **5.** Sample answer: The quadrilateral has two sides that are congruent.

7. Sample answer: The opposite sides of the quadrilateral are parallel.

Pages 372–373   Lesson 9-4

1. Sample answer: A chalkboard is a rectangle and a piece of floor tile is a square.   **3.** Jonathan; All squares are parallelograms since both sets of opposite sides of a square are parallel. Victoria is incorrect. Some parallelograms are not squares since some parallelograms do not have right angles.
5. equilateral triangle; yes   **7.** Alike: All sides are congruent. All angles are congruent.   Different: pentagon has 5 sides and square has 4 sides

9.

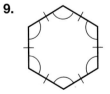

11. pentagon; yes   **13.** square; yes   **15.** Alike: all sides congruent; all angles congruent Different: triangle has three sides and square has four sides

17. Alike: quadrilaterals; at least 1 pair of parallel sides   Different: square has all sides congruent and four right angles

19.   **21.**

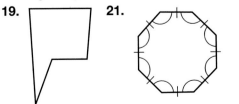

Page 373   Mid-Chapter Self Test

1. 125°; obtuse   **3.** 90°; right   **5.** about 30°

7.   **9.**

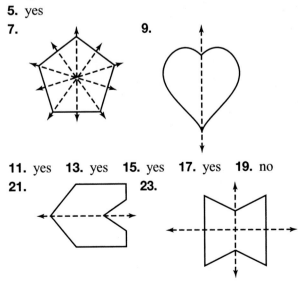

Page 374   Lesson 9-4B

1. The second net did not cover the cube. The "top" and "bottom" covers are on the same side of the cube.
3. Sample answer: Each net contains six squares. Each net has four squares lined up in a row. Each net has two squares positioned on opposite sides of the row of four squares.

Pages 376–378   Lesson 9-5

5. yes

7.   **9.**

11. yes   **13.** yes   **15.** yes   **17.** yes   **19.** no

21.   **23.**

Selected Answers   **629**

25. 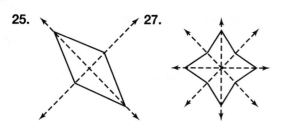 **27.**

29. yes **31.** no **33.** yes **35.** 8 **37.** 1 **41.** $6\frac{1}{3}$
43. 525 people

Pages 380–382   Lesson 9-6

1. The angles are congruent, they have the same shape, and they are different sizes. **5.** congruent
7. 3 **9.** neither **11.** neither **13.** congruent
15a. 2.5 m **15b.** $\overline{NM}$ **17.** A and D; A and F; A and G; B and D; B and F; B and G; F and D; A and B; D and G; F and G

19.

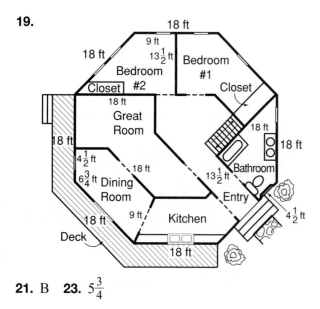

21. B **23.** $5\frac{3}{4}$

Pages 384–385   Lesson 9-6B

1. 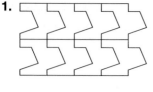 **3.**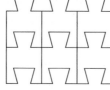

5. 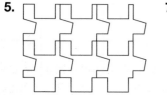 **7.**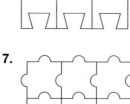

Pages 386–389   Study Guide and Assessment

1. f **3.** a **5.** e **7.** b **9.** d **11.** If an angle measures less than 90°, it is an acute angle. If an angle measures 90°, it is a right angle. If an angle measures between 90° and 180°, it is an obtuse angle. **13.** 42°; acute **15.** acute **17.** acute
19.

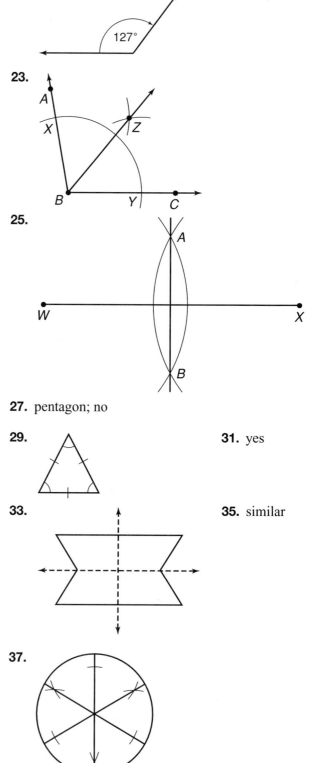

21. about 90°

23.

25.

27. pentagon; no

29.

31. yes

33.

35. similar

37.

1. C   **3.** D   **5.** A   **7.** B   **9.** A   **11.** C
13. 126 feet   **15.** $\frac{1}{8}$ **17.** 26 inches by 18 inches

Chapter 10
Geometry: Understanding Area and Volume

Pages 396–397   Lesson 10-1A
7–11. Sample answers are given.   **7.** 5 square units
9. $4\frac{1}{2}$ square units   **11.** The areas are the same.

Pages 400–401   Lesson 10-1
1. A rectangle is a special parallelogram.   **3.** If the lengths of the bases and the heights are the same, the areas will be the same.   **5.** 13.6 sq m   **7.** 54 in^2
9. 38.8 cm^2 **11.** 70 m^2 **13.** 42.7 cm^2
15. 4,611 cm^2 **17a.** The area doubles.
17b. The area is increased four times.   **19.** 2, 3, 6

Pages 404–405   Lesson 10-2
1. Two congruent triangles form a parallelogram.
3. Ebony is correct. If 5.8 is used as the height, then the base is the length of the side to which the height is drawn, the 7 inch side.   **5.** 35 mm^2 **7.** 1,136 m^2
9. 13.5 units^2 **11.** 51 ft^2 **13.** 41.3 cm^2
15. 1,400 in^2 **17.** 128 cm^2 **19.** 14 ft^3; 16 ft
21. Sample answer: An elephant's ears expands its overall surface area, allowing it more space to release heat and cool itself.   **25.** 320 mm^2
27. 46.4 ounces

Pages 407–409   Lesson 10-3
1. The circle can be cut and arranged into a figure like a parallelogram to find the area.   **5.** 254.3 m^2
7. 1.4 ft^2 **9.** 28.3 in^2 **11.** 615.4 ft^2 **13.** 95.0 in^2
15. 379.9 in^2 **17.** 63.6 m^2 **19.** 379.9 mm^2
21. 95.0 m^2 **23.** about 63.6 m^2 **25.** The area is multiplied by 4.   **27.** A

Page 411   Lesson 10-3B
1. Sample answer: The circle graph, because it allows a visual comparison of the proportions.
3. Sample answers: data that doesn't equal 100%; a bar or line graph

5.   **World Population**

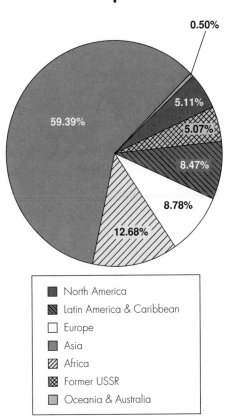

North America
Latin America & Caribbean
Europe
Asia
Africa
Former USSR
Oceania & Australia

7. The area of the circle is divided into parts to represent the portion of the whole for each category.

Pages 413–414   Lesson 10-4
1. square pyramid   **5.** cone   **7.** 5; 9; 6
9. cylinder   **11.** triangular prism   **13.** cone
15. 4; 6; 4   **17.** none; none; none   **19.** 8; 18; 12
21a. 8; 12; 6   **21b.** yes   **23.** 1,519.76 in^2 **25.** B

Page 414   Mid-Chapter Self Test
1. 3,850 ft^2 **3.** 12.6 in^2 **5.** cylinder

Page 415   Lesson 10-4B
1. the sides that are rectangles 3 units long and 2 units wide   **2.** yes   **3.** Draw the hexagonal base first. Then from each vertex, draw a line to represent the depth. Then connect the ends of the segments.
5.

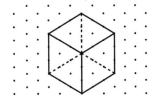

Page 417   Lesson 10-5A
1. Sample answer: It allowed them to build a small pyramid with easy-to-move boxes first.   **3.** Emilio

made sure that the number of boxes they planned to use was the most possible for the type of pyramid they were building.   **5.** 55 cubic inches   **7.** 20 boxes   **9.** Sample answer: 4 by 2 by 1   **11.** 501

Pages 419–420   Lesson 10-5

1a. Because each of the three dimensions being multiplied is expressed in a unit of measure.
3. Sample answer: a prism 1 cm high, 2 cm wide, and 9 cm long   **5.** 588 cm³   **7.** 14,400 in³
9. 1,536 ft³   **11.** 2,737.9 m³   **13.** 90 mm³
15a. 3,750 m³   **15b.** 3,750,000 liters   **19.** B

Pages 423–424   Lesson 10-6

1a. length   **1b.** area   **1c.** volume   **1d.** length
1e. volume   **1f.** area   **3.** Sample answers: determining the correct amount of paint for a room; or choosing enough wrapping paper for a package
5. 142 cm²   **7.** 312 in²   **9.** 286 m²   **11.** 96 cm²
13. 648 ft²   **15.** 718 m²   **17.** 1,240 in²   **21.** 328 ft²
23. 6 units on each side   **25.** 9 hours 9 minutes

Page 425   Lesson 10-6B

1.

	A	B	C	D	E
1	LENGTH	WIDTH	HEIGHT	VOLUME	SURFACE AREA
2	3	3	7	63	102
3	10.2	4.1	1.6	66.912	129.4
4	4	4	4	64	96
5	8	4	4	128	160
6	8	8	4	256	256
7	8	8	8	512	384

3. Delete the height column. Change "volume" to "area" and change the formulas in column D to A2*B2, A3*B3, and so on. Change "surface area" to "perimeter". Change the formulas in column E to 2*A2 + 2*B2, 2*A3 + 2*B3, and so on.   **5.** When one dimension is doubled, the volume doubles. When two dimensions are doubled, the volume is multiplied by four. When all three dimensions are doubled, the volume is multiplied by eight. The volumes are changed in this way because if any one dimension is multiplied by 2, the volume is multiplied by 2.

Pages 426–429   Study Guide and Assessment

1. height   **3.** edges   **5.** sphere   **7.** volume
9. surface area   **11.** 14 m²   **13.** 87.5 cm²

15. 72 cm²   **17.** $62\frac{7}{16}$ ft²   **19.** 153.9 in²
21. 40.3 in²   **23.** rectangular prism   **25.** 6; 12; 8
27. $103\frac{1}{8}$ yd³   **29.** 7,280 m²   **31.** 25.3 mm²
33a. $85\frac{2}{3}$ ft²   **33b.** 4 bushels   **35.** 9 m³

Pages 430–431   Standardized Test Practice

1. A   **3.** B   **5.** D   **7.** B   **9.** A   **11.** $6.35
13. 4; 6; 4   **15.** 718 ft²

Chapter 11
Algebra: Investigating Integers

Pages 435–436   Lesson 11-1

1. −1 and +2   **3.** Sample answer: The yards lost or gained in one down of a football game. For example, −3 would represent a loss of 3 yards, and +4 would represent a gain of 4 yards.
5.

7. +6   **9.** −4   **11.** −345   **13.** −178
15.

17.

19.

21.

23. +75   **25.** −100   **27.** +100   **29.** −23
31. −45   **33.** −77   **35.** +110
37.

39. +7,000   **43.** C   **45.** 135%   **47.** 673.0

Pages 438–439   Lesson 11-2

1. Graph the numbers on a number line. The number to the left is less than the number to the right.
3. Cordelia; since −5 is to the left of −3 on the number line, −5 < −3.   **5.** <   **7.** −3, −2, 0, 5
9. >   **11.** <   **13.** <   **15.** >   **17.** <   **19.** 41, 10, 3, 0, −10, −20   **21.** −17   **23a.** −458, −435, −361, −38, 450, 1,542, 1,763, 1,947, 2,795
23b. solid   **23c.** 450   **25.** All negative integers are to the left of 0 on a number line while all positive integers are to the right. The number to the left is always less than the number to the right.   **27.** C

Page 440   Lesson 11-3A

1. none   **3.** They are opposites.   **5.** 0   **7.** Place 4 yellow counters on the mat to represent +4. Place 3 red counters on the mat to represent −3. Remove as many zero pairs as possible. The 1 yellow counter left represents the sum (+1).

Pages 443–444   Lesson 11-3

1. $-5 + 2 = -3$ **3.** negative   **5.** 2   **7.** −2
9. −8   **11.** negative   **13.** positive   **15.** negative
17. −9   **19.** 0   **21.** −3   **23.** 4   **25.** −8   **27.** 10
29. 8   **31.** −3   **33.** −123°C   **35a.** Sample answer: $-2 + (-3) = -5$ **35b.** Sample answer: $0 + (-5) = -5$ **35c.** Sample answer: $-8 + 3 = -5$
37. 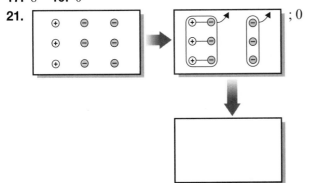   **39.** D

Pages 447–448   Lesson 11-4

1. $4 - (-2) = 6$ **3.** Donnell; you can subtract 8 from 5 by using negative integers. $5 - 8 = -3$
5. 0   **7.** −2   **9.** −2   **11.** −5   **13.** 3   **15.** −24
17. 8   **19.** 0
21.

; 0

23a. 18 ft   **23b.** 19 ft   **25.** Yes; sample examples: $8 - 3 = 5, 0 - (-2) = 2$, and $-3 - (-5) = 2$.
27. $\frac{1}{14}$

Pages 451–452   Lesson 11-5

1. $-2 \times (-2) = 4$ **3.** −3   **5.** −16   **7.** 21
9. 18   **11.** 28   **13.** 0   **15.** −18   **17.** −24
19. −25   **21.** −42   **23a.** 12 lb   **23b.** 68 lb
25. the number's opposite

27.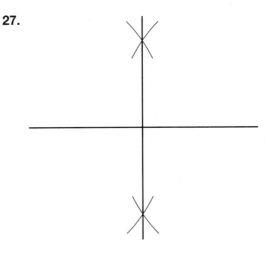

Page 452   Mid-Chapter Self Test

1. −282   **3.** <   **5.** −10   **7.** −4   **9.** −18

Pages 454–455   Lesson 11-6A

1. Callie and Nora used a calendar to locate April 15. Then they counted backward from that date the number of weeks they wanted to spend on each activity.   **3.** 7:30 A.M.   **5.** Sample answer: Sue has a dental appointment at 3:30 P.M. She wants to get there 10 minutes early. If it takes 40 minutes to drive to the dentist's office, what time should she leave her house for the appointment?   **7.** 24 ways
9a. 7 ft   **9b.** 28 ft
11.

−2	3	−4
−3	−1	1
2	−5	0

13. $29.39

Pages 457–458   Lesson 11-6

1. $-9 \div 3 = -3$ **3a.** Sample answer: $-5 \times (-3) = 15; 15 \div (-5) = -3$ and $15 \div (-3) = -5$
3b. Sample answer: $4 \times (-6) = -24; -24 \div 4 = -6$ and $-24 \div (-6) = 4$ **5.** −2   **7.** 9   **9.** 2
11. −4   **13.** −5   **15.** 3   **17.** −9   **19.** −6
21a. −5   **21b.** 4   **23.** loss of 4 yd per play
25. Sample answers: $14 \div (-2) = -7, -21 \div 3 = -7, 28 \div (-4) = -7, -35 \div 5 = -7$
27. Sample answer: 5

Pages 460–461   Lesson 11-7

1. Start at 0. Move 3 units to the left on the x-axis. Then move 6 units up parallel to the y-axis to locate the point. Place a dot at this location and label the point.   **3.** (3, 1)   **5.** (−3, −2)

7.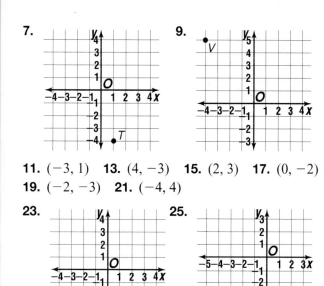

9.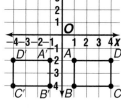

11. $(-3, 1)$ **13.** $(4, -3)$ **15.** $(2, 3)$ **17.** $(0, -2)$
19. $(-2, -3)$ **21.** $(-4, 4)$

23. 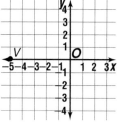 **25.**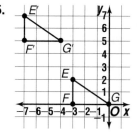

27. **29.**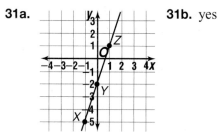

31a. 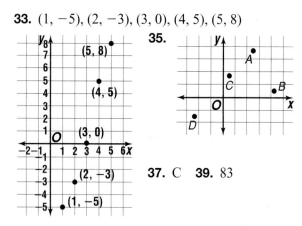 **31b.** yes

33. $(1, -5)$, $(2, -3)$, $(3, 0)$, $(4, 5)$, $(5, 8)$

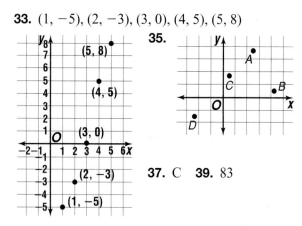

35.

(5, 8)
(4, 5)
(3, 0)
(2, −3)
(1, −5)

37. C **39.** 83

Pages 462–463 Lesson 11-8A

1. The position of the triangle is changed. **3.** The new figure is the mirror image of triangle *C*.

Pages 466–467 Lesson 11-8

1. Both translations and reflections are transformations. In translations and reflections, the image is the same size and shape as the figure. The translation image has the same orientation as the figure, but the reflection image is a flip of the figure.

3a. 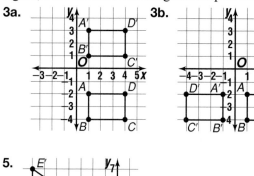 **3b.**

5.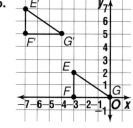

7. Reflection; the mountains are reflected in the lake. **9.** reflection

11. 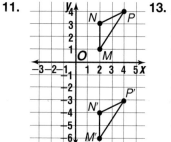 **13.**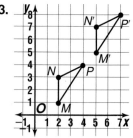

15. 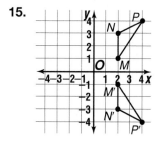 **17.** Translations; the winged horses are translations of each other. **19.** The image is a reflection of the polygon over the *y*-axis.
21. A

Pages 468–471 Study Guide and Assessment

1. false; positive **3.** true **5.** false; sometimes
7. false; never **9.** true **11.** false; reflection

13.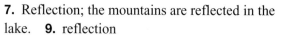

15.

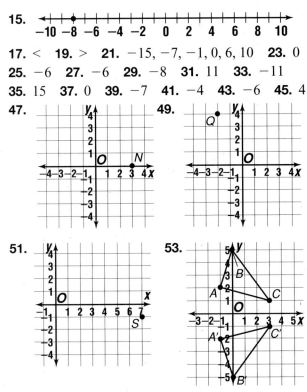

17. $<$ **19.** $>$ **21.** $-15, -7, -1, 0, 6, 10$ **23.** 0
25. -6 **27.** -6 **29.** -8 **31.** 11 **33.** -11
35. 15 **37.** 0 **39.** -7 **41.** -4 **43.** -6 **45.** 4
47.

49.

51.

53.

55. 8 yd **57.** reflection

Pages 472–473 Standardized Test Practice
1. C **3.** D **5.** A **7.** A **9.** B **11.** B **13.** -8
15. $(3, 3)$ **17.** ∃ **19.** $28\frac{1}{2}$

Chapter 12
Algebra: Exploring Equations

Pages 477–479 Lesson 12-1
1. Sample answer: The 5 positive counters and the cup represent $x + 5$ and the 8 negative counters represent -8. **3.** 6 **5.** 7 **7.** -9 **9.** -6
11. -5 **13.** -1 **15.** 2 **17.** -6 **19.** -20
21. -43 **23.** -125 **25.** -2 **27.** -8 **29.** 10
31. -6 **33.** Sample answer: $37 + (-9) = 28$
35. 21 **37.** 9%

Pages 482–483 Lesson 12-2
1. Sample answer: The 3 negative counters and the cup represent $y - 3$ and the 4 positive counters represent 4. **3.** Sample answer: If the equation involves addition, you subtract counters. If the equation involves subtraction, you add counters.
5. -2 **7.** 7 **9.** 16 **11.** -2 **13.** -2 **15.** -1

17. -8 **19.** 5 **21.** 7 **23.** 3 **25.** 2 **27.** -9
29a. $-140 + r = -35$ **29b.** 105 feet **31.** B

Pages 486–487 Lesson 12-3
1. $4x = 8; x = 2$ **3.** Sample answer: They're both correct. Dividing by 2 is the same as multiplying by one-half. **5.** 8, 8, 2 **7.** 5 **9.** -10 **11.** 2
13. -3 **15.** -9 **17.** $\frac{1}{4}$ **19.** 5 **21.** 24 **23.** 38
25. 2 **27.** -36 **29.** 8.2 **31.** $4 million
33.

E
Change per month
D2/10
D3/10
D4/10
D5/10
D6/1
D7/10

35. 3 **37.** C

Page 487 Mid-Chapter Self Test
1. 6 **3.** 21 **5.** -15 **7.** -2 **9.** 3 gallons

Pages 490–491 Lesson 12-4
1. $3x + 2 = -1; x = -1$ **3.** subtract 2 from each side; 4 **5.** add 4 to each side; 20 **7.** 4 **9.** 3
11. -9 **13.** -32 **15.** -20 **17.** 32
19. 16 inches **21.** 8 **23.** 44, 46.5

Pages 492–493 Lesson 12-4B
1. Sample answer: no; Ed's information may be more useful because it contains all the known amounts. **3.** Sample answer: Subtract 8 from the solution. **5.** Sample answer: You could use the guess and check strategy. **7.** Sample answer: In solving a two-step equation, you undo the operations in reverse order of the order of operations. **9.** about 12 buses **11.** $39.99 **13.** 24 combinations
15. 4 ways

Page 495 Lesson 12-5A
1. 8 **3.** 13 **5a.** 12 **5b.** 18 **5c.** 21 **5d.** 11
7. 1, 7, 13, 19

Pages 497–499 Lesson 12-5
1.

input (y)	output ($y-2$)
-4	-6
-1	-3
0	-2
3	1

3. Sample answer: Juanita is correct. If you use 4, 5, and 6 for inputs, the outputs are 1, 2, and 3. Each output is 3 less than the input.

5.

input (n)	output ($\frac{1}{4}n$)
4	1
8	2
12	3

7. $n \div 2$ **9a.** $0.75c$ **9b.** $37.50

11.

input (n)	output ($3n$)
0	0
3	9
4	12

13.

input (n)	output ($\frac{1}{8}n$)
−8	−1
0	0
12	$1\frac{1}{2}$

15. $5n$ **17.** $n \div -2$ **19.** n^2 **21.** -22 **23.** $n \div 5$
25. 100, 60, 80, 70   **27.** $2n + 2$ **29.** -12

Pages 501–503   Lesson 12-6

1. Sample answer: Let the input values represent the x-coordinates and let the output values represent the corresponding y-coordinates.   **3.** Sample answer: Make a table recording the input and output of the given function. Write the input/output as an ordered pair. Record at least three ordered pairs on the table. Graph these ordered pairs and draw a line connecting the graphed points.

5.

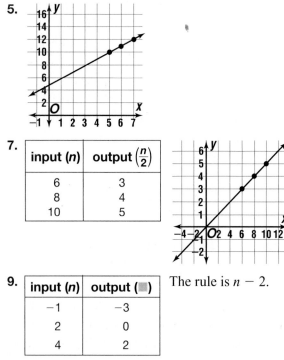

7.

input (n)	output ($\frac{n}{2}$)
6	3
8	4
10	5

The rule is $n - 2$.

9.

input (n)	output (■)
−1	−3
2	0
4	2

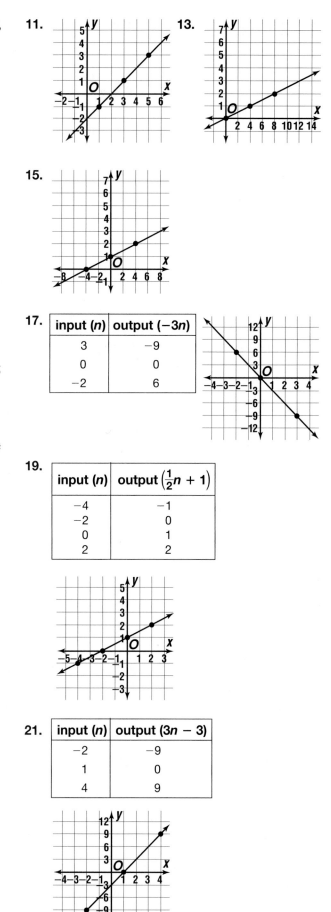

11.

13.

15.

17.

input (n)	output ($-3n$)
3	−9
0	0
−2	6

19.

input (n)	output ($\frac{1}{2}n + 1$)
−4	−1
−2	0
0	1
2	2

21.

input (n)	output ($3n - 3$)
−2	−9
1	0
4	9

23.

input (*n*)	output (■)
0	0
4	2
8	4

The rule is $\frac{n}{2}$

25a. Jane: $25w - 20$; Julie: $20w$

25b.
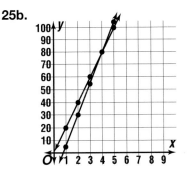

25c. The point of intersection represents when the total earnings are the same for both girls.

27a.

input	output
1	61
2	62
3	63
4	64
5	65
6	66
7	67
8	68
9	69
10	70
11	71
12	72

27b.
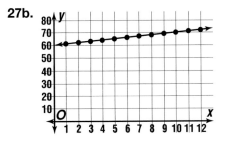

27c. In 5 months, the card will be worth $65.

29.

input (*n*)	output (*n* + 7)
−5	2
0	7
4	11

31. about 31.4 meters

Pages 504–507 Study Guide and Assessment

1. variable **3.** subtracted from **5.** function table
7. -18 **9.** $2x + 3 = -9$ **11.** 3 **13.** -13
15. -3 **17.** 4 **19.** 2 **21.** 0 **23.** 4 **25.** -55
27. -5 **29.** -20 **31.** 6 **33.** -9 **35.** $-3, 0, 6$
37. -5 **39.** $n - 4$
41. 1, 3, 5

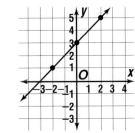

43. 8 laps **45.** $n + 3.50$

Pages 508–509 Standardized Test Practice

1. B **3.** D **5.** D **7.** E **9.** D **11.** E
13. 240 m³ **15.** about 15 people
17. circle

····································

Chapter 13
Using Probability

Page 514 Lesson 13-1A

1. Sample answer: The number of even sums tossed and the number of odd sums tossed should be about the same. The number of even products tossed should be greater than the number of odd products tossed. **3.** Sample answer: For the game involving sums, the bar graph resembled an isosceles triangle and is symmetrical about the axis which contains the sum of 7. For the products game, the bar graph reaches a maximum height at products 6 and 12 and the overall graph is not symmetrical. **5.** Sample answer: A sum of 12 is less likely to occur since out of 36 results, a sum of 12 can only occur one way.

Pages 517–518 Lesson 13-1

1. Sample answer:

3. Sample answer: Complementary events are two events in which either one or the other can occur, but not both. The sum of their probabilities is 1. An

example is the probability of snow and the probability of no snow.   **5.** $\frac{2}{3}$; The event is very likely to occur.   **7.** $\frac{1}{3}$; The event is not too likely to occur.   **9.** 0; The event cannot occur.   **11.** $\frac{1}{4}$; The event is not too likely to occur.   **13.** $\frac{1}{2}$; The event is equally likely to occur.   **15.** $\frac{3}{26}$; The event is not too likely to occur.   **17.** $\frac{21}{26}$; The event is very likely to occur.   **19.** $\frac{1}{2}$; not odd; $\frac{1}{2}$ **21.** 1; not an integer; 0   **23.** $\frac{2}{5}$; not greater than 18; $\frac{3}{5}$ **25a.** $\frac{1}{5}$
25b. Sample answer: Sometimes, eliminating the ones that cannot be correct leaves the correct answer.
25c. increase; There would be fewer choices.
29.

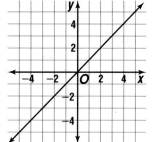

31. 47.36

Pages 520–521   Lesson 13-2A

1. Sample answer: Yes; surveying every fourth member provides an adequate sample.   **3.** Sample answer: The information is easy to locate and it is easier to draw conclusions from organized data.
5. Sample answer: It enables you to determine a reasonable interval width for the given data.
7. Sample answer: Ashley organized the results of the survey by making a table. The table makes it easy to find information such as the most preferred and the least preferred pet.   **9.** 8 days
11. Sample answer: $20 − $6 or $14.   **13.** D

Pages 524–525   Lesson 13-2

1. Sample answer: Population is the entire group while random sample is a subset of the entire group.
5. yes; There is probably a diverse group of people at a shopping center.   **7a.** $\frac{3}{5}$, or 60%   **7b.** about 9
9. no; You would probably survey people whose hobby is model trains.   **11.** yes; A mall would have a diverse group of people.   **13.** no; You would probably find many people who prefer the type of cars sold by the dealership.   **15.** no; A lot of chocolate lovers would be in a chocolate factory.
17. no; No Republicans were included in the survey.
19. about 369   **21.** C   **23.** 180 inches

Page 525   Mid-Chapter Self Test

1. $\frac{1}{6}$ **3.** $\frac{1}{2}$ **5a.** $\frac{3}{50}$, or 6%   **5b.** about 900

Pages 527–529   Lesson 13-3

1. $\frac{4}{9}$ **5.** $\frac{1}{8}$ **7.** about 19   **9.** $\frac{2}{5}$ **11.** $\frac{1}{8}$ **13.** $\frac{4}{25}$
15. about 80   **17.** about 32   **19.** about 44
21. about $\frac{1}{40}$, or 2.5%   **23c.** Sample answer: In the experiment, it is easier to see the black and white circles but it is hard to see the newsprint circles. The newsprint circles represent the camouflage trait.
25. No; people in a jeans store may prefer the jeans sold by the store.   **27.** 988 ft^2

Page 530   Lesson 13-3B

1. Sample answer: $\frac{27}{50}$ **3.** Sample answer: The fractions should be about the same.   **5.** Sample answer: In the program, change each 10 to 20.

Pages 533–534   Lesson 13-4

1. Sample answer: At a restaurant, a customer can choose either a chicken or beef dinner and one of three beverages. How many possible outcomes are there?   **3.** Sample answer: Tree diagrams are useful for organizing the outcomes. One disadvantage is that there could be too many outcomes.   **5.** blue shirt, blue shorts; blue shirt, black shorts; blue shirt, red shorts; red shirt, blue shorts; red shirt, black shorts; red shirt, red shorts

7a.

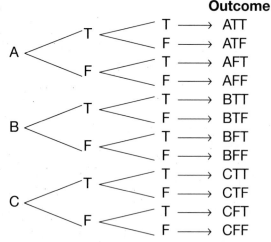

7b. $\frac{1}{2}$ **9.** leather, purple; leather, green; leather, black; leather, brown; nylon, purple; nylon, green; nylon, black; nylon, brown   **11.** blue, red; blue, green; blue, white; blue, black; orange, red; orange,

green; orange, white; orange, black; yellow, red; yellow, green; yellow, white; yellow, black
13. tower, wood, 60; tower, wood, 100; tower, wood, 250; tower, plastic, 60; tower, plastic, 100; tower, plastic, 250; spinner, wood, 60; spinner, wood, 100; spinner, wood, 250; spinner, plastic, 60; spinner, plastic, 100; spinner, plastic, 250 **15.** 10 ways **17.** 15 ways **19.** Sample answer: This is not a fair game. Only in two of the eight outcomes are all chips different. It could be made a fair game by giving Amiri 3 points or changing the second counter from red and white to green and white. **21.** C

Page 535   Lesson 13-4B
1. Sample answer: $\frac{22}{50}$ **3.** They are about the same.

Pages 538–539   Lesson 13-5
1. Multiply the probability of the first event by the probability of the second event. **3.** Flora is correct. The probability of a 4 or 5 is $\frac{2}{6}$, and the probability of a 6 is $\frac{1}{6}$. $\frac{1}{3} \times \frac{1}{6} = \frac{1}{18}$. **5.** $\frac{1}{10}$ **7a.** $\frac{2}{5}$ **7b.** $\frac{3}{10}$ **7c.** $\frac{3}{25}$ **9.** $\frac{1}{4}$ **11.** 0 **13.** $\frac{1}{3}$ **15.** $\frac{3}{8}$ **17.** $\frac{5}{18}$ **19a.** 0.48 **19b.** 0.08 **21.** 20 ways **23.** D

Pages 540–543   Study Guide and Assessment
1. e **3.** g **5.** b **7.** c **9.** d **11.** Sample answer: Multiply the probability of the first event by the probability of the second event. **13.** $\frac{1}{3}$; The event is not too likely to occur. **15.** $\frac{5}{6}$; The event is very likely to occur. **17.** $\frac{2}{3}$; The event is very likely to occur. **19.** no; At a football game, you would most likely find people who prefer football. **21a.** $\frac{7}{20}$ or 35% **21b.** 147 students **23.** about $\frac{5}{16}$ or 31.25% **25.** black jeans–tapered; black jeans–straight; black jeans–baggy; blue jeans–tapered; blue jeans–straight; blue jeans–baggy **27.** ball game–Friday; ball game–Saturday; amusement park–Friday; amusement park–Saturday; concert–Friday; concert–Saturday; **29.** $\frac{1}{4}$ **31.** $\frac{1}{6}$ **33a.** 11 **33b.** 32 **35.** purple skirt–yellow vest; purple skirt–purple sweater; white skirt–yellow vest; white skirt–purple sweater; yellow skirt–yellow vest; yellow skirt–purple sweater

Pages 544–545   Standardized Test Practice
1. D **3.** C **5.** B **7.** D **9.** A **11.** C **13.** C **15a.** $\frac{1}{6}$ **15b.** $\frac{4}{75}$

Photo Credits

Cover (bkgd)Index Stock, (hologram)Glencoe photo, (inset)Paul L. Ruben; **iv** Mark Burnett, (bl)courtesy of Arthur Howard; **ix** (l)Hickson & Assoc., (lc)AT&T, (r)Mark Burnett, (rc)Bob Mullenix; **v vii** Mark Burnett; **x** Thomas Kitchin/Tom Stack & Assoc.; **xi** Aaron Haupt; **xii** Don Valentine/Tony Stone Images; **xiii** courtesy Binney & Smith Co.; **xiv** Elisabeth Weiland/Photo Researchers; **xix** Steve Kaufman/Peter Arnold, Inc.; **xv** William Blake/Corbis; **xvi** Doug Martin; **xvii** Bill Bachman/Photo Researchers; **xviii** SuperStock; **xx** Tom McHugh/Photo Researchers; **xxi** GK & Vikki Hart/The Image Bank; **xxii** (l)Frans Lanting/Tony Stone Images, (r) Ann Duncan/Tom Stack & Assoc.; **xxiii** Kenji Kerins; **xxiv** Doug Martin; **xxv** (l)Matt Meadows, (r) David R. Frazier Photo Library; **xxvi** (l)SuperStock, (r) Fred Bavendam/Peter Arnold, Inc.; **xxvii** Van Bucher/Photo Researchers; **xxviii** Mark Steinmetz; **1** (t)KS Studio, (b)Matt Meadows; **2** (t)Matt Meadows, (others)Aaron Haupt; **4** (l)Hickson & Assoc., (r)AT&T; **5** (tl)Bob Mullenix, (tr)Mark Burnett, (b)Aaron Haupt; **7** (l)B & C Alexander/Photo Researchers, (r)Charles Krebs/Tony Stone Images; **8** (tl)Bill Ivy/Tony Stone Images, (tr)Hans Pfletshinger/ Peter Arnold, Inc., (b)Stephen Dalton/Photo Researchers, **9** Mark Steinmetz; **10** Aaron Haupt; **12** Eastcott/Momatuik/ Tony Stone Images; **13** Kathi Lamm; **16** SuperStock; **17** Tom Brakefield/The Stock Market; **19** SuperStock; **22** Aaron Haupt; **23** Spencer Swanger/Tom Stack & Assoc.; **25 26** Aaron Haupt; **27** (bkgd)Onne Van Der Wal/ Stock Newport, (l)Andy Freeberg, (r)KS Studio; **29** Aaron Haupt; **32** KS Studio; **34** (t)Library of Congress/Corbis, (c)Philadelphia Museum of Art/ Corbis, (b)courtesy Chicago Historical Society; **36 37 41** Aaron Haupt; **44** (t)Frans Lanting/Tony Stone Images, (bl)Monika Smith/Cordaiy Photo Library Ltd./Corbis, (br)Ted Horowitz/The Stock Market; **46** Aaron Haupt; **47** David Muench/Tony Stone Images; **48** Aaron Haupt; **49** KS Studio; **50** Aaron Haupt, (inset)KS Studio; **52** KS Studio; **53** Thomas Kitchin/ Tom Stack and Assoc.; **56** Ken Frick; **58** (t)KS Studio, (b)Matt Meadows; **60** StudiOhio; **62** (b)BLT/Brent Turner Productions; **64** David Lees/Corbis; **65** Focus on Sports; **66** Doug Martin; **70** David Madison/Bruce Coleman, Inc.; **71** KS Studio; **76** Aaron Haupt; **78** ChromeSohm Inc./Corbis-Bettmann; **81** Hy Peskin/ FPG; **82** *Close to Home* © 1996 reprinted with permission of Universal Press Syndicate. All rights reserved; **83** (t)StudiOhio, (b)KS Studio; **85** (t)Len Rue Jr./Photo Researchers, (b)Fritz Polking/ Frank Lane Pictures Agency; **92** (bkgd)Terry Donelly/Tony Stone Images, (t)Murry Sill, (c)Aaron Haupt, (b)Gerald & Buff Corsi/Tom Stack and Assoc.; **95** (t)Barry Rosenthal/FPG, (b)Doug Martin; **96** Robert A. Tyrell; **97** R. Scheiber; **98** By permission of Bud Blake, King Features Syndicate; **99** (bkgd)Stuart Westmorland/ Corbis, (t)Tim Wright/Corbis, (bl)Jeffery Salters/SABA, (br)KS Studio; **100** Doug Martin; **102** (t)Focus on Sports, (b)from ©Hammond's Road Atlas; **103** Mike Bacon/Tom Stack & Assoc.; **107** Focus on Sports; **109** (t)The Orange County *Register*, (b)Aaron Haupt; **110** courtesy KTXQ; **112** Michal Heron/The Stock Market; **113** courtesy Gleim Jewelers; **114** Matt Meadows; **115** David Muench/ Corbis, (inset)Matt Meadows; **116** Matt Meadows; **128** Focus on Sports; **128-129** image ©1998 PhotoDisc, Inc.; **129** Focus on Sports; **130** (bkgd)Dick Luria/FPG, (l)Ed Taylor Studio/FPG, (r)Francis Lepine/Earth Scenes; **131** Dick Luria/FPG; **133** KS Studio; **136** Doug Martin; **138** KS Studio; **139** SuperStock; **141** Dominic Oldershaw; **145** KS Studio; **146** Doug Martin; **149** Gary Bumgarner/Tony Stone Images; **150** (t br)KS Studio, (bl)Aaron Haupt; **152** Doug Martin; **153** KS Studio; **159** Ken Brate/Photo Researchers; **160** (bkgd)David Barnes/The Stock Market, (t)J. Pickerell/FPG, (bl)John Storey/©1997 *People Weekly*, (br)KS Studio; **161** Matt Meadows; **163** Don Valentine/Tony Stone Images; **164** (t)Rick Stewart/AllSport, (b)Peter Angelo Simon/The Stock Market; **166** Tom McHugh/Photo Researchers; **167** Chuck Savage/The Stock Market; **176** Amanita Pictures; **178** (t)courtesy Binney & Smith Co., (b)Bob Mullenix; **179** KS Studio; **182** Ron Chapple/FPG; **185** (bkgd)Westlight, (l)courtesy Kathryn Sharar Prusineski, (r)Aaron Haupt; **186** KS Studio; **188** Dominic Oldershaw; **189** M&C Werner/Comstock; **190** (l)Doug Martin, (r)J.T. Miller/The Stock Market; **193** Doug Martin; **195** KS Studio; **196** Doug Martin; **198** (t)KS Studio, (b)Elaine Shay; **199** Gunter Ziesler/Peter Arnold, Inc.; **200** Richard Mackson/Time Inc. Picture Collection; **202** Aaron Haupt; **205** (t)Matt Meadows, (b)PEANUTS © United Features Syndicate. Reprinted by Permission; **206 207 208** Doug Martin; **210** Gerard Fritz/Tony Stone Images; **212** SuperStock; **213** (t)Doug Martin, (b)Larry Lefever from Grant Heilman; **214** Aaron Haupt; **215** SuperStock; **216** Aaron Haupt; **217** Doug Martin; **218** David Woodfall/Tony Stone Images; **226-227** Karen Leeds/The Stock Market; **226** (bkgd)Karen Leeds/The Stock Market, F. Stuart Westmoreland/Photo Researchers; **228** Doug Martin; **230** Doug Martin, (b)Robert Maler/Animals Animals; **231** UPI/Corbis-Bettmann; **233** Doug Martin; **234** Erickson Productions/The Stock Market; **235** (bkgd)KS Studio, (l)USGS Photographic Library, Denver CO, (r)Aaron Haupt; **236** (t br) Aaron Haupt, (bl)Stan Obert/ USGS; **238** (t)Matt Meadows, (b)Elaine Shay; **239** Doug Martin; **241** (l)Aaron Haupt, (r)Doug Martin; **242** Doug Martin; **246** Aaron Haupt; **248** Bill Meng/ Wildlife Conservation Society; **250** Chris Noble/Tony Stone Images; **253** Baron Wolman/Tony Stone Images; **254** NASA/ Science Source/Science Photo Library/Photo Researchers; **256** Tim Courlas; **264 264-265** David R. Frazier Photo Library; **266** (bkgd)William Blake/ Corbis, Doug Martin; **268** Matt Meadows; **277** Matt Meadows; **279** PEANUTS © United Features Syndicate. Reprinted by Permission; **280** UPI/Corbis-Bettmann; **281** Library of Congress/Corbis; **282** Jack Zehrt/ FPG; **283** Corbis-Bettmann; **285** Leonard Lessin/Peter Arnold, Inc., **288** (l)Latent Image, (r)Glencoe file; **289** Paul Berger/ Tony Stone Images; **290** KS Studio; **292** Peter Welmann/ Animals Animals; **293** John S. Dunning/Photo Researchers; **296** Matt Meadows; **297** (l)Doug Martin, (r)KS Studio; **298** Inge King/Peter Arnold, Inc., **305** Doug Martin; **308** (bkgd) Index Stock, (t)Aaron Haupt; **312** Aaron Haupt; **313** Manoj Shah/Tony Stone Images; **314** KS Studio; **317 319** Aaron Haupt; **320** Mike Timo/Tony Stone Images; **322** (t)Doug Martin, (c b)KS Studio; **324** (t)SuperStock, (b)Catherine Ursillo/Photo Researchers; **325** (l)UPI/Corbis-Bettmann, (r)OSCAR MAYER is a registered ™ of Kraft Foods Inc. and is used with permission; **327** SuperStock; **328** (t)David R. Frazier Photo Library, (bl)courtesy Carol Larson, (other)KS Studio; **333** Mark Scott/FPG; **337** Amanita Pictures; **350** (bkgd)Mark C. Burnett/Photo Researchers, (l)Mark Steinmetz; **350-351** Matt Meadows; **352** KS Studio; **356** Matt Meadows; **358** Bill Bachman/Photo Researchers; **360** Doug Martin; **364** (t)Scala/ Art Resources, (b)KS Studio; **365** (t)Aaron Haupt, (b)Matt Meadows; **367** (bkgd)Andrea Pistolesi/The Image Bank, (t)courtesy Southern Poverty Law Center, (b)Michael Marsland/ Yale University Office of Public Affairs; **370** J. Du Boisberran/ The Image Bank; **372** J. Sekowski; **375** Mark C. Burnett/Photo Researchers; **377** Fred Bavendam/Peter Arnold, Inc.; **379** Jeff

Glossary

acute angle (352) An angle with a measure greater than 0° and less than 90°.

algebra (22) A mathematical language that uses letters along with numbers. The letters stand for numbers that are unknown. $10x - 3 = 17$ is an example of an algebra equation.

algebraic expression (22) A combination of variables, numbers, and at least one operation.

angle (352) Two rays with a common endpoint form an angle.

angle *BAC* or ∠*BAC*

area (146) The number of square units needed to cover a surface enclosed by a geometric figure.

average (71) The sum of two or more quantities divided by the number of quantities; the mean.

bar graph (54) A graph using bars to compare quantities. The height or length of each bar represents a designated number.

bar notation (218) In repeating decimals, the line or bar placed over the digits that repeat. For example, 2.63 indicates the digits 63 repeat.

base (28) In a power, the number used as a factor. In 10^3, the base is 10. That is, $10^3 = 10 \times 10 \times 10$.

base (398) Any side of a parallelogram.

base (412) The faces on the top and bottom of a three-dimensional figure.

bisect (364) To separate something into two congruent parts.

box-and-whisker plot (76) A diagram that summarizes data using the median, the upper and lower quartiles, and the extreme values. A box is drawn around the quartile value and whiskers extend from each quartile to the extreme data points.

cell (144) The basic unit of a spreadsheet. A cell can contain data, labels, or formulas.

center (280, 413) The given point from which all points on a circle or a sphere are the same distance.

centimeter (100) A metric unit of length. One centimeter equals one-hundredth of a meter.

circle (280) The set of all points in a plane that are the same distance from a given point called the center.

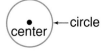

circle graph (60) A graph used to compare parts of a whole. The circle represents the whole and is separated into parts of the whole.

circumference (280) The distance around a circle.

clustering (113) A method used to estimate decimal sums and differences by rounding a group of closely related numbers to the same whole number.

combination (532) An arrangement or listing of objects in which order is not important.

common multiples (206) Multiples that are shared by two or more numbers. For example, some common multiples of 2 and 3 are 6, 12, and 18.

compass (362) An instrument used for drawing circles or parts of circles.

compatible numbers (133, 268) Two numbers that are easy to divide mentally. They are often members of fact families.

complementary (353) Two angles are complementary if the sum of their measures is 90°.

complementary events (516) Two events in which either one or the other must take place, but they cannot both happen at the same time. The sum of their probabilities is 1.

composite number (181, 182) Any whole number greater than 1 that has more than two factors.

cone (413) A three-dimensional figure with curved surfaces, a circular base and one vertex.

congruent angles (363) Angles that have the same angle measure.

congruent figures (379) Figures that are the same size and shape. The symbol ≅ means *is congruent to*.

congruent segments (362) Segments having the same length.

coordinate grid (459) Another name for a coordinate system.

coordinate system (82, 459) A plane in which a horizontal number line and a vertical number line intersect at their zero points.

cross products (318) The products of the terms on the diagonals when two ratios are compared. If the cross products are equal, then the ratios form a proportion. In the proportion $\frac{3}{6} = \frac{4}{8}$, the cross products are 3×8 and 6×4.

cubed (28) The product in which a number is a factor three times. Two cubed is 8 because $2 \times 2 \times 2 = 8$.

cup (292) A customary unit of capacity equal to 8 fluid ounces.

cylinder (413) A three-dimensional figure with all curved surfaces, two circular bases and no vertices.

D

data (46) Numerical information gathered for statistical purposes.

decagon (370) A polygon having ten sides.

degree (352) The most common unit of measure for angles.

diameter (280) The distance across a circle through its center.

distributive property (137) For any numbers a, b, and c, $a(b + c) = ab + ac$ and $(b + c)a = ba + ca$.

E

edge (352, 412) The intersection of faces of a three-dimensional figure.

equation (34) A mathematical sentence that contains the equal sign, $=$.

equilateral triangle (371) A triangle with three congruent sides.

equivalent fractions (193) Fractions that name the same number. $\frac{3}{4}$ and $\frac{6}{8}$ are equivalent.

evaluate (22) To find the value of an expression by replacing variables with numerals.

event (515) A specific outcome or type of outcome.

experimental probability (197) An estimated probability based on the relative frequency of positive outcomes occurring during an experiment.

exponent (28) In a power, the number of times the base is used as a factor. In 5^3, the exponent is 3. That is, $5^3 = 5 \times 5 \times 5$.

extreme value (76) The least value or the greatest value in a set of data.

F

face (412) The flat surfaces of a three-dimensional figure.

factor (28) A number that divides into a whole number with a remainder of zero. 5 is a factor of 30.

factor tree (183) A diagram showing the prime factorization of a number. The factors branch out from the previous factors until all the factors are prime numbers.

fair game (514) A game in which players have an equal chance of winning.

fluid ounce (292) A customary unit of capacity.

foot (202) A customary unit of length equal to 12 inches.

frequency table (46) A table for organizing a set of data that shows the number of times each item or number appears.

Glossary **643**

function (496) A relation in which each element of the input is paired with exactly one element of the output according to a specified rule.

function machine (494) A machine that uses a number called the input, performs one or more operations on it, and produces a result called the output.

function table (496) A table organizing the input, rule, and output of a function.

gallon (292) A customary unit of capacity equal to 4 quarts.

gram (164) A unit of mass in the metric system.

greatest common factor (GCF) (188) The greatest of the common factors of two or more numbers. The GCF of 24 and 30 is 6.

height (398) The shortest distance from the base of a parallelogram to its opposite side.

hexagon (370) A polygon having six sides.

improper fraction (198) A fraction that has a numerator that is greater than or equal to the denominator.

inch (202) A customary unit of length. Twelve inches equal one foot.

independent events (536) Two or more events in which the outcome of one event does *not* affect the outcome(s) of the other event(s).

input (494) Information or data given to a function machine to produce output or results.

integer (434) The whole numbers and their opposites. $\ldots, -3, -2, -1, 0, 1, 2, 3, \ldots$

interval (50) The difference between successive values on a scale.

kilogram (164) A metric unit of mass. One kilogram equals one thousand grams.

leaf (68) The units digit written to the right of the vertical line in a stem-and-leaf plot.

least common denominator (LCD) (210) The least common multiple of the denominators of two or more fractions.

least common multiple (LCM) (206) The least of the common multiples of two or more numbers. The LCM of 2 and 3 is 6.

like fractions (238) Fractions with the same denominator.

line graph (54) A graph used to show change and direction of change over a period of time.

line of symmetry (375) A line that divides a figure into two halves that are reflections of each other.

line segment (362) Two endpoints and the straight path between them. A line segment is named by its endpoints. A representation of line segment ST ($\overline{ST}$) is shown below.

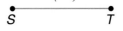

liter (164) The basic unit of capacity in the metric system. A liter is a little more than a quart.

lower quartile (77) The median of the lower half of a set of data.

M

mean (71) The sum of the numbers in a set of data divided by the number of pieces of data.

measures of central tendency (71) Numbers or pieces of data that can represent the whole set of data. These common measures of central tendency are *mean*, *median*, and *mode*.

median (71) The middle number in a set of data when the data are arranged in numerical order. If the data has an even number, the median is the mean of the two middle numbers.

meter (102) The basic unit of length in the metric system.

metric system (102) A base-ten system of weights and measures. The meter is the basic unit of length, the gram is the basic unit of weight, and the liter is the basic unit of capacity.

mile (202) A customary unit of length equal to 5,280 feet or 1,760 yards.

milligram (164) A metric unit of mass. One milligram equals one-thousandth of a gram.

milliliter (164) A metric unit of capacity. One milliliter equals one-thousandth of a liter.

millimeter (100) A metric unit of length. One millimeter equals one-thousandth of a meter.

mixed number (198) The sum of a whole number and a fraction. $1\frac{1}{2}$, $2\frac{3}{4}$, and $4\frac{5}{8}$ are mixed numbers.

mode (71) The number(s) or item(s) that appear most often in a set of data.

multiple (206) The product of the number and any whole number.

negative integer (434) Integer that is less than zero.

net (374) The shape that is formed by "unfolding" a three-dimensional figure. The net shows all the faces that make up the surface area of the figure.

obtuse angle (352) Any angle that measures greater than 90° but less than 180°.

octagon (370) A polygon having eight sides.

opposite (435) Two integers are opposites if they are represented on the number line by points that are the same distance from zero, but in opposite directions from zero. The sum of opposites is zero.

ordered pair (82, 459) A pair of numbers used to locate a point in the coordinate system. The ordered pair is written in this form: (*x*-coordinate, *y*-coordinate).

order of operations (16, 28) The rules to follow when more than one operation is used.
1. Do all powers before other operations.
2. Do all multiplications and divisions from left to right.
3. Then do all additions and subtractions from left to right.

origin (82, 459) The point of intersection of the *x*-axis and *y*-axis in a coordinate system.

ounce (292) A customary unit of weight. 16 ounces equals one pound.

outcomes (515) Possible results of a probability event. For example, 4 is an outcome when a number cube is rolled.

output (494) The result of input that has had one or more operations performed on it in a function machine.

parallel (371) Lines going in the same direction and always being the same distance apart. Parallel lines never meet or cross each other.

parallelogram (371) A quadrilateral that has both pairs of opposite sides equal and parallel.

pentagon (370) A polygon with five sides.

percent (330) A ratio with a denominator of 100.

perimeter (145) The distance around any closed figure.

pint (292) A customary unit of capacity equal to two cups.

place value (95) A system for writing numbers in which the position of the digit determines its value.

polygon (368, 370) A simple closed figure in a plane formed by three or more line segments.

population (522) The entire group of items or individuals from which the samples under consideration are taken.

positive integer (434) Integer that is greater than zero.

pound (292) A customary unit of weight equal to 16 ounces.

power (28) A number that can be written using an exponent. The power 3^2 is read *three to the second power*, or *three squared*.

prime factorization (183) Expressing a composite number as the product of prime numbers. For example, the prime factorization of 63 is $3 \times 3 \times 7$.

prime number (181, 182) A whole number greater than 1 that has exactly two factors, 1 and itself.

prism (412) A three-dimensional figure that has two parallel and congruent bases in the shape of polygons and at least three lateral faces shaped like rectangles. The shape of the bases tells the name of the prism.

proportion (317) An equation that shows that two ratios are equivalent, $\frac{a}{b} = \frac{c}{d}$, $b \neq 0$, $d \neq 0$.

protractor (352) An instrument used to measure angles.

pyramid (412) A solid figure that has a polygon for a base and triangles for sides. A pyramid is named for the shape of its base.

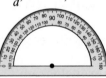

Q

quadrant (459) One of the four regions into which two perpendicular number lines separate the plane.

quadrilateral (368, 370) A polygon with four sides.

quart (292) A customary unit of capacity equal to two pints.

R

radius (280) The distance from the center of a circle to any point on the circle.

random (522) When an outcome is chosen without any preference the outcome occurs at random.

range (72) The difference between the greatest number and the least number in a set of data.

rate (312) A ratio of two measurements having different units.

ratio (191, 193) A comparison of two numbers by division. The ratio of 2 to 3 can be stated as 2 out of 3, 2 to 3, 2:3, or $\frac{2}{3}$.

ray (363) A part of a line that extends indefinitely from one point in one direction. A representation of ray *DE* ($\overrightarrow{DE}$) is shown below.

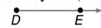

reciprocals (285) Any two numbers whose product is 1. Since $\frac{5}{6} \times \frac{6}{5} = 1$, the reciprocal of $\frac{5}{6}$ is $\frac{6}{5}$.

rectangle (371) A quadrilateral with four congruent angles.

rectangular prism (412) A three-dimensional figure with six rectangular shaped faces. A rectangular prism has a total of six faces, twelve edges, and eight vertices.

reflection (375, 464) A type of transformation where a figure is flipped over a line.

regular polygon (370) A polygon having all sides congruent and all angles congruent.

repeating decimal (218) A decimal whose digits repeat in groups of one or more. 0.181818... can also be written $0.\overline{18}$. The bar above the digits indicates those digits repeat.

right angle (352) An angle that measures 90°.

S

sample (522) A randomly-selected group that is used to represent a whole population.

sample space (515) The set of all possible outcomes.

scale (50) The set of all possible values of a given measurement, including the least and greatest numbers in the set, separated by the intervals used.

scale drawing (324) A drawing that is similar but either larger or smaller than the actual object.

sequence (298) A list of numbers in a certain order, such as, 0, 1, 2, 3, or 2, 4, 6, 8.

sides (145) Line segments that enclose a polygon.

similar figures (379) Figures that have the same shape but different sizes. The symbol ~ means *is similar to*.

simplest form (194) The form of a fraction when the GCF of the numerator and the denominator is 1. The fraction $\frac{3}{4}$ is in simplest form because the GCF of 3 and 4 is 1.

simulation (535) The process of acting out a problem.

solution (34) Any number that makes an equation true. The solution for $12 = x + 7$ is 5.

solve (34) To replace a variable with a number that makes an equation true.

sphere (413) A three-dimensional figure with no faces, bases, edges, or vertices. All of its points are the same distance from a given point called the center.

center

spreadsheet (144) A tool used for organizing and analyzing data.

square (371) A parallelogram with all sides congruent and all angles congruent.

squared (28) A number multiplied by itself; 4×4, or 4^2.

square pyramid (412) A pyramid with a square base.

statistics (46) The study of collecting, analyzing, and presenting data.

stem (68) The greatest place value common to all the data that is written to the left of the line in a stem-and-leaf plot.

stem-and-leaf plot (68) A system used to condense a set of data where the greatest place value of the data forms the stem and the next greatest place value forms the leaves.

straightedge (358) Any object with a straight side that can be used to draw a straight line.

supplementary (353) Two angles are supplementary if the sum of their measures is 180°.

surface area (421) The sum of the areas of all the surfaces (faces) of a three-dimensional figure.

T

terminating decimal (214) A decimal whose digits end. Every terminating decimal can be written as a fraction with a denominator of 10, 100, 1000, and so on.

theoretical probability (515) The ratio of the number of ways an event can occur to the number of possible outcomes.

three-dimensional figure (412) A figure that encloses a part of space.

ton (292) A customary unit of weight equal to 2,000 pounds.

transformation (464) Movement of geometric figures to new points in a coordinate system.

translation (384, 464) One type of transformation where a figure is slid horizontally, vertically, or both.

tree diagram (531) A diagram used to show the total number of possible outcomes in a probability experiment.

triangle (368, 370) A polygon with three sides.

U

unfair game (514) A game in which players do not have an equal chance of winning.

upper quartile (77) The median of the upper half of a set of numbers.

variable (22) A symbol, usually a letter, used to represent a number.

vertex (352) A vertex of a polygon is a point where two sides of the polygon intersect.

vertex (412) The point where the edges of a three-dimensional figure intersect.

volume (418) The amount of space that a three-dimensional figure contains. Volume is expressed in cubic units.

x-**axis** (82, 459) The horizontal line of the two perpendicular number lines in a coordinate plane.

x-**coordinate** (82, 459) The first number of an ordered pair.

yard (202) A customary unit of length equal to 3 feet, or 36 inches.

y-**axis** (82, 459) The vertical line of the two perpendicular number lines in a coordinate plane.

y-**coordinate** (82, 459) The second number of an ordered pair.

zero pair (441) The result of pairing one positive counter with one negative counter.

Spanish Glossary/Glosario

SPANISH GLOSSARY/GLOSARIO

acute angle / ángulo agudo (352) Ángulo que mide más de 0° y menos de 90°.

algebra / álgebra (22) Lenguaje matemático que usa letras y números. Las letras representan números desconocidos.

$10x - 3 = 17$ es un ejemplo de ecuación algebraica.

algebraic expression / expresión algebraica (22) Combinación de variables, números y al menos una operación.

angle / ángulo (352) Dos rayos con un extremo común forman un ángulo.

ángulo *BAC* ó ∠*BAC*

area / área (146) Número de unidades cuadradas que se requieren para cubrir la superficie cerrada por una figura geométrica.

average / promedio (71) Suma de dos o más cantidades dividida entre el número de cantidades; la media.

B

bar graph / gráfica de barras (54) Gráfica que usa barras para comparar cantidades. La altura o longitud de cada barra representa un número específico.

bar notation / notación de barras (218) En los decimales periódicos, la línea o barra que se coloca encima de los dígitos que se repiten. En $2.\overline{63}$, la barra encima de 63 indica que el bloque de dos dígitos, 63, se repite indefinidamente.

base / base (28) Número que se usa como factor en una potencia. En 10^3, la base es 10, es decir, $10^3 = 10 \times 10 \times 10$.

base / base (398) Cualquier lado de un paralelogramo.

base / base (412) Caras inferior y superior de una figura tridimensional.

bisect / bisecar (364) Separar algo en dos partes congruentes.

box-and-whisker plot / diagrama de caja y patillas (76) Diagrama que resume información usando la mediana, los cuartiles superior e inferior y los valores extremos. Se dibuja una caja alrededor de los cuartiles y se trazan patillas uniendo los cuartiles a los valores extremos respectivos.

C

cell / celda (144) Unidad básica de una hoja de cálculos. Las celdas pueden contener datos, rótulos o fórmulas.

center / centro (280, 413) Un punto dado del cual equidistan todos los puntos de un círculo o de una esfera.

centimeter / centímetro (100) Unidad métrica de longitud. Un centímetro es igual a la centésima parte de un metro.

circle / círculo (280) Conjunto de todos los puntos en un plano que equidistan de un punto dado llamado centro.

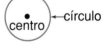

circle graph / gráfica circular (60) Gráfica que se usa para comparar las partes de un todo. El círculo representa el todo y aparece dividido en las partes en las que el todo ha sido separado.

circumference / circunferencia (280) La distancia alrededor de un círculo.

clustering / agrupamiento (113) Método que se usa para estimar sumas y restas de decimales, redondeando al mismo número entero un grupo de números estrechamente relacionados.

combination / combinación (532) Arreglo o lista de objetos en la que el orden no es importante.

common multiples / múltiplos comunes (206)
Múltiplos compartidos por dos o más números. Por ejemplo, algunos múltiplos comunes de 2 y 3 son 6, 12 y 18.

compass / compás (362) Instrumento que se utiliza para trazar círculos o partes de círculos.

compatible numbers / números compatibles
(133, 268) Dos números que son fáciles de dividir mentalmente. A menudo son miembros de la misma familia de factores.

complementary / complementarios (353) Dos ángulos son complementarios si la suma de sus medidas es 90°.

complementary events / eventos
complementarios (516) Dos eventos tales, que uno de ellos debe ocurrir, pero ambos no pueden ocurrir simultáneamente. La suma de sus probabilidades es siempre 1.

composite number / número compuesto
(181, 182) Cualquier número entero mayor que 1 que posee más de dos factores.

cone / cono (413) Figura tridimensional con superficies curvas, una base circular y un vértice.

vértice

congruent angles / ángulos congruentes (363)
Ángulos que tienen la misma medida.

congruent figures / figuras congruentes (379)
Figuras que tienen la misma forma y tamaño. El símbolo ≅ significa *es congruente a*.

congruent segments / segmentos congruentes
(362) Segmentos que tienen la misma longitud.

coordinate grid / plano de coordenadas (459)
Otro nombre para el sistema de coordenadas.

coordinate system / sistema de coordenadas
(82, 459) Plano en el cual se han trazado dos rectas numéricas, una horizontal y una vertical, las cuales se intersecan en sus puntos cero.

eje *x*
eje *y*
O
origen

cross products / productos cruzados (318)
Productos que resultan de la comparación de los

términos de las diagonales de dos razones. Las razones forman una proporción si y sólo si los productos son iguales. En la proporción $\frac{3}{6} = \frac{4}{8}$, los productos cruzados son 3×8 y 6×4.

cubed / al cubo (28) Producto de un número por sí mismo, tres veces. Dos al cubo es 8 ya que $2 \times 2 \times 2 = 8$.

cup / taza (292) Unidad de capacidad del sistema inglés que equivale a 8 onzas líquidas.

cylinder / cilindro (413) Figura tridimensional que tiene superficies curvas, dos bases circulares y que carece de vértices.

D

data / datos (46) Información numérica que se recoge con fines estadísticos.

decagon / decágono (370) Polígono de diez lados.

degree / grado (352) La unidad de medida angular más común.

diameter / diámetro (280)
La distancia a través de un círculo pasando por el centro.

diámetro

distributive property / propiedad distributiva
(137) Para números *a*, *b* y *c* cualesquiera, $a(b + c) = ab + ac$ y $(b + c)a = ba + ca$.

E

edge / arista (352, 412) Intersección de las caras de una figura tridimensional.

equation / ecuación (34) Enunciado matemático que contiene el signo de igualdad ($=$).

equilateral triangle / triángulo equilátero
(371) Triángulo cuyos lados son congruentes entre sí.

equivalent fractions / fracciones equivalentes
(193) Fracciones que designan el mismo número. $\frac{3}{4}$ y $\frac{6}{8}$ son fracciones equivalentes.

evaluate / evaluar (22) Calcular el valor de una expresión sustituyendo las variables por números.

event / evento (515) Resultado específico o tipo de resultado de un experimento probabilístico.

experimental probability / probabilidad experimental (197) Probabilidad de un evento que se estima basándose en la frecuencia relativa de los resultados favorables al evento en cuestión, que ocurren durante un experimento probabilístico.

exponent / exponente (28) Número de veces que la base de una potencia se usa como factor. En 5^3, el exponente es 3, o sea, $5^3 = 5 \times 5 \times 5$.

extreme value / valor extremo (76) El valor mínimo o máximo de un conjunto de datos.

F

face / cara (412) Las superficies planas de una figura tridimensional.

factor / factor (28) Número entero que divide otro número entero con un residuo de 0. 5 es un factor de 30.

factor tree / árbol de factores (183) Diagrama que sirve para encontrar la factorización prima de un número. Los factores se ramifican de los factores anteriores hasta que todos los factores son números primos.

fair game / juego justo (514) Juego en el que los jugadores tienen la misma oportunidad de ganar.

fluid ounce / onza líquida (292) Unidad de capacidad del sistema inglés de medidas.

foot / pie (202) Unidad de longitud del sistema inglés de medidas que equivale a 12 pulgadas.

frequency table / tabla de frecuencia (46) Tabla que se utiliza para organizar un conjunto de datos y que muestra cuántas veces aparece cada ítem o dato.

function / función (496) Relación en que cada elemento de entrada es apareado con un único elemento de salida, según una regla específica.

function machine / máquina de funciones (494) Máquina que utiliza un número llamado entrada y que ejecuta una o más operaciones en el número, produciendo un resultado llamado salida.

function table / tabla de funciones (496) Tabla que organiza la entrada, la regla y las salidas de una función.

G

gallon / galón (292) Unidad de capacidad del sistema inglés de medidas que equivale a 4 cuartos de galón.

gram / gramo (164) Unidad de masa del sistema métrico.

greatest common factor (GCF) / máximo común divisor (MCD) (188) El mayor factor común de dos o más números. El MCD de 24 y 30 es 6.

H

height / altura (398) La distancia más corta desde la base de un paralelogramo hasta su lado opuesto.

hexagon / hexágono (370) Polígono de seis lados.

I

improper fraction / fracción impropia (198) Fracción cuyo numerador es mayor que o igual a su denominador.

inch / pulgada (202) Unidad de longitud del sistema inglés de medidas. Doce pulgadas equivalen a un pie.

independent events / eventos independientes (536) Dos o más eventos en que el resultado de uno de ellos *no* afecta el resultado de los otros eventos.

input / entrada (494) Información o datos que se le proporcionan a una máquina de funciones para que produzca información de salida o resultados.

integer / entero (434) Los números enteros y sus opuestos.

$$\ldots, -3, -2, -1, 0, 1, 2, 3, \ldots$$

interval / intervalo (50) La diferencia entre valores sucesivos de una escala.

kilogram / kilogramo (164) Unidad métrica de masa. Un kilogramo equivale a mil gramos.

leaf / hoja (68) El dígito de las unidades que se escribe a la derecha de la línea divisoria vertical en un diagrama de tallo y hojas.

least common denominator (LCD) / mínimo común denominador (mcd) (210) El menor múltiplo común de los denominadores de dos o más fracciones.

least common multiple (LCM) / mínimo común múltiplo (mcm) (206) El menor múltiplo común de dos o más números. El mcm de 2 y 3 es 6.

like fractions / fracciones con igual denominador (238) Fracciones que tienen el mismo denominador.

line graph / gráfica lineal (54) Gráfica que se usa para mostrar cambio y la dirección del cambio, durante un período de tiempo.

line of symmetry / eje de simetría (375) Recta que divide una figura en dos mitades que son reflexiones una de la otra.

eje de simetría

line segment / segmento de recta (362) Dos puntos y la senda rectilínea entre ellos. Un segmento de recta recibe el nombre de los puntos que la unen. A continuación se muestra una representación del segmento de recta ST ($\overline{ST}$).

$$S \bullet \!\!\!-\!\!\!-\!\!\!-\!\!\!-\!\!\!-\!\!\!- \bullet T$$

liter / litro (164) Unidad fundamental de capacidad del sistema métrico. Un litro es un poco más de un cuarto de galón.

lower quartile / cuartil inferior (77) La mediana de la mitad inferior de un conjunto de datos.

mean / media (71) La suma de los números en un conjunto de datos dividida entre el número total de datos.

measures of central tendency / medidas de tendencia central (71) Números o piezas de datos que pueden representar el conjunto completo de datos. Las medidas comunes de tendencia central son la *media*, la *mediana* y la *modal*.

median / mediana (71) El número central de un conjunto de datos, una vez que los datos se han ordenado numéricamente. Si hay un número par de datos, la mediana es la media de los dos datos centrales.

meter / metro (102) Unidad fundamental de longitud del sistema métrico.

metric system / sistema métrico (102) Sistema de pesos y medidas de base diez. El metro es la unidad fundamental de longitud; el gramo es la unidad fundamental de masa; y el litro es la unidad fundamental de capacidad.

mile / milla (202) Unidad de longitud del sistema inglés que equivale a 5,280 pies ó 1,760 yardas.

milligram / miligramo (164) Unidad métrica de masa. Un miligramo equivale a la milésima parte de un gramo.

milliliter / mililitro (164) Unidad métrica de capacidad. Un mililitro equivale a la milésima parte de un litro.

millimeter / milímetro (100) Unidad métrica de longitud. Un milímetro equivale a la milésima parte de un metro.

mixed number / número mixto (198) La suma de un entero y una fracción. $1\frac{1}{2}$, $2\frac{3}{4}$ y $4\frac{5}{8}$ son números mixtos.

mode / modal (71) Número(s) o ítem(es) de un conjunto de datos que aparece(n) más frecuentemente.

multiple / múltiplo (206) El múltiplo de un número entero es el producto del número por cualquier otro número entero.

negative integer / entero negativo (434) Entero que es menor que cero.

net / red (374) Forma que se obtiene al "desdoblar" una figura tridimensional. La red muestra todas las caras que integran la superficie de una figura.

obtuse angle / ángulo obtuso (352) Cualquier ángulo que mide más de 90° pero menos de 180°.

octagon / octágono (370) Polígono de ocho lados.

opposite / opuestos (435) Dos enteros son opuestos si, en la recta numérica, están representados por puntos que equidistan de cero, pero en direcciones opuestas. La suma de opuestos es cero.

ordered pair / par ordenado (82, 459) Par de números que se utiliza para ubicar un punto en un plano de coordenadas. Se escribe de la siguiente forma: (coordenada *x*, coordenada *y*).

order of operations / orden de las operaciones (16, 28) Reglas a seguir cuando hay más de una operación involucrada.
1. Ejecuta todas las potencias antes que cualquier otra operación.
2. Ejecuta todas las multiplicaciones y divisiones de izquierda a derecha, en el orden que aparezcan.
3. Luego ejecuta todas las sumas y restas de izquierda a derecha, en el orden que aparezcan.

origin / origen (82, 459) Punto de intersección axial en un plano de coordenadas.

ounce / onza (292) Unidad de peso del sistema inglés de medidas. 16 onzas equivalen a una libra.

outcomes / resultados (515) Resultados posibles de un experimento probabilístico. Por ejemplo, 4 es un resultado posible cuando se lanza un dado.

output / salida (494) Resultado de una entrada sobre la que se han realizado una o más operaciones, en una máquina de función.

parallel / paralelas (371) Rectas que se extienden en la misma dirección y que se mantienen siempre a la misma distancia una de la otra. Las rectas paralelas nunca se intersecan o cruzan.

parallelogram / paralelogramo (371) Cuadrilátero cuyos pares de lados opuestos son congruentes y paralelos.

pentagon / pentágono (370) Polígono de cinco lados.

percent / tanto por ciento (330) Razón cuyo denominador es 100. Por ejemplo, 7% y $\frac{7}{100}$ indican el mismo número. 7% se lee 7 *por ciento*.

perimeter / perímetro (145) Medida del contorno de una figura cerrada.

pint / pinta (292) Unidad de capacidad del sistema inglés de medidas que equivale a dos tazas.

place value / valor de posición (95) Sistema de escritura de números en el que la posición de cada dígito de un número determina su valor.

polygon / polígono (368, 370) Figura simple cerrada en un plano, formada por tres o más segmentos de recta.

population / población (522) El grupo total de artículos o individuos del cual se toman las muestras bajo estudio.

positive integer / entero positivo (434) Entero que es mayor que cero.

pound / libra (292) Unidad de peso del sistema inglés de medidas que equivale a 16 onzas.

power / potencia (28) Número que puede escribirse usando un exponente. La potencia

3^2 se lee *tres a la segunda potencia* o *tres al cuadrado.*

prime factorization / factorización prima (183) Forma de expresar un número compuesto como el producto de números primos. La factorización prima de 63, por ejemplo, es $3 \times 3 \times 7$.

prime number / número primo (181, 182) Número entero mayor que 1 que sólo tiene dos factores, 1 y sí mismo.

prism / prisma (412) Figura tridimensional que posee dos bases paralelas y congruentes con forma de polígono, y por lo menos, tres caras laterales con forma de rectángulo. La forma de las bases identifica el prisma.

proportion / proporción (317) Ecuación que demuestra la igualdad de dos razones, $\frac{a}{b} = \frac{c}{d}$, $b \neq 0, d \neq 0$.

protractor / transportador (352) Instrumento que se usa para medir ángulos.

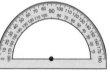

pyramid / pirámide (412) Figura tridimensional que tiene una base poligonal y caras triangulares. Las pirámides se clasifican según la forma de su base.

Q

quadrant / cuadrante (459) Una de las cuatro regiones en que dos rectas perpendiculares dividen un plano.

quadrilateral / cuadrilátero (368, 370) Polígono de cuatro lados.

quart / cuarto de galón (292) Unidad de capacidad del sistema inglés de medidas que equivale a dos pintas.

R

radius / radio (280) Distancia del centro de un círculo a cualquier punto del mismo.

radio

random / aleatorio (522) Un resultado ocurre aleatoriamente o al azar si se escoge sin preferencia alguna.

range / rango (72) Diferencia entre los valores máximo y mínimo de un conjunto de datos.

rate / tasa (312) Razón de dos medidas que tienen distintas unidades de medida.

ratio / razón (191, 193) Comparación de dos números mediante división. La razón de 2 a 3 puede escribirse como 2 de cada 3, 2 a 3, 2:3 ó $\frac{2}{3}$.

ray / rayo (363) Parte de una recta que se extiende indefinidamente en una dirección. A continuación se muestra una representación del rayo DE ($\overrightarrow{DE}$).

reciprocals / recíprocos (285) Dos números cuyo producto es 1. Como $\frac{5}{6} \times \frac{6}{5} = 1$, el recíproco de $\frac{5}{6}$ es $\frac{6}{5}$ y viceversa.

rectangle / rectángulo (371) Cuadrilátero cuyos cuatro ángulos son congruentes entre sí.

rectangular prism / prisma rectangular (412) Figura tridimensional que posee seis caras rectangulares. Un prisma rectangular tiene en total seis caras, doce aristas y ocho vértices.

reflection / reflexión (375, 464) Transformación en que a una figura se le da vuelta de campana por encima de una recta.

regular polygon / polígono regular (370) Polígono cuyos lados, así como sus ángulos, son todos congruentes.

repeating decimal / decimal periódico (218) Decimal en el cual los dígitos, en algún momento, comienzan a repetirse en bloques de uno o más números. 0.181818... puede también escribirse $0.\overline{18}$. La barra encima de los dígitos indica que estos dígitos se repiten.

right angle / ángulo recto (352) Ángulo que mide 90°.

S

sample / muestra (522) Grupo escogido al azar o aleatoriamente que se usa para representar la población entera.

sample space / espacio muestral (515) Conjunto de todos los resultados posibles de un experimento probabilístico.

scale / escala (50) Conjunto de todos los posibles valores de una medida dada, el cual incluye los valores máximo y mínimo del conjunto, separados mediante los intervalos que se han usado.

scale drawing / dibujo a escala (324) Dibujo que es semejante, pero más grande o más pequeño que el objeto real.

sequence / sucesión (298) Lista de números en cierto orden como, por ejemplo, 0, 1, 2, 3 ó 2, 4, 6, 8.

sides / lados (145) Segmentos de recta que encierran un polígono.

similar figures / figuras semejantes (379) Figuras que tienen la misma forma, pero no necesariamente el mismo tamaño. El símbolo ~ se lee *es semejante a.*

simplest form / forma reducida (194) Forma de una fracción en que el MCD de su numerador y denominador es 1. La fracción $\frac{3}{4}$ está escrita en forma reducida pues el MCD de 3 y 4 es 1.

simulation / simulación (535) Proceso de representar un problema.

solution / solución (34) Cualquier número que satisface una ecuación. La solución de $12 = x + 7$ es 5.

solve / resolver (34) Proceso de encontrar el número o números que satisfagan una ecuación.

sphere / esfera (413) Figura tridimensional que carece de caras, bases, aristas o vértices. Conjunto de todos los puntos en el espacio que equidistan de un punto dado llamado centro.

centro

spreadsheet / hoja de cálculos (144) Herramienta que se usa para organizar y analizar datos.

square / cuadrado (371) Paralelogramo cuyos lados, así como sus ángulos, son todos congruentes.

squared / al cuadrado (28) Número multiplicado por sí mismo; 4×4 ó 4^2.

square pyramid / pirámide cuadrada (412) Pirámide cuya base es un cuadrado.

statistics / estadística (46) Estudio de la recolección, análisis y presentación de datos.

stem / tallo (68) El mayor valor de posición común a todos los datos, el cual se escribe a la izquierda de la línea divisoria vertical en un diagrama de tallo y hojas.

stem-and-leaf plot / diagrama de tallo y hojas (68) Sistema que se usa para condensar un conjunto de datos y en el cual el mayor valor de posición de los datos forma el tallo y el segundo mayor valor de posición de los datos forma las hojas.

straightedge / regla (358) Cualquier objeto con un lado recto que se usa para trazar rectas.

supplementary / suplementarios (353) Dos ángulos son suplementarios si la suma de sus medidas es 180°.

surface area / área de superficie (421) Suma de las áreas de todas las superficies de una figura tridimensional.

T

terminating decimal / decimal terminal (214) Decimal cuyos dígitos terminan. Todo decimal terminal puede escribirse como una fracción con un denominador de 10, 100, 1,000, etc.

theoretical probability / probabilidad teórica (515) La razón del número de maneras en que puede ocurrir un evento al número total de resultados posibles.

three-dimensional figure / figura tridimensional (412) Figura que encierra parte del espacio.

ton / tonelada (292) Unidad de peso del sistema inglés que equivale a 2,000 libras.

transformation / transformación (464) Movimiento de figuras geométricas en un sistema de coordenadas a otros puntos del sistema.

translation / traslación (384, 464) Tipo de transformación en que una figura se desliza horizontalmente, verticalmente o de ambas maneras.

tree diagram / diagrama de árbol (531) Diagrama que se utiliza para encontrar y mostrar el número total de resultados posibles de un experimento probabilístico.

triangle / triángulo (368, 370) Polígono de tres lados.

unfair game / juego injusto (514) Juego en que los jugadores no tienen la misma oportunidad de ganar.

upper quartile / cuartil superior (77) La mediana de la mitad superior de un conjunto de números o datos.

variable / variable (22) Un símbolo, por lo general, una letra que se usa para representar números.

vertex / vértice (352) Un vértice de un polígono es el punto de intersección de dos lados del polígono.

vertex / vértice (412) Punto en que se intersecan las aristas de una figura tridimensional.

volume / volumen (418) Cantidad de espacio que encierra una figura tridimensional. Se expresa en unidades cúbicas.

x*-axis / eje *x (82, 459) La recta horizontal de las dos rectas numéricas perpendiculares, en un plano de coordenadas.

x*-coordinate / coordenada *x (82, 459) Primer número de un par ordenado.

Y

yard / yarda (202) Unidad de longitud del sistema inglés que equivale a 3 pies ó 36 pulgadas.

y*-axis / eje *y (82, 459) La recta vertical de las dos rectas numéricas perpendiculares, en un plano de coordenadas.

y*-coordinate / coordenada *y (82, 459) Segundo número de un par ordenado.

zero pair / par nulo (441) Resultado de aparear una ficha positiva con una negativa.

Index

48, 51, 56, 59, 62, 79, 84, 97, 103, 106, 110, 134, 138, 142, 147, 158, 161, 162, 193, 217, 295, 360, 435, 443, 447, 451, 457, 458, 482, 490, 497, 501, 517, 533, 538

real objects, 20, 31, 47, 60, 73, 75, 76, 103, 132, 140, 149, 166, 264, 359, 374, 376, 379, 384, 392–393, 402, 406, 409, 410–411, 421, 444, 462–463, 464, 465, 494–495, 526, 535, 539

    base-ten blocks, 94, 95, 96, 123, 156

    centimeter cubes, 9, 418

    compass, 362, 363, 364, 410–411

    counters, 15, 121, 440, 441–443, 445–447, 449–451, 456–457, 476–477, 480–482, 484–486, 488–490, 517, 535

    dot paper, 526

    geoboard and geobands, 271–272, 284, 368–369, 380

    graph paper, 128

    grid paper, 106, 132, 140, 149, 238, 239, 383, 384, 396–397, 398–399, 421, 464, 465, 530

    integer mat, 440, 441–443, 445–447, 449–451, 456–457, 476–477, 480–482, 484–486, 488–490

    isometric dot paper, 415

    meterstick, 121

    number cubes, 121, 514, 533

    protractor, 352, 353, 358, 359, 365, 410–411

    ruler, 104, 123, 191, 364, 396–397, 398–399, 401, 410–411, 415, 421

    spinner, 75, 515, 517, 532, 533, 538

    straightedge, 358, 359, 362, 363, 364, 365, 530

    tape measure, 60, 100, 103, 166, 202

technology, 3, 17, 26, 27, 31, 32, 45, 73, 75, 93, 99, 107, 119, 121, 128, 129, 131, 135, 160, 166, 131, 144, 146, 154, 207, 410–411, 421, 425, 530

See also problem solving

Math Journal, 6, 14, 56, 69, 79, 114, 119, 138, 147, 162, 179, 212, 230, 255, 275, 290, 313, 338, 360, 413, 423, 435, 458, 501, 517, 533

Math in the Media, 98, 205, 279, 382, 479

Mean, 71–74, 88, 396
    game, 75

Measurement
    capacity, 164–166
    centimeters, 100–104
    cup, 292
    customary system, 202–205, 229, 292–296
    foot, 202–205
    gallon, 292
    gram, 164–166
    inch, 202–205
    kilogram, 164–166
    kilometers, 100–104
    length, 100–104, 202–205
    liter, 164–166
    mass, 164–166
    meters, 100–104
    metric system, 100–104, 164–166, 167–169
    mile, 202–205
    milligram, 164–166
    milliliter, 164–166
    millimeters, 100–104
    to the nearest unit, 203, 229
    nonstandard units, 203
    pint, 292
    pound, 292
    quart, 292
    time, 254–257
    ton, 292
    yard, 202–205
    See also Applications, Connections, and Integration Index, pages xxii–1

Measures of central tendency, 71, 79, 80

    average, 71
    mean, 71–75, 88
    median, 71–75, 88
    mode, 71–75, 88
    range, 72–74, 88

Median, 71–74, 88
    game, 75

Mental math, 35, 36, 40, 239
    study hints, 145, 157, 167, 214, 285, 298, 318, 331, 334, 335

using the distributive property, 138–139

Metric system, 100–104, 164–166, 167–169
    capacity, 164–166
    centimeter, 100–104, 122, 167–169
    changing units, 167–169
    gram, 164–166, 167–169
    kilogram, 164–166, 167–169
    liter, 164–166, 167–169
    mass, 164–166
    meter, 100–104, 122, 167–169
    milligram, 164–166, 167–169
    milliliter, 164–166, 167–169
    millimeter, 100–104, 122, 167–169

Mid-Chapter Self Test, 19, 67, 111, 155, 201, 249, 283, 327, 373, 414, 452, 487, 525

Mile, 202–205

Milligram, 164

Milliliter, 164

Mini-lab, 9, 17, 47, 48, 51, 60, 73, 106

Mixed numbers, 96, 198–201
    adding, 246–249
    as decimals, 218
    dividing, 289–291
    as improper fractions, 198–199, 250
    multiplying, 277–279
    renaming, 250
    subtracting, 246–249, 250–252

Mode, 71–74, 88
    game, 75

Multiples, 206, 210
    common, 206–209
    least common, 207–209

Multiplication
    decimal by whole numbers, 132–136
    of decimals, 140–143
    equations, 484–487
    estimating products, 13, 14, 39, 268–270
    of fractions, 271–276
    of integers, 449–452
    of mixed numbers, 277–279
    in order of operations, 16, 28
    of whole numbers, 13, 14
    using the distributive property, 137–139

solve a simpler problem,
150–151
use an equation, 492–493
use a graph, 58–59
use logical reasoning, 356–357
use a pattern, 8–11, 39
work backward, 454–455

Problem-Solving Study Hints,
47, 51, 207, 281, 299, 337, 353,
358, 447

Projects *See* Chapter Projects and
Interdisciplinary Investigations

Properties
distributive, 137–139

Proportional Reasoning
area, 382
conversions, 309, 321, 324–327
examples, 309, 317–321,
324–327
geometry, 379–382
measurement, 324–327
number, 60–63, 191–192, 309
percent, 60–63, 329–343
perimeter, 380–382
predictions, 319, 320
probability, 316
proportion, 309, 317–321,
324–327
ratio, 191–192, 194–196,
309–316, 324
sequences, 298–300
surface area, 424
volume, 418, 420
See Applications, Connections,
and Integration Index on
pages xxii–1

Proportions, 309, 317–321
models, 317
scale drawings, 324–327
solving, 318–320, 324–327
spreadsheets, 321

Protractor, 352

Pyramid, 412
modeling, 416
square, 412

Quadrants, 459

Quadrilateral, 368, 369, 370–373

Quartile, 77

Radius, 280

Random sample, 522
survey, 522–525

Range, 72–74, 89

Rates, 312–315
as fractions, 313

Rational Numbers *See* Fractions
and Mixed Numbers

Ratios, 191–192, 309, 310–311,
312–315
as fractions, 312
geometry, 310–311, 317
measurement, 309, 310–311
models, 310–311
number, 191–192
probability, 316
scale, 324
simplest form, 194–196

Ray, 363

Reading Math Study Hints, 23, 34,
96, 105, 137, 168, 199, 203, 281,
286, 324, 330, 352, 353, 370,
379, 413, 434, 464, 481, 516, 531

Reasonable answers, 116–117,
141, 152–153, 156–157, 168

Reciprocals, 285

Rectangles, 371–372
area of, 146–149
perimeter of, 145–149, 155
sides of, 145

Rectangular prism, 412
surface area of, 421–424, 425
volume of, 418–420, 425

Reflection, 376, 462–463,
464–467
image, 464
line of, 376

Regular polygon, 370

Repeating decimal, 218–219
bar notation, 218

Right angle, 352

Rounding
decimal quotients, 153–155, 162
decimals, 109–111
in estimation, 12–15, 39
fractions, 228–231

using a number line, 12–13
whole numbers, 12

Sample, 522
random, 522
very large, 526

Sample space, 515, 531
using combinations, 532
using lists to show, 532
using tree diagram to show,
531–533

Scale
for bar graph, 54
for box-and-whisker plot, 76
for frequency table, 50–53
for line graph, 55

Scale drawings, 324–327

School to Career
architecture, 367
arts and crafts, 328
aviation, 99
biological science, 519
business, 160
cartography, 235
graphic design, 185
mail-order entrepreneur, 27
statistics, 453

Scientific calculator *See*
calculator

Sequences, 298–300, 301

Similar figures, 379–382

Simplest form
of fractions, 194–196, 244
of ratios, 194–196

Simulation, 535

Slide, 384–385

Solve a simpler problem,
150–151

Solving equations, 34–37, 40, 85,
121, 124, 213, 245, 274, 276,
283, 287, 289, 290, 291, 353,
354, 476–493, 487
using guess and check, 34–35,
36, 40
using mental math, 35–36, 40

Span, 203

Spatial Reasoning
make a model, 416–417
surface area, 421–425

three-dimensional figures, 412–415
volume, 418–419, 425

Sphere, 413
center, 413

Spreadsheets, 3, 93, 144, 154, 177, 265, 267, 301, 309, 321, 351, 392, 395, 425, 433, 475, 478, 511, 513, 610–611
cells, 144
formulas, 144

Square, 371–372

Squared, 38

Square pyramid, 412

Standardized Test Practice, 42–43, 90–91, 126–127, 174–175, 224–225, 262, 263, 306–307, 348–349, 390–391, 430–431, 472–473, 508–509, 544–545

Statistics
average, 71, 75, 86
bar graph, 54
box-and-whisker plot, 76–77
circle graph, 60–61
extreme value, 76
frequency table, 46–47, 50–51
interval, 50
leaf, 68
line graph, 55
making predictions from, 64–65
mean, 71–75
median, 71–75
misleading, 78–81
mode, 71–75
quartile, 77
range, 72–77
scale, 50
stem, 68
stem-and-leaf plot, 68–69

Stem-and-leaf plots
intervals, 68
leaves, 68
making, 68–70
making box-and-whisker plot from, 77
stems, 68

Straightedge, 358–359, 364–365

Study Guide and Assessment, 39–41, 86–89, 122–125, 170–173, 220–223, 258–261, 302–305, 344–347, 386–390,
426–429, 468–471, 504–507, 540–543

Study Hints
Estimation, 119, 153, 239, 407
Mental Math, 145, 157, 167, 214, 285, 298, 318, 331, 334, 335
Problem Solving, 47, 51, 207, 281, 299, 337, 353, 358, 447
Reading Math, 23, 34, 96, 105, 137, 168, 199, 203, 281, 286, 324, 330, 352, 353, 370, 379, 413, 434, 464, 481, 516, 531
Technology, 55, 29, 119, 194, 274

Subtraction
of decimals, 13, 14, 112–115
equations, 480–483
estimating, 13, 14, 112–115
estimating differences of fractions, 232–234
of integers, 445–448
of like fractions, 239–241, 243
meaning of, 239
measures of time, 254
of mixed numbers, 246–249, 250–253
in order of operations, 16, 28
of unlike fractions, 242–245
of whole numbers, 14

Supplementary angles, 353–354

Surface area, 421
of rectangular prisms, 422–424, 425

Survey, 45, 46, 522–525

Tables, 3, 7, 9, 10, 11, 16, 18, 22, 45, 53, 60, 62, 67, 71, 79, 80, 84, 85, 93, 101, 105, 107, 111, 112, 113, 115, 116, 118, 125, 136, 149, 159, 166, 177, 196, 201, 213, 216, 223, 245, 249, 264, 288, 292, 298, 312, 342, 395, 411, 418, 439, 475, 535, 536
frequency, 46–49, 50–51, 54–56, 87, 88, 520, 521, 543
function, 495–503, 518

Tally mark, 46, 47

Tangram, 310

Techniques *See* Mathematical Techniques

Technology Labs
Evaluating Expressions, 26

Probability, 530
Proportions, 321
Reading Spreadsheets, 144
Sequences, 301
Surface Area and Volume, 425

Technology Mini-Labs, 17

Technology Study Hints, 55, 29, 119, 194, 274

Technology tips, 128, 177, 227, 264, 267, 309

Temperature, 483

Ten-thousandths, 96

Tenths, 95

Terminating decimal, 214
as mixed number, 214–215

Test Practice, 11, 15, 19, 25, 30, 33, 37, 49, 53, 57, 63, 67, 70, 74, 81, 85, 98, 104, 108, 111, 115, 121, 136, 143, 148, 155, 159, 163, 166, 169, 180, 184, 190, 196, 201, 205, 209, 213, 216, 219, 231, 234, 237, 241, 245, 249, 253, 257, 270, 276, 279, 283, 288, 291, 294, 297, 300, 355, 361, 366, 373, 378, 382, 401, 405, 409, 414, 417, 420, 424, 436, 439, 444, 448, 452, 455, 458, 461, 467, 479, 483, 487, 491, 493, 499, 503, 518, 521, 525, 529, 534, 539

Test-Taking Tips, 43, 91, 127, 175, 225, 263, 307, 349, 391, 431, 473, 509, 545

Theoretical probability, 515

Thinking Labs
Draw a Diagram, 322–323
Eliminate Possibilities, 236–237
Guess and Check, 32–33
Look for a Pattern, 296–297
Make a List, 186–187
Make a Model, 416–417
Make a Table, 520–521
Reasonable Answers, 116–117
Solve a Simpler Problem, 150–151
Use an Equation, 492–493
Use a Graph, 58–59
Use Logical Reasoning, 356–357
Work Backward, 454–455

Thousandths, 95

INDEX

Number and Operations

$+$	plus or positive
$-$	minus or negative

$\left. \begin{array}{l} a \cdot b \\ a \times b \\ ab \text{ or } a(b) \end{array} \right\}$ a times b

$\div$	divided by
$\pm$	positive or negative
$=$	is equal to
$\neq$	is not equal to
$<$	is less than
$>$	is greater than
$\leq$	is less than or equal to
$\geq$	is greater than or equal to
$\approx$	is approximately equal to
$\%$	percent
$a{:}b$	the ratio of a to b, or $\frac{a}{b}$

Geometry and Measurement

$\cong$	is congruent to
$\sim$	is similar to
$^{\circ}$	degree(s)
$\overleftrightarrow{AB}$	line AB
$\overline{AB}$	segment AB
$\overrightarrow{AB}$	ray AB
$\llcorner$	right angle
$\perp$	is perpendicular to
$\parallel$	is parallel to
AB	length of $\overline{AB}$, distance between A and B
$\triangle ABC$	triangle ABC
$\angle ABC$	angle ABC
$\angle B$	angle B
$m\angle ABC$	measure of angle ABC
$\odot C$	circle C
$\overset{\frown}{AB}$	arc AB
π	pi $\left(\text{approximately } 3.14159 \text{ or } \frac{22}{7}\right)$
(a, b)	ordered pair with x-coordinate a and y-coordinate b
$\sin A$	sine of angle A
$\cos A$	cosine of angle A
$\tan A$	tangent of angle A

Algebra and Functions

a'	a prime
a^n	a to the nth power
a^{-n}	$\frac{1}{a^n}$ (one over a to the nth power)
$\lvert x \rvert$	absolute value of x
$\sqrt{x}$	principal (positive) square root of x
$f(n)$	function, f of n

Probability and Statistics

$P(A)$	the probability of event A
$n!$	n factorial
$P(n, r)$	permutation of n things taken r at a time
$C(n, r)$	combination of n things taken r at a time